Water for Energy and Fuel Production

GREEN CHEMISTRY AND CHEMICAL ENGINEERING

Series Editor: Sunggyu Lee
Ohio University, Athens, Ohio, USA

Proton Exchange Membrane Fuel Cells: Contamination and Mitigation Strategies
Hui Li, Shanna Knights, Zheng Shi, John W. Van Zee, and Jiujun Zhang

Proton Exchange Membrane Fuel Cells: Materials Properties and Performance
David P. Wilkinson, Jiujun Zhang, Rob Hui, Jeffrey Fergus, and Xianguo Li

Solid Oxide Fuel Cells: Materials Properties and Performance
Jeffrey Fergus, Rob Hui, Xianguo Li, David P. Wilkinson, and Jiujun Zhang

Efficiency and Sustainability in the Energy and Chemical Industries: Scientific Principles and Case Studies, Second Edition
Krishnan Sankaranarayanan, Jakob de Swaan Arons, and Hedzer van der Kooi

Nuclear Hydrogen Production Handbook
Xing L. Yan and Ryutaro Hino

Magneto Luminous Chemical Vapor Deposition
Hirotsugu Yasuda

Carbon-Neutral Fuels and Energy Carriers
Nazim Z. Muradov and T. Nejat Veziroğlu

Oxide Semiconductors for Solar Energy Conversion: Titanium Dioxide
Janusz Nowotny

Lithium-Ion Batteries: Advanced Materials and Technologies
Xianxia Yuan, Hansan Liu, and Jiujun Zhang

Process Integration for Resource Conservation
Dominic C. Y. Foo

Chemicals from Biomass: Integrating Bioprocesses into Chemical Production Complexes for Sustainable Development
Debalina Sengupta and Ralph W. Pike

Hydrogen Safety
Fotis Rigas and Paul Amyotte

Biofuels and Bioenergy: Processes and Technologies
Sunggyu Lee and Y. T. Shah

Hydrogen Energy and Vehicle Systems
Scott E. Grasman

Integrated Biorefineries: Design, Analysis, and Optimization
Paul R. Stuart and Mahmoud M. El-Halwagi

Water for Energy and Fuel Production
Yatish T. Shah

Water for Energy and Fuel Production

Yatish T. Shah

CRC Press
Taylor & Francis Group
Boca Raton London New York

CRC Press is an imprint of the
Taylor & Francis Group, an **informa** business

CRC Press
Taylor & Francis Group
6000 Broken Sound Parkway NW, Suite 300
Boca Raton, FL 33487-2742

First issued in paperback 2017

Version Date: 20140319

ISBN 13: 978-1-138-07606-8 (pbk)
ISBN 13: 978-1-4822-1618-9 (hbk)

Library of Congress Cataloging-in-Publication Data

Shah, Yatish T.
Water for energy and fuel production / author, Yatish T. Shah.
pages cm. -- (Green chemistry and chemical engineering)
Summary: "Water in all its forms may be the most important solvent in the development of the new "Energy Economy". This book illustrates that as energy and fuel industries diversify, we are transitioning to an economy where water will play a more and more important role in the supply of energy and fuels. It discusses the role of water in the production of raw fuels such as oil, gas, coal, uranium, and biomass. It also describes methods for how supercritical water and steam are vital for the conversion of raw fuels to synthetic fuels. "-- Provided by publisher.
Includes bibliographical references and index.
ISBN 978-1-4822-1618-9 (hardback)
1. Water--Industrial applications. 2. Energy industries--Materials. 3. Hydraulic machinery. 4. Water as fuel. 5. Green chemistry. I. Title.

TJ840.S435 2014
621.042--dc23 2014002521

Visit the Taylor & Francis Web site at
http://www.taylorandfrancis.com

and the CRC Press Web site at
http://www.crcpress.com

Contents

Series Preface

GREEN CHEMISTRY AND CHEMICAL ENGINEERING

A Book Series by CRC Press/Taylor & Francis

The subject and discipline of chemistry and chemical engineering have encountered a new landmark in the way of thinking about, developing, and designing chemical products and processes. The revolutionary philosophy, termed "green chemistry and chemical engineering," focuses on the design of products and processes that are conducive to reducing or eliminating the use and/or generation of hazardous substances. In dealing with hazardous or potentially hazardous substances, there may be some overlaps and interrelationships between environmental chemistry and green chemistry. While environmental chemistry is the chemistry of the natural environment and the pollutant chemicals in nature, green chemistry proactively aims to reduce and prevent pollution at its very source. In essence, the philosophies of green chemistry and chemical engineering tend to focus more on industrial applications and practice rather than academic principles and phenomenological science. However, similar to the chemistry and chemical engineering philosophy, the green chemistry and chemical engineering derives from and builds on organic chemistry, inorganic chemistry, polymer chemistry, fuel chemistry, biochemistry, analytical chemistry, physical chemistry, environmental chemistry, thermodynamics, chemical reaction engineering, transport phenomena, chemical process design, separation technology, automatic process control, and so on. In sum, green chemistry and chemical engineering is the rigorous use of chemistry and chemical engineering for pollution prevention and environmental protection.

The Pollution Prevention Act of 1990 in the United States established a national policy to prevent or reduce pollution at its source whenever feasible. Adhering to the spirit of this policy, the Environmental Protection Agency (EPA) launched its Green Chemistry Program to promote innovative chemical technologies that reduce or eliminate the use or generation of hazardous substances in the design, manufacture, and use of chemical products. The global efforts in green chemistry and chemical engineering have recently gained a substantial amount of support from the international communities of science, engineering, academia, industry, and government in all phases and aspects.

Some of the successful examples and key technological developments include the use of supercritical carbon dioxide as a green solvent in separation technologies; application of supercritical water oxidation for destruction of harmful substances; process integration with carbon dioxide sequestration steps; solvent-free synthesis of chemicals and polymeric materials; exploitation of biologically degradable materials; use of aqueous hydrogen peroxide for efficient oxidation; development of hydrogen proton exchange membrane (PEM) fuel cells for a variety of power generation needs;

advanced biofuel productions; devulcanization of spent tire rubber; avoidance of the use of chemicals and processes causing generation of volatile organic compounds (VOCs); replacement of traditional petrochemical processes by microorganism-based bioengineering processes; replacement of chlorofluorocarbons (CFCs) with nonhazardous alternatives; advances in design of energy-efficient processes; use of clean, alternative, and renewable energy sources in manufacturing; and so on. Even though this list is only a partial compilation, it is undoubtedly growing exponentially.

This book series "Green Chemistry and Chemical Engineering" by CRC Press/Taylor & Francis is designed to meet the new challenges of the twenty-first century in the chemistry and chemical engineering disciplines by publishing books and monographs based on the cutting-edge research and development to the effect of reducing adverse impacts on the environment by chemical enterprise. In achieving this, the series will detail the development of alternative sustainable technologies that will minimize the hazard and maximize the efficiency of any chemical choice. The series aims at delivering the readers in academia and industry with an authoritative information source in the field of green chemistry and chemical engineering. The publisher and its series editor are fully aware of the rapidly evolving nature of the subject and its long-lasting impact on the quality of human life in both the present and the future. As such, the team is committed to making this series the most comprehensive and accurate literary source in the field of green chemistry and chemical engineering.

Sunggyu Lee

Preface

The need for energy at both the individual level and the societal level is essential and rapidly growing. Throughout the history of the human race, man has found the sources of energy to make life more comfortable, convenient, and progressive. Energy is needed for all walks of life including industrial growth, transportation, manufacturing of products, and improvement of the quality of life. The modernization of society, increase in industrial productivity, increase in population, and man's constant desire for freedom to travel all contribute to the increasing demands on energy.

Fundamentally, there are 10 known sources of energy and fuel: oil, gas, solid fuels such as coal, oil shale, and bitumen (commonly known as solid fossil fuels), uranium (or nuclear), biomass, waste, solar, wind, geothermal, and water. Over the years, once the new sources of energy are discovered and harnessed, they have to compete for their share of the market. The demand for any given source of energy and fuel has gone up and down depending on the supply, competitive price, usability, and its influence to the environment. For example, in the 1800s, the source of energy (largely for heating and cooling) was biomass because oil, gas, coal, and uranium had not been discovered and the technologies to harness and use other sources of energy and fuels had not been developed. Over the years that followed, the percentage use of biomass steadily declined because of the availability, the need for use, and the pricing structure of other sources of energy and fuels. Over the past 100 years, our society and economy have been predominantly fossil fuel based.

The sources of energy and fuels have always been challenged by the need for better environment and economics. This is also a challenge for the development of new energy technologies. One source of energy and fuel cannot replace the other unless there are strong arguments based on the environment, economics, usability, and improvement of the quality of life. Often these forces compete with each other, and then local politics and social acceptance make the final decisions. One example is the recent rapid change in the use of renewable energy in favor of fossil energy due to a well-accepted notion that fossil energy, in general, affects our environment in more negative ways than the effects of renewable energies. This has led to the so-called green energy revolution. Even within fossil energy sources, it is generally believed that natural gas is less harmful to the environment than coal. This, along with changing supplies of gas, has led to the installation of more gas-driven power plants than those using coal. In our history, the changes in the sources of energy and fuels have been gradual because most technological developments in energy industries have been evolutionary and not revolutionary in their impacts.

Out of all sources of energy and fuels, one source that is most abundant, most green, most compatible with all other sources of energy and fuels, and most acceptable to the environmentalists is water. Nearly four-fifths of the world is surrounded by water. Water is everywhere underground and closely tied to the development of our fossil energy resources. Water is not only essential for all lives on this planet, but it is also the best under-tapped source of energy and fuels.

The main theme of this book is to convey the message that as we continue to find new sources of energy and fuels and continue the development of new technologies to meet our growing needs of energy, we need to be more vigilant in pursuing the role of water in the future energy landscape. Water can play a vital role in creating new energy sources that will keep our environment intact. For one thing, water is free and in plentiful supply. For the other thing, it carries the most important green source of energy, namely, hydrogen. In fact, there is a general belief that the best long-term solution for energy is hydrogen due to its least impact on environment.

We have used water without its appreciation due to its abundance. However, until now, we have not exploited all the versatility of water in its use to provide energy and fuels. New technological development over the past several decades have brought forward some of the most fascinating characteristics of water that was unknown before. These characteristics offer an even wider role for water in the future production of energy and fuel.

The book outlines, with concrete examples, five separate roles of water in the energy and fuel industry, which are as follows:

1. ***Benign roles of water in the production of energy and raw fuels***
   Water has played an important role in the recovery and purification of oil, gas, coal, oil shale, tar sand, and uranium. This role will become even more important as we (1) try to recover more unconventional sources of gas and oils, (2) use enhanced oil recovery methods more vigorously to work hard to get the remaining sources of oil, and (3) develop new technology such as "fracking process" to improve our efficiency of both conventional and unconventional oil and gas recovery. The use of water will also expand significantly with a substantial growth of recovery of bitumen from tar sand.

   The water has also played a significant role as a thermal energy carrier and a reactor moderator in the nuclear power industry. With the commercialization of the next generation of nuclear reactors using "supercritical water," the importance of water in the nuclear industry will further increase. The importance of water as a benign energy carrier will also increase significantly because of the aggressive pursuit of enhanced geothermal systems across the world. An increased use of solar energy will also require more use of water as a thermal energy carrier. Finally, it is predicted that by 2040, electrical energy will be about 40% of our total energy usage. This growth of electrical energy will require more steam-driven turbines that will drive more electric generators.
2. ***Role of steam as a reactant***
   While steam has been used as a reactant for a variety of gasification and reforming processes in the past, this role will significantly increase because of (1) large growth in natural gas production (by unconventional gas recovery), which can be used for reforming (and hydrogen generation) and power plants; (2) the increased use of biomass gasification and reforming as it is a good source of hydrogen and it is renewable and environmentally friendly;

and (3) large growth in waste conversion industry to generate biogas that can also be reformed to produce hydrogen. In sum, the role of steam to produce hydrogen will significantly increase in outgoing years.

3. ***Role of water as a reactant, a reaction medium, and a catalyst***
Recent developments showing the significant changes in the properties of water at elevated temperatures and pressures have galvanized a significant amount of R&D in making synthetic solid, liquid, and gaseous fuels from a variety of feedstock using water. At elevated temperatures and pressures, water possesses unique organic, liquid-like properties. The hydrothermal processes at high temperatures and pressures accelerate the conversion of biomass to more coal-like solids, oil-like liquids (biocrudes), and gases containing methane and hydrogen. This technology has a very bright and growing future.

   Water has also been used as a medium to carry out the catalytic aqueous-phase reforming of selective carbohydrates to produce hydrogen, syngas, alkanes, and other monofunctional products that can be further upgraded to produce a variety of fuels, fuel additives, and chemicals. This selective catalytic process has resulted in the birth of a new "Bioforming process," which also has a significant growth potential.

   There have been other interesting developments in the use of water as a reactant. The acid-catalyzed hydrolysis has been successfully used to produce active and versatile chemicals such as levulinic acid, furfural, gamma-valerolactone from cellulose, and carbohydrates. These chemicals can provide platforms for many fuels, fuel additives, and useful chemicals. This technology has a bright future and has resulted in the development of the "Biofine process."

   Anaerobic digestion of waste to produce hydrogen and methane has been practiced for a long time. This is one of the most efficient energy conversion processes. An increase in waste production worldwide (e.g., municipal solid waste that will reach close to one billion tons per year in few years) will increase the application of this process. Water is an important part of this biochemical process.

   The use of water in hydrolysis and fermentation processes will also grow significantly as more emphasis on conversion of lignocellulose to ethanol and higher alcohols is mandated all over the world. The production of ethanol will significantly grow and increase the role of water in the fermentation industry.

4. ***Role of supercritical water in the production of synthetic fuels***
Since the late 1970s, the use of supercritical water to convert all carbonaceous materials to synthetic fuels and chemicals has been rapidly increasing. Supercritical water possesses some very unique properties, which allow easy conversion of many carbon-based feedstock to liquid fuels, methane, and hydrogen. In recent years, this technology has been widely exploited to produce hydrogen. The growth in this technology will be accompanied by a significant growth in the use of water.

5. ***Role of water as a direct source for fuels and energy***
Hydrogen can be generated by water dissociation. This subject has been very heavily researched in recent years. Technologies such as electrolysis,

photocatalytic and photobiological dissociation of water and thermochemical dissociation of water can provide some important breakthroughs for the production of hydrogen. The uses of solar and nuclear energies for this purpose have also been heavily examined.

Gas hydrates provide methane from water-based clathrate molecules. Gas hydrates are loosely bound methane molecules in water (or ice) cage. These naturally occurring compounds are the largest source of carbon in the world. The recovery of methane from the hydrates has been heavily researched all over the world.

Water has also been a direct source of energy by its use in dams for hydroelectricity (potential energy). This is one of the cleanest sources of energy and is rapidly growing all over the world. Besides hydroelectricity, in the recent years, other sources such as hydrokinetic energy that are imbedded in all types of moving water such as in tidal waves, offshore waves, undercurrents, and inland waterways are being harnessed. Tidal waves can provide both potential and kinetic energies for power. These are renewable, predictable, and clean sources of energy. Water can also provide power near the equator using ocean thermal energy conversion technologies that use the difference in surface temperature and temperature at the high depth to drive heat engines.

The potentials for each of these five applications are enormous. So far, we have only scratched the surface. In fact, together it appears that the future of energy and fuel landscape is moving toward the expansion of the water industry with all other sources of energy and fuels as supporting players. The concept of water refinery is not unimaginable. The above-mentioned five roles of water are described in detail in 12 chapters of this book. The content of each chapter is described in the "Introduction" chapter.

The expanded use of water for energy and fuel production leads to another societal issue that will have to be managed. Clean water is essential for human and animal life. The expanded use of water for energy and fuel production may create a problem for the available amount of drinkable water for human and animal needs and useable water for agricultural needs. Just as there are tensions between the use of food materials such as corn, soybeans, maize, and other carbohydrates for fuels (such as ethanol), tensions will also be created by the use of water for energy and fuel production, and the need of clean water for human and animal needs as well as for agricultural purposes. The strategic management of water will be the next important societal issue. The treatment of water used in energy and fuel industries will become an independent industry by itself. This industry will have to manage the overall societal needs for the water to maintain the required strategic distribution of water among its different usages. Along with energy and fuel, a prudent use of water will be the next societal challenge. Water, however, will be the centerpiece of future energy and fuel landscape.

This book is very useful to all academic, industry, and government personnel who are engaged in R&D, pilot-scale development, and commercialization of new technologies for energy and fuel.

Y.T. Shah
Norfolk State University

Author

Dr. Y.T. Shah received his BS in chemical engineering from the University of Michigan and MS and ScD in chemical engineering from Massachusetts Institute of Technology. He has more than 40 years of academic and industrial experience in energy-related areas. He was chairman of the Department of Chemical and Petroleum Engineering at the University of Pittsburgh, dean of the College of Engineering at the University of Tulsa and Drexel University, chief research officer at Clemson University, and provost at Missouri University of Science and Technology, University of Central Missouri, and Norfolk State University. He was also a visiting scholar at Cambridge University in the United Kingdom and a visiting professor at the University of California, Berkley, and Institut für Technische Chemie I der Universität Erlangen, Nürnberg, Germany. Currently, he is a professor of engineering at Norfolk State University. Dr. Shah has written four books related to energy: *Gas-Liquid-Solid Reactor Design* (published by McGraw-Hill), *Reaction Engineering in Direct Coal Liquefaction* (published by Addison-Wesley), *Cavitation Reaction Engineering* (published by Plenum Press), and *Biofuels and Bioenergy—Processes and Technologies* (published by CRC Press). He has also published more than 250 refereed reviews, book chapters, and research technical publications in the areas of energy, environment, and reaction engineering. He is an active consultant to numerous industries and government organizations in the energy areas.

1 Introduction

The energy and fuel landscape is changing faster than ever. We need energy and fuels for heat, power, electricity, and transportation. Generally, the major sources of raw fuels are oil, natural gas, coal, biomass, waste, and uranium. These raw fuels can provide either chemical or nuclear energy to generate power or raw materials for synthetic fuels and a host of chemicals and materials. Other major sources of energy for power production are solar, wind, geothermal, and water. Together, these 10 sources satisfy our needs for raw energy and fuels for residential, industrial, and transportation purposes. They also provide raw materials for important chemicals and materials.

The term "synthetic fuels," or "synfuels," generally refers to hydrogen, syngas (a mixture of hydrogen and carbon monoxide), methane, or a mixture of methane with other hydrocarbon gases, liquid alkanes (such as gasoline, diesel fuel, heating oil, jet fuel, naphtha, alcohols, biocrudes, and many other chemical additives), and intermediates as well as refined solid fuels (such as refined coal or oil shale, hydrochar, and biochars) that are all capable of direct use (as a source of chemical energy) to provide heat, electricity, or a source of energy for transportation purposes. These synthetic fuels also allow the storage of energy for an extended period of time.

1.1 GLOBAL ENERGY LANDSCAPE: PAST, PRESENT, AND FUTURE

In the twenty-first century, the energy industry is forced to diversify to address the three issues: (1) depletion of worldwide oil productions and reserves as well as refining capacity, (2) a stronger global demand for energy due to increased needs by China, India, and other developing nations, and (3) the effect of carbon emissions (in the form of volatile hydrocarbons and carbon dioxide) on global warming and climate change. During the last two decades, along with further expansion of clean energy sources such as wind, solar, hydroelectric, and geothermal, the development of renewable biomass energy (often called "bioenergy") has increased its momentum. Unlike fossil energy, bioenergy is proven to be carbon neutral because of the use of carbon dioxide by plants for photosynthesis. Renewable and carbon-free energy sources such as nuclear, solar, wind, and hydroelectric have also gained significant momentum due to push on "green energy."

Today, the power and fuel need is supplied by five separate types of energy industry. The largest supplier of power and fuel (>85%) is still "fossil energy," which is organic hydrocarbon based, and the fossil energy industry is attempting to reduce carbon emissions to the environment and its effect on global climate change. In this regard, more and more coal-based power plants are being replaced by the power plants that use the natural gas. In general, natural gas is more environmentally friendly than coal. The supply of natural gas is increasing because of the success in recovering new unconventional gas resources. The carbon-emitting fossil energy industry is still facing significant political pressures due to its harmful influence on the environment.

Over the last several decades (particularly in Europe and South America), much effort has been made to use renewable bioenergy, which uses biomass (and cellulosic waste), because this type of energy is carbon neutral in its natural life cycle. Carbon dioxide emitted by the use of biofuels is captured by plants and vegetables for their regenerations. Unlike fossil energy industry, bioenergy industry extensively uses aqueous processes (such as hydrothermal liquefaction, anaerobic digestion, hydrolysis, fermentation, aqueous-phase reforming, and supercritical gasification) along with the traditional thermochemical processes used in the fossil energy industry. This book illustrates the extensive use of water as a reactant to generate power and fuels in the bioenergy industry.

The remaining three types of energy industry that are also gaining an additional momentum in the recent years are all carbon-free energy industries and include (1) nuclear energy; (2) solar, wind, geothermal, hydroelectric, and hydrokinetic energies; and (3) hydrogen. These sources of energy do not emit any carbon in the environment. Nuclear energy has been in existence for a while, but its acceptance has been politically hindered because of safety issues. It is projected that its use will moderately grow over the next 30 years. Solar, wind, geothermal, hydroelectric, and hydrokinetic energy sources, which are time and/or location dependent, are renewable and are dependent on the natural elements. While each of these sources of energy will be an industry by itself, renewable nature and carbon-free characteristics unite them.

Hydrogen is the most abundant source of energy on this earth and it is the cleanest and most likely the solution to the energy needs of the world in the long term. Hydrogen economy may dominate "energy economy" in the long term, and it is slowly becoming an industry by itself. Unfortunately, hydrogen is found only in the compound form and its recovery as pure hydrogen requires fossil, biofuel, and water resources. While the generation of hydrogen from fossil and biofuels may cause carbon emissions, the use of hydrogen is carbon free. Water is the most abundant source of hydrogen and the generation of hydrogen from water can be carbon free. In the recent years, significant efforts have been made to recover hydrogen from water by innovative water dissociation technologies, most of which are outlined in this book. The research and development (R&D) in hydrocarbon-based, cellulose-based, carbohydrate-based, nuclear, and carbon-free energy sources will continuously change the future landscape of the energy industry.

Because of our quest to accommodate the growing needs of energy by the developing countries such as China, India, Brazil, Russia, and many African nations as well as to satisfy the need for reduced carbon emission to the environment, the global supply and demand picture will considerably change over the next several decades. ExxonMobil has carried out supply and demand projections for energy and fuels up to 2040 [1]. Similar reports (with somewhat different projections) have also been published by other oil companies (such as BP). Here we briefly summarize some of the important conclusions of the ExxonMobil report (EMR) [1].

1. Since 1800, the energy and fuel landscape has been constantly changing as new sources are developed and the old sources either dry up, or become relatively more expensive, less usable, or environmentally more unacceptable. The energy and fuel landscape from 1800 to 2040 has been graphically depicted by the EMR [1]. The best calculations from this graph are presented in Table 1.1. The data show that renewable fuel (biomass) was the sole source of energy in 1800. Over the last more than two centuries, its use has declined significantly because of the discovery of oil, gas, and coal reserves. For nearly a century, our energy production has largely been fossil fuel based. However, the renewable energy will be coming back due to more technological developments and favorable environmental impacts. Within fossil fuels, over the next several decades, the use of natural gas will significantly increase because of the new technological developments and less-favorable environmental impacts by coal. The table indicates that even in 2040, 75% of energy usage will still be provided by oil, gas, and coal and that oil and gas will supply 60% of the total energy demand. The extent of growth in the use of renewable energies (solar, wind, biomass, geothermal, and hydroelectric/hydrokinetic) will be significant. The level of increased percentage contribution by the renewable energies to overall energy consumption will depend on the rate of commercialization of these technologies.
2. The growth in energy demand projection during a 50-year (1990–2040) span by several countries is illustrated in Table 1.2 [1]. It is predicted that the growth in energy demand in Organisation for Economic Co-operation and Development (OECD) countries—most of the developed countries of the world—will be very small (around 18.5%), whereas the growth in each non-OECD country (except Russia/Caspian) will be >200%. Overall growth in non-OECD countries will be about 174%. The energy demand in Russia/Caspian region will actually go down during this period. This

TABLE 1.1
Energy Landscape 1800–2040 Calculated from EMR

Year	Fossil (%)	Nuclear (%)	Renewables (%)
1800	2	0	98
1850	10	0	90
1900	50	0	50
1950	76	0	24
2000	82	7	11
2040	75	9	16

Source: ExxonMobil, "The outlook for energy: A view to 2040," US Edition, ExxonMobil Report, ExxonMobil, Irving, TX, 2012. With permission.

Note: The numbers in the table are approximate calculations from the graphical data reported in the EMR.

TABLE 1.2
Growth in Energy Demand Projections for OECD and Some Non-OECD Countries (1990–2040)

Country	Energy Demand (Quadrillion Btu's) and Growth			
	1990	2015	2040	Growth (1990–2040) (%)
OECD countries	189	225	224	18.5
Non-OECD countries	171	315	469	174
China	33	105	138	318
India	13	35	61	369
Latin America	15	29	45	200
Middle East	11	30	51	364
Africa	17	29	62	265
Russia/Caspian	57	43	43	−24.6

Source: ExxonMobil, "The outlook for energy: A view to 2040," US Edition, ExxonMobil Report, ExxonMobil, Irving, TX, 2012. With permission.

Note: The numbers in this table are best calculations/estimations from the graphical and tabulated data reported in the EMR.

projected growth will be due to (1) an increase in population, particularly in India and Africa, (2) increased demands for electricity in poor countries, and (3) increased industrial and transportation activities because of increased gross domestic product (GDP) and improved quality of life in these countries. Non-OECD countries will largely contribute to the total growth in energy demand in the world.

3. The growth in end-use demand in sectors and regions during the period 1990–2040 along with the growth in the CO_2 emissions is illustrated in Table 1.3 [1]. These results again show that energy demand in each sector (residential/commercial, industrial, and transportation) will significantly go up in non-OECD (except Russia) countries, whereas it will either increase moderately or go down in OECD countries. The energy demand in Russia will go down in every sector. The increased demand in non-OECD countries is for the same reasons outlined above. The largest growth across the world will be in electricity demand. In fact, the report predicts that in 2040, electricity demand will be about 40% of the total energy consumption in the world. OECD countries will focus on energy efficiency and environmental issues. Carbon dioxide emission in OECD countries will go down, whereas this will significantly go up in non-OECD countries. In 2040, 70% of the total CO_2 emission will be generated by non-OECD countries and the three major sources for CO_2 emissions will be industrial, transportation, and electricity [1]. While energy-related CO_2 emission per capita will go down in the United States and Europe, it will go up in India.

TABLE 1.3
Growth in End-Use Demand and CO_2 Emissions in Sectors and Regions (1990–2040; 50-Year Span)

Sector	Growth (1990–2040) (%)						
	Region						
	NA	Europe	Russia/Caspian	Africa	Asia/Pacific	LA	Middle East
Residential/ commercial	33.3	17.6	−33	214	90	100	800
Transport	24	36	−17	350	36	225	266
Industrial electricity	6.6	−12	−25	183	225	200	300
Demand	18.2	56	20	800	614	600	500
CO_2 emission	−7	−27	−41	343	226	186	271

Source: ExxonMobil, "The outlook for energy: A view to 2040," US Edition, ExxonMobil Report, ExxonMobil, Irving, TX, 2012. With permission.
Note: The numbers are calculated from the tabulated and graphical data presented in the EMR.
LA, Latin America; NA, North America.

Other important changes will also occur in the energy landscape over the next 30 years. More shift from hydrocarbon-based energy to non-hydrocarbon-based energy supply will occur. The use of hydrogen for energy will become more prominent. R&D efforts to produce hydrogen from water and other sources using novel techniques will increase. The efforts to use "solar fuels" will become more prominent. As mentioned earlier, the renewable energy will play a larger role in the energy portfolio. In short, energy economy will become less and less dependent on the hydrocarbon industry. The EMR also gives more details on other factors such as the nature of transportation fuel, the nature of vehicle use, and their energy consumptions [1].

1.2 THE THEME AND OUTLINE OF THE BOOK

While the world is craving for more sources of energy and fuels, one source of energy that is most abundant and environmentally acceptable is water. Water in all its forms (i.e., subcritical, supercritical, steam, heavy water) is the most important solvent in the development of new "energy economy." Four-fifths of the earth's surface is covered with water, which plays a very significant role in the generation of various forms of energy and fuels. The premise of this book is that water is essential not only for human health and environment but also for the development of a broad-based "energy economy."

The central theme of this book is to illustrate that as energy and fuel industries diversify, we are transitioning from predominantly oil- and fossil fuel-based economies to the economy where water plays a more and more important role in the supply of energy and fuels. The book shows that water contributes to the production of energy and fuels in at least 12 different ways. Each of these methods will become more important as the need for energy and fuels grows and diversifies. The role of water is distinctly different in these 12 methods of water usage for the production of energy and fuel, which is discussed below. Each method is separately described in Chapters 2–13. The content of each chapter is briefly described below.

While the rest of the book is divided into 12 separate chapters based on 12 separate roles of water (in all its forms) in energy and fuel industries, these roles can also be classified in terms of the following five major functions played by the water:

1. The benign role of water in the production of raw fuels such as oil, gas, coal, uranium, and biomass as well as the benign role of water as a means for energy carrier in the form of hot water and steam (Chapters 2 and 3).
2. Role of steam as a reactant in the conversion of raw fuels to synthetic fuels, which can then be a direct source of energy (Chapter 4).
3. Role of water as a reactant, reaction medium, and catalyst in the conversion of raw fuels such as coal, biomass, and waste to useful synthetic fuels. Aqueous water under subcritical conditions can react with various feedstock thermochemically in the presence of a catalyst or biochemically to produce gaseous, liquid, or solid synthetic fuels (Chapters 5–9).
4. Role of supercritical water in the conversion of fossil- and bio-based feedstock to synthetic fuels in the presence and absence of a catalyst. Supercritical water provides an excellent medium for many organic reactions (Chapter 10).
5. Water as a direct source of energy and fuels. Hydrogen can be generated by water dissociation. Gas hydrates provide methane from water-based clathrate molecules. Water can also be a direct source of energy when stored in dams (potential energy) or other sources of hydrokinetic energy such as in tidal waves, offshore sea and ocean waves, undercurrents, and inland waterways. In the equator region, power can also be generated using ocean thermal energy conversion technology, which harnesses the temperature difference between the ocean surface and the underwater to drive heat engine (Chapters 11–13).

1.2.1 Chapter 2: Water for Raw Fuel Production

Water plays a very important role in the generation of renewable source of energy, namely bioenergy and biofuels. The growth of all types of plants, vegetables, crops, and trees requires water. The growth of algae, which is one of the richest sources of oil, requires wetlands. Thus, water is a basic necessity for the production of biomass, a raw material for biofuels and bioenergy.

Water, however, also plays a vital role in the recovery of raw fuels such as oil, gas, coal, and uranium. The amount of water required (often denoted as "produced water") to recover raw fuels is so large that the treatment, disposal, and management of produced water has become a rapidly growing industry by itself. Some people consider fossil fuel industry as water industry with oil, gas, coal, and uranium as byproducts. For every barrel of oil produced, six to eight barrels of water is needed or produced, which may grow to 12 barrels of water as "enhanced oil and gas recovery methods" become more important.

Chapter 2 illustrates the growing importance of water in raw fuel productions by examining the use of water in four specific cases: (1) the role of underground water and produced water in the recovery of oil and gas, and in particular an increased water usage for the recovery of coal bed methane and gas from geopressurized zones; (2) an increased use of water in enhanced oil recovery methods; (3) an important role of water in the newly developed and promising method called "fracking" in the recovery of unconventional gas (shale gas and tight gas); and (4) the use of water in mining, preparation, and extraction of coal, tar sands and heavy oils, uranium, and oil shale. The large expansion of tar sand industry will particularly enhance the use of water.

The four cases examined are the basis of existing and all future growth and diversification of raw fuel industries. While the large use of water for cases 1 and 4 described earlier is already known, it is the expanded use of water for all cases that will increase the importance of water in raw fuel production industries. As the existing oil wells age, the recovery of last remains from the oil wells will require more and more use of enhanced oil recovery methods. These methods use a number of techniques such as surfactant, polymer, alkaline flooding, and steam injection, which involve water or steam. In the case of unconventional oil recovery, pressurized steam is often used. Steam combustion is also used to increase fluidity of trapped oils.

The recoveries of unconventional gas such as shale gas, tight gas, and coal bed methane are the game changers in the natural gas industries.

Water has been used as a material for creating underground fractures in various geological structures to recover unconventional gas. This process is called fracking, and it opens up the tight geological structures by the injection of a high-pressure water in the ground either horizontally or at an angle. The fracking process increases the porosity of the tight geological structures and therefore releases the trapped gas. The pressurized water also contains additives (e.g., surfactants) and other chemicals. New unconventional gas industry will thus use significantly more water than old conventional natural gas industry. Chapter 2 briefly examines these and other relevant issues.

1.2.2 Chapter 3: Water as Energy Carrier

Water also plays a benign but vital role in the recovery of various forms of energy. Chapter 3 illustrates this with four important applications of water and steam to recover energy: (1) nuclear reactor, (2) geothermal sources, (3) solar energy, and (4) thermal energy generated from the combustion of various types of fuels. In each case, water or steam plays a vital role in converting energy and carrying the thermal energy to generate electricity and heat.

Water has always been an effective "energy carrier." For example, in a nuclear reactor, water carries energy generated by the nuclear fission process in the form of steam (thermal energy) to convert it into electrical energy. Water also plays a role of reactor moderator, ensuring safety. More than 80% of current nuclear reactors use water as an energy carrier and/or reactor moderator. New nuclear reactors may use supercritical water as the next generation of reactor coolant and thermal energy carrier.

Water has also been used as a medium to transfer geothermal energy to heat pumps, air conditioners, or electrical devices by carrying geothermal energy in the form of steam for a subsequent energy conversion process. In future, the development of "enhanced geothermal systems" will require water not only as a geothermal energy carrier but also as a fluid required to open up deep compressed geological structures that carry geothermal heat. This dual role of water will make its use larger and more important in the recovery of geothermal energy. New geothermal recovery systems will also use existing underground infrastructure for oil and gas recovery for geothermal energy recovery.

Water can also be an effective "energy storage" device for renewable energy sources such as wind and solar energy that are time and location dependent. The electricity generated from these sources can be stored in the form of hydrogen through water dissociation, and this hydrogen can then be used to generate electricity during "off-time" periods. The stored energy can also be used for "peak energy" needs. While conversion of electrical energy into hydrogen is not the most efficient process, it provides another option for storing electrical energy instead of using conventional power grids, batteries, or capacitors. Unlike these conventional sources, once the electrical energy is stored by hydrogen, it will not dissipate over time. Water can also be used for thermal storage of excess electricity. Solar energy has also been used for heating and cooling homes and industrial buildings through the use of water. Water is once again an important thermal energy carrier for this use of solar energy.

Finally, steam turbine has been an effective device that uses steam to drive turbine which in turn generates electricity. Steam in this process is often produced using combustion heat generated from the burning of coal, oil shale, biomass, waste, and so on. Once again, steam is an effective energy carrier in the combustion processes to generate power. As mentioned earlier, the EMR [1] has predicted that by 2040, nearly 40% of our energy consumption will be in electric power. This expanded use of electricity will require a significant growth in the use of steam turbine. All of these benign roles of water as an energy carrier are briefly examined in Chapter 3.

1.2.3 Chapter 4: Steam for Synthetic Gas Production

While the "steam gasification and reforming" process has been in place since the beginning of the fossil fuel industry, steam gasification and reforming of coal was not as popular and productive as steam reforming of natural gas. In the recent years, steam gasification and reforming of biomass has become more popular and productive for hydrogen generation.

The use of steam to recover gaseous synthetic fuels of different compositions is outlined in Chapter 4. Steam gasification and reforming of carbonaceous fuels (fossil as well as biomass) either alone or in combination with air (or oxygen), carbon dioxide, or hydrogen is a commercially accepted process. Depending on the nature of feedstock and operating conditions, the process generates gaseous fuel largely consisting of methane, carbon monoxide, carbon dioxide, water, and hydrogen. Minor amounts of other volatile hydrocarbons and nitrogen as well as impurities such as sulfur, nitrogen and chlorinated compounds may also be present in the product gases. For fossil fuels such as coal, shale oil, bitumen, tar sand, and crude oil, the gasification by steam alone has not been as effective as gasification by steam with oxygen, carbon dioxide, and hydrogen. The thermodynamics of steam gasification of coal are not very favorable [2]. Generally, such a gasification process predominantly generates pure syngas at temperatures higher than 1000°C–1200°C. The required temperature is, however, lower for biomass and low-rank coals. With excess steam and at high temperatures, the most dominant product is hydrogen with some carbon dioxide.

When steam gasification is carried out in the presence of catalysts, such as alkalis, and Ni-based or other supported noble metal catalysts (e.g., Ru, Rh catalysts), both gasification and reforming occur simultaneously. Along with steam gasification, steam reforming has been used for a long time to generate hydrogen needed for the ammonia and urea productions, petroleum refining, and other hydrogenation reactions producing chemicals. Ammonia is an important raw material for the fertilizer industry. Steam reforming is, to date, the most economical method for hydrogen production. In the recent years, steam reforming has been carried out along with dry reforming and partial oxidation reactions to generate syngas of various hydrogen–carbon monoxide compositions.

Chapter 4 evaluates various aspects of steam gasification and reforming technologies (SGRT) such as (1) the mechanism and kinetics of steam gasification and reforming processes in the presence and absence of other gases, (2) catalysis and reactors for steam gasification and reforming processes, and (3) effects of feedstock and operating conditions on the product distributions. The chapter also examines underground gasification and combustion and multistage processes for steam gasification and reforming. Finally, the effects of water gas shift reaction and simultaneous presence of dry reforming and partial oxidation reactions (i.e., tri-reforming) on the SGRT are also assessed.

In the recent years, novel approaches to steam reforming and gasification such as solar reforming and gasification and microwave-assisted reforming have also been investigated. Since steam reforming is an endothermic process, the use of solar energy for heating makes the process more energy efficient. Chapter 4 examines this and other novel steam gasification and reforming processes.

1.2.4 Chapter 5: Synthetic Fuel Production by Water under Subcritical Conditions

In recent years, the use of water under high-temperature and high-pressure conditions (in subcritical region) to carry out various thermochemical transformations has been increasing due to recognition that the properties of water change significantly with increase in temperature and pressure. These changes allow a number of organic reactions to occur in the aqueous medium. Water becomes nonpolar as temperature increases. Chapter 5 describes in detail this new-found role of water as a reaction medium for transformation of raw fuels such as coal, biomass, waste, and others to produce a variety of gaseous, liquid, and solid synfuels.

The chapter first examines the changes in the properties of water as temperature and pressure increase. One of the most important transformations that occur is that water becomes nonpolar at higher temperatures. For example, the properties of water at 370°C are similar to those of acetone at 25°C. This transformation allows many hydrocarbon reactions to occur in the aqueous phase.

It is well known that fossil fuels, such as coal, oil, and gas, are the results of slow geological transformations of biomass waste and human and animal remains buried underground. These transformations have taken millions of years, resulting in fossil fuels such as coal and oil with higher carbon content and lower hydrogen and oxygen content than the original feedstock of biological nature.

Even within coal and oil, there are gradations of properties. For example, the highest ranking (i.e., longest geological age) anthracite coal contains the highest amount of carbon and the lowest amount of hydrogen and oxygen compared to younger coals such as bituminous, subbituminous, and lignite. The youngest lignite coal contains hydrogen/carbon ratio (H/C) and oxygen/carbon ratio (O/C) similar to that of biomass and peat. Just like biomass, lignite coal contains high oxygen concentration and high moisture content. All this is well illustrated by the famous Van Krevelen's plot [2], which shows H/C versus O/C for various types of fossil fuels and biomass. The plot shows that, in general, fossil fuels contain lower hydrogen and oxygen and higher carbon contents compared to those found in the biomass.

The chapter illustrates that the aging and geological transformation of biomass can be accelerated by the hydrothermal conversion processes. High thermal and pressure forces exerted during hydrothermal conversion processes rapidly convert biomass into more coal (hydrochar), oil (biocrude), or syngas similar to natural gas. Biomass is easily transformed to these products in high-temperature and high-pressure water because of strong thermochemical interactions between the biomass and the water.

Hydrothermal carbonization brings solid biomass properties closer to that of coal [3]. The hydrochar produced during this process exhibits properties that are closer to a subbituminous coal. Biomass produces cleaner hydrochar, which is an important raw material in the production of numerous types of gaseous and liquid fuels.

At higher temperatures and pressures under subcritical conditions, biomass can also be converted to oils. This process of hydrothermal liquefaction produces biocrude, which is similar to crude oil, and just like crude oil, it can also be upgraded. Finally, hydrothermal gasification produces methane, hydrogen, or syngas. The hydrothermal carbonization, liquefaction, and upgrading of biocrude as well as hydrothermal gasification processes have been successfully examined in the recent years, and some new commercial processes have been evolved based on these concepts. Chapter 5 briefly illustrates these topics.

The chapter also evaluates the coal–water chemistry in three different areas: (1) the effect of pretreatment of coal by water on coal liquefaction, (2) the liquefaction of coal in high-temperature and high-pressure water, and (3) the use of coal–water slurry in various combustion processes. The low-rank lignite coal is amenable to liquefaction in high-temperature water; however, the liquid product that is generated is of poor quality and requires significant upgrading. Coal–water slurry can be used as a fuel for combustion in boilers, gas turbines, and diesel engines. While the affinity of coal for water is not as pronounced as that of biomass, the chapter illustrates that coal–water chemistry still needs to be further explored, particularly under high-temperature and high-pressure conditions.

1.2.5 Chapter 6: Production of Synthetic Fuels by Aqueous-Phase Reforming

Along with the hydrothermal conversion of biomass, an aqueous-phase reforming in subcritical water also plays an important role in the production of synfuels from a variety of oxygenated compounds in biomass. An aqueous-phase reforming process carries out selective conversion of sugar-based reactants such as glucose and fructose to hydrogen, syngas, or liquid alkanes and monofunctional groups depending on the nature of catalysts and other operating conditions. These compounds can be subsequently upgraded to liquid fuels using a variety of conventional refining operations.

The use of an aqueous-phase reforming process using a suitable catalyst to generate selective fuel products from various organic compounds is a relatively new and exciting technology. In this process, both catalytic materials and the nature of support are equally important. Chapter 6 gives a detailed and up-to-date account of the use of selective catalysis for the production of a variety of synfuels and/or useful platform chemicals in an aqueous-phase environment.

In the recent years, significant development work has been carried out to produce liquid fuels by upgrading (through a variety of condensation reactions) of monofunctional groups produced by aqueous-phase reforming process. This has led to the development of a "bioforming process" by Virent Inc., Madison, Wisconsin. The process is highly energy efficient and produces selective hydrocarbons that can be useful for specialized jet fuel, diesel, and other transportation fuel materials. The chapter outlines our present state of knowledge of this important and novel use of water chemistry to produce hydrogen and selective liquid transportation fuels.

1.2.6 Chapter 7: Production of Synthetic Fuels and Chemicals by Hydrolysis Followed by Selective Catalytic Conversions

Chapter 7 deals with another method of producing liquid fuels, fuel additives, and chemicals from a variety of feedstock using water. The method involves acid hydrolysis of a variety of carbohydrates, cellulose waste, and biomass to produce important platform chemicals such as furfural, levulinic acid (LA), and gamma-valerolactone (GVL). These chemicals can be catalytically upgraded to produce a variety of fuels, fuel additives, and useful chemicals.

The chapter describes the "biofine hydrolysis process," which fractionates lignocellulose into various fractions such as cellulose, hemicellulose, and lignin by hydrolysis and produces six-carbon (glucose) and five-carbon sugars (xylose). Instead of reforming these oxygenated compounds, as described in Chapter 6, five- and six-carbon sugars are catalytically converted to intermediate platform chemicals such as furfuryl and hydroxymethylfurfuryl (HMF) for five-carbon sugars and LA for six-carbon sugars. LA can also be further converted to GVL, another important platform chemical that can also be converted to a number of fuels, fuel additives, and chemicals. Formic acid and ligneous char are produced as byproducts for the biofine hydrolysis process.

The chapter shows how different types of fuels, fuel additives, and chemicals for various industrial applications can be made using different upgrading strategies for the platform chemicals. The process can handle a variety of feedstock and is proven to be economical. A small commercial plant for this process is already in operation in Italy and larger commercial plants are being pursued in Ireland, the United Kingdom, and the United States.

1.2.7 Chapter 8: Production of Hydrogen and Methane by Anaerobic Digestion of Aqueous Waste

Aqueous-phase conditions are well known for carrying out biological reactions. An anaerobic digestion of aqueous agricultural and other biological waste can produce methane and hydrogen using suitable enzymes or consortia of microorganisms. Water can thus biochemically react with biomass to generate methane and hydrogen. Such reactions generate "landfill gas" (which is predominantly methane [about 55%] and carbon dioxide) from cellulosic waste. Landfill gas is an important raw material for power generation or for the production of other gaseous and liquid fuels via reforming and Fischer–Tropsch (FT) technologies.

Chapter 8 deals with the production of methane and hydrogen by biochemical anaerobic digestion of biomass and waste in aqueous environment. The chapter illustrates biochemical mechanisms to convert cellulosic waste into methane and hydrogen. Although landfill gas is a prime example of such conversion, the aqueous waste from numerous other types of waste such as animal and human manure, agricultural waste, forestry, and plant waste can also be converted to methane and hydrogen (commonly known as "biogas"). Anaerobic digestion of biological waste is one of the most energy-efficient and fastest growing industries in the world. The chapter examines various operational issues related to this industry.

1.2.8 Chapter 9: Production of Ethanol by Aqueous-Phase Fermentation

The biochemical conversion of sugar, glucose, fructose, and so on to ethanol and other alcohols has been a long-standing industry. Although the fermentation process has been used for the production of beers, liquors, and so on, its application for the transportation fuels and their additives has become more important in the recent years because of an increased emphasis on renewable energy.

Fuel-grade ethanol can be produced from corn, starch, barley, or sugarcane by hydrolysis and fermentation processes. This has been commercialized for a long time. In the recent years, more emphasis has been placed on the conversion of lignocellulosic materials to ethanol by hydrolysis and fermentation processes. Recent research on new methods of pretreatments, acid and enzyme hydrolysis, and discovery of new microorganisms for fermentation has allowed this biological process to be applied to a broad range of lignocellulosic materials. New developments have also led to the production of higher alcohols such as butanol, which has a higher fuel value. Chapter 9 briefly examines our current state of art for these technologies and processes.

The future development of alcohol production from lignocellulosic materials will continue to require better methods of pretreatment, hydrolysis, and fermentation.

New microorganisms and enzymes will have to be developed to make these processes more efficient and economical. The fermentation process can generate specific types of products and will require basic understanding of the applications of genomics and proteomics to different types of lignocellulosic materials.

1.2.9 Chapter 10: Production of Synthetic Fuels by Supercritical Water

Water under supercritical conditions behaves very differently than at room temperature. Water has high solubility for many organic and cellulosic compounds under these conditions. The density, viscosity, and other properties facilitate the conversion of a variety of feedstock to fuels such as hydrogen, methane, and syngas. Since the pioneering work of Modell at Massachusetts Institute of Technology (MIT), this technology has made enormous progress and now many pilot-scale operations for the application of this technology to generate synfuels have become a reality.

In recent years, the interest in the use of supercritical water for the production of fuels and chemicals has been rapidly expanding. As mentioned earlier, the main reason is that the unique properties of supercritical water allow a variety of organic reactions to be carried out in the supercritical phase. In these reactions, water not only plays a benign role of solvent but also plays a role as an active reactant or a catalyst. Properties of water under supercritical conditions ensure that important organic reactions can be carried out in a homogeneous medium.

Supercritical water can play five different functions: (1) a medium in which numerous types of organic chemical synthesis occur, (2) a medium for partial or complete oxidation of numerous hazardous or nonhazardous materials, (3) a medium in which complex materials decompose and produce liquids and gases, (4) a medium for thermal or catalytic gasification of simple and complex materials to produce fuels such as methane and hydrogen, and (5) a medium to generate hydrogen by catalytic gasification and reforming of various carbonaceous materials. Chapter 10 outlines the role of supercritical water in each of these functions with a special emphasis on the functions that generate synthetic fuels.

Collectively, Chapters 4–10 illustrate various thermochemical, catalytic, and biotechnological options to convert coal, biomass, waste, and their mixtures to a variety of synthetic gaseous, liquid, and solid fuels. In all of these cases, water provides an important role of a reaction medium, a reactant, or a catalyst.

1.2.10 Chapter 11: Production of Hydrogen by Water Dissociation

Water can also be a direct source of fuel. Hydrogen can be generated from water by its dissociation. Hydrogen is the cleanest form of energy and may be the only long-term solution to our energy needs.

Chapter 11 examines three basic methods to dissociate water to produce hydrogen: electrolysis, photocatalytic or photobiological dissociation, and thermal or thermochemical dissociation. Various ramifications of each of these methods are also briefly examined. The chapter also briefly assesses other novel methods for the production of hydrogen from water.

One of the major issues with all these technologies for water dissociation is the low conversion efficiency. Various modifications of electrolysis, photocatalysis, and thermochemical methods have been tested in the literature. These are briefly assessed in the chapter as well. The use of solar and nuclear energy to dissociate water has also been extensively examined in the literature. These technologies are also surveyed in the chapter. A significant breakthrough in water dissociation technology can significantly change the energy landscape and push us more close to the hydrogen economy.

1.2.11 Chapter 12: Production of Methane from Gas Hydrates

Chapter 12 deals with another direct source of fuel from water, that is, gas hydrates. Methane gas hydrates are naturally occurring methane trapped in water. These gas hydrates are unique substances that are found at the bottom of the sea and in arctic conditions such as in Alaska and Siberia. These naturally occurring hydrates require right temperature and pressure conditions for their stable formation. While hydrates are in general unstable, highly dispersed, and difficult to recover, there is more carbon in methane gas hydrates than in all other fossil fuels combined.

Gas hydrates are of great importance for a number of reasons. Naturally occurring methane gas clathrates contain enormous amounts of strategic energy reserve. In offshore hydrocarbon drilling and production operations, gas hydrates cause major and potentially hazardous flow assurance problems. Gas hydrates also pose potential danger to deep water drilling installations, pipelines, and subsea cables. The recovery of gas hydrates by carbon dioxide provides an opportunity to dispose carbon dioxide by sequestration. Gas hydrates also provide an increasing awareness of the relationship between hydrates and subsea slope stability. Finally, it creates long-term considerations with respect to hydrate stability, methane release, and global climate change. Some of these topics along with numerous methods for the recovery are briefly discussed in Chapter 12.

1.2.12 Chapter 13: Water as a Direct Source of Energy

Water is also a direct source for energy and power. This is accomplished by three different methods: hydroelectricity, hydrokinetic energy, and ocean thermal energy conversion.

The generation of power (hydroelectricity) with the potential energy from waterfalls using dams has been long known, and many dams across the world generate a significant amount of electricity from waterfalls. This is one of the cleanest sources of power and is practiced globally. The industry can be broken into large, small, mini-, micro-, and pico-plants depending on the level of the electricity generation. The use of this technology is continuing to grow all over the world.

More recently, more efforts have been made to harness the kinetic energy of the moving water in rivers, seas, and oceans. This method captures energy from sea and ocean waves and undercurrents, tidal waves, and inland waterways. New modern technologies are introduced that can generate hydrokinetic power using devices that can handle high-amplitude waves and fast currents. The chapter examines these different technologies and their progress in the commercialization. The method is applicable globally and is gaining rapid acceptance.

The third method, ocean thermal energy conversion, is only applicable within 20° of the equator. In this method, the temperature difference (about 20°C–25°C) between the surface of ocean and the underwater is used to drive a heat engine, which in turn drives turbine to generate electricity. Although the economics of this method are not as favorable as the previous two methods, it has a number of side benefits in its use for desalination, aquaculture, seafood, hydrogen production, and other industries.

Chapter 13 examines our current state of art in all three methods. It is clear that a strong and growing global demand for energy and fuel will require technological developments for all sources of energy. New developments must be economical, usable, and environmentally acceptable. The energy and fuel landscape may change rapidly depending on the success of the new technology developments.

The 12 chapters outlined here demonstrate an important role of water in the development of future energy landscape. The chapters not only illustrate the versatility of water and its role as a solvent, energy carrier, reactant, catalyst, and a direct source of fuel and energy, but also show how water can help the growth of energy and fuel industry with a least environmental impact. As the production of energy and fuel diversifies, water will continue to play an increasingly important role in the new energy economy.

1.3 WATER-BASED REFINERY AND WATER MANAGEMENT FOR THE FUTURE

Besides numerous roles of water outlined in this book, the concept of water-based refinery may also be not unreal. The refinery, by definition, refines the crude feedstock into useful fuels (or fuel additives) and chemicals. Over many decades, petroleum refineries have converted crude oil of different compositions into various kinds of fuels and chemicals that meet the required industry standards. Analogous to petroleum refineries, coal conversion plants have also converted coal into useful gaseous and liquid fuels. The conversion of natural gas into syngas has also been a part of many refineries.

Unlike petroleum refineries for fossil fuel, biorefineries will be more versatile in that while parts of biorefinery can be integrated with the existing oil refineries, other parts will require more water-based processes. For example, gasification of biomass and conversion of biosyngas to liquid fuels can be integrated with the existing coal gasification and conventional FT process. Similarly, steam gasification and reforming of biomass would be similar to steam gasification of coal and reforming of gasification products. However, five other major technologies outlined in this book—hydrothermal conversions under sub- and supercritical conditions, bioreforming and biofine processes and water dissociation technologies—are largely water based and the use of these technologies will require water-based refining processes. The use of water to obtain the hydrogen required for the refining operations will become more important as water dissociation technologies advance. The five technologies mentioned earlier and discussed in detail in this book show that a water-based refinery will be capable of generating all kinds of gaseous and liquid synfuels from biomass that is currently produced from fossil fuel in oil-based refinery.

Besides coal, oil, gas, and biomass, an important raw material for future energy and fuel industry is waste. The United States generates about 250 million tons of waste per year. The worldwide waste production will exceed 1 billion tons per year during this decade. Since it is also connected to the growth in world population, waste production is one of the fastest growing industries. The landfills around the world are full, and the new paradigm is that landfills are only temporary storage places for the waste and waste to energy and products should be more actively pursued. Since most waste contains between 50% and 85% cellulose, its conversion using water-based processes will become more important.

The future energy industry will also call for refineries that can process multiple feedstock, if possible. The use of mixed feedstock (such as coal and biomass, coal, and waste) may be the new reality of the future because of (1) the location- and time-dependent availability of various raw materials, (2) the desire to reduce carbon emission in the atmosphere, and (3) the cost reduction of building targeted refinery based on the feedstock. Some of the technologies such as gasification followed by gas-to-liquid conversion and supercritical water processing may handle such mixtures.

The expanded use of water for energy and fuel production leads to another societal issue that will have to be managed. Clean water is essential for human and animal life. The expanded use of water for energy and fuel production may create a problem for the available amount of drinkable water for human and animal needs and useable water for agricultural needs. Just as there are tensions between the use of food materials such as corn, soybeans, maize, and other carbohydrates for fuels (such as ethanol), tensions will also be created for the use of water for energy and fuel production and the need of clean water for human and animal needs as well as for agricultural purposes. The strategic management of water will be the next important societal issue. The treatment of water used in energy and fuel industries will become an independent industry by itself. This industry will have to manage the overall societal need for the water to maintain the required strategic distribution of water among its different usages. Along with energy and fuel, a prudent use of water will be the next societal challenge.

Water, however, will be the centerpiece of future energy and fuel landscape.

REFERENCES

1. ExxonMobil, "The outlook for energy: A view to 2040," US Edition, ExxonMobil Report, ExxonMobil, Irving, TX (2012).
2. Lee, S. and Shah, Y., *Biofuels and Bioenergy: Processes and Technologies*. Taylor & Francis, Boca Raton, FL (2012).
3. Funke, A. and Zeigler, F., "Hydrothermal carbonization of biomass: A summary and discussion of chemical mechanisms for process engineering," *Biofuel, Bioproducts and Biorefining*, 4, 160–177 (2010).

2 Role of Water in Recovery and Production of Raw Fuels

2.1 INTRODUCTION

We know that all biological products require water to survive and grow. Thus, bioenergy coming from different types of biomass requires the supply of water. Even the most oil-storing species such as algae requires wetland environment. This chapter, however, shows that water plays a very important role in the recovery and production of various types of fossil fuels and uranium (for nuclear energy). Water will continue to play an increasingly important role in the recovery and production of raw fuels such as conventional and unconventional gas and oils, and solid fuels such as coal, oil shale, tar sands, and uranium. In fact, the role of water in these processes is so important that the production, use, treatment, recycling, and management of water associated with the recovery of fossil fuels is becoming one of the fastest growing independent industries. This chapter illustrates the fact that the use of water and steam is not only essential for the recovery and production of various types of fossil fuels and uranium, but it will significantly increase over the next several decades [1].

There are those who say that fossil energy industry is effectively water industry with oil, gas, and coal as byproducts. This may be particularly true for oil and gas that generally reside with water in underground reservoirs. In North American onshore oil industry, eight barrels (bbl) of water are brought to surface for every bbl of oil. This produced water is often highly saline and contaminated by hydrocarbons: It is hazardous and requires treatment, disposal, and potential recycling. Handling this produced water is an integral part of the oil and gas industries.

In a recent global water intelligence report, the projected growth rate of the produced water from oil and gas industries was illustrated [2]. These produced volume of water forecast data are summarized in Table 2.1 [2]. The data shown in this table indicate that the produced water for oil and gas industries will grow from present 25 billion bbl per year to about 35 billion bbl per year by 2025. The largest growth in produced water will be due to the growth in unconventional gas and oil recoveries and more use of enhanced oil recovery (EOR) methods for the recovery of conventional gas and oil.

The growth in produced water volume will accompany an increase in produced water market activity. The report projects that the dollar value of the produced water market will grow from $5 billion in 2010 to $9.9 billion in 2025—nearly 100%

TABLE 2.1
Produced Volume of Water and Percentage Increase (since 2007) Forecast in Oil and Gas Industries over the Next 15 Years

Year	Coal Bed Methane/Shale Gas	Oil Sand	On/Offshore	All Other Gas	Total
2007	4.5	2.8	14.5	2.8	24.6
2010	4.5 (0%)	2.8 (0%)	14.9 (2.8%)	2.8 (0%)	25.0 (1.6%)
2015	4.7 (4.4%)	3.2 (14.3%)	16.7 (15.2%)	2.9 (3.6%)	27.5 (11.8%)
2020	4.9 (8.9%)	4.5 (60.7%)	18.4 (26.9%)	3.0 (7.1%)	30.8 (25.2%)
2025[a]	5.0 (11.1%)	5.5 (96.4%)	21.5 (48.3%)	3.0 (7.1%)	35.0 (42.3%)

Source: *Global Water Intelligence*, 12, 2–8, 2011. With permission.
Note: All numbers are in billion gallons of water per year.
[a] By 2025, the total increase in produced water will be more than 10 billion gallons per year.

growth in 15 years. The dollar value of the produced water equipment market will grow from $693 million in 2010 to $2.9 billion in 2025 [2]. This growth will accompany a significant growth in various water purification technologies.

While the use of water is essential for recovery of fossil fuels and uranium, the chapter illustrates at least four reasons why this usage will grow significantly (Table 2.1) over the next several decades. These four reasons are as follows:

1. The increased use of water for the recovery of unconventional gas such as coal bed methane, deep gas, and gas trapped in geopressurized zones. Water plays an important role in the underground storage of conventional oil as well as methane in coal bed. The removal of water from coal beds to release trapped methane will become more and more important. The water produced from this unconventional gas recovery will be more than that required for conventional gas capture [3–12]. Future growth in gas industry significantly depends on the recovery of these forms of gases. The chapter will briefly illustrate why more water will be required and produced to recover these gases.
2. More use of EOR methods to recover conventional and unconventional oils. The EOR methods (also often called tertiary oil recovery methods) heavily use water and steam. The thermal recovery methods such as steam flooding, cyclic steam stimulation, and *in situ* combustion, and the water flooding methods that include chemical flooding such as polymer flooding, micellar–polymer flooding, and alkaline flooding as well as microbial flooding and cyclic microbial recovery all use water. These methods have been very successfully applied to improve oil recovery efficiency of both conventional and unconventional oils. In future, the water-to-oil ratio in the EOR processes will further increase due to aging oil wells and expansion of more difficult recovery processes for unconventional oils [13–17]. By 2025, this ratio is likely to reach 12 bbl of water per each bbl of oil from the current number of eight.

3. The new "fracking" technique to recover unconventional shale gas will require a significantly large use of water. In recent years, the recoveries of shale and tight gases have revolutionized the gas industry. Shale deposits were formed about 350 million years ago. Shale is a very fine-grained sedimentary rock, which is easily breakable into thin parallel layers. It is a very soft rock, but it is impermeable to water in its natural state. The shales can contain natural gas usually when two thick black shale deposits "sandwich" a thinner area of shale. Due to the nonporous and impermeable properties of these shales, the extraction of natural gas from shale formation is more difficult.

   Similarly, the "tight gas" is a gas that is stuck in a very tight and unusually impermeable hard rock or in a sandstone or limestone formation that is unusually impermeable and nonporous (tight sand). Unlike extracting conventional natural gas, a significant more effort has to be put into extracting gas from a tight formation. Recently, both shale and tight gases have been successfully recovered using a new technique of fracking, by which impermeable rocks in both cases are fractured using a horizontal or directional drilling technique, which uses high-pressure water solutions (with many chemical additives) to fracture impermeable and tight shale or sand matrix. The dramatic expansion of shale gas industry will require water (with chemicals) as "fracking fluid" to fracture impermeable shale rocks. Additional water will also be required to capture tight gas by the "fracking process" [18–37]. A significant increase in water requirement for this process may cause local conflicts due to competing needs for water. We will briefly discuss the role of water in this successful but somewhat controversial process.
4. Additional water will be required for the recovery of solid fuels such as coal, oil shale, tar sand, and uranium as these sources become harder to recover due to their locations and recovery methods. Both surface mining and deep mining are used in the recovery process. The extraction processes can be carried out outside the mine or *in situ*. A significant amount of water is used for these processes and water is essential for recovery and treatment of these fuels [38–42]. The use of water in the recovery of tar sands will grow very rapidly as more and more tar sands are discovered deeper into the ground and at a lower concentration in the sand.

2.2 INCREASED WATER USAGE FOR RECOVERY OF COAL BED METHANE AND GAS FROM GEOPRESSURIZED ZONES

Recovery of methane from coal beds is an attractive prospect for development because of the ability of coal bed to retain a large amount of methane gas; coal is able to store six to seven times more gas than an equivalent volume of rock common to conventional gas reservoirs. In most regions of the United States, coal bed methane wells produce between 100 and 500 thousand cubic feet of methane per day. The amount of methane in a coal deposit depends on the quality and depth of deposit. In general, the higher the energy value of the coal and the deeper the coal bed, the more methane in the deposit [3–12]. Like the United States, the extraction of coal bed methane is gaining a significant momentum in Canada as well [8]. In principle, coal bed methane will be very important for all parts of the world where significant coal deposits are found.

In coal bed methane, the gas is trapped along with water in the porous coal bed. The pressure of water keeps methane loosely attached to coal and therefore preserves methane in coal deposits. The water pressure also allows bacterial actions between microorganisms in the water and coal to continue to generate new methane. The water needs to be released to allow methane to escape the coal bed. When the water is removed from the coal bed by pumping, the gas pressure is decreased and this allows methane to detach from coal and flow up to the gas well. As shown in Figure 2.1 [5,9], in the initial production stage of methane from the coal bed, the well mostly produces water. Once the water is nearly removed, the production of gas increases. Depending on the geological conditions, it may take several years to achieve a full-scale gas production. Generally, the well with the deeper coal bed produces gas with a short initial time lag due to the presence of less water. In general, the water produced from the coal bed is much higher than that obtained from conventional wells [6,7,12].

Coal bed methane wells are drilled using techniques similar to those used for conventional wells (Figure 2.2) [42]. When coal beds are shallow, less expensive modified water well drilling rigs can be used. In general, however, hydraulic fracturing (or fracking) is used as a primary means of stimulating gas flow in coal bed methane wells. The gas can also be stimulated using a cavitation technique [5,6,11,12]. In this technique, pressure in the reservoir is increased by the injection of water and air (or foam) into the well. The pressure is then suddenly released and this causes a violent blowout of gas, water, coal, and rock fragments from the well. This "surging action" can be repeated several times leaving larger holes and more fractures within coal seams, which in turn causes a faster rate of gas release. The quality of the produced water in both hydraulic fracturing and cavitation techniques mainly depends on the geology of the coal formation. Typically, saltier water is produced

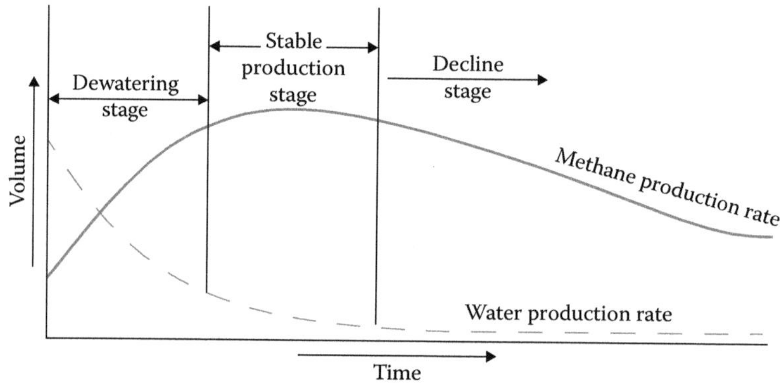

FIGURE 2.1 **(See color insert.)** Typical production curve for a coal bed methane well showing relative methane and water production. (Adapted from Rice, D., "Coal bed methane—An untapped energy resource and environment concern," US Geological Survey, Energy Resource Surveys Program, USGS Fact Sheet FS-019-97, 1997.)

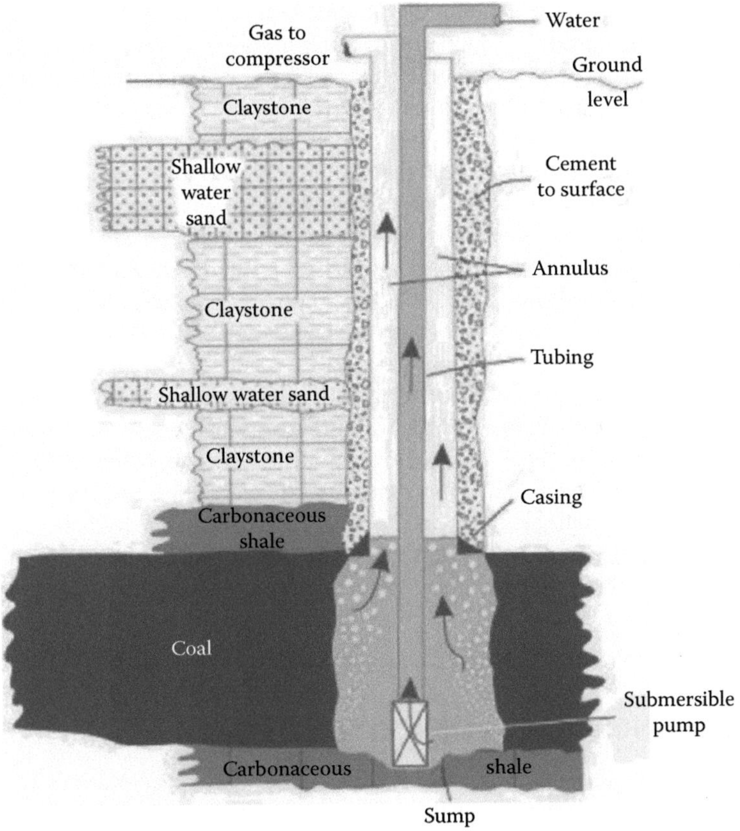

FIGURE 2.2 **(See color insert.)** Simplified illustration of a coal bed methane production well. (From Huth, E., Sule, M., Todman, L., Brant, J., and Templeton, M., "Treatment and reuse of coalbed methane produced water using pervaporation irrigation," 22nd Annual Produced Water Society Conference, January 17–19, 2012. With permission.)

from deeper coal formations. The produced water may contain nitrate, nitrite, chlorides, other salts, benzene, toluene, ethyl benzene, other minerals, metals, and high levels of the total dissolved solids [6,12]. The method of disposal of the produced water depends on (1) the quality of water and (2) the geographical location of the coal bed. Sometimes, the produced water can be an important source for (1) drinking water or (2) water used for the irrigation purposes [6].

Unlike in coal bed methane, in geopressurized zones, confinement of water causes thermal built-up partly because the rate of upward movement of water is not great enough to carry away geothermal heat added to the system and partly because water has a high-specific heat and a low thermal conductivity [6,11]. This thermal built-up further increases the pressure in the geopressurized zones. Water salinity is also increased with depth in the sand bed aquifers within geopressurized zones.

Geopressurized zones are underground natural formations that are at unusually high pressures for their depth. These zones are formed by the deposition of clays over more porous sand or silt and the gradual compaction of these clays over years. This compaction squeezes water and gas out of clay into more porous sand or silt. The gas in the sand under so-called geopressure is usually found at the depth of 10,000–25,000 ft. Thus, it carries some similarity with "deep gas" [6,7,11]. A combination of high depth and high pressure makes the extraction of gas from such zones very difficult. However, of all unconventional gas resources, geopressurized zones hold the highest amount of gas reserve. Just like deep gas, geopressurized zones are mostly found in the Gulf Coast region. It is estimated that the amount of natural gas in the geopressurized zones can be anywhere between 5,000 and 49,000 Tcf. This presents an incredible opportunity because at present the total technically recoverable gas resource is about 1100 Tcf (see the National Energy Technology Laboratory [NETL] website for unconventional gas).

Gas in the geopressurized zones is usually dissolved in hot brine solution (about 150°C–200°C) under pressure. The high pressure makes gas recovery easy when the gas is tapped from these zones. However, this gas is accompanied by water that will have to be removed. The geopressurized zones contain three types of energy: (1) the unconventional gas reserve, (2) the high-pressurized fluids that can impart mechanical energy, and (3) the hot brine solution that may provide geothermal energy. The recovery of these energy resources will require high investment costs and large amount of water production, treatment, and usages [6,11,12] (see the NETL website).

2.3 ENHANCED OIL RECOVERY (EOR) PROCESS

EOR is defined as the incremental ultimate oil that can be economically recovered from a petroleum reservoir over oil that can be recovered from the same reservoir by conventional primary and secondary methods. The intent of EOR is to increase the effectiveness of oil removal from the pores of the rock (displacement efficiency) and to increase the volume of the rock contacted by injected fluids (sweep efficiency) [16,17]. The oil remaining after conventional recovery operations is retained in the pore space of reservoir rock at a lower concentration than originally existed. This residual oil is found as either droplets trapped in the individual pores or cluster of pores or films partly coating the pore walls. Entrapment of this residual oil is predominantly due to capillary and surface forces as well as due to pore geometry.

Bypassing of oil in the reservoir occurs due to a number of reasons: (1) nonhomogeneity of the reservoir rock causing inefficient sweeping by the displacement fluids; (2) simultaneous effects of viscous, gravity, and capillary forces; and (3) high mobility of displacing fluid compared to that of oil. The net effect depends on the conditions at individual locations. In general, gravity forces cause vertical segregation of the fluids and water tends to underrun the oil-containing rock [16,17].

The recovery of conventional oil from a reservoir requires pressure gradient to push oil out from the reservoir to the surface. Initially, gas and water that accompany oil provide this pressure, and because of that as indicated earlier, when oil comes out of the ground, a significant amount of water (eight bbl of water per each bbl of oil) accompanies it. Initially, this process may be facilitated by pumping the fluid out of the ground by a pump. This is generally known as primary oil recovery process.

This process generally recovers only 5%–15% of the oil from a well. Secondary oil recovery process, often involving pumping water down (water flooding) the well to maintain pressure on the oil, may increase the recovery to 30%. EOR techniques can increase the proportion of the oil brought to the surface to 60%. When the well is aged, the recovery of the remaining oil particularly requires the implementation of EOR techniques [16,17].

Fundamentally, three types of EOR processes are currently being used: (1) miscible displacement processes that use miscible hydrocarbons, carbon dioxide, or inert gas; (2) chemical processes that use surfactant polymer, polymer, or caustic solutions in water; or (3) thermal processes that use steam stimulation, steam flooding, hot water injection, or *in situ* combustion. In this chapter, we focus on the latter two processes because they use water or steam. Various methods used for EOR processes are described in two excellent books and numerous articles by Speight [16,17]. The present description closely follows his work along with other works reported in few additional publications [13–15].

2.3.1 Chemical Processes

In general, due to their high cost, complex technology, and high risk, chemical flood processes account for <1% of the total tertiary recovery. The successful chemical processes for oil recovery require floodwater of precise salinity. Fundamentally, three types of chemical solutions—surfactant–polymer solution, polymer solution, and caustic alkaline solution—are used. Here we briefly describe the effectiveness of each of these processes.

2.3.1.1 Surfactant–Polymer Solution (Microemulsion Flooding)

The injection of surfactant–polymer solution is a two-step process [16,17]. The first step is the injection of a surfactant slug commonly referred as either micellar solution or microemulsion. The purpose of the surfactant is to displace oil that cannot be displaced by water alone. The second step is the injection of polymer mobility buffer. The polymer provides mobility control for a more piston-like displacement.

In microemulsion flooding process, a stable solution of oil, water, electrolytes of salts, and one or more surfactants are injected into the formation that is then displaced by mobility buffer solution, which in turn is displaced by injection of water. Two approaches can be used in microemulsion flooding. In one approach, a relatively low-concentration (2–4 wt%) surfactant microemulsion is injected in large pore volumes of 15%–60%. In the second approach, a high-concentration (8–12 wt%) surfactant microemulsion is injected in a relatively small pore volume from 3% to 20%. As the time passes, the second approach merges with the first approach due to the dilution effect. Mobility control is important in the success of this process [16,17].

Microemulsion technique can be applied over a wide range of operating conditions. In microemulsion flooding, the slug must be designed for specific reservoir conditions of temperature, resident water salinity, and crude oil type. The success of the microemulsion–polymer flooding in a given reservoir depends on the proper assessment of the characteristics such as the nature of oil and water content, relative permeability, mobility ratios, formation fractures and variations in permeability, porosity, formation continuity, and rock mineralogy [16,17].

2.3.1.2 Polymer Solution

The addition of polymers in water increases the solution viscosity, thereby increasing the sweep efficiency. Two classes of polymers are normally used: polyacrylamides (PACs) and polysaccharides (PSAs). PAC (normally used in 50–1000 ppm concentration) decreases the mobility of the injected fluid by decreasing the permeability of the reservoir rock. Addition of PSA, however, increases the viscosity with a very low level of permeability reduction in the reservoir rock. The high viscosity of both solutions compared to water results in a significant long-range oil production.

Polymer flooding is most effective for heterogeneous reservoirs because they respond favorably to the improved vertical sweep efficiency. This technique is also preferred over microemulsion flooding for recovery of more viscous oils. Currently, polymer flooding is being used in a significant number of commercial field projects [16,17].

The injection of polymer solutions may face stability problems due to oxygen contamination. For PAC solutions, this may be alleviated using sodium hydrosulfite in low concentrations. In general, the degradation of polymer due to mechanical, chemical, thermal, and microbial reasons can be totally prevented by using specialized equipment or techniques [16,17].

2.3.1.3 Caustic Alkaline Solution

This method is inexpensive and preferred, but it provides low productivity. In this method, inorganic alkaline chemicals such as sodium hydroxide, sodium carbonate, or sodium orthosilicate are added to the water to enhance oil recovery by one or more of the following mechanisms: (1) interfacial tension reduction, (2) spontaneous emulsification, or (3) wettability alteration. For an efficient oil recovery, the pH range 12–13 is optimum. Sometimes the mobility ratio can be improved by an addition of polymer in the alkaline solution [16,17].

2.3.2 Thermal Processes

Thermal EOR processes are very popular and gaining more use due to the fact that they can be applied to both conventional and unconventional oil recovery. Three types of thermal processes are currently used: steam stimulation, hot water injection, and *in situ* combustion. We briefly examine each of these processes in Sections 2.3.2.1–2.3.2.3.

2.3.2.1 Steam Stimulation

Steam stimulation is a general term used when steam is injected into a well and then produced back out of the same well. The method is also referred as cyclic steam injection, steam soak, or huff and puff. The process uses up to 1000 bbl of water per day (in the form of superheated steam) for 10–30 days and then the well is shut in for about 1–4 weeks to allow the steam to soak in the well [16,17]. During this soaking period, heat dissipates into the reservoir and reduces the viscosity of oil and expands the volume of oil causing fluid movement. This action facilitates oil recovery from the reservoir. The well begins its production level after the soaking period until the flow of oil slows down. Once this occurs, the process is repeated. After much of oil is recovered, cyclic steam flood is converted to steam injection. In this process, steam is injected from one well and the oil is recovered from other nearby wells [16,17].

Steam flood in some reservoirs results in a dramatic increase in oil production. This method which accounts for about 20% of EOR processes, has gained more popularity in recovery of unconventional oils such as heavy oils, tar sands, and shale oil [16,17]. The technique works well because of the crude oil expansion due to increased temperature, which continues as pressure drops. Steam flood wells are drilled on an ~5-acre spacing and require a reservoir depth of ~10 ft or more. The method is most effective in wells no deeper than 5000 ft.

2.3.2.2 Hot Water Injection

Besides steam injection, hot water injection is often pursued. This method is, however, not as effective because of heat loss and resulting fingering phenomenon and loss of sweep efficiency [16,17].

2.3.2.3 *In Situ* Combustion

In situ combustion can be dry (only using air) or wet or partially wet. In this process, fire is ignited by injecting compressed air into the injection wells and driven across the reservoir. The heat from fire reduces oil viscosity leading to drop in pressure and expansion of oil. The vapor can also be collected from the well and condensed at the well mouth. The process is not very efficient because a large amount of heat is not utilized.

The *in situ* combustion can be forward combustion or reverse combustion. In the forward combustion process, fire is ignited in the formation near the injection well, and with continuing air injection, the fire and produced oil are driven toward nearby producing wells. In the reverse combustion process, the fire is started near the compressed air injection well and allowed it to progress toward nearby wells. The process is then reversed; the air is injected from nearby wells and the original air injection well is used to collect the oil that is produced. The reverse combustion is often considered to be more efficient than the forward combustion. The combustion process is more effective for heavy crude oil. In wet and partially wet injection processes, steam accompanies air [16,17].

As the existing oil well ages and new oil wells are discovered in more difficult locations, the use of EOR techniques to recover oils will significantly increase in future. This will require an additional usage of water or steam.

2.4 ROLE OF WATER IN THE FRACKING PROCESS

The natural gas can occur in the oil fields (known as associated gas), in coal seams (known as coal bed methane), in sandstone (tight and deep gas) or shale (shale gas), or in the fields not associated with oil or coal (i.e., nonassociated gas). The gas is also found in geopressurized zones. The most important game changer in recent years is shale gas largely due to the novel but controversial process of fracking [18–36].

The drilling and hydraulic fracturing of horizontal gas wells on average requires 3.5 million gallons of water. This water generally comes from surface water bodies or from groundwater, private water sources, municipal water, and reused produced water. Water, however, plays a very significant role in recovering unconventional gas such as deep gas, tight gas, gas from geopressurized zones, shale gas, and coal bed methane. Section 2.2 showed the increased usage of water for recovery of coal bed methane and the gas from geopressurized zones. In this section, we briefly discuss the additional usage of water required to recover shale gas and tight gas using the fracking process. This hydraulic fracturing process to release trapped gas in tight and nonporous geological matrix requires a substantial amount of water and chemical additives.

The recovery of shale gas and tight gas uses horizontal drilling and hydraulic fracturing (i.e., fracking). Over the next 25 years, the unconventional gas production will increase from 15% of the total current production to about 77% of the total gas production. Unconventional natural gas reserves are located at varying depths below the ground. In Texas' Barnett and Haynesville/Bossier plays, for example, the natural gas-producing areas are 1,000–12,000 ft below the ground. In Michigan and Illinois, natural gas-producing areas are much shallower ranging from hundreds of feet to 2000 ft below the ground. All of these reservoirs now use horizontal drilling and fracking process [17,29,35,36].

The drilling and hydraulic fracturing of a horizontal shale gas well uses large volumes of water. The EPA reports [30,31,34] that fracturing shale gas wells requires between 2.3 and 3.8 million gallons of water per well. An additional 40,000 to 1 million gallons of water is required to drill the well. This is considerably more water than is required for conventional gas because the wells to access shale gas and tight gas are deeper. The data [26,33,35] show that for shale gas in Marcellus region, 4.5 million gallons of water per well is required. Water requirements within Texas Eagle Ford shale area can be even greater where the fracking process can use up to 13 million gallons of water per well with additional water requirement for drilling the wells [17,18,22,36]. Within the state of Texas, the water requirement does depend on shale plays; for example, water requirement in Barnett shale play can range from 1 to 8 million gallons of water per well; in Haynesville and Bossier shale play, it can range from 1 to 10 million gallons per well; and in Woodford, Pearsall, and Barnett-PB shale plays, it can be as low as 1–5 million gallons per well [17–19,22,36].

The water used in the fracking process also contains some harmful chemicals such as acids, scale inhibitors, iron control agents, surfactants, friction reducing agents, corrosion inhibitors, gelling agents, and bactericide/biocide compounds. Once the fracking process is complete, the fracking fluid is withdrawn from the well, but this withdrawal rate can be as low as 20% of the injection rate. The remaining fluid can have a harmful effect on the underground water aquifers [20–22,24,30–32] due to its chemical contents. The water withdrawn from the well is accumulated in the surface pond and treated before it is reused or reinjected in the well. The fracking process not only requires a large amount of water but also creates a number of water-related issues such as (1) water withdrawals, (2) groundwater contamination associated with well drilling and production, (3) waste water management, (4) surface spills and leaks, and (5) stormwater management. Water supply, purification, treatment, and management are very important and essential parts of the fastest growing shale gas industry.

2.5 WATER REQUIREMENT FOR MINING, PREPARATION, AND EXTRACTION OF SOLID FUELS

Mining of solid fuels such as coal, oil shale, tar sand (and heavy oil), and uranium, and their subsequent preparation and extraction require a large quantity of water [37–41,43–48]. We briefly address this issue in this section.

2.5.1 Oil Shale Industry

The mining and retorting of oil shale consumes a large amount of water. Above-ground retorting typically consumes between one and five bbl of water per each bbl of produced shale oil, depending on the technology [43–46,48]. Water is normally used for spent oil shale cooling and oil shale ash disposal. *In situ* processing, according to one estimate, uses about one-tenth as much water [47]. Also, water must be pumped out of oil shale mines. The resulting fall in water table may have negative effects on nearby arable lands and forests [43]. A 2008 programmatic environmental impact statement issued by the US Bureau of Land Management stated that surface mining and retort operations produce 2–10 US gallons of waste water per 1 short ton of processed oil shale [46].

2.5.2 Tar Sand and Heavy Oil Industries

Water requirement in tar sand industry is well reviewed by Speight [16,17,44] in his numerous books and other publications. Just like oil shale, tar sands and heavy oil are obtained from strip (surface) mining or underground mining. The mining is generally accompanied by extraction process to recover bitumen or heavy oil. In terms of bitumen separation and recovery, the "hot water process," to date, is the only successful commercial process to be applied to bitumen recovery from mined tar sand/oil sand in North America.

The hot water process utilizes the linear variation of bitumen density and the nonlinear variation of water density with temperature so that bitumen that is heavier than water at room temperature becomes lighter than water at 80°C. Surface-active materials in the tar sand also contribute to the process. The two most important steps in the process are "conditioning" and "separation."

In the conditioning step, the tar sand is heated and mixed with water to form a pulp of 60%–85% by weight of solids at 80°C–90°C. First the tar sand lumps are reduced in size by ablation. The conditioned pulp is screened through a double-layer vibrating screen. Water is added to the screened material and the pulp then enters the "separation" cell. The bulk of the sand settles in the cell and is removed from the bottom as tailings, but the majority of the bitumen floats to the surface and is removed as froth. A middling stream containing water and fine solids and some bitumen from the midway up the side of the cell wall is also recovered.

Froth from the hot water process may be mixed with a hydrocarbon diluent such as coker naphtha and centrifuged. The Suncor process described by Speight [16,17,44] employs a two-stage centrifuging operation. The final bitumen product contains 1–2 wt% mineral and 5–15 wt% water. About 2 tons of tar sands is required to produce one bbl of oil. Roughly 75% of the bitumen can be recovered from sand. More details on oil recovery from tar sands are given in excellent reviews of Speight [16,17,44].

Relatively large amount of water is required to process tar sands. Currently, tar sand extraction and processing require several bbl of water for each bbl of oil produced, though some of the water can be recycled. *In situ* production methods are used on bitumen deposits buried too deep for mining to be economically recovered. These techniques include steam injection, solvent injection, and firefloods (see various EOR methods in Sections 2.3.1 and 2.3.2).

2.5.3 Uranium Mining and Leaching

Uranium, a substance essential for nuclear energy, is recovered from the ground by the extraction process [41]. In 2009, a worldwide production of uranium amounted to 50,572 tons [41]. While this number is small compared to that for coal, oil shale, and tar sands, this mining process also requires a large amount of water. As with other types of hard rock mining, uranium is extracted by the three main methods: box cut mining, open pit mining, and *in situ* leaching (ISL). While water requirement for open pit mining or underground mining of uranium is similar to that of other minerals, coal, oil shale, and tar sand, the major water usage in uranium mining is in the implementation of the ISL process.

The ISL process is also known as solution mining, which involves leaving the ore where it is in the ground and recovering uranium from it by dissolving it and pumping the pregnant solution to the surface where the uranium is recovered. This process has a little surface disturbance and no waste is generated. Uranium ISL uses the native groundwater in the ore body that is fortified with a complexing agent and in most cases an oxidant such as hydrogen peroxide. In many cases, the complexing agent used is sulfuric acid. It is then pumped through the underground ore body to recover the minerals in it by leaching. Once the pregnant solution is returned to the surface, uranium is recovered in much the same way as in other uranium plants [41].

Often, the use of oxidant is replaced by high concentration of acid solution. In the United States, ISL mines use an alkali leach due to the presence of significant quantities of acid-consuming minerals such as gypsum and limestone in the host aquifers. Any more than two to five percent carbonate minerals means that alkali leach must be used in preference to the more efficient acid leach.

In uranium mine near Moab, Utah, uranium deposits were formed when oxygenated groundwater, which had leached uranium from crystal rocks, flowed down into aquifers, where it was reduced to form precipitate uraninite; the main ore of uranium. This corresponds to oxidized and reduced conditions in groundwater redox chemistry. The rocks formed in the oxidizing conditions are reduced by a reducing fluid. The reduced fluid carries uranium-bearing minerals [41].

While the uranium concentration in sea water is low at 3.3 mg per cubic meter of seawater, the total amount is large. Several countries such as the United Kingdom, France, Germany, and Japan are exploring the recovery of uranium from the seawater using inorganic adsorbents such as titanium oxide. Japan is also exploring the production of adsorbents by irradiation of polymer fiber. Uranium adsorption by the polymer adsorbent is about ten-fold high compared to that of conventional titanium oxide adsorbent. In 2012, Oak Ridge National Laboratory (ORNL) researchers announced the successful development of a new adsorbent material dubbed "HiCap" that vastly outperforms the previous best adsorbents. They showed that their adsorbents can extract five to seven times more uranium at uptake rates, which are seven times faster than the world's best adsorbents. HiCap also effectively removes toxic metals from water.

2.5.4 Coal Mining and Preparation

While coal mining also requires significant use of water and produces acid drainage that can affect local aquifers, the large use of water in the coal industry is also in the coal preparation plants [40]. Coal preparation plant requires washing, crushing, and removal of various impurities from coal. One area where significant water is used is in the removal of inorganic sulfur (iron pyrites) from coal by the floatation process. In this process, finely pulverized coal goes through a floatation process in which iron pyrite particles are removed from coal by gravity separation due to the density difference between pyrites and coal particles. The removal of sulfur, other impurities (like ash), and metals is important for the downstream operations for coal. For example, once the coal is finely pulverized and all ash and metals are removed, coal-water slurry becomes an important fuel for combustion purposes. The subject of coal-water slurry combustion is discussed in Chapter 5. The conversion of coal to oil by direct or indirect coal liquefaction processes also uses a significant amount of water. For 50,000 barrels per stream day (BPSD), water requirement can vary from 7,300 to 10,500 gallons per minute (GPM) depending on the nature of coal. For coproduction of Fischer–Tropsch (FT) liquids plus electric power of 25,000 BPSD and 1,250 MW plant, water requirement can be 20,800 GPM [40].

REFERENCES

1. Duda, J., "Emerging issues for fossil energy and water-investigation of water issues related to coal mining, coal to liquids, oil shale, and carbon capture and sequestration," DOE/NETL-2006/1233 (June 2006).
2. "Water's growing role in oil and gas," *Global Water Intelligence* (Market Profile), 12 (3), 2–8 (2011).
3. "Coal bed methane," *Energy Justice Network*, Report published by Oil and Gas Accountability Project (2012).
4. Rice, D. and Nuccio, V. *Water Produced with Coal Bed Methane*. USGS Publication, US Geological Survey, Denver, CO (1999), http://pubs.usgs.gov/fs/fs-0156-00/fs-0156-00.pdf.
5. Kumar, H. and Mathews, J., "An overview of current coal bed methane technologies," Report from EMS Energy Institute, Pennsylvania State University, University Park, PA (2010).

6. ALL Consulting, *Handbook on Coal Bed Methane Produced Water: Management and Beneficial Use Alternatives*, prepared for Ground Water Protection Research Foundation, DOE, Morgantown, WV (2003).
7. "NETL: Future supply—Coal bed natural gas," Netl.doe.gov (2011).
8. Squarek, J. and Dawson, M., "Coal bed methane expands in Canada," *Oil & Gas Journal*, 104, 37–40 (2006).
9. Rice, D., "Coal bed methane—An untapped energy resource and environment concern," US Geological Survey, Energy Resource Surveys Program, USGS Fact Sheet FS-019-97 (1997).
10. "Coal bed methane," Wikipedia, the free encyclopedia (2012).
11. Jones, P.H., "Role of geopressure in hydrocarbon and water system," in Roberts, W.H., III, and Cordell, R.J. (eds.), *SG 10: Problems of Petroleum Migration*, Studies in Geology. American Association of Petroleum Geologists, Tulsa, OK, 207–216 (1980).
12. Veil, J.A., Puder, M.G., Elcock, D., and Redweik, R.J., Jr., *A White Paper Describing Produced Water from Production of Crude Oil, Natural Gas, and Coal Bed Methane*, prepared by Argonne National Laboratory, Argonne, IL, prepared for US Department of Energy, National Energy Technology Laboratory, Morgantown, WV (2004), http://www.evs.anl.gov/pub/dsp_detail.cfm?PubID=1715.
13. Muslimov, R.Kh., "The growing role of enhanced oil recovery in replenishment of oil reserves," *International Journal of Science*, 2 (10), 2–5 (2007).
14. Al-Mjeni, R., Arora, S., Cherukupalli, P., Wunnik, J., Edwards, J., Felber, B., Gurpinar, O. et al., "Has the time come for EOR," *Outfield Review Winter*, 22 (4), 16–37 (2010/2011).
15. "Enhanced oil recovery," Report by Teledyne ISCO, a Teledyne Technologies, Thousand Oaks, CA (November, 2007).
16. Speight, J.G., *Enhanced Oil Recovery Handbook: A Guide to Heavy Oil*. Gulf Publishing Company, New York (2009).
17. Speight, J.G., *Enhanced Recovery Methods for Heavy Oil and Tar Sands*. Gulf Publishing Company, New York (2009).
18. Nicot, J.P. and Scanlon, B.R., "Water use for shale-gas production in Texas, US," *Environmental Science & Technology*, 46, 3580–3586 (2012).
19. Cooley, H. and Donnelly, K., "Hydraulic fracturing and water resources: Separating the frack from fiction," 25 years Pacific Institute Report (June 2012).
20. American Petroleum Institute, *Freeing up Energy—Hydraulic Fracturing: Unlocking America's Natural Gas Resources* (2010), http://www.api.org/~/media/Files/Policy/Exploration/HYDRAULIC_FRACTURING_PRIMER.ashx.
21. Osborn, S.G., Vengosh, A., Warner, N.R., and Jackson, R.B., "Methane contamination of drinking water accompanying gas-well drilling and hydraulic fracturing," *Proceedings of the National Academy of Sciences*, 108 (20), 8172–8176 (2011).
22. Arthur, J.D., Bohm, B., Coughlin, B.J., and Layne, M., *Evaluating the Environmental Implications of Hydraulic Fracturing in Shale Gas Reservoirs* (2008), http://www.all-llc.com/publicdownloads/ArthurHydrFracPaperFINAL.pdf.
23. Groat, C.G. and Grimshaw, T.W., *Fact-Based Regulation for Environmental Protection in Shale Gas Development*. The Energy Institute, University of Texas at Austin, Austin, TX (2012).
24. Ground Water Protection Council (GWPC) and ALL Consulting. *Modern Shale Gas Development in the United States: A Primer* (2009), http://www.all-llc.com/publicdownloads/ShaleGasPrimer2009.pdf.
25. Hammer, R. and Van Briesen, J., "In fracking's wake: New rules are needed to protect our health and environment from contaminated wastewater," Natural Resources Defense Council Document D: 12-05-A, NRDC, New York (May 2012).

26. Jackson, R.B., Pearson, B.R., Osborn, S.G., Warner, N.R., and Vengosh, A., *Research and Policy Recommendations for Hydraulic Fracturing and Shale-Gas Extraction.* Center on Global Change, Duke University, Durham, NC (2011).
27. Kargbo, D.M., Wilhelm, R.G., and Campbell, D.J., "Natural gas plays in the marcellus shale: Challenges and potential opportunities," *Environmental Science & Technology*, 44 (15), 5679–5684 (2010).
28. McAllister, E. and Gardner, T., "EPA may truck water to residents near fracking site," *Reuters* (January 6, 2012), http://www.reuters.com/article/2012/01/06/us-usa-fracking-epa-idUSTRE8041YE20120106.
29. Myers, T., "Potential contaminant pathways from hydraulically fractured shale to aquifers," *Ground Water* (2012), doi:10.1111/j.1745-6584.2012.00933.x.
30. Nicot, J.P., Hebel, A.K., Ritter, S.M., Walden, S., Baier, R., Galusky, P., Beach, J., Kyle, R., Symank, L., and Breton, C., *Current and Projected Water Use in the Texas Mining and Oil and Gas Industry*, prepared for the Texas Water Development Board, Austin, TX (2011), http://www.twdb.state.tx.us/rwpg/rpgm_rpts/0904830939_MiningWaterUse.pdf.
31. USEPA, *Plan to Study the Potential Impacts of Hydraulic Fracturing on Drinking Water Resources.* Office of Research and Development, Washington, DC (2011a), http://www.epa.gov/hfstudy/HF_Study__Plan_110211_FINAL_508.pdf.
32. USEPA, *Investigation of Ground Contamination near Pavillion, Wyoming Draft.* Office of Research and Development (2011b), http://www.epa.gov/region8/superfund/wy/pavillion/EPA_ReportOnPavillion_Dec-8-2011.pdf.
33. US Government Accountability Office (GAO), *Information on the Quantity, Quality, and Management of Water Produced during Oil and Gas Production*, GAO-12-256. GAO, Washington, DC (2012).
34. Veil, J.A., *Water Management Technologies Used by Marcellus Shale Gas Producers*, prepared by the Environmental Science Division, Argonne National Laboratory, Argonne, IL, prepared for US Department of Energy, Office of Fossil Energy, National Energy Technology Laboratory, Morgantown, WV (2010), http://www.evs.anl.gov/pub/doc/Water%20Mgmt%20in%20Marcellus-final-jul10.pdf.
35. Wiseman, H., "Regulation of shale gas development," in *Fact-Based Regulation for Environmental Protection in Shale Gas Development.* The Energy Institute, The University of Texas at Austin, Austin, TX (2012), http://energy.utexas.edu/images/ei_shale_gas_regulation120215.pdf.
36. Veil, J., "Water requirements for shale gas activities in the Marcellus and Fayetteville shale," Report by Veil Environmental, LLC (2010).
37. Chan, M., Duda, J., Forbes, S., Rodosta, T., Vagnetti, R., and McIlvried, H., "Emerging issues for fossil energy and water-investigation of water issues related to coal mining, coal to liquids, oil shale and carbon capture and sequestration," DOE/NETL Report No. 1233, DOE, Washington, DC (June, 2006).
38. "Heavy crude oil," Wikipedia, the free encyclopedia (2012).
39. Speight, J.G., "Extraction technologies for natural bitumen and heavy oil," *Coal, Oil Shale, Natural Bitumen, Heavy Oil and Peat*, 2, 1–7 (2007).
40. "Coal mining," Wikipedia, the free encyclopedia (2012).
41. "Uranium mining," Wikipedia, the free encyclopedia (2012).
42. Huth, E., Sule, M., Todman, L., Brant, J., and Templeton, M., "Treatment and reuse of coalbed methane produced water using pervaporation irrigation," *22nd Annual Produced Water Society Conference*, Houston, TX, January 17–19, 2012.
43. Francu, J., Harvie, B., Laenen, B., Siirde, A., and Veiderma, M., "A study on the EU oil shale industry viewed in the light of the Estonian experience," Report by EASAC to the Committee on Industry, Research and Energy of the European Parliament (PDF). European Academies Science Advisory Council, 12–13, 18–19, 23–24, 28 (May 2007).

44. Speight, J.G., *Synthetic Fuels Handbook: Properties, Process, and Performance.* McGraw-Hill, New York, 13, 182, 186 (2008).
45. Bartis, J.T., LaTourrette, T., Dixon, L., Peterson, D.J., and Cecchine, G., *Oil Shale Development in the United States. Prospects and Policy Issues*, prepared for the National Energy Technology Laboratory of the United States Department of Energy. The RAND Corporation, Santa Monica, CA, x, 15–18, 50 (2005).
46. US Department of the Interior, "Effects of oil shale technologies," in *Proposed Oil Shale and Tar Sands Resource Management Plan Amendments to Address Land Use Allocations in Colorado, Utah, and Wyoming and Final Programmatic Environmental Impact Statement.* Bureau of Land Management, Lakewood, CO, FES 08-32 Vol. 2: Chapters 5 and 6 (September 2008).
47. Fischer, P.A., "Hopes for shale oil are revived," *World Oil Magazine* (Gulf Publishing Company, Houston, TX), P. Kulkarni (ed.) archived from the original on June 17, 2008 (August 2005).
48. "Shale oil extraction," Wikipedia, the free encyclopedia, 1–11 (2013).

3 Energy Recovery by Benign Hydrothermal Processes

3.1 INTRODUCTION

Water has been used as a benign thermal energy carrier for recovery of energy from various other sources. While the examples of use of water and steam as benign thermal energy carrier are numerous, this chapter focuses on four important applications of water and steam as energy carrier. Although water and steam are used for heating and cooling in all refineries and other process industries, this chapter focuses on the use of water and steam to carry energy created by other sources such as nuclear energy, geothermal energy, solar energy, and different types of fuel burning to generate electric power, and/or heating and cooling purposes. Four applications that are considered here include (1) the role of water as a coolant and a thermal energy carrier for the nuclear reactor, (2) the use of water and steam in the recovery of geothermal energy, (3) the use of water to store heat produced from solar energy, and (4) the role of steam to drive turbine for power generation from various types of fuel burning. In each of these applications, water or steam plays a benign but very important role for various types of energy conversion and recovery processes.

3.2 ROLE OF WATER IN PRODUCTION OF NUCLEAR POWER

As discussed in Chapter 2, the extraction of uranium requires a large quantity of water. This section illustrates the important role of water in the nuclear reactor. There are several types of nuclear reactors, that is, water cooled, gas cooled, fast neutron, and so on, currently used in commercial practice. However, as shown in Table 3.1 [1–4], a majority of the commercial nuclear reactors (>80%) currently in operation use water as a coolant and an energy carrier. In fact, water is not only a coolant but also a moderator of the nuclear reactors and provides both energy carrier and safety functions for the reactors. Some of the water-based nuclear reactors are briefly described below.

3.2.1 LIGHT WATER REACTOR

The light water reactor uses light water (hence enriched uranium) and this category contains two different types of reactor: pressurized water reactor (PWR) and boiling water reactor (BWR) [2]. The light water also combines the functions of coolant and moderator. In both PWR and BWR, the water flows through the reactor core,

TABLE 3.1
Water-Based Nuclear Power Plants (NPPs) in Commercial Operation

Reactor Type	Total NPP (%)	GWs	Coolant/Moderator
PWR	65	270	Water/water
BWR	20	81	Water/water
PHWR (CANDU)	11.5	27	Heavy water/heavy water
RBMK (light water)	2.5	10	Water/graphite
Other	1	0.04	Water/graphite

Source: *Nuclear Engineering International Handbook*, 2011.
Note: The total number of reactors in the world is 438 with 399.3 GW.
PHWR, pressurized heavy water reactor.

a zone containing a large array of fuel rods where it picks up the heat generated by the fission of U^{235} present in the fuel rods. The coolant transfers heat to turbine and returns back to the reactor core. This loop is called primary circuit. It is the pressure at which coolant flows through the reactor core that makes the distinctions between PWR and BWR. In both types of light water reactors, about one-third of the fuel is replaced every year (implying life cycle for any given rod to be about three years) because by that time the concentration of fission fragments produced as a result of fission reactions absorbs enough neutrons to interfere with the chain reaction. Thus, before U^{235} is exhausted, fuel rods are periodically replaced to maintain high efficiency of the fission process.

3.2.2 Boiling Water Reactor

The BWR does not have a steam generator [1–3]. Instead, water in the BWR boils inside the pressure vessel and the steam–water mixture is produced when very pure water (reactor coolant) moves upward through the core absorbing heat. The uranium core in the reactor vessel creates heat. The control rods enter the reactor from below. The water boils and produces steam that is passed through a turbine, which in turn drives the electric generator. While the BWR has many similarities to the PWR, there is only one circuit with water at lower pressure (about 75 atm) in the BWR so that it boils in the core at about 285°C. About 12%–15% of water is in the upper part of the core as steam and this has a lower moderating effect. BWR units can operate in the load-following mode more readily than PWR. The steam passes directly to the turbines before being condensed and recycled. Both water and steam are thus a part of a close reactor circuit. The entire close loop along with the BWR is illustrated in Figure 3.1 [2].

As shown in the figure, the core of the BWR contains 3.5–4.0 m-long fuel rods (90–100) and assemblies (up to 750), which hold up to 140 tons of uranium. In most nuclear reactors, the fuel is enriched ceramic uranium oxide (UO_2 with melting point of 2800°C). The fuel pellets (usually about 1 cm in diameter and 1.5 cm long) are typically arranged in a long zirconium alloy tube to form a fuel rod. A fuel assembly is an open lattice that can be inserted and withdrawn from the reactor core.

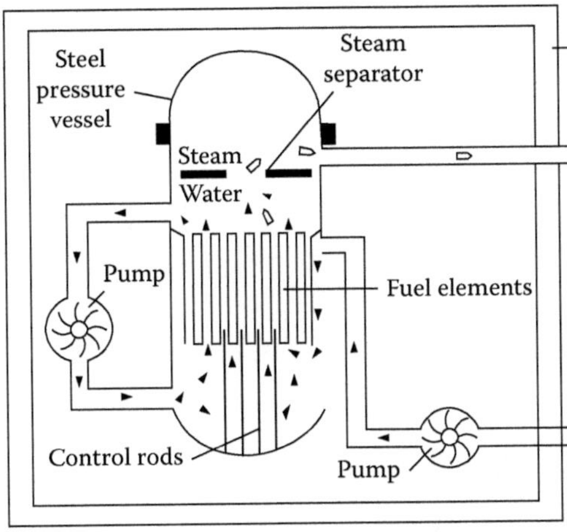

FIGURE 3.1 Schematic of a typical BWR. (Adapted from "Nuclear power reactors," a document of World Nuclear Association, July 2013.)

The BWR has a secondary control system that restricts the flow through the core so that the amount of steam in the upper part of the reactor can be adjusted. This is important because steam has a lower moderating effect and the steam produced in the fission zone is slightly radioactive, mainly due to short-lived activation products. The turbine is therefore housed in the same reinforced building as the reactor.

3.2.3 Pressurized Water Reactor

About 60% of the world commercial nuclear power reactors are PWRs [1–3]. A graphical illustration of this type of reactor and the attached cooling system is shown in Figure 3.2 [2]. Similar to the BWR, the PWR has a core where fission reactions occur and a containment structure; unlike the BWR, the PWR system has a primary cooling system and a secondary steam generation circuit.

A core in the PWR contains 80–100 tons of uranium in 150–250 fuel assemblies, each with 200–300 vertical rods. Each fuel rod contains a stack of pellets of enriched uranium oxide packed in a sealed tube of Zircalloy. The control rods containing neutron-absorbing materials such as boron or cadmium are used to fine-tune the reactor operation and shut down the reactor in an abnormal operation or in an event of a malfunction. Boric acid fluid is used as a secondary shutdown system.

As shown in Figure 3.2, the reactor vessel, the primary cooling system, and the steam generator for the secondary steam circulation system are enclosed in a meter-thick concrete and steel containment structure to protect the reactor and provide seal for any radiation leakage. The escape of fission products that are formed during fission is prevented by (1) high melting temperature ceramic pellets themselves, as fission products are trapped in small pores, and (2) Zircalloy cladding that is corrosion resistant to low neutron absorption. Any fission gas that escapes from the pellets is accommodated in the small space at the top of the fuel rod.

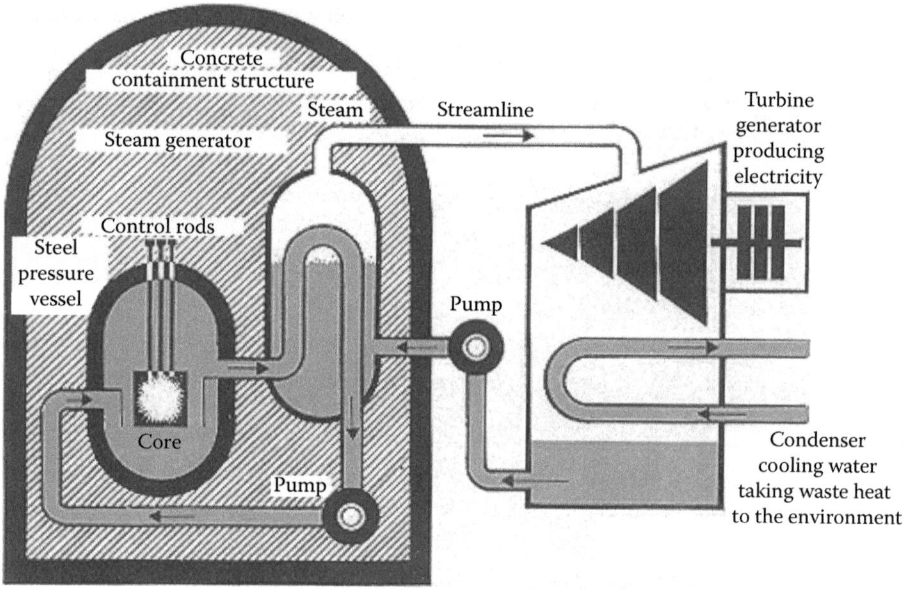

FIGURE 3.2 **(See color insert.)** PWR—A common type of LWR. (Adapted from "Nuclear power reactors," a document of World Nuclear Association, July 2013.)

In the PWR, ordinary water is used as a moderator and as a coolant. In a primary cooling system, water flows freely between the fuel rods, while being directed through fuel assembly in a predetermined fashion. Water is kept at 325°C under 150 atm pressure. As shown in Figure 3.2, the pressure is maintained by steam in a pressurizer. The water in the primary cooling system also serves as a moderator by controlling the negative feedback effect of steam production and the resulting slowdown of the fission reaction. Thus, generation of steam reduces heat generated by the fission reaction causing steam to condense back to water. This negative feedback effect provides one of the safety features of the reactor. An addition of boron to the primary cooling system can also be used as a secondary shutdown system. Thus, closed loop water recirculation system used for primary cooling is operated at 325°C and 150 atm.

The secondary steam generation circuit is operated under low pressure, and water in this circuit boils in the secondary steam generators that are towers containing long narrow tubes. The generated steam drives turbine, which in turn generates electricity. The condensed steam from the turbine is returned to the heat exchangers in contact with the primary circuit. The PWR thus differs from the BWR in that the steam to run the turbine is produced in a steam generator in the secondary steam generation circuit [1–3].

3.2.4 Pressurized Heavy Water Reactor (CANDU)

CANada Deuterium Uranium (CANDU) is a pressurized heavy water nuclear reactor operated on natural uranium fuel U^{238} and uses heavy water (D_2O) as a coolant and a moderator [2]. The CANDU reactor is capable of online refueling during operation. In this type of reactor, the heat from primary coolant is once again transferred to a secondary boiling water loop. The main difference between this reactor and the light water reactor is the use of heavy water in CANDU.

3.2.5 Graphite-Moderated, Direct Cycle (Boiling Water) Pressure Tube Reactor

The graphite-moderated, direct cycle (boiling water) pressure tube reactor (RBMK) was designed in the former Soviet Union [2]. The reactor uses ordinary boiling water as a coolant and graphite as a moderator. This type of reactor is also capable of online fueling. Both CANDU and RBMK circulate water through pipes rather than in a pressure vessel surrounding the entire reactor. In both of these reactors, fuel rods can be reached while the reactor is in operation, and refueling takes place almost on a continuous basis. RBMK is, however, not currently used.

3.2.6 Supercritical Water-Cooled Reactor

Twenty percent of all electricity in the United States uses light water nuclear reactors [3,4]. The next-generation reactors, called supercritical water reactors, promise to increase reactor energy efficiency by as much as 13% while simplifying plant design. Water at about 374°C and about 220 atm pressure becomes supercritical where a phase difference between gas and liquid disappears. Heat produced by fission can also be converted into electricity in a reactor cooled by supercritical water. The supercritical state of water offers some distinct advantages of physical, thermal, and chemical properties for an efficient energy transformation operation in a nuclear reactor. The building of such reactors will, however, require materials that withstand high temperature and pressure. The commercial use of this type of reactor is still in its infancy.

It is clear from the above descriptions that water plays an essential role as an energy carrier and a reactor safety moderator in the nuclear power industry.

3.3 HYDROTHERMAL PROCESSES FOR RECOVERY OF GEOTHERMAL ENERGY

Geothermal energy is thermal energy generated and stored in the Earth [5–20]. This energy of the Earth's crust originates from the original formation of the planet (20%) and from the decay of radioactive minerals (80%). The difference in temperature between the core of the Earth and its surface drives a continuous conduction of heat from the core to the surface. The temperature of the Earth increases with an increased depth from the surface. The core of the Earth is believed to be over 5000°C due to radioactive decay.

The hot water and steam generated by the geothermal heat can be used for power generation. Approximately 10,715 MW of geothermal power is collected in 24 different countries [19]. The worldwide installed geothermal electric capacity is illustrated in Table 3.2 [19]. While the United States has more geothermal capacity than any other nation in the world, it has also been extensively explored in other parts of the world because geothermal power is renewable, reliable, sustainable, environmentally friendly, and cost effective [5,6]. For example, Philippines obtain >25% of its electricity from geothermal energy. The United States produces more than 3000 MW of power from

TABLE 3.2
Global Geothermal Capacity with Greater Than 50 MW

Country	Capacity as of 2010 (MW)
United States	3086
Philippines	1904
Indonesia	1197
Mexico	958
Italy	843
New Zealand	628
Iceland	575
Japan	536
Iran	250
El Salvador	204
Kenya	167
Costa Rica	166
Nicaragua	88
Russia	82
Turkey	82
Papua New Guinea	56
Guatemala	52

Source: "Geothermal energy," Wikipedia, the free encyclopedia, 2012.

geothermal energy largely used in eight states including California, Alaska, Oregon, and Nevada. California leads the nation with 80% of the total US energy consumption [19].

Conventional geothermal energy is generally limited to the areas near tectonic plate boundaries—the regions that are seismically active. Earthquakes and magma movement break up the rock covering allowing water to circulate. As the water rises to the surface, natural hot springs and geysers occur with water temperature as high as 200°C. Besides power, geothermal heat pump also uses the steady temperatures just underground to heat and cool buildings cleanly and inexpensively. About 28 GW of direct geothermal heat capacity is used for heating, spas, industrial processes, desalination, and agricultural applications [5].

The most common current way of capturing the geothermal energy is to tap into naturally occurring "hydrothermal convection" systems where cooler water seeps into the Earth's crust, is heated up by geothermal energy, and then rises to the surface. The hot water coming to the surface can be captured as steam, which in turn can drive turbine to generate electricity. The steam can also be effectively captured from holes that are drilled with a careful design.

Three methods—dry steam, flash steam, and binary cycle—are used to operate the power plants by geothermal energy [9,14]. While all of them use steam and hot water from the ground, in dry steam approach steam goes directly through the turbine then into a condenser where steam is condensed into water, which is returned to the ground. The recycling of water prolongs the life of the heat source. This method is schematically described in Figure 3.3a [14]. In the second approach shown in Figure 3.3b [14], very hot water is depressurized or "flashed" into steam that can then be used to drive turbine. In the third "binary system" approach, the hot water is passed through a heat exchanger, where it heats a second fluid, such as isobutene, in a closed loop. The isobutene boils at a lower temperature than water, so it is more easily converted into steam to run the turbine. This method is schematically described in Figure 3.3c [14].

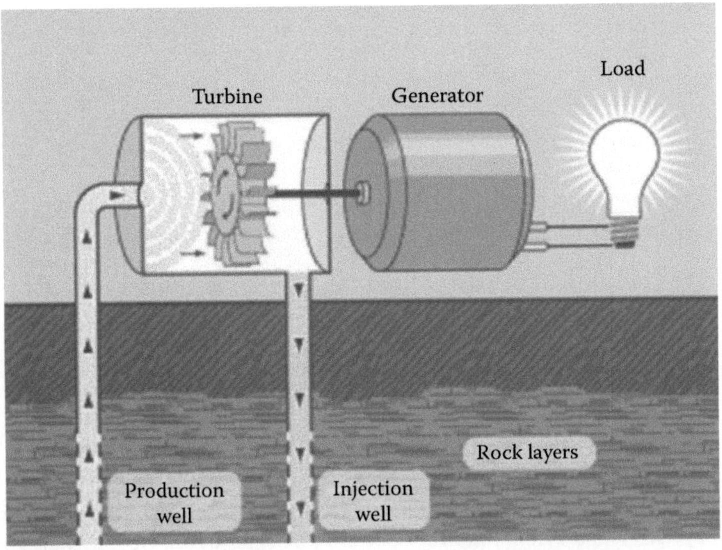

(a)

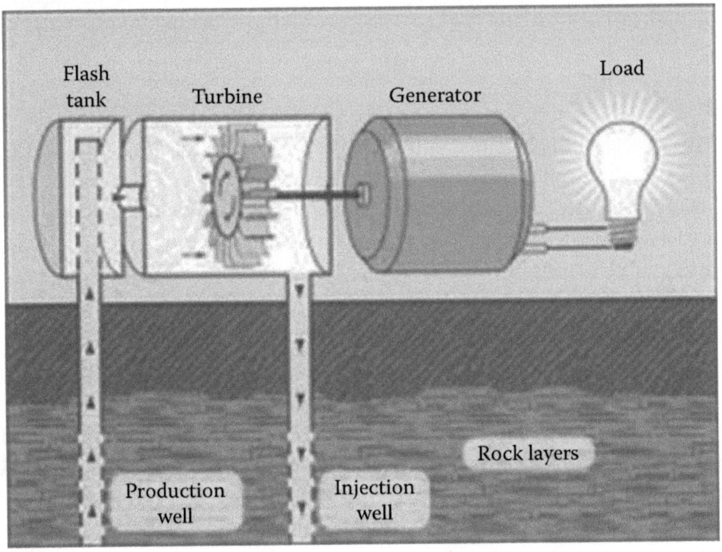

(b)

FIGURE 3.3 **(See color insert.)** Three methods of recovering geothermal energy: (a) dry steam, (b) flash steam. (Adapted from Union of Concerned Scientists, *How Geothermal Energy Works*, Union of Concerned Scientists, Cambridge, MA, 2012; National Renewable Energy Laboratory. Planta Solar 20. http://www.nrel.gov/csp/solarpaces/project_detail.cfm/projectID=39.)

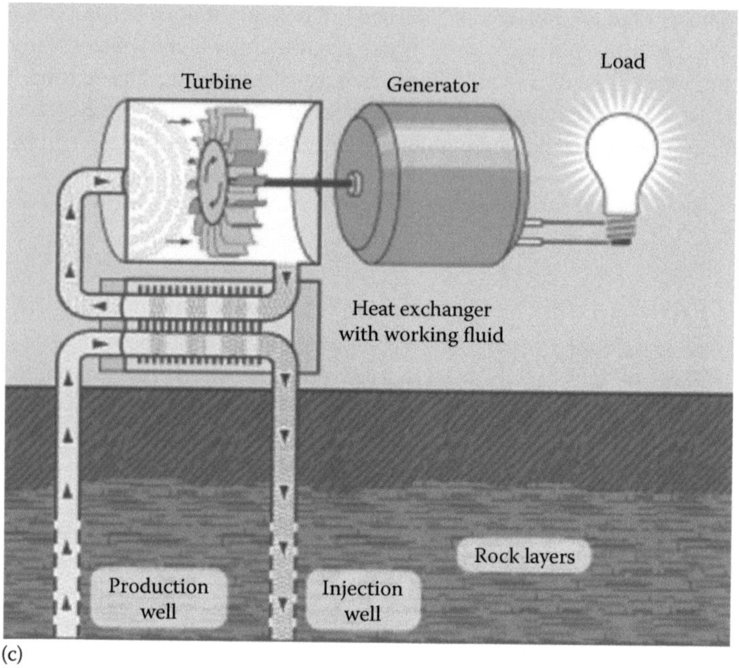

(c)

FIGURE 3.3 **(See color insert.)** (Continued) Three methods of recovering geothermal energy: (c) binary cycle. (Adapted from Union of Concerned Scientists, *How Geothermal Energy Works*, Union of Concerned Scientists, Cambridge, MA, 2012; National Renewable Energy Laboratory. Planta Solar 20. http://www.nrel.gov/csp/solarpaces/project_detail.cfm/projectID=39.)

In general, the choice of the approach depends on the nature of the geothermal resource. If water comes out as steam, the dry steam approach is used. If water comes out as hot water at high enough temperature, the flash steam approach is used. Since most resources produce hot waters, an exchanger design (third approach) is more prevalent in common practice. The largest geothermal system now in operation is a steam-driven plant in an area called Geysers, north of San Francisco, California [13,14].

Geothermal energy (steam) can sometimes be accompanied by impurities that are harmful to the environment. The open systems such as geysers can emit some air pollutants such as hydrogen sulfide and trace amounts of arsenic and minerals along with steam. For the power plant run by hot water system such as Salton Sea reservoir in Southern California, a significant amount of salt can be built up in the pipes, which must be removed. This salt can be either put into landfill or reinjected back into the ground. The closed loop binary cycle system has no emission problem because everything is returned back to the ground.

Besides electricity, the geothermal heat can also be used to heat and cool homes, heat greenhouses, dry out fish and de-ice roads, improve oil recovery, and heat fish farms, spas, and local resorts among other applications. In Iceland, almost every building is heated with hot spring water. A convenient way to tap geothermal energy is to use heat pump, which supplies heating and cooling using geothermal energy. In this method, either air or antifreeze liquid is pumped through pipes that are buried underground and reentered into the buildings. In the summer, the liquid moves heat from the building into the ground, and in the winter, the opposite process takes place providing air and water to the heating system of the building. The system can be simple, in which heating and cooling using the ground source can occur by tubes running from outside the air, under the ground, and into the house ventilation system, or more complex, in which compressor and pumps are used as an electric air-conditioning system to optimize the heat transfer. These ground source heating and cooling systems are the most environmentally clean systems. The Department of Energy has pointed out that heat pumps operated by geothermal energy are more efficient and save more money than any other electrical systems. Currently, more than 600,000 homes in the United States use geothermal energy-driven heat pumps, and this number is increasing at the rate of 60,000 homes per year, with the largest growth in rural areas [6,14,19,20].

3.3.1 Enhanced Geothermal Systems

While geothermal heat can be obtained anywhere under the surface of the Earth, the conditions that make water circulate to the surface are found only in <10% of the Earth's surface [5,9,13,18,20]. A method to capture geothermal heat from dry areas is known as enhanced geothermal system (EGS) or "hot dry rock." The systematic steps demonstrating how EGSs work are graphically illustrated in Figure 3.4 [5,11,14]. As shown in this figure, the hot dry rock reservoirs, typically at greater depths below the Earth's surface than conventional sources, are first broken up by pumping high-pressure water through them. Once the rock is perforated (by the hydraulic fracturing process), additional water not only expands perforations in the rock but also captures heat from the open rock. This steam is collected by a production well and brought to the surface, and it powers turbine to generate electricity. Finally, the cooled water is returned to the reservoir by injection wells to complete the circulation loop. The system can further be optimized by employing carefully designed multiple production wells. Plants that use a closed loop binary cycle described earlier release no fluids or heat-trapping emissions other than water vapor, which may be used for cooling. As indicated in the figure, water and steam play a key role in recovering geothermal energy from deep dry rocks.

The EGS process does carry some risk as hydraulic horizontal fracturing (fracking) used in the recovery of unconventional gas such as "shale gas" allows permeation of carbon dioxide or "fracking fluid" to water aquifers. The EGS can induce seismic activity that might occur from hot dry rock drilling and development, although the likelihood of this occurrence is low, when projects are located at an appropriate distance away from the major fault lines and properly monitored. Appropriate site selection, assessment, and monitoring of rock fracturing and seismic activity during and after construction are very critical. The EGS can produce a continuous power and it is feasible anywhere in the world, depending on the economic limits of the drilling depth. Good locations are over deep granite covered by a 3–5 km layer of insulating sediments that reduce the heat loss [6,7]. The EGS

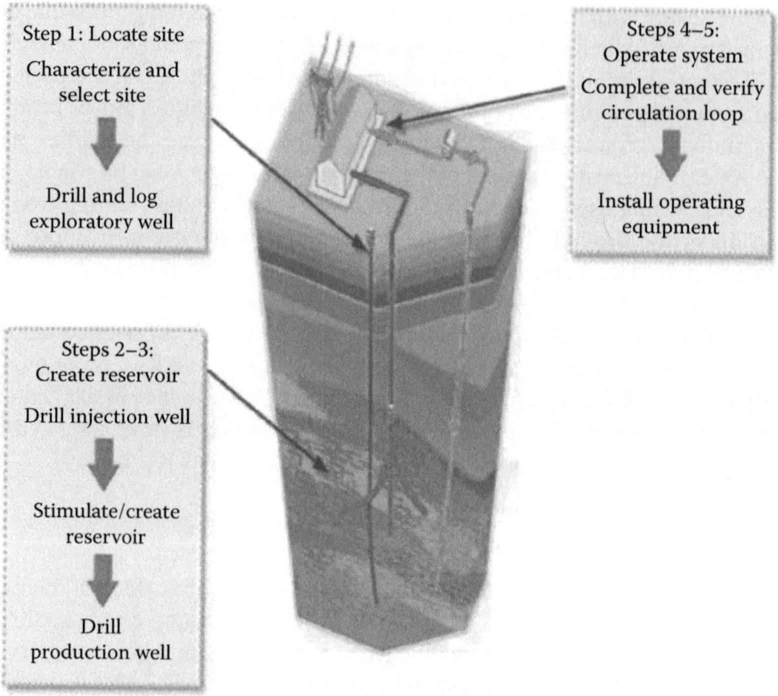

FIGURE 3.4 **(See color insert.)** Steps taken to recover geothermal energy via the EGS. (Adapted from Union of Concerned Scientists, *How Geothermal Energy Works*, Union of Concerned Scientists, Cambridge, MA, 2012; Office of Energy Efficiency and Renewable Energy, *An Evaluation of Enhanced Geothermal Systems Technology*, US Department of Energy, Washington, DC, 2008.)

wells are expected to have a useful life of 20–30 years. The EGSs are currently being developed in many countries including France, Australia, Japan, Germany, the United States, and Switzerland. The largest current EGS project is being developed in Cooper Basin in Australia with a capacity of 25 MW [12]. A summary of the current commercial EGS projects under development is described in Table 3.3 [20].

3.3.2 Coproduction of Geothermal Electricity in Oil and Gas Wells

Geothermal energy can also be captured by using the existing oil and gas wells. In many existing oil and gas reservoirs, a significant amount of high-temperature water and/or high-pressure conditions prevails, which will allow the production of electricity along with the production of oil and gas. In some cases, exploiting these sources can even enhance the extraction of oil and gas itself. A Massachusetts Institute of Technology (MIT) study indicates that in southern and southeastern states there is a potential for developing 44,000 MW of geothermal capacity by 2050 by coproducing electricity, oil, and natural gas at oil and gas fields [5,9,17,18]. The study also suggests that such advanced geothermal systems could supply 10% of

TABLE 3.3
Current Commercial EGS Projects under Development

Project	Country	Size (MW)	Plant Type	Depth (km)
Landau[a]	Germany (EU)	3	Binary	3.3
Aardwarmte Den Haag[a]	Netherland (EU)	6	Thermal	2.0
Paralana (phase 1)	Australia	7–30	Binary	4.1
Cooper Basin	Australia	25–50	Kalina	4.3
United Downs, Redruth	United Kingdom	10	Binary	4.5
Eden project	United Kingdom	3	Binary	3.0–4.0

Source: "Enhanced geothermal systems," Wikipedia, the free encyclopedia, 1–9, 2012.
[a] These projects are operational.

the US base load electricity by that year. Besides conventional oil and gas wells, horizontal wells created in deep oil shale rocks to recover shale gas by "fracking process" can also be helpful in recovering geothermal energy in oil shale rocks. The horizontal wells created for deep and tight gas as well as gas in geopressurized zones and coal bed methane reservoirs can also be useful for the recovery of geothermal energy along with the recovery of unconventional gas. Water and steam play a very critical role in these advanced geothermal systems [5,9,17,18].

3.4 ROLE OF WATER IN STORAGE OF SOLAR ENERGY

The use of solar energy for home heating has been in existence for a long time [21–35]. In this method, solar panels installed on the top of the roof of the houses or buildings can absorb solar heat and this heat is stored in water and steam circulating under the solar panels. This heat can be stored and used continuously for the environmental control in residential houses and industrial buildings.

While solar energy can be stored and used in a number of different ways, water plays an important role in harnessing solar energy. Solar hot water systems use sunlight to heat water. In low geographical latitudes, domestic hot water use at moderate temperatures can be provided by solar water heating systems [21]. There are at least three types of solar heaters: evacuated tube collectors (most widely used), glazed flat plate collectors (used for domestic water heating), and unglazed plastic collectors (mainly used for the swimming pools in the United States) [22–24]. In 2007, the capacity of solar water heater systems installed worldwide was 154 GW, led by China (70 GW), Israel, and Cyprus [21–25].

Water is also heavily used in solar energy-driven heating, ventilation, and air-conditioning (HVAC) systems in residential home as well as in industrial building. Water can be a good solar energy storage device that can be used to provide heating and cooling on a needed basis for daily and seasonal durations. Solar distillation can be used to make saline and brackish water potable. Solar energy can be used for water disinfection and water stabilization pond to treat wastewater. Solar concentrating technologies such as parabolic dish, trough, and Scheffler reflectors can provide process heating for commercial and industrial applications. For example, 50% of process heating, air-conditioning, and electrical requirement for a clothing factory in Shenandoah, Georgia, is provided by a solar energy project [21–25].

Finally, power generated by solar energy using photovoltaic (PV) systems needs to be stored. Off-grid PV systems have traditionally used rechargeable batteries to store excess electricity. Another approach is the use of pumped storage of hydroelectricity that stores energy in the form of water pumped when the energy is available from a lower elevation reservoir to a higher elevation one. The energy is recovered when demand is high by releasing the water to run through a hydroelectric power generator [25–35]. Solar energy can also be stored by producing solar fuels such as hydrogen using numerous techniques described in Chapter 11. The production of solar fuels mostly involves dissociation of water. Hydrogen can also be produced using solar reforming of fossil and biofuels using steam. Different techniques required to accumulate concentrated solar power are described in numerous references [25–35].

3.5 STEAM TURBINE

A steam turbine is a device that extracts thermal energy from pressurized steam and uses it to do mechanical work on a rotating output shaft [36]. Because turbine generates rotary motion, it is particularly suitable to drive an electrical generator. In 1996, about 90% of all electricity generation in the United States used steam turbine [36]. The steam turbine is a form of heat engine that derives much of its improvement in thermodynamic efficiency through the use of multiple stages in the expansion of steam. Steam turbines are made in a variety of sizes ranging from <0.75 kW used for mechanical drives for pumps and compressors to 1.5 million kW used for electricity generators.

Basically five types of steam turbines are used: condensing, noncondensing, reheat, extraction, and induction [36]. *Condensing* turbines are most commonly found in the electric power plants. In this type, steam coming out of turbine is condensed (about 90%). Process steam applications mostly use back-pressure *noncondensing* steam turbine (commonly used in paper and pulp operations, refineries and desalination plants, etc.) in which exhaust pressure is controlled to suit the needs of the steam pressure. *Reheat* turbines are exclusively used in the electric power plants. Here, steam returns to the boiler from turbine, picks up more superheated steam, and returns back to turbine to continue its expansion. *Extraction* turbines are common in all applications. In this case, steam is released from the various stages of the turbine and used for industrial process needs and sent to boiler feedwater heaters to improve an overall cycle efficiency. *Induction* turbines introduce low-pressure steam at an intermediate stage to produce additional power.

Steam turbines are very valuable because they can be used for any fuel. For example, in a nuclear reactor, nuclear energy is converted to thermal energy by generating steam, and the steam can then be used to generate power by steam turbines. In combustion processes using coal, waste, biomass, or other fuels, heat generated by combustion is absorbed by water to generate steam, and the steam is then used to generate power via the use of a steam turbine. Steam is thus a very benign vehicle for energy conversion and heat and power generation.

REFERENCES

1. "Nuclear reactor," Wikipedia, the free encyclopedia (2012).
2. "Nuclear power reactors," a document of World Nuclear Association (July 2013).
3. Duffey, R.B. and Pioro, L.L., "Supercritical water-cooled nuclear reactors: Review and status," *Nuclear Energy Materials and Reactors*, II, 1–14 (2002).
4. "GEN-4 Technology, "Supercritical-water-cooled reactor system," www.gen-4.org/Technology/systems/scwr.htm (2012).
5. Tester, J., Anderson, B.J., Batchelor, A.T., Blackwell, D.D., DiPippo, R., Drake, E.M., Garnish, J. et al., *The Future of Geothermal Energy: Impact of Enhanced Geothermal Systems (EGS) on the United States in the 21st Century*. MIT Press, Cambridge, MA (2006).
6. Geothermal Energy Association (GEA), US Geothermal Power Production and Development Update (2009).
7. Williams, C.F., Reed, M.J., Mariner, R.H., DeAngelo, J., and Galanis, S.P., Jr., *Assessment of Moderate and High-Temperature Geothermal Resources of the United States*, US Geological Survey fact sheet 2008-3082, 4. US Department of the Interior, Washington, DC (2008).
8. Department of Energy, Oak Ridge National Laboratory (ORNL), "Geothermal (ground-source) heat pumps: Market status, barriers to adoption, and actions to overcome barriers," Report No. ORNL/TM-2008/232, DOE publications, Oak Ridge, TN (2008).
9. Office of Energy Efficiency and Renewable Energy (EERE), *An Evaluation of Enhanced Geothermal Systems Technology*. US Department of Energy, Washington, DC (2008).
10. Lund, J.W., "Characteristics, development and utilization of geothermal resources," *Geo-Heat Centre Quarterly Bulletin*, 28 (2), 1–9 (2007).
11. Duchane, D. and Brown, D., "Hot dry rock (HDR) geothermal energy research and development at Fenton Hill, New Mexico," *Geo-Heat Centre Quarterly Bulletin*, 23 (4), 13–19 (2002).
12. Beardsmore, G., "The burgeoning Australian geothermal energy industry," *Geo-Heat Centre Quarterly Bulletin*, 28 (3), 20–26 (2007).
13. AltaRock Energy, "Alta Rock EGS Demonstration Project Status with NCPA at the Geysers," www.altarockenergy.com.
14. Union of Concerned Scientists, *How Geothermal Energy Works*. Union of Concerned Scientists, Cambridge, MA (2012).
15. Kaieda, H., Ueda, A., Wakahama, H., Mito, S., Sugiyama, K., Ozawa, A., Kuroda, Y. et al., "Field experiments for studying on CO2 sequestration in solid minerals at the Ogachi HDR geothermal site, Japan," *Proceedings of the Thirty-Fourth Workshop on Geothermal Reservoir Engineering*, February 9–11, Stanford University, Stanford, CA (2009).
16. NewEnergyFocus, "Plans unveiled for 'UK's first' commercial-scale geothermal plant," *NewEnergyFocus* (October 20, 2009).
17. Martin, D., "Geothermal power plant that could run 5,000 British homes to be built in Cornwall," *Daily Mail* (June 2, 2009).
18. US Department of Energy, Energy Efficiency and Renewable Energy, *Geothermal Technologies Program: Enhanced Geothermal Systems*, DOE/GO-102004-1958. US Department of Energy, Washington, DC (2004).
19. "Geothermal energy," Wikipedia, the free encyclopedia (2012).
20. "Enhanced geothermal systems," Wikipedia, the free encyclopedia, 1–9 (2012).
21. Morrison, G., Budihardjo, I., and Behnia, M., "Water-in-glass evacuated tube solar water heaters," *Solar Energy*, 76 (1), 135–140 (2004).
22. Duffie, J.A. and Beckman, W.A., *Solar Engineering of Thermal Processes*, 3rd ed. John Wiley & Sons, Hoboken, NJ (2006).

23. Mills, D.R. and Morrison, G.L., "Compact linear Fresnel reflector solar thermal power-plants," *Solar Energy*, 68 (3), 263–283 (2000).
24. Wald, M.L., "In the desert, harnessing the power of the sun by capturing heat instead of light," *New York Times* (July 17, 2007), http://www.nytimes.com/2007/07/17/business/17thermal.html?_r = 1.
25. Areva, "Our technology and features," http://www.areva.com/EN/solar-198/arevasolarour-technology.html#tab = tab2.
26. Ummadisingu, A. and Soni, M., "Concentrating solar power—Technology, potential, and policy in India," *Renewable and Sustainable Energy Reviews*, 15 (9), 5169–5175 (2011).
27. Kaygusuz, K., "Prospect of concentrating solar power in Turkey: The sustainable future," *Renewable and Sustainable Energy Reviews*, 15 (1), 808–814 (2011).
28. US Environmental Protection Agency, *Final Report: Design and Fabrication of a Reduced Cost Heliostat*, http://cfpub.epa.gov/ncer_abstracts/index.cfm/fuseaction/display.abstractDetail/abstract/9033/report/F.
29. Pitz-Paal, R., Botero, N.B., and Steinfeld, A., "Heliostat field layout optimization for high-temperature solar thermochemical processing," *Solar Energy*, 85 (2), 334–343 (2011).
30. Kolb, G.J., Jones, S., Donnelly, M., Gorman, D., Thomas, R., Davenport, R., and Lumia, R., *Heliostat Cost Reduction Study*, Report No. SAND2007-3293. Sandia National Laboratories, Livermore, CA (2007).
31. National Renewable Energy Laboratory. Planta Solar 20. http://www.nrel.gov/csp/solarpaces/project_detail.cfm/projectID=39.
32. Solar Reserve, "Crescent Dunes," http://www.solarreserve.com/what-we-do/csp- projects/crescent-dunes/.
33. Abengoa Solar, PS20, the largest solar power tower worldwide. http://www.abengoasolar.com/corp/web/en/nuestras_plantas/plantas_en_operacion/espana/PS20_la_mayor_torre_comercial_del_mundo.html.
34. Ávila-Marín, A.L., "Volumetric receivers in solar thermal power plants with central receiver system technology: A review," *Solar Energy*, 85 (5), 891–910 (2011).
35. Pavlović, T.M., Radonjić, I.S., Milosavljević, D.D., and Pantić, L.S., "A review of concentrating solar power plants in the world and their potential use in Serbia," *Renewable and Sustainable Energy Reviews*, 16 (6), 3891–3902 (2012).
36. "Steam turbine," Wikipedia, the free encyclopedia (2012).
37. *Nuclear Engineering International Handbook,* Nuclear Engineering International Organization, ESCO, Braintree, Essex (2011).

4 Steam Gasification and Reforming Technologies

4.1 INTRODUCTION

All carbon-based materials (i.e., coal, crude and heavy oil, shale oil, bitumen, tar sand, plastics, biomass, organic waste, etc.) can be converted to carbon monoxide, hydrogen, carbon dioxide, and methane in the gasification process by a controlled amount of oxygen and/or steam at temperatures higher than about 700°C. The product distribution generally depends on the temperature, pressure, residence time, catalyst, and the nature of the feedstock. The gas produced from steam gasification (with or without oxygen) is often called syngas (either synthesis gas or synthetic gas) or producer gas, both of which are fuel themselves that generate heat and energy. Unlike direct combustion of original raw materials, syngas (hydrogen and carbon monoxide) is much more versatile in that it can be used for combustion at much higher temperatures. It can be used in fuel cells (FCs) and can also serve as raw materials for the production of numerous chemicals and liquid fuels. The gasification also produces gaseous fuels that do not contain corrosive ash elements such as chloride and potassium [1–10].

Water in its gaseous form (i.e., steam) plays a very important role in overall gasification process. During gasification, carbonaceous material undergoes several important processes: (1) at temperatures around 100°C, the dehydration results in the generation of steam in the gas phase, and (2) further dehydration and pyrolysis of carbonaceous materials occur at temperatures around 200°C–300°C resulting in the loss of raw materials up to about 70% of their original weight. The nature of char produced by this reaction depends on the nature of the feedstock; (3) the volatile materials from char react with oxygen to produce carbon dioxide and carbon monoxide; (4) the char also reacts with steam to produce hydrogen and carbon monoxide; and (5) at higher temperatures, the water–gas shift reaction between carbon monoxide and steam produces hydrogen and carbon dioxide [1–10].

The equilibrium constants for various reactions involving carbon and intermediate products are illustrated in Figure 4.1 [1–3]. In the real process, at high temperatures, steam gasification predominantly produces hydrogen and carbon dioxide because of the dominance of water–gas shift reaction. Both carbon monoxide and carbon dioxide are favored during direct combustion [1,3,5].

To some extent, gasification and reforming are overlapping phenomena in that gasification involves the transformation of solid (and liquid) raw materials to the gas-phase products through a series of thermal reactions. Catalytic reforming involves the transformation of these intermediate products to hydrogen, carbon monoxide, and carbon dioxide by steam reforming, dry reforming,

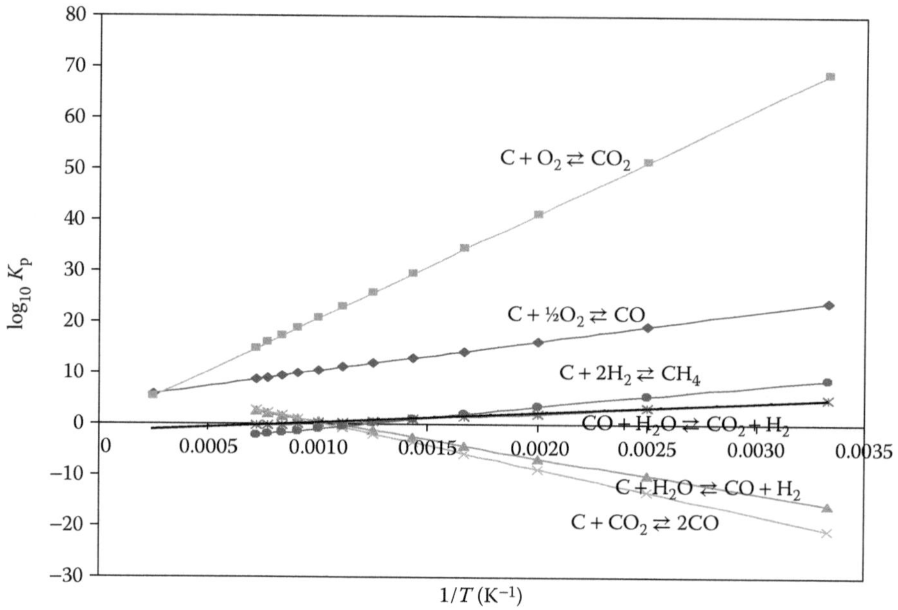

FIGURE 4.1 Equilibrium constant-temperature relations for carbon reactions with oxygen, steam, hydrogen, and carbon dioxide. (Adapted from Lee, S., Speight, J.G., and Loyalka, S.K., *Handbook of Alternative Fuel Technologies*. Taylor & Francis, Boca Raton, FL, 2007.)

partial oxidation, and water–gas shift reactions aided by a suitable catalyst. High-temperature gasification can also produce syngas; however, reforming catalyst allows the productions of hydrogen, carbon monoxide, and carbon dioxide at a lower temperature and at a faster rate. Generally, hydrocarbon-free product distribution by gasification requires temperatures in excess of 1000°C–1200°C. The catalytic reforming process can achieve the same type of product distribution at temperatures around 800°C.

Steam reforming is the oldest and most widely used technology available to convert hydrocarbons into a gaseous product containing hydrogen and carbon dioxide. The reaction between steam and hydrocarbons is an endothermic reaction and is carried at high temperatures (somewhere between 400°C and 1000°C) in the presence of a catalyst. Generally, Ni catalyst is used; however, in recent years, several other types of catalysts have been investigated. The stoichiometry of hydrocarbon reforming for maximum hydrogen production is described by the following reactions [11–17]:

$$C_nH_m + 2nH_2O \rightarrow nCO_2 + \left(\frac{2n+m}{2}\right) \times H_2 \tag{4.1}$$

$$C_nH_m + nH_2O \rightarrow nCO + \left(\frac{2n+m}{2}\right) \times H_2 \tag{4.2}$$

These two reactions are generally accompanied by the water–gas shift reaction as [18]

$$(X_n)CO + H_2O \rightleftarrows CO_2 + H_2 \tag{4.3}$$

The initial steps in steam reforming are the dissociative adsorption of the hydrocarbons on the metal sites of the catalyst and the reaction of the adsorbed C_xH_y species with the adsorbed H_2O-derived species to produce CO and H_2 (Equation 4.2). With an active catalyst at temperatures below 600°C, reforming of hydrocarbons is irreversible with no intermediates and the only byproduct is carbon that forms on the catalyst. Besides the above reactions, the following reactions also occur at varying degrees:

$$CO + 3H_2 \rightleftarrows CH_4 + H_2O \tag{4.4}$$

$$2CO \rightarrow CO_2 + C \tag{4.5}$$

$$CH_4(g) \rightleftarrows 2H_2(g) + C(s) \tag{4.6}$$

Equation 4.1 is the combination of reforming and water–gas shift reactions. Equations 4.3, 4.4, and 4.6 are equilibrium-limited reactions. Under normal conditions, Equations 4.5 and 4.6 dominate, and they together produce coke on the catalyst. In general, both methanation and disproportionation reactions are equilibrium limited. The reformed fuel contains carbon monoxide that must be reduced to a low level (except for the use in high-temperature FCs [HTFCs]). To reduce carbon monoxide concentration at the desired level of <10 ppm, the reforming reaction is followed by a high-temperature and a low-temperature water–gas shift reaction, both of which are exothermic [18]. The residual carbon monoxide can be further reduced by its preferential oxidation. The mixture of hydrogen and carbon dioxide coming out of the preferential oxidation process can then undergo a separation process to remove carbon dioxide and generate pure hydrogen. The separation process can be a physical (absorption by molecular sieves), a chemical (absorption in an amine solution), or a membrane separation (usually Pd membrane) at high temperature. The separated carbon dioxide is used with ammonia to produce urea. The purified hydrogen is used in the production of ammonia and a host of other refining and chemical production operations.

The catalysts for steam reforming of hydrocarbons are mainly nickel based on oxide support to obtain high thermal stability. Nickel catalysts are preferred because of their low cost, reasonable thermal stability, and high activity [19–23]. At low temperatures (425°C–500°C), iron catalyst promoted with chromium oxide is sometimes used to enhance oxidation reaction. More details on the catalysis of steam and tri-reforming reactions are given in Sections 4.2.3, 4.3, and 4.4.

Currently, more than 65% of hydrogen production uses steam reforming of fossil fuel technology because it is a mature and reasonably inexpensive technology (compared to other processes). It does not require a new infrastructure. It also reduces the need for transport and storage of hydrogen. The disadvantages of the steam reforming process are as follows: (1) reformers are complex, large, and expensive; (2) reformers have high warm-up period; and (3) reformers introduce additional losses into the energy conversion process. The reforming process can also create pollutants such as carbon dioxide, carbon monoxide, unburned original hydrocarbons, and nitrous oxide that can be generated by oxidation with air.

While steam gasification and reforming has huge potential to generate hydrogen, as discussed earlier, steam gasification of coal is more difficult than that of biomass. Biomass has the potential to accelerate the realization of hydrogen as a major fuel of the future. It is more acceptable than coal because it is renewable and consumes atmospheric carbon dioxide during its growth, thus having a small net CO_2 impact compared to fossil fuels. However, hydrogen produced from biomass has major challenges. There are few commercial plants. While biomass is more reactive to steam compared to coal, the yield of hydrogen is low from biomass since the hydrogen content of biomass is low to begin with (~6% vs. 25% for methane) and the energy content is low due to 40% oxygen content of biomass. Since over half of the hydrogen from biomass comes from splitting water in the steam reforming reaction, the energy content of the feedstock is an inherent limitation of the process. Due to high oxygen content, the yield of hydrogen per unit weight of biomass is low. The low yield of hydrogen on a weight basis is, however, misleading since the energy conversion efficiency is high. For example, the steam reforming of bio-oil at 825°C with a fivefold excess of steam has an energy efficiency of 56% [24–26]. The cost of growing, harvesting, and transporting biomass is, however, high. Thus, even with reasonable energy efficiencies, it is not currently economically competitive with natural gas steam reforming for stand-alone hydrogen without the advantage of high-value coproducts. One way to make steam reforming of biomass more competitive is to use coal–biomass mixture as the feedstock [5,27–33]. This mixture will increase the hydrogen production rate, maintain the overall reactivity between steam and feedstock, and also reduce the emission of carbon dioxide. A 70/30 mixture of coal and biomass is CO_2 neutral for environmental purposes [2].

4.2 MECHANISMS, KINETICS, AND CATALYSIS OF STEAM GASIFICATION AND REFORMING

4.2.1 Mechanism of Steam Gasification

Steam gasification is an endothermic reaction and requires heat to move the reaction in the forward direction. Generally, excess steam is also required to promote the gasification reaction. However, excess steam affects the thermal efficiency of the process. The reaction can be expressed as [1]

$$C(s) + H_2O(g) = CO(g) + H_2(g) \qquad \Delta H^0_{298\,K} = 131.3 \text{ kJ/mol} \tag{4.7}$$

The equilibrium constant for this reaction is illustrated in Figure 4.1 and compared with the equilibrium constants for gasification with other gasifying agents such as oxygen, hydrogen, and carbon dioxide [1]. It is clear that the pure steam gasification is not as favored as the one with oxygen. Often steam is accompanied by oxygen and hydrogen to get more favorable rate of reaction and better product distribution. The steam gasification produces hydrogen and carbon monoxide. Its relative ratio depends on synthesis chemistry and process engineering as well as the presence of other gasifying agents. Two reaction mechanisms for carbon–steam reactions over a wide range of gasification conditions are proposed [1,3]:

Mechanism A:

$$C_f + H_2O = C(H_2O)_A \tag{4.8}$$

$$C(H_2O)_A \rightarrow CO + H_2 \tag{4.9}$$

$$C_f + H_2 = C(H_2)_B \tag{4.10}$$

In this mechanism, C_f designates unoccupied carbon sites, $C(H_2O)_A$ and $C(H_2)_B$ are adsorbed species of water and hydrogen, respectively. Equations 4.8 and 4.10 are reversible, whereas Equation 4.9 is irreversible. This mechanism is referred to as inhibition by hydrogen adsorption because the adsorbed hydrogen prevents the adsorption of steam molecules and thus prevents the gasification reaction [1,3]

Mechanism B:

$$C_f + H_2O = C(O)_A + H_2 \tag{4.11}$$

$$C(O)_A \rightarrow CO \tag{4.12}$$

Here gasification rate is affected by the competitive reaction of chemisorbed oxygen with hydrogen, and this competition for sites limits the conversion of the adsorbed oxygen to carbon monoxide. Thus, this mechanism is often called "inhibition by oxygen exchange" [1,3].

Both of these mechanisms can express the gasification rate in the form as follows:

$$R = \frac{k_1 p_{H_2}}{1 + k_2 p_{H_2} + k_3 p_{H_2O}} \tag{4.13}$$

where:

R is the rate of gasification
p_{H_2} and p_{H_2O} are partial pressures of hydrogen and water
k_1 is the kinetic constant
k_2 and k_3 are adsorption constants

and the expression fits the experimental data well. The rate expression can be derived, assuming pseudo-steady states for adsorbed species.

The above discussion assumes pure carbon and steam as a gasifying agent. The rate analysis gets more complex when coal and other gasifying agents are taken into the considerations. Carbon in coal is distributed within the coal matrix and its concentration depends on the rank of coal. The gasification of coal must therefore consider transport processes within the coal matrix and accessibility of carbon within the matrix. When other gasifying agents such as oxygen and hydrogen are added, their roles on overall gasification reactions must also be considered. Alkali metal salts act as catalysts for steam gasification reactions. The order of catalyst activity of alkali metals on coal gasification reaction is Cs > Rb > K > Na > Li. For coal gasification, coke deposition during gasification on active catalyst sites also affects the catalyst life. The coking can be reduced by increasing the steam-to-carbon ratio [1–3].

4.2.2 Mechanism of Steam Reforming

As mentioned earlier, the steam reforming of methane is very attractive because methane contains the largest hydrogen/carbon (H/C) ratio in any hydrocarbon. Unfortunately, methane molecule is very stable with C–H bond energy of 439 kJ/mol. Such high bond energy makes methane resistant to many reactants and reactions. Furthermore, C–H bond in methane is very strong. Methane molecule can be, however, activated by group 8–10 transition metals and can be oxidized to produce syngas. Further conversion of CO by water–gas shift reaction generates the final product with large concentrations of hydrogen and carbon dioxide. Once the carbon dioxide is removed from the mixture of carbon dioxide and hydrogen by adsorption, absorption, or membrane separation process, pure hydrogen is obtained. The adsorption process allows the purity of hydrogen of about 999.999% at 25 bar feedstock pressure. In the recent years, the use of ceramic ion transport membranes (ITMs) with reformers has opened up the possibilities of the production of high-quality and low-cost hydrogen [5,6,33–35] (Barrio et al., 2012, pers. comm.).

Methane reforming by steam is an endothermic reaction and favored at lower pressures. While noble metal catalysts have been tested and used in the past, most commercial operations use nickel catalyst because of its low cost and high activity, stability, and selectivity. The activity of the catalyst depends on the catalyst surface area and the temperature (around 400°C–1000°C) for steam pressure up to 30 atm. The activity of the catalyst is usually described by the turnover frequency (TOF) that is generally 0.5 s^{-1} at around 450°C. This number corresponds to about 10% methane conversion. High conversion rate demands higher temperature because the reaction is limited by thermodynamics that is favored at higher temperature. Very high conversion requires the reactor to be operated at temperatures higher than around 900°C. Often the catalysts in the reformer are poorly used because heat transfer between gas and solid is a limiting factor in the reaction. The reactor design plays an important role in the performance of the reactor, which will be discussed later.

Numerous studies on mechanism of methane reforming have been reported and these are well reviewed by Wei and Iglesia [36], Rostrup-Nielsen et al. [37], and Bradford and Vannice [22,23]. The following discussion closely follows these reviews. Wei and Iglesia [36] have shown that the rate-limiting step for steam reforming is C–H bond activation. They proposed the following mechanism:

$$H_2O + * \rightarrow O^*(a) + H_2(g) \tag{4.14}$$

$$CH_4(g) + 2* \rightarrow CH_3*(a) + H*(a) \tag{4.15}$$

$$CH_3*(a) + * \rightarrow CH_2*(a) + H*(a) \tag{4.16}$$

$$CH_2*(a) + * \rightarrow CH*(a) + H*(a) \tag{4.17}$$

$$CH*(a) + O*(a) \rightarrow CO*(a) + H*(a) \tag{4.18}$$

$$CO^{*}(a) \rightarrow CO(g) + * \quad (4.19)$$

$$2H^{*}(a) \rightarrow H_2(g) + 2^{*} \quad (4.20)$$

In the above equations, "*" denote Ni (or catalyst in general) surface atom. In this mechanism, methane adsorbs dissociatively on the Ni surface producing methyl group and water molecule reacts with Ni surface atoms to produce adsorbed oxygen and gaseous hydrogen. The methyl group goes through further stepwise dehydrogenation steps. The final product of this dehydrogenation CH– reacts with adsorbed oxygen to produce syngas (CO and H_2).

Along with the main reactions outlined above, the reforming reactions are accompanied by the carbon formation reactions:

$$2CO \rightarrow C + CO_2 \qquad \Delta H^0_{298\,K} = -172.5 \text{ kJ/mol} \quad (4.21)$$

$$CH_4 \rightarrow C + 2H_2 \qquad \Delta H^0_{298\,K} = 74.9 \text{ kJ/mol} \quad (4.22)$$

These two reactions deposit carbon on the catalyst in the form of filaments that ultimately deactivate catalyst. The carbon formation reactions are also counterbalanced by carbon-consuming reactions:

$$C + CO_2 \rightarrow 2CO \quad (4.23)$$

$$C + H_2O \rightarrow CO + H_2 \quad (4.24)$$

Both of these reactions also depend on the operating conditions and the nature of the reactor design. Generally, at low temperatures, the Ni catalyst surface is covered with hydrocarbons, which degrades into a polymeric layer. However, at high temperatures, cracking of olefinic and aromatic hydrocarbons produces coke that deposits on the catalyst surface. Since NiC is not stable, carbon is formed in the form of filaments that grow on the catalyst surface. The size of Ni particles has a direct bearing on the location of filaments on the Ni surface. Smaller and more dispersed Ni particles reduce the formation of carbon filaments. Thus, Ni dispersion is an important variable on the catalyst activity and stability (degradation). The literature has shown that the size and location of Ni particle ensemble is an important variable for controlling the coke formation on the catalyst [34–40] (Barrio et al., 2012, pers. comm.). The coke formation can also be controlled by controlling the carbide formation. While alloys reduce carbide formation, they hide the active sites of nickel for the reforming reactions. The literature has also shown that the addition of a small amount of dopants (e.g., Sn) reduces coking without affecting the activity for the reforming reaction [34–40] (Barrio et al., 2012, pers. comm.). Carbon formation can also be reduced by the alloys of copper–nickel, sulfur–nickel, nickel–tin, and nickel–rhenium [34–40] (Barrio et al., 2012, pers. comm.).

4.2.3 Catalysts for Steam Gasification

Catalysts can be added to the steam gasification process in two forms: (1) as active bed additives or (2) as separate heterogeneous catalysts that are used in the steam reforming reactions [28,41–57]. The active additives are used to (1) reduce the amount of tar formation; (2) promote several other chemical reactions to change the production rate, composition, and heating value of the gas; (3) promote char gasification; (4) prevent active agglomeration of the feedstock, char, and tar that can lead to reactor choking; and (5) remove carbon dioxide through the active adsorption process. The steam reforming catalysts also reform tar and produce gas of high quality.

4.2.3.1 Dolomite, Olivine, and Alkali Metal-Based Catalysts

These are generally cheap and disposable catalysts. Dolomite is a magnesium ore with the general formula $MgCO_3$. $CaCO_3$ is considered to be a good catalyst for biomass gasification. Dolomite is also a good adsorbent for carbon dioxide and capable of removing tar very efficiently. It is, however, a very fragile substance and may quickly attrite in highly turbulent conditions within a fluidized bed. CaO additive was studied by Dalai et al. [41], who showed that the use of this additive reduced the gasification temperature to about 150°C to get the same level of gas production. Both carbon conversion and hydrogen production increased with impregnation of CaO in cellulose, cedar, and aspen. The production rates of gas and hydrogen also depended on the nature of feedstock; cedar and aspen performed better than cellulose.

Hu et al. [42] tested calcined olivine and dolomite in a fixed-bed reactor and found higher activities of calcined catalysts compared to those of natural catalysts. Other literature also showed that in the presence of olivine, tar conversion increased with an increase in temperature from 800°C to 900°C, and at 900°C and higher, all water-soluble heterocyclic compounds get converted [21] (Barrio et al., 2012, pers. comm.). With 17 wt% olivine in the sand at 900°C, the conversion of heavy polyaromatics increased from 48% to 71%. Calcined dolomite, however, increased the conversion up to 90%. Aznar et al. [44–46] showed that dolomite was very effective in removing tar coming from a blend of plastic waste with pinewood sawdust and coal in the temperature range of 750°C–880°C. Xu et al. [47] demonstrated that at 700°C, hydrogen concentration in the product increased by the use of CaO as an adsorption agent for carbon dioxide.

Monovalent alkali metals such as Li, Na, K, Rb, Cs, and Fr were also found to be catalytically active in steam gasification. Both K and Na are a part of biomass and accumulate in the ash, which in turn can act as a catalyst. This solves the problem of ash handling and the ash reduces the tar content in the gas phase. The ash catalytic activity, however, can be lost due to particle agglomeration. Sutton et al. [43] pointed out that direct addition of alkali metals can require (1) expensive recovery of catalyst, (2) increased char content after gasification, and (3) ash disposal problems. Lee [48] and Lee et al. [49] found that the addition of Na_2CO_3 enhances the catalytic gasification of rice straw over a nickel catalyst and the additive increases the gas formation. They also found that the gas production rate is affected by the nature of the additive and follows the order: Na > K > Cs > Li. The use of activated alumina as a secondary catalyst was found to be effective by Simell et al. [50–52]; however, this catalyst deactivated faster due to coking compared to dolomite. Sami et al. [29] showed that both zirconia and alumina promoted toluene and ammonia conversions at lower temperatures, indicating enhanced oxidation activity of zirconia with alumina. Furthermore, H_2S had a little effect on the activity of alumina-doped zirconia.

4.2.3.2 Nickel-Based Catalysts

As mentioned earlier, the gasification and reforming in the presence of steam are overlapping reactions. Tar and lower hydrocarbons produced by the gasification can be simultaneously reformed in the presence of a suitable catalyst. Rostrup-Nielsen et al. [37] presented a very good review of applicability of transition metals (group VIII) and noble metal catalysts to steam gasification/reforming process. While a number of noble metal catalysts such as Ru and Rh have superior performance for steam reforming, the cost and easy availability of these catalysts compared to that of nickel, made the latter choice more practical. The literature has convincingly demonstrated the usefulness of nickel catalysts for biomass gasification [19–23,36,37,43]. Olivares et al. [53] showed that nickel reforming catalysts display 8–10 times more reactivity than calcined dolomite. Nickel catalysts can be, however, deactivated by the poisons such as sulfur, chlorine, and alkali metals. They can also be deactivated by the formation of coke. The coke deposition can be reduced by increasing steam/biomass ratio; however, this increases the energy cost and changes the gas-phase composition of the product. In general, Ni-gamma-alumina catalyst gave higher conversion and lower deactivation compared to Ni-alpha-alumina catalysts. The MgO/CaO addition to alumina also gives the catalyst more stability. Lanthanum-based pervoskite support was also found to be very effective. The topics of coking, catalyst deactivation, and effective support for the nickel are discussed in Sections 4.2.4, 4.3, and 4.4. Suffice to say that nickel-based catalysts have gained a significant support for steam gasification and reforming.

4.2.4 Catalysts for Steam Reforming

In general, two types of sites are required for the steam reforming catalysts: the catalytic sites for hydrogenation and dehydrogenation and the acidic sites [22,23,28,36,37,41–56]. The acidic sites promote the formation of carbonium ions. For aromatization and isomerization reactions, the two types of sites are necessary. While, as mentioned earlier, Ni catalysts on oxide supports have been most extensively used in the industry, recent studies show that bimetallic catalysts such as Ni/Ru and Pt/Re have been more effective catalysts. Again, due to economical reasons, one of the catalysts needs to be nickel. Trimetallic catalysts of noble metal alloys have also been tested. In general, bi- and trimetallic catalysts give better stability (with low sintering at high temperatures) and increased catalyst activity and stability. Coke deposition on the catalysts has been the main reason for catalyst decay; however, coke can be removed by the oxidation at high temperatures. The coke deposition can vary from 15% to 25% on the catalyst [22,23,28,36,37,41–56].

The coke formation can occur by one or more of the following reactions:

$$CH_4(g) \rightleftarrows 2H_2 + C(s) \tag{4.25}$$

$$2CO(g) \rightleftarrows CO_2(g) + C(s) \tag{4.26}$$

$$CO(g) + 2H_2(g) \rightleftarrows H_2O(g) + C(s) \tag{4.27}$$

$$CO_2(g) + 2H_2(g) \rightleftarrows 2H_2O(g) + C(s) \tag{4.28}$$

Equation 4.25 is the famous Boudouard reaction. Coke can be formed from CO, CO_2 as well as CH_4. Coke can also be formed from ethylene through the polymerization reaction as

$$C_2H_4 \rightarrow \text{Polymers} \rightarrow \text{Coke} \tag{4.29}$$

The coke deposition at a sustained level should be avoided because it leads to several undesirable side reactions, loss in catalyst activity, and poor heat transfer between the catalyst and the gas phase. If the coke deposition becomes very extensive, it can block the open surface area causing an excessive pressure drop within the reactor and it can also cause localized “hot spots” that can induce “runaway” conditions for the reactor. Coke formation can be minimized by the use of an excess steam. The catalyst can also be regenerated periodically, by burning off the deposited coke through the oxidation reactions.

4.3 DRY REFORMING

While steam reforming has been extensively used to produce hydrogen, it is generally not a desirable process to make syngas of a diverse composition that may be needed for the downstream conversion of syngas to a variety of fuels and chemicals by Fischer–Tropsch (FT) and other processes [57–62]. To generate syngas of different composition, steam reforming is often coupled with dry reforming and partial oxidation. A combination of steam reforming, dry reforming, and partial oxidation is called “tri-reforming.” Since both dry reforming and tri-reforming have strategic fuel values, they are briefly described in this and the following section.

Depending on the reaction conditions, steam reforming and water–gas shift reaction can also be accompanied by “dry reforming” reaction in which carbon dioxide produced from reverse water–gas shift reaction can react with hydrocarbons according to the following reactions:

$$CO_2 + H_2 \rightarrow CO + H_2O \qquad \Delta H^0_{298\,K} = 41.2\ \text{kJ/mol} \tag{4.30}$$

and

$$CH_4 + CO_2 \rightarrow 2CO + 2H_2 \qquad \Delta H^0_{298\,K} = 247.4\ \text{kJ/mol} \tag{4.31}$$

Here, the second “dry reforming” reaction is illustrated for methane. This reaction was first studied by Fischer and Tropsch in 1928. It is briefly covered here because of its close alignment with the steam reforming reaction. While the kinetic mechanisms for dry reforming and steam reforming reactions on conventional catalysts are very similar, generally steam reforming is faster and dry reforming requires higher temperature and is accompanied by more coke formation than steam reforming. While dry reforming provides a mechanism for chemical use of greenhouse gas “carbon dioxide,” it is not the solution for the complete removal of carbon dioxide due to stoichiometry of various reactions occurring simultaneously.

Generally, nickel and other noble metal catalysts, alone or composite bimetallic or trimetallic form, work well. The literature has shown that coking and catalyst deactivation issue can be partially addressed by the use of pervoskite and hydrotalcite (HT) catalyst supports with ABO_3 functionalities. Lanthanum- and strontium-based supports were found to be more effective. These catalysts reduce the degree of coking and the resulting catalyst deactivation. Dry reforming reaction produces H_2/CO ratio of 1, which is lower than 2 for partial oxidation and 3 for steam reforming.

Dry reforming gives good conversion generally at high temperature (around 850°C). Dry reforming is more endothermic than steam reforming and must be carried out at high temperature and low pressure to achieve maximum conversion. Besides noble metals, transition metal carbides (especially Mo) are also effective, but these catalysts are stable only at high temperatures.

There are at least two examples of the commercial process for the dry reforming of methane [57–62]. The industrial caloric process (CALCOR), which has been developed using nickel-based catalysts, is used for the production of CO-rich synthesis gas from natural gas or liquefied petroleum gas using a large excess of CO_2. Pure carbon monoxide is an important chemical feedstock, for example, in the production of acetic acid and phosgene, and it is important to produce on-site due to the transportation risks caused by its toxicity. This multistage process was developed by Caloric GmbH. In the first stage of the process, carbon dioxide and methane are reacted together producing a mixture of carbon monoxide, hydrogen, carbon dioxide, and water. The heat for the endothermic reaction is provided by the burning of fuel in a similar manner to the steam reforming process. In the next stage, carbon dioxide in the effluent stream is removed and recycled to the reformers. Carbon monoxide is separated from hydrogen (which can be used as a fuel and sold) and methane leaving only a very small amount of methane remaining in the product stream and giving purities up to 99.95%. The coking is prevented by packing the reactor with the catalyst that has varying activities and shapes.

The dry reforming of methane has also been practiced by the SPARG (sulfur-passivated reforming) process created by Haldor Topsøe [57–62]. It was commercialized at Sterling Chemicals Inc., Houston, Texas, in 1987. The process produces a variety of syngas compositions [57–62] and reduces H_2/CO ratio from 2.7 to 1.8 without modification in steam reforming facility [38–43]. The process is operated at 915°C–945°C and coke deposition on Ni catalyst is reduced due to the treatment of the catalyst by sulfur. The process uses mixtures of CO_2 and H_2O, and thus, it is a combined dry and steam reforming process. Impurities such as methane, hydrogen, or other hydrocarbons in the feed stream decrease the mechanical strength of polycarbonates produced from syngas via phosgene reaction path. Higher hydrocarbons are therefore removed in the pre-reforming step to reduce the product impurities as well as coke deposition on the catalyst. Sulfur in the product may require additional purification steps.

The literature indicates that the mechanism for dry reforming of C_1–C_3 hydrocarbons is somewhat different from that of higher hydrocarbons [57]. The same holds for steam reforming reaction. The general route in the cases of C_1–C_3 alkanes involves the dissociation of hydrocarbons and subsequent oxidation of carbon fragments; oxidative dehydrogenations of ethane and propane also proceed partially. The catalysts are in more reduced state and the activation of the hydrocarbon is the rate-controlling step. In the case of C_4 and higher hydrocarbons, the first step of the process is direct hydrogenation of alkanes. Activation of carbon dioxide, but not the activation of hydrocarbon, is the rate-controlling step. Hydrogen formed interacts with carbon dioxide and shifts the equilibrium of the dehydrogenation reaction.

Just as for methane, a required condition for dry reforming of higher hydrocarbon is that the catalyst system adsorbs and activates carbon dioxide. The acidic property of CO_2 necessitates the choice of a catalyst with basic properties. However, alkali metal and alkaline earth oxides are ineffective because of strong carbonate formation. Oxides of a moderate basicity are necessary, and moreover, they must participate in the redox process with CO_2 reduction. While MnO was used in the earlier studies, its modification by oxides of K, Na, Cr, and La influences both its acceptor function and the degree of surface oxidation. It controls the mechanism of hydrocarbons and alcohol transformations. Possible other good candidates are La_2O_3, cesium oxides, and praseodymium oxides. La_2O_3 showed the greatest interactions among CO_2, hydrocarbons, and alcohols. Binary oxide-based support system and dual metals can improve the performance. Promoters and the method of catalyst preparation also have an effect on the catalyst performance [57].

As mentioned earlier, the dry reforming of hydrocarbons leads to a variety of products and the transformation to syngas with different degrees of success depending on the operating conditions and the nature of the catalyst. The major issues with dry reforming are (1) endothermic nature of reaction requiring high-energy input for the reaction process, (2) difficulty in igniting the reaction at low temperature (lower than about 500°C), and (3) requiring very high temperature (>650°C) to reduce coke deposition on the catalyst. In sum,

1. The success of the CO_2 conversion depends on three factors: catalyst activity, catalyst stability (which depends significantly on the coke formation and the nature of the coke), and efficient heat transfer operations. While there are numerous catalysts examined in the literature, it is clear that nickel catalyst is still the most practical from an economic point of view at the commercial scale. Noble metal catalysts such as Rh, Ru, and Pt are more active and perhaps more stable, but they are too expensive to be of commercial value. Future research should be focus on bimetallic catalysts such as Ni—Ru. Ru is about 40–50 times less expensive than Rh, and therefore, it will carry more practical viability for the commercial process [57–62].
2. The nature of the catalyst support is also very important [57]. The support often interacts with metals, and because of that, it is often considered as part of the catalyst. The best situation is the uniform distribution of very active metals in small sizes distributed along the support and they do not migrate or sinter during high-temperature reforming process. Perovskite support offers special attraction because in this case metals are uniformly and tightly distributed in the support lattice. The catalyst must be a basic

in nature, but literature has shown that too much basicity does not help the reforming process. Along with Al_2O_3, lanthanum, cerium, and zirconium oxides need to be examined. Just like mixed metals, mixed supports should also be considered.

In more recent investigations on dry reforming, the overall objectives have been to devise (1) a process that has less coke deposition on the catalyst such that the catalyst is active and stable for a long period, (2) a process in which the catalyst ignites at as low temperature as possible, (3) a process that is heat efficient, (4) a process in which high conversion of CO_2 and hydrocarbons is achieved, and (5) a process in which major products are carbon monoxide and hydrogen. As indicated earlier, the last objective is a particular problem without deep dehydrogenation when the hydrocarbons contain two or more carbon numbers.

4.4 TRI-REFORMING

Fundamentally, there are three types of high temperature reforming processes: stream reforming, dry reforming, and partial oxidation [57,63–65]. The term "tri-reforming" is applied to the process in which all of these reforming processes are combined in a single use. The three reforming processes are expressed by the following set of chemical reactions for methane:

1. Steam reforming: $CH_4 + H_2O \rightarrow CO + 3H_2$ $\quad \Delta H^0_{298\,K} = 206$ kJ/mol
2. Dry reforming: $CH_4 + CO_2 \rightarrow 2CO + 2H_2$ $\quad \Delta H^0_{298\,K} = 247$ kJ/mol
3. Partial oxidation: $CH_4 + O_2 \rightarrow CO + 2H_2$ $\quad \Delta H^0_{298\,K} = -38$ kJ/mol

The water–gas shift reaction always accompanies these three reactions. The major technical problem of conducting steam reforming alone is carbon deposition on the catalysts that can lead to rapid deactivation and breakup of the catalyst. Carbon deposition can be substantially reduced by the use of an excess of water and a temperature of about 800°C. Other drawbacks of stream reforming are (1) the expensive generation of superheated steam (in excess) at high temperature; (2) the production of a significant amount of CO_2 in the product gas causing the onset of reverse water–gas shift reaction ($CO_2 + H_2 \rightarrow CO + H_2O$) particularly at high temperature; and (3) the H_2-to-CO ratio is higher than the optimum required for the downstream synthesis gas conversion to methanol, acetic acid, or hydrocarbons.

Partial oxidation offers some advantages over steam reforming. The reaction produces extremely high yields of syngas by an exothermic reaction, and therefore, the reactor would be more economical to heat. Oxygen is often used in steam reforming to provide heat and high methane conversion. Partial oxidation also gives a better ratio of hydrogen to carbon monoxide for subsequent conversion processes. The product gases from the reaction are low in carbon dioxide that must often be removed before the syngas can be used. Steam reforming and partial oxidation produce syngas of more moderate H_2/CO ratio (of about 2). This makes the direct use of syngas more versatile.

The dry reforming has the added advantage that it simultaneously consumes two greenhouse gases: hydrocarbons and CO_2. The best reducing agent for CO_2

is hydrogen. While dry reforming converts carbon dioxide and hydrocarbons into useful syngas, the "tri-reforming" allows the process to produce the syngas with varying H_2/CO ratios. The H_2/CO ratio in syngas is very important for its further use for a variety of chemical products. Syngas can be converted to acetone, acetic acid, and ethylene by an exothermic reaction [22], while pure CO can be used for the production of acetic acid, formic acid, polyurethane, polycarbonates, methyl acrylates, and so on. Both dry reforming and steam reforming reactions are endothermic. The heat generated from partial oxidation reduces the need for expensive external heating. Both dry and steam reforming reactions require very high temperatures (>600°C) to reduce the cooking. While steam reduces carbon deposition, an addition of oxygen provides necessary heat that can jump start dry and steam reforming reactions and maintain the catalyst in a clean and carbon-free state through oxidation of coke on the catalyst surface. The extent to which oxygenates are added to the reforming reactions is strictly determined by the process conditions and the catalyst employed. Since dry reforming produces water, steam reforming always accompanies dry reforming, making these studies relevant for tri-reforming.

Finally, since the real systems where tri-reforming (a combination of steam reforming, dry reforming, and partial oxidation) will be applied consist of hydrocarbon mixtures, it is important to compare how different types of hydrocarbons will perform under the same operating conditions. Puolakka [65] made one such comparison and his results indicate that propane and ethanol give favorable product distributions compared to toluene and dodecane. Such results may help optimizing the composition of the mixed streams to get the best syngas (with desired H_2/CO ratio) production by tri-reforming. More work on "tri-reforming" is currently being pursued.

4.5 EFFECTS OF FEEDSTOCK AND OPERATING CONDITIONS ON PRODUCT DISTRIBUTIONS

4.5.1 Steam Gasification

4.5.1.1 Coal

Corella et al. [66] used the following model for steam gasification of coal at low–medium (600°C–800°C) temperatures with simultaneous CO_2 capture in a fluidized bed at an atmospheric pressure. The study also examines the effect of inorganic species on the gasification process.

The gasification of coal with steam follows the following set of reactions [66]:
First, fast pyrolysis of coal follows the reactions:

$$\begin{aligned} \mathrm{Coal}(\mathrm{C}_x\mathrm{H}_y\mathrm{O}_z\mathrm{ISs}) &\rightarrow \mathrm{Tar\,1} + \mathrm{Char\,1} \rightarrow \mathrm{Tar\,2}\ (\mathrm{CH}_{0.85}\mathrm{O}_{0.17}) + \\ &\mathrm{Char\,2}\ (\mathrm{CH}_{0.2}\mathrm{O}_{0.13}\mathrm{ISs}) + \mathrm{H}_2 + \mathrm{CO} + \mathrm{CO}_2 + \\ &\mathrm{CH}_4 + \mathrm{C}_2\mathrm{H}_4 + \cdots \end{aligned} \tag{4.32}$$

Here, ISs are the inorganic species in the coal. The conversion of tarl and charl is a thermal reaction. Tar2 and char2 further react with steam and carbon dioxide as

$$\text{Tar}\,2(CH_{0.85}O_{0.17}) + H_2O \rightarrow CO + H_2 + \cdots \tag{4.33}$$

$$\text{Char}\,2(CH_{0.2}O_{0.13}\text{ISs}) + H_2O \rightarrow CO + H_2 + \text{Char}\,3(C_{xx}H_{yy}O_{zz}) + \text{Ashes(ISs)} \tag{4.34}$$

$$\text{Char}\,2(CH_{0.2}O_{0.13}\text{ISs}) + CO_2 \rightarrow CO + H_2 + \text{Char}\,3(C_{xx}H_{yy}O_{zz}) + \text{Ashes(ISs)} \tag{4.35}$$

The above reactions are not in stoichiometric proportions. Steam reforming of methane and light hydrocarbons that may occur simultaneously is expressed as

$$CH_4 + H_2O \rightarrow CO + 3H_2 \tag{4.36}$$

$$C_2H_4 + H_2O \rightarrow 2CO + 3H_2 \tag{4.37}$$

Along with the shift reaction,

$$CO + H_2O \rightleftarrows H_2 + CO_2 \tag{4.38}$$

All inorganic species with possible catalytic effects are designated as ISs. For example, iron-based species (Fe_2O_3, Fe_3O_4, etc.) affect the rate of overall steam gasification reaction. Some of the reactions, in particular, reforming and water–gas shift reactions, are catalyzed by nickel. Inorganic species such as indium can also have a catalytic effect. Finally, alkaline and alkaline earth metallic species (sometimes called American Academy of Environmental Medicine [AAEM] species) such as K, Ca, Na, Cs, and Mg significantly influence the overall gasification process. AAEM species can be either parts of char generated or additives in the gasification process. AAEM species affect (1) reactivity of coal and char; (2) product distribution of H_2, CO_2, and CH_4; and (3) tar content in the product gas. One of the AAEM species, CaO is a good absorbent for carbon dioxide to form calcium carbonate. During the gasification reaction, coke can be generated on CaO that can be removed by steam or carbon dioxide reactions with coke producing hydrogen and carbon monoxide. Thus, steam gasification of coal is often carried out with an addition of CaO so that the gasifier simultaneously removes carbon dioxide during the gasification process.

In general, carbon conversion and char gasification in a fluidized bed reactor increase with temperature between 600°C and 900°C. While tar yield (or tar content) and CO_2 capture decrease with an increase in the temperature in the same range. High contents of alkalis during gasification can also cause the problems of agglomeration, sintering, and melting, all of which are harmful to the smooth operation of the gasifier. Besides CaO, often calcined dolomites (CaO–MgO) and magnesium-based minerals silicates such as serpentine [$Mg_3Si_2O_5(OH)_4$], olivine ($Mg_{0.9}Fe_{0.1}SiO_4$), and calcine limestones or calcites have also been tested.

The study showed that for a clean and efficient steam gasification of coal in a fluidized bed at low/medium temperatures, at an atmospheric pressure, and with simultaneous capture of CO_2, the CaO/coal ratio is (1) a key parameter to obtain optimal product distribution, (2) a free parameter to be decided by the process designer, and (3) required to have relatively high values, clearly higher than 2 and perhaps as high as 80 or more. The type of coal or the types of ISs in the coal have some influence in the reaction network existing in the gasifier, but its influence is less than the effect of the temperature. Both product distribution and the usefulness of CaO are more controlled by temperature than AAEM, IS, and the nature of CaO.

Recently, Sharma [67,68] outlined a stepwise scheme to improve steam gasification reactivity of coal. In this scheme, coal is first refined using coal-derived solvents such as anthracene oil and paraffin oil. The refined coal has a higher amount of inorganic materials that can act as catalyst for the steam gasification to produce chemicals and char. The particle size of coal has no effect on the gasification reactivity, and the catalytic effects of minerals follow the order: Na > K > Ca > Ni [67]. The char is further subjected to steam gasification to produce syngas that can be further refined using steam reforming reaction. According to Sharma [67], the simplified set of reactions occurs during coal/char gasification as follows:

$$C_{90-120-240}H_{6-9-20}O_xS_yN_z + O_2 + H_2O \rightarrow C_nH_m + \text{Other products } (CO,\ CO_2,\ \text{etc.}) \tag{4.39}$$

$$C_nH_m \rightarrow nC + mH \tag{4.40}$$

$$C + O_2 \rightarrow CO_2 \qquad \Delta H = -40.59 \text{ kJ/mol} \tag{4.41}$$

$$C + O_2 \rightarrow 2CO \qquad \Delta H = 159.7 \text{ kJ/mol} \tag{4.42}$$

$$C + H_2O \rightarrow CO + H_2 \qquad \Delta H = 118.9 \text{ kJ/mol} \tag{4.43}$$

$$CO + H_2O \rightarrow CO_2 + H_2 \qquad \Delta H = -40.9 \text{ kJ/mol} \tag{4.44}$$

$$C + 2H_2 \rightarrow CH_4 \qquad \Delta H = -87.4 \text{ kJ/mol} \tag{4.45}$$

Shift reaction takes place only at high concentration of steam. The last reaction is important under pressure. Sharma [67] concluded that the main factors for the steam reactivity of gasification are (1) refining of coal that increases the surface area of coal, (2) volatile matters in residual coal and char (the more the volatile matter the more the reactivity), and (3) the concentration of mineral matter in coal and char. Sharma [68] also studied steam gasification reactions that can be useful for the reactor design. Exxon examined steam gasification of coal liquefaction residue [7]. Exxon technology utilized steam to sequentially gasify and hydrogenate both raw coal and carbon residue left in coal gasification. The study also used calcium hydroxide or a similar alkaline earth metal compound as possible catalysts for the process.

4.5.1.2 Biomass

In the recent years, the steam gasification of biomass is gaining more importance because it produces gaseous fuel with high hydrogen content that can either produce electricity with high efficiency or provide a feedstock for various chemical and fuel productions. Steam gasification also (1) provides gases with high heating value, (2) reduces the diluting effect of nitrogen from air, and (3) eliminates the need for expensive oxygen separation plant. Catalytic gasification in a fluidized bed allows (1) lower temperature, (2) a variety of particle sizes, and (3) a variety of feedstock.

A serious issue in the broad implementation of steam gasification is the generation of unwanted materials such as tars, particles, nitrogen compounds, and alkali metals. Tar is a mixture of one- to five-ring aromatic hydrocarbons that can plug the reactor. Its removal is essential, which can be done either in the gasifier or by hot gas cleaning after the gasification process. Within the gasifier, tar can be reduced by choosing the appropriate operating parameters, inserting additive catalyst, or changing the gasifier design so that it cannot plug the reactor. The removal of tar thermally requires the operation of the gasifier at a temperature above 1000°C. The prevention of ash agglomeration however requires the gasifier at a temperature below 700°C. Ash frequently contains various oxides of Ca, K, Mg, P, Si, Na, and S that can agglomerate, deposit on the surface, and contribute to erosion and corrosion of the gasifier. Alkali metals can also react with silica to form silicates or with sulfur to form alkali sulfates, both of which are sticky and can cause sintering and defluidization [28,41–49,53–56,69–74] (Encinar et al., 2010, pers. comm.). Reforming tar using a Ni catalyst is an effective method for removing tar. The coke deposition in a reforming reaction can be reduced using excess steam. Catalytic steam gasification of biomass is a complete network of heterogeneous reactions [28,41–49,53–56,69–74] (Encinar et al., 2010, pers. comm.). The reactions can be described as follows:

Primary reactions:

$$C_xH_yO_z + H_2O \rightarrow C(x-1)\,CO + \left(\frac{y}{2}+1\right)H_2 \quad (4.46)$$

$$(C_xH_yO_z + H_2) \rightarrow (\text{Heat})H_2 + CO + CO_2 + CH_4 + C_nH_{2m} + C(s) + \text{Tars} \quad (4.47)$$

Secondary reactions:

$$C_nH_{2m} + nH_2O \rightleftarrows nCO + (n+m)H_2 \quad (4.48)$$

Additional gas-phase reactions:

$$C + H_2O \rightleftarrows H_2 + CO \quad (4.49)$$

$$C + CO_2 \rightleftarrows 2CO \quad (4.50)$$

$$C + 2H_2 \rightleftarrows CH_4 \quad (4.51)$$

$$CO + H_2O \rightleftarrows H_2 + CO_2 \quad (4.52)$$

$$CH_4 + H_2O \rightleftarrows CO + 3H_2 \quad (4.53)$$

$$CH_4 + CO_2 \rightleftarrows 2CO + 2H_2 \quad (4.54)$$

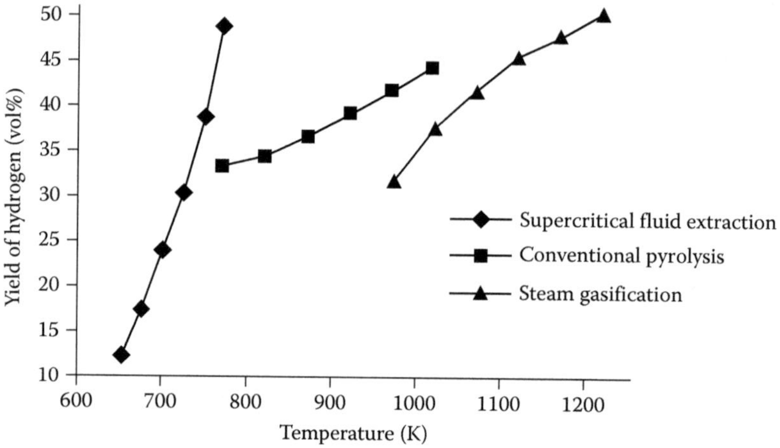

FIGURE 4.2 A comparison of yields of hydrogen from beechwood at different temperatures via pyrolysis, supercritical water extraction, and steam gasification (with steam/solid = 2). (Adapted from Demirbas, M., *Energy Sources, Part A*, 28, 245–252, 2006.)

In order to operate gasification in the temperature range of 600°C–700°C, gasification is generally operated with reforming in the same reactor or in two stages.

Demirbas [72] compared the hydrogen production from conventional pyrolysis, steam gasification, and supercritical extraction. A comparison of hydrogen yield as a function of temperature for these three processes is illustrated in Figure 4.2 [72]. While the results described in this figure are for beech wood, similar results were obtained for corncob, olive waste, and wheat straw. The results show that an increase in the steam-to-biomass ratio increases the hydrogen production. At low temperature, supercritical extraction is the best process, whereas steam gasification produces the best results at higher temperatures. Inayat et al. [69] presented a model for steam gasification accompanied by CO_2 adsorption by CaO in a fluidized bed reactor. The model indicated that high steam-to-biomass ratio gave higher hydrogen production. While an increase in temperature gave an increased hydrogen production, at a very high temperature, reverse water–gas shift reaction changes the trend. The model showed that at a temperature of 950 K and a steam-to-biomass ratio of 3, hydrogen production was maximum. Demirbas [12,70,72,73] also studied other types of biomass such as hazelnut shell, tea waste, and spruce wood, and again showed that at higher temperatures, steam gasification gave higher hydrogen yield than conventional pyrolysis. Higher steam-to-biomass ratio also gave higher hydrogen production. Similar results for mosses and algae were reported by Demirbas [70]. Specific samples examined were *Polytrichum commune*, *Thuidium tamarascinum*, *Cladophora fracta*, *Chlorella protothecoides*, beech wood, and spruce wood. A kinetic model for steam gasification of a cellulose surrogate was presented by Salaices et al. [74].

Li et al. [71] examined catalytic steam gasification of municipal solid waste (MSW) in a combined (two-stage) fixed-bed reactor. The catalyst used was a trimetallic catalyst (nano-Ni–La–Fe/γ-Al_2O_3) and the MSW contained kitchen garbage, wood and leaves, paper, textile, and plastics. The syngas composition was measured as functions of temperature, steam-to-MSW ratio, and catalyst-to-MSW ratio at an atmospheric pressure. The results showed >99% tar removal at 800°C with a significant production of hydrogen. The catalyst significantly improved hydrogen production. Higher temperature gave higher gas and hydrogen yields. While higher steam-to-MSW ratio gave better results, an excessive steam-to-MSW ratio lowered the gasification temperature and degraded the product quality. The optimum value of Steam/MSW ratio was found to be 1.33 under the operating conditions. The optimum value of the catalyst-to-MSW ratio was found to be about 0.5. A two-stage (pyrolysis followed by catalytic steam gasification) process for olive waste was studied by Encinar et al. (2010, pers. comm.). The catalyst used was dolomite. The two-stage process produced gas, liquid, and solid, the yields of which were strongly dependent on the temperature and the amount of catalyst. Higher temperature and catalyst amount gave higher amount of gases and the presence of steam gave higher amount of hydrogen and carbon dioxide.

Hofbauer et al. [75–79] used a fast internally circulating fluidized bed (CFB; at a pilot scale) to gasify biomass with steam. Using a natural catalyst as bed material, and at a temperature of 750°C, tar content was significantly reduced and gas with high hydrogen content was obtained. The internal circulating bed allowed the flexibility in varying residence times needed to lower tar concentration. Herguido et al. [80] studied gasification of pine sawdust, pinewood chips, cereal straw, and thistles from energy crops in the presence of steam in a fluidized bed reactor. The product gases were hydrogen, CO, and CO_2, and their amount and composition varied with the nature of biomass in the temperature range of 650°C–780°C.

A novel two-stage fluidized bed approach was used by Pfeifer et al. [81] in which the first stage carried out steam gasification of solid biomass to generate heat and power as well as provide raw materials for downstream chemical synthesis. The residual biochar from the first stage is combusted in the second stage and the hot bed materials from the second stage provide the heat needed for the first stage. This concept was also analyzed by Gopalakrishnan [82] and Matsuoka et al. [83]. The latter study showed that separating the combustion zone from the gasification zone resulted in high-efficiency gasification. They used γ-alumina as particles for bed materials and tested two different types of sawdusts. Since the residence time of the bed material can be controlled in the gasifier of the circulating dual bubbling fluidized bed system, the tars captured by the porous alumina particles (coke) as well as chars were effectively gasified. Since coke was preferentially gasified compared with char, higher carbon conversion and hydrogen yield can be achieved in this type of dual bed system than in the conventional CFB.

In the studies described earlier, the process generated gases with about 40 vol% hydrogen. Furthermore, an addition of carbonate adsorbed carbon dioxide and moved carbon dioxide from the gasification to the combustion zone (they called it adsorption-enhanced reforming [AER]). This concept has been successfully adapted by an 8 MW combined heat and power (CHP) plant in Güssing, Austria, since 2002. A new pilot plant of 100 kW has also been built to see the effect of the AER concept in improving hydrogen concentration to 75 vol% in the product gases. The possibilities of getting high hydrogen concentration, operating the reactors at low temperature, and thereby improving energy conversion efficiency make this concept very attractive. As mentioned earlier, Salaices et al. [74] presented a very workable kinetic model for catalytic steam gasification of cellulose surrogate with Ni/alpha-alumina catalyst in a CFB with a riser. The model successfully predicted the production of various gases such as hydrogen, carbon dioxide, carbon monoxide, water, and methane.

While a significant number of studies have investigated steam gasification in the presence of air (or oxygen) to improve carbon conversion and energy efficiency of the steam gasification process, Barrio et al. (2012, pers. comm.) examined the effect of hydrogen on the steam gasification process. They found that hydrogen inhibits the steam gasification reaction. They also concluded that the nature of char coming from beech or birch wood did not significantly affect the final results.

While a major effort on steam gasification is focused at a low temperature using a catalyst, Donaj et al. [84] and Gupta and Cichonski [85] examined the effectiveness of high-temperature steam gasification. Donaj et al. [84] examined the steam gasification of straw pellets at temperatures between 750°C and 950°C. The effect of the steam-to-feed ratio on carbon conversion was marginal (below 850°C), and in general, higher steam-to-feed ratio gave higher hydrogen production. Gupta and Cichonski [85] examined the steam gasification of paper, cardboard, and wood pellets in the temperature range of 700°C–1100°C. Once again in all cases, hydrogen production increased with the temperature and the steam-to-biomass ratio.

Lucas et al. [25] examined the high-temperature air and steam gasification (HTAG) of densified biofuels. The experiments were carried out in a fixed-bed updraft gasifier. The results showed that an increase in the feed temperature reduced the production of tars and soot and char residues, and also increased the heating value of the dry fuel gas produced. Butterman and Castaldi [86] showed that an increase in CO_2 feed rate enhanced the char conversion and the production of CO. The experiments produced a low concentration of methane and a high concentration of hydrogen (above 500°C for the herbaceous and nonwood samples and above 650°C for the wood biomass). The experiments also showed similarities between the gaseous products from biomass and MSW. The mass decomposition rates and the gas evolution profiles showed two distinct regions with transition around 400°C. Large pyrolysis char volumes correlated well with higher lignin compositions. The biomass fuels examined included woods, grasses, and other lignocellulosic samples. These included oak, sugar maple, poplar, spruce, white pine, Douglas fir, alfalfa, cordgrass, beachgrass, maple bark, pine needles, blue noble fir needles, pecan shells, almond shells, walnut shells, wheat straw, and green olive pit. The complete gasification occurred around 900°C–1000°C.

Aznar et al. [87] examined biomass gasification with steam–O_2 mixtures followed by a catalytic steam reformer and a CO-shift system. The use of two CO-shift converters downstream from a fluidized bed biomass gasifier, using steam–O_2 mixtures and a catalytic steam reformer, generated an exit gas with 73% hydrogen (by volume) on a dry basis and only 2.6% CO. The remaining gas contained CO_2, O_2, and CH_4. The results showed that the H_2O/CO ratio in the gas phase at the inlet of the high-temperature shift (HTS) reactor is a very important parameter in the system. CO conversion up to 90% was obtained, but to get this conversion, the steam/CO ratio greater than 2 at the inlet of the HTS reactor was needed. Due to low tar content in the inlet gas to the HTS reactor, a significant less deactivation of the catalyst in the shift reactor occurred.

4.5.1.3 Mixed Feedstock

In recent years, significant efforts have been made to gasify the mixtures of coal and biomass, coal and waste, or biomass and waste in the presence of steam. Seo et al. [30,88,89] used the successful two-stage fluidized bed model described earlier for coal–biomass blend in the temperature range of 750°C–900°C and the steam/fuel ratio of 0.5–0.8. Biomass-to-coal ratio was varied from 0 to 1. The study showed that the product gas yield, carbon conversion, and cold gas efficiency increased with increasing temperature and steam/fuel ratio. These parameters were higher for biomass gasification than those for coal gasification. A synergistic effect on gas yields was observed with a larger surface area, pore volume, and presence of micropores at a biomass/total feed ratio of 0.5. The calorific values of the product gas at 800°C were 9.89–11.15 MJ/m^3 with the coal, 12.10–13.19 MJ/m^3 with the biomass, and 13.77–14.39 MJ/m^3 with the coal–biomass blend. The maximum cold efficiency was 0.45 with a biomass/total feed ratio of 0.5. Sun et al. (2001, pers. comm.) examined various kinetic models for the gasification of biomass blended with waste filter carbon at temperatures around 850°C. Once again, high temperature and high flow of steam increased the gasification rate; the gasification rate of filter carbon was lower than that of wood chip. The data were taken for the steam pressure of 0.5 atm. A modified volume reaction kinetic model best fit all the data.

Kumabe et al. [27] showed that at 900°C, the mixture of woody biomass and coal in the presence of steam and air gave favorable results. The results of this study were similar to those described above; increase in biomass gave more gases and more hydrogen was produced at higher steam-to-feedstock ratio. Higher amount of biomass also gave lower amount of char and tar. The study produced gas with composition that was favorable to methanol, hydrocarbon fuels, and dimethyl ether (DME) under high biomass feed conditions. The co-gasification was carried out in a downdraft fixed-bed reactor and it provided cold gas efficiency ranging from 65% to 85%. Demirbas et al. [90], Demirbas and Caglar [91], and Demirbas [92,93] studied hydrogen production from various biomass samples, black liquor, biomass/coal, and biomass/heavy oil mixtures. In a most recent study, Demirbas [94] studied the effects of co-firing MSW with pulverized coal in a bubbling fluidized bed combustor. The results showed that the mixture produced less NO_x and SO_x in direct proportions to the MSW concentration in the mixture. Similarly, mixture produced less CO_2 than coal alone. The mixture burning can, however, bring the problems with chlorine impurities in MSW that can lead to corrosion problems and inorganic impurities such as Si, Al, Ti, Fe, Ca, Mg, Na, K, S, and P that can significantly change the composition of ash and its melting and agglomeration characteristics. This change in ash characteristics may limit the market for its downstream use.

Numerous other investigators have also examined the steam gasification (some in the presence of air or oxygen) of a variety of coal–biomass mixtures. Chmielniak and Sciazko [95] produced syngas from steam gasification of coal–biomass mixture that was subsequently transformed to methanol, DME, ethylene, and gasoline.

Yamada et al. [31] also produced useful syngas from a mixture of coal and biomass briquettes. Kumabe et al. [27] examined steam gasification (with air) of a mixture of Japanese cedar and mulia coal and obtained useful syngas for the production of DME. Their gasification results were very similar to those described earlier and they obtained cold gas efficiency of 65%–85% during the gasification process. Pan et al. [96] examined steam gasification of residual biomass and poor coal blends. Pine chips from Spain were used as biomass, and the two types of coal—black coal (low grade) from Escatron, Spain, and Sabero coal from Sabero, Spain—represented poor-grade coals. Once again, reasonable quality of syngas was produced with an overall thermal efficiency of about 50%. Satrio et al. [97] examined steam gasification of coal–biomass mixture with the specially designed catalyst pellets with outside shell consisting of nickel on alumina and core consisting of calcium and magnesium oxides that can adsorb carbon dioxide. This catalyst design gave higher production of hydrogen. Finally, Ji et al. [32] studied steam gasification of a mixture of low-rank fuel mixture of biomass, coal, and sludge in a fluidized bed reactor at 900°C temperature. Just like other studies, higher temperature gave more gas and hydrogen but not high heating value of gas. The calorific value of syngas produced from sludge mixture, sludge, wood chips, and lignite was 13, 10, 6.9, and 5.7 MJ/m^3, respectively.

An excellent review of problems associated with co-firing of coal and biomass fuel blends was given by Sami et al. [29]. This review critically assesses the effectiveness of this mixed feedstock for combustion and pyrolysis—two extreme cases of gasification. While they specifically do not discuss steam gasification and reforming, significant parts of their analysis are applicable to the process of steam gasification and reforming. Indrawati et al. [98] examined partial replacement of fossil energy by renewable sources such as rice husk, palm kernel shell, sawdust, and municipal waste in the cement production. While this study also does not specifically address steam gasification and reforming of mixed feedstock, the study points to another application of the mixed feedstock.

4.5.1.4 Tar

As indicated earlier, formation of tar is a major issue with steam gasification. Tar is a complex mixture of condensable hydrocarbons and it can contain one-ring to five-ring aromatic compounds with other oxygen containing hydrocarbon species [99–103]. Generally, tar is defined as C_6^+ aromatic organics produced under gasification conditions. Tar is a problem during gasification because (1) it can deposit on the outlet pipes of the gasifier and also on the particulate filters; (2) it can clog fuel lines and injectors in the internal combustion engine; and (3) it reduces the gasifier's efficiency to produce additional useful fuel products such as hydrogen, carbon monoxide, carbon dioxide, and methane.

Baker et al. [99] illustrated the conceptual relationship between tar disappearance and the temperature during thermal steam gasification of carbonaceous materials. They also divided tar components in four different categories (Equation 4.55) [99]. The first category is easiest to crack and the fourth category (which mostly contains polycyclic aromatic hydrocarbons) is the most difficult to crack. Their analysis showed that at low temperatures (400°C), a significant amount of tar is produced, and at temperatures higher than around 1000°C, very little tar is produced. The literature has shown that as temperature increases, the nature of tar undergoes the following transformation [99–103]:

$$\begin{aligned}&\text{Mixed oxygenates (400°C) (primary)} \rightarrow \\ &\text{Phenolic ethers (500°C) (secondary)} \rightarrow \\ &\text{Alkyl phenolics (600°C) (tertiary-alkyl)} \rightarrow \\ &\text{Heterocyclic ethers (700°C) (tertiary-PNA)} \rightarrow \\ &\text{PAH * (800°C)} \rightarrow \text{Larger PAH (900°C)}\end{aligned} \quad (4.55)$$

In the above reaction, PAH* is high-molecular-weight polynuclear aromatic hydrocarbons. Along with the temperature, tar concentration depends on the reaction time, the amount of oxygen, and the presence of a suitable catalyst during steam gasification. Higher oxygen concentration generally reduces tar concentration through the processes of cracking and oxidation among others. The conventional steam gasification operated at 700°C–800°C produces tar with naphthalenes, acenaphthylenes, fluorenes, phenanthrenes, benzaldehydes, phenols, naphthofurans, and benzanthracenes. While high-temperature steam gasification operating between 900°C and 1000°C produces tar that contains naphthalenes, acenaphthylenes, phenanthrenes, fluranthenes, pyrenes, acephenanthrylenes, benzanthracenes, benzopyrenes, 226 MW (molecular weight) polycyclic aromatic hydrocarbons (PAHs), and 276 MW PAHs. Milne et al. [100] further characterized tar in terms of primary, secondary, and tertiary products based on molecular beam mass spectroscopy. Some of the details of the constituents of primary, secondary, and tertiary products and their behavior with temperature are described by Milne et al. [100].

4.5.1.5 Black Liquor

Huang and Ramaswamy [104] examined steam gasification of black liquor coming out of the paper and pulp industry at temperatures as high as 1500°C. Their results were in agreement with other reports. The carbon conversion was nearly complete at temperatures higher than about 750°C. Hydrogen concentration first increased with temperature but showed a maximum at high temperatures because of the dominance of reverse water–gas shift reaction. Higher steam gave higher hydrogen concentration in the product gas. Operating with a $0.3 < \text{SBR} < 0.6$ in combination with high pressure of 30 atm, high temperature of 1000°C appears to be the most beneficial for obtaining smelt with no C(s) and maximizing Na and S capture in the melt. Here SBR is steam-to-dry black liquor ratio.

Black liquor gasification can be used to substitute the existing combustion process for potential higher energy efficiency, lower greenhouse gas emissions, and more safety. The steam gasification of black liquor technology can help the current paper and pulp mills technology to be extended into future biorefineries. In general, the equilibrium model examined by Huang and Ramaswamy [104] indicates that the hydrogen concentration in the product increased with a decrease in pressure and an increase in SBR, and it showed a maximum with an increase in temperature. Li and Heiningen [38] also illustrated the conversion data for a black liquor via steam gasification with and without catalysts. Whitty [105] examined steam gasification of black liquor char under pressurized conditions.

4.5.1.6 Lignin

The gasification of lignin has been investigated by a number of investigators [106–108]. These and other studies have investigated various characteristics of pyrolysis and gasification of lignin, the effect of alkali addition on gasification and production of hydrogen and medium heating value gas during steam gasification of lignin. Most studies have examined lignin from paper and pulp industries as well as Westvaco Kraft lignin. In the latter category, Kraft-1, Kraft-2, and Alcell were gasified in the presence of steam at 600°C–800°C and they produced gases with 30–50 vol% hydrogen. Most studies used a fixed-bed reactor.

4.5.2 Steam Reforming

4.5.2.1 Ethanol

As discussed in a subsequent chapter 9, alcohols and in particular ethanol can be easily obtained by the process of fermentation of sugar, glucose, fructose, and many lignocellulosic biomass [109–116]. In Brazil, ethanol is extensively produced using sugarcanes. Ethanol is easier and safer to store and transport because of its low toxicity and volatility and biodegradable characteristics. Ethanol can also be produced from various energy plants, waste materials from agro industries, or forestry residue materials as well as cellulosic and organic fractions of MSW. Easy availability of ethanol makes it a good candidate for steam reforming to produce hydrogen.

Unlike methanol and gasoline derived from fossil fuel sources, ethanol derived from biosources is carbon neutral to the environment. The carbon dioxide produced from the steam reforming of ethanol can be used to regenerate additional biomass. Bioethanol, generally containing about 12% ethanol in an aqueous solution, can be directly subjected to steam reforming, thus eliminating the distillation step required to produce pure ethanol. Since both water and ethanol can be converted to hydrogen, the process of steam reforming avoids the separation stage. The thermal efficiency of steam reforming of aqueous ethanol solution is very high (>85%) and this makes the process economically very attractive. The steam reforming of ethanol is carried out by the following reaction:

$$C_2H_5OH + 3H_2O \rightleftarrows 2CO_2 + 6H_2 \tag{4.56}$$

This reaction follows a number of steps that involve the dehydrogenation of ethanol to form acetaldehyde, which in turn decomposes to produce methane and carbon monoxide. Further reforming of methane and water–gas shift reaction leads to the formation of hydrogen. Since ethanol has high hydrogen content, the process produces a significant amount of hydrogen. There are, however, side reactions such as dehydration and decomposition of ethanol which produce methane, diethyl ether, and acetic acid that reduce the production of hydrogen. These side reactions can be minimized by the use of selective catalysts. In addition, the formation of large amounts of carbon monoxide reduces the hydrogen yield and it also requires complex gas cleanup process. Overall, ethanol is still one of the best raw materials for steam reforming to produce hydrogen.

An extensive number of studies to develop different types of catalysts for ethanol steam reforming have been reported in the literature [109–116]. Mas et al. [109] used Ni(III)–Al(III) lamellar double hydroxide as catalyst precursor. They developed a Langmuir–Hinshelwood type of kinetic model for steam reforming of ethanol for this catalyst. A general model was found to be valid for a wide range of water/ethanol feed ratio and temperatures. Biswas and Kunzru [110] examined the effects of copper, cobalt, and calcium doping on Ni–CeO_2–ZrO_2 catalysts for steam reforming of ethanol. The data were obtained in the temperature range of 400°C–650°C. The nickel loading was kept fixed at 30 wt%, whereas Cu and Co loading was varied from 2 to 10 wt% and Ca loading was varied from 5 to 15 wt%. For Cu- and Ca-doped catalysts, the activity increased significantly; however, Co-doped catalysts showed poor activity. The catalyst activity was in the order: Ni > $NiCu_5$ > $NiCa_{15}$ > $NiCo_5$. For steam reforming reaction, the highest hydrogen yield was obtained on the undoped catalyst at 600°C. With calcium doping, in the temperature range of 400°C–550°C, higher hydrogen yield was obtained compared to those for undoped catalysts. Akdim et al. [111] compared the steam reforming of non-noble metal (Ni–Cu) with noble metals (Rh or Ir) supported over neutral SiO_2, amphoteric Al_2O_3, and redox CeO_2. The data showed that for each domain of temperature, quite different mechanistic routes were governing for the three tested systems. The data suggested some methods that improved the catalyst formula for the steam reforming of ethanol. Finally, the effect of support on catalytic behavior of nickel catalysts in the steam reforming of ethanol for hydrogen production was investigated by Fajardo et al. [112]. They studied Al_2O_3-, MgO-, SiO_2-, and ZnO-supported nickel catalysts and showed that the catalyst behavior can be influenced by the experimental conditions and chemical composition of the catalysts.

The steam reforming of ethanol by different types of Co catalysts was investigated by Sekine et al. [113], Song et al. [115], and He et al. [116]. Sekine et al. [113] examined steam reforming of ethanol over Co/$SrTiO_3$ with an addition of another metal: Pt, Pd, Rh, Cr, Cu, or Fe. Ethanol conversion and H_2 yield improved significantly by adding Fe and Rh at 823 K; however, Rh addition promoted CH_4 formation. Within Fe loading of 0.33–1.33 mol%, Fe addition increased the selectivity of steam reforming of ethanol. The addition of Fe on Co/SiO_2 catalyst was not very effective. High activity of Fe/Co/$SrTiO_3$ catalyst came from interaction among Fe, Co, and $SrTiO_3$. Song et al. [115] showed that the use of novel synthesis methods such as solvothermal decomposition, colloidal crystal templating, and reverse microemulsion to prepare CeO_2-supported Co catalysts gave better performance than the catalysts prepared using conventional incipient wetness impregnation method for ethanol steam reforming. The improvement can be attributed to a better cobalt dispersion and a better Co–CeO_2 interaction for the catalysts prepared using these novel methods. He et al. [116] examined a series of Co–Ni catalysts prepared from HT-like materials by coprecipitation for steam reforming of ethanol. The results showed that the particle size and reducibility of the Co–Ni catalysts are influenced by the degree of formation of HT-like structure and increasing Co content. All catalysts were active and stable at 575°C. The activity decreased in the order: 30Co–10Ni > 40Co–20Ni > 20Co > 10Co–30Ni > 40Ni. The 40Ni showed the strongest resistance to deactivation, whereas all Co-containing catalysts showed higher activity than 40Ni catalyst. The highest hydrogen yield was found for 30Co–10Ni catalyst in which XCo and Ni are intimately mixed and dispersed in the HT-derived support.

Dong et al. [114] examined hydrogen production by steam reforming of ethanol using potassium-doped 12CaO–7Al_2O_3 catalysts. The conversion of ethanol and H_2 yield over $C_{12}A_7O^-/x\%K$ catalyst mainly depended on the temperature, K-doping amount, steam-to-carbon ratio, and contact time. Based on numerous types of catalyst analysis, the authors concluded that the active oxygen species and doped potassium play important roles in the steam reforming of ethanol over $C_{12}A_7–O^-/27.3\%K$ catalyst.

As shown earlier, the steam reforming of ethanol undergoes several reaction pathways depending on the catalysts and the reaction conditions. Therefore, the choice of the catalyst plays a vital role in the reforming process. Navarro et al. [13] pointed out that the reactions to avoid are C_4 and C_2H_4 inductive of carbon deposition on the catalyst surface. Thus, the catalysts that selectively produce hydrogen must (1) dehydrogenate ethanol, (2) break the carbon–carbon bonds of surface intermediates to produce CO and CH_4, and (3) reform these C_1 products to generate hydrogen. As shown earlier, various oxide catalysts, metal-based catalysts (Ni, Co, Ni/Cu), and noble metal-based catalysts (Pt, Pd, Rh) have proven to be active for steam reforming of ethanol. The metallic function and the acid-based properties play an important role in the steam reforming. A good review of hydrogen selectivity and coking resistance of various types of catalysts is given by Navarro et al. [13].

4.5.2.2 Methanol

Methanol is an abundant chemical often produced from fossil fuels as well as biomass [117–134]. Industrially, it is produced at 250°C–300°C temperature and 80–100 atm pressure using a copper–zinc-based oxide catalyst. Methanol is an important feedstock for the production of hydrogen and hydrogen-rich syngas. While methanol can be decomposed as

$$CH_3OH \rightarrow CO + 2H_2 \qquad \Delta H^0_{298\,K} = 90.1\ \text{kJ/mol} \tag{4.57}$$

and this reaction is endothermic and can be catalyzed by a number of catalysts including Ni and Pd, in this chapter, we mainly focus on steam reforming of methanol. Methanol is a good feedstock because of its easy availability, high-energy density, and easy storage and transportation. Currently, a significant work is being carried out for low-temperature steam reforming to produce high-purity hydrogen for power generation in FC in automobiles. The steam reforming of methanol follows the reaction:

$$CH_3OH + H_2O \rightarrow CO_2 + 3H_2 \qquad \Delta H^0_{298\,K} = 49.4\ \text{kJ/mol} \tag{4.58}$$

While a number of catalysts have been examined, commercial Cu/ZnO water–gas shift reaction and methanol synthesis catalysts have been found to be effective for steam reforming of the methanol. Copper on ZrO_2 support prepared by a numerous different methods including precipitation, microemulsion, formation of amorphous aerogels, CuZr alloys, and so on have been successfully attempted. For this catalyst, a large surface area of the active metals needs to be maintained to avoid rapid deactivation. For this, zirconia support should be in the amorphous state under the calcination and reaction conditions. $Cu/ZnO/ZrO_2$ catalyst has been found to be active at a temperature as low as 170°C, but the catalyst deactivates rapidly at temperatures above 320°C. The deactivation can, however, be reduced by the incorporation of Al_2O_3 that increases the temperature of crystallization of ZrO_2, which remains amorphous at the reaction temperature. The incorporation of alumina also increases both the copper and the Brunauer, Emmett, and Teller (BET) surface area, thereby increasing the catalyst activity.

Henpraserttae and Toochinda [117] examined a novel preparation technique of Cu/Zn catalyst over Al_2O_3 for methanol steam reforming. The study focused on the preparation methods of active Cu/Zn-based catalysts with and without urea by incipient wetness impregnations to lower the metal loading and the catalyst cost. The experimental data for methanol steam reforming were obtained in a fixed-bed reactor in the temperature range of 453–523 K to lower the energy costs. The data showed that the activity in the hydrogen production from the catalysts with urea was higher than that from the catalysts without urea. The impregnated catalysts can show the activity at temperatures as low as 453 K. The Cu/Zn catalysts prepared with an incipient wetness impregnation over Al_2O_3 with urea can give a hydrogen yield of about 28%. Thus, the impregnated catalysts could be alternative catalysts for hydrogen production from methanol reforming with a lower cost of the catalyst compared with the co-precipitation method used in the commercial operation. More details on methanol synthesis technology from various raw materials are given by Lee [118].

The partial oxidation of methanol is attractive because it is an exothermic reaction and it follows the reaction:

$$CH_3OH + \frac{1}{2}O_2 \rightarrow CO_2 + 2H_2 \qquad \Delta H^0_{298\,K} = -192.2\ \text{kJ/mol} \tag{4.59}$$

The above reaction starts at the temperature as low as 215°C. Both the reaction rate and the selectivity for hydrogen increase very rapidly with temperature. The carbon monoxide formation in the entire temperature range is low. The literature has shown that production of hydrogen and carbon dioxide increases with copper content and it reaches the maximum with 40/60 atomic percentage of copper and zinc [117–134]. Unreduced copper–zinc oxide catalysts display very low activities and produce only carbon dioxide and water with very little hydrogen. The catalysts, however, become eventually reduced under high-temperature reaction conditions. The apparent activation energy and the TOF are higher at lower copper content and slightly decrease with an increase in the copper content and then achieve a constant value. These and some other similar data show that the reaction depends on both ZnO and CuO phases. Methanol conversion increases with oxygen partial pressure up to 0.063 atm. A further increase in oxygen partial pressure precipitously decreases methanol conversion. The incorporation of Al_2O_3 (up to 15% Al) to the Cu/ZnO system results in a lower activity, implying that aluminum has an inhibiting effect on the partial oxidation of methanol.

Besides Cu/ZnO catalyst, Pd/ZnO catalyst has also been effective in methanol partial oxidation reaction. For 1 wt% Pd/ZnO catalyst, methanol conversion reaches 40–80% within the 230°C–270°C range. Methanol conversion and H_2 selectivity increase with an increase in temperature. The nature of support also affects the kinetics. Pd/ZrO_2 catalyst, while producing hydrogen and carbon dioxide, also shows a significant increase in the decomposition reaction.

A combination of steam reforming and partial oxidation results in an auto-thermal operation. Under this condition, the following reaction

$$CH_3OH + (1 - 2n)\,H_2O + nO_2 \rightarrow CO_2 + (3 - 2n)H_2 \quad (0 < n < 0.5) \qquad (4.60)$$

with copper-based catalysts also perform well. On Cu/ZnO catalyst, initially methanol is combusted by oxygen and water is produced. When oxygen is depleted, methanol conversion and the production of hydrogen and carbon monoxide increase, and the water production goes down. When Al_2O_3 is added to the catalyst, better performance for steam reforming is obtained. Purnama et al. [119] also found the beneficial effect of oxygen addition to the feed during steam reforming of methanol on Cu/ZrO_2 catalysts. In the auto-thermal operation, the relative ratio of oxygen, methanol, and steam plays an important role on hydrogen production. For Cu–ZnO (Al) catalyst, the best feed ratio of oxygen/methanol/steam was found to be 0.3/1/1. In general, oxy reforming of methanol is complex, but it also strongly interacts with water–gas shift reaction.

The auto-thermal operation of methanol for FC application in vehicles has been adopted by DaimlerChrysler, Toyota, and Nissan. Small-scale hydrogen production by reforming methanol is also commercialized. For its application in refueling station, hydrogen purification step is needed. This is generally carried out by either pressure swing adsorption (PSA) or membrane separation technology. In general, the cost of hydrogen production from methanol reforming is higher than that from methane reforming. The Mercator project funded by the European Commission is an integrated methanol steam reformer and selective oxidation system. The FC contains a series of catalytic plates with combustion of anode off-gas on one side and steam reforming of methanol on the other side.

A number of studies examined the metal-supported catalyst systems for steam reforming of methanol for FC applications [120–134]. Such catalysts overcome the slow heat transfer of packed-bed systems by integrating endothermic steam reforming with exothermic hydrogen combustion. A wash-coated aluminum heat exchanger showed the best performance using a suspension of commercial reforming catalysts. With an aluminum foam, 90% methanol conversion was achieved for a sustainable period of time (about 450 h). Lindström [120], Lindström and Pettersson [121–124,126,127], Lindström et al. [125,129], and Kolb et al. [128] examined methanol reforming over copper-based catalysts for FC applications.

A novel technology of steam reforming of methanol accompanied by palladium membrane separation to produce pure hydrogen was investigated by Pan and Wang [131,132] and Pan et al. [133]. This technique provides a possibility for bypassing the technical problems of storage and delivery of hydrogen by delivering methanol to forecourt hydrogen-dispensing stations and on-site hydrogen productions. Li et al.

[134] examined a strategy in which a coal-derived methanol is used as a hydrogen carrier. The steam reforming of methanol can generate hydrogen at the desired place.

4.5.2.3 Liquid Hydrocarbons

Besides methane, methanol, and ethanol, gasoline, diesel, and jet fuel can also be important feedstock for the steam reforming to produce hydrogen [135–142] (Sun et al. 2001, pers. comm.). These three types of fuels contain a variety of hydrocarbons and sulfur. While these components themselves can be important feedstock for steam reforming, they are not as readily available on a large scale as various other fuels. The technical problems associated with these hydrocarbons include (1) the catalyst deactivation by sulfur in the feedstock and (2) the significant amount of coke deposition on the catalyst that eventually results in its deactivation. Along with steam reforming, in the recent years, catalytic partial oxidation of high hydrocarbons using short contact time (milliseconds) and high temperatures (850°C–900°C) over noble metal catalysts on porous monolithic ceramic supports have been examined [135–142] (Sun et al. 2001, pers. comm.). These reactions can be represented by a generalized reaction:

$$C_nH_m + \frac{n}{2}O_2 \rightarrow nCO + \frac{m}{2}H_2 \tag{4.61}$$

The above reaction is about two times faster than the steam reforming reaction and the heat of reaction generated by this reaction depends on the oxygen-to-fuel ratio. Unlike steam reforming and partial oxidation of methane, methanol, and ethanol, steam reforming and partial oxidation of fuels involve dehydrogenation, C–C bond cleavages, total oxidation, steam reforming, CO_2 reforming, hydrocarbon cracking, methanation, and water–gas shift reaction all occurring simultaneously. In addition, these reactions occur for all different component hydrocarbons at different rates. Thus, the process is very complex and not clearly understood. In general, aromatics are less reactive and are more prone to the reaction producing cokes than aliphatic components and olefins. Through a complex set of reactions, fuels also produce hydrogen, carbon dioxide, carbon monoxide, and water along with a significant amount of lower hydrocarbons. The final product distribution depends on the temperature and the residence time. Several catalysts including nickel, platinum, rhodium, and bimetallic have been tested for hydrocarbons such as n-octane, n-heptane, and n-hexane [135–142] (Sun et al. 2001, pers. comm.). In general, ceria and zirconia supports or a mixture of ceria–zirconia supports has been found to be reasonably effective in averting coke deposition [135–142] (Sun et al. 2001, pers. comm.).

A combination of steam reforming, partial oxidation, and water–gas shift reaction has been tested to obtain an auto-thermal operation. Generally, partial oxidation and steam reforming are carried out in separate zones, with the first one controlled by the oxygen-to-carbon ratio and the second one by the steam-to-carbon ratio. The adiabatic temperature and the amount of hydrogen produced depend on the relative amounts of energy released in these two steps. Higher steam-to-carbon ratio reduces the carbon monoxide concentration in the product. For diesel fuel, thermodynamic equilibrium can be achieved at an oxygen-to-carbon ratio of 1 and a steam-to-carbon ratio of 1.25 at 700°C temperature.

The advanced thermal recycling (ATR) process requires catalysts and supports with high resistance to coking at high temperatures. Excess steam and/or oxygen helps avoid coking. Also at high temperature, sulfur is less of a problem. The noble metal catalysts (Pt, Rh, Ru) supported on ceria or zirconium or their mixtures work well. In the recent years, applications of pervoskite oxides (ABO_3) for steam reforming of higher hydrocarbons and various fuels have been extensively examined [135–142] (Sun et al. 2001, pers. comm.). A group of six metal carbides has also shown a good success.

4.5.2.4 Glycerol

Glycerol has been a byproduct of a number of conversion processes, particularly transesterification of used oil, algae, and crop oils (there are about 350 of them) to produce diesel fuel [143–155] (Cheng et al., 2012, pers. comm.). This byproduct can also be effectively utilized to produce hydrogen by steam reforming process. Steam reforming of glycerol involves a complex set of reactions, numerous intermediates, and hydrogen that is accompanied by several other products. The hydrogen yield depends on the steam-to-glycerol ratio and follows the reactions:

$$C_3H_8O_3 \rightarrow 3CO + 4H_2 \quad (4.62)$$

$$CO + H_2O \rightleftarrows CO_2 + H_2 \quad (4.63)$$

With an overall reaction as

$$C_3H_8O_3 + 3H_2O \rightarrow 3CO_2 + 7H_2 \quad (4.64)$$

Simonetti et al. [148] showed that at about 275°C, glycerol can be catalytically converted to H_2/CO mixture. Because of this low temperature, the endothermic steam reforming process can be combined with an exothermic FT process to make the overall process energy efficient for fuel generation from glycerol. The primary products for steam reforming of glycerol are hydrogen, methane, carbon dioxide, carbon monoxide, carbon, and unreacted water and glycerol. The formation of methane competes with the formation of hydrogen. According to steam reforming and decomposition reactions,

$$C_xH_yO_x + xH_2O \rightarrow xCO_2 + \left(\frac{x+y}{2}\right)H_2 \quad \text{Steam reforming} \quad (4.65)$$

$$C_xH_yO_x \rightarrow xCO + \frac{y}{2}H_2 \quad \text{Decomposition reaction} \quad (4.66)$$

The maximum hydrogen concentration in the product can be either 77% or 57%. A study by Adhikari et al. [151,153] showed that at about 680°C, the upper limit of moles of hydrogen per mole of glycerol produced is six at an atmospheric pressure and at a steam-to-glycerol ratio of nine.

While nickel on alumina is a workable catalyst for steam reforming of glycerol, the effects of numerous promoters such as Ce, La, Mg, and Zr were examined at 600°C [143–155] (Cheng et al., 2012, pers. comm.). These results indicate that all promoters improved the production of hydrogen with zirconium giving the best results. The increase in hydrogen production can be due to an increased nickel concentration, an increased capacity to activate steam, and the stability of nickel phase.

Recent studies [143–155] (Cheng et al., 2012, pers. comm.) investigated various noble metal catalysts on a variety of supports at 500°C–600°C, an atmospheric pressure, and a steam-to-carbon molar ratio of 3.3. The results indicated the activity order: Ru = Rh > Ni > Ir > Co > Pt > Pd > Fe. Among Y_2O_3, ZrO_2, CeO_2, La_2O_3, SiO_2, MgO, and Al_2O_3 supports, Y_2O_3 (along with ZrO_2 and CeO_2) support gave the best glycerol conversion and hydrogen production. These studies also demonstrated that at low conversion and low temperature (225°C–275°C), Pt/C and Pt–Re/C gave stable results. For CeO_2 support, Zhang et al. [145] showed that at 400°C, Ir/CeO_2 gave the best glycerol conversion with 85% hydrogen selectivity, whereas Co/CeO_2 and Ni/CeO_2 gave 88% and 75% hydrogen selectivity at 425°C and 450°C, respectively. Glycerol has a higher tendency for coke formation compared to methane and this coke formation can be significantly reduced by increasing the steam-to-glycerol ratio in the feed. The catalytic steam reforming of glycerol (both conversion of glycerol and selectivity of hydrogen) is affected by the operating parameters such as the reaction temperature, the pressure, the steam-to-glycerol ratio, and the oxygen-to-glycerol ratio.

In a recent study, Maciel and Ishikura [144] have given an outstanding review of steam reforming of renewable feedstock for the production of hydrogen. They have considered methanol, ethanol, glycerol, glucose, and biomass as potential raw materials for steam reforming. Their overall analysis led to the following conclusions: (1) reforming should be carried out at lower temperatures and an atmospheric pressure to reduce the operating costs; (2) the catalyst should provide high selectivity to hydrogen and inhibit CO and byproduct formation such as methane; and (3) the catalyst must resist coke formation that reduces the number of active sites and hence the reaction rates, and implies a regeneration process that is costly. Feedstock issues such as supply, cost, logistics, and the value of byproducts are major factors in cost-effectiveness of steam reforming process.

4.5.2.5 Biomass

Just like methane and other hydrocarbons, biomass can also undergo partial oxidation and steam reforming in the presence of oxygen and steam at temperatures above around 725°C yielding gaseous products and chars [156–171]. The char can also be converted to gaseous products such as hydrogen, carbon dioxide, carbon monoxide, and methane under high-temperature conditions. The overall reaction is as follows:

$$C_xH_yO_z + H_2O + O_2 \rightarrow H_2 + CO_x + CH_4 + HCs + char \tag{4.67}$$

The hydrogen production for a variety of biomass under different operating conditions has been examined in the literature [156–171]. The literature data indicate that in a fluidized bed reactor, under suitable operating conditions, as high as 60 vol% hydrogen can be produced from biomass.

The major drawback of steam reforming of biomass is the tar formation that is not easily amenable to steam reforming process. The tar formation can be minimized by suitable operating conditions (i.e., operating at very high temperature), suitable gasifier design (i.e., entrained bed reactor), or incorporation of additives or promoters to the catalysts. At temperatures above around 1000°C, tar can be cracked, and for temperatures above around 1200°C, pure syngas can be obtained. Higher residence time can also help cracking of the tar. The additives such as dolomite and olivine to the nickel catalyst help to reduce the tar formation. Alkaline metal oxides are also used to reduce the tar formation.

Another important issue with biomass gasification and reforming is the formation of ash that can cause slagging, fouling, and agglomeration. The inorganic impurities in biomass can be removed by biomass pretreatment using leaching and extraction processes. The literature has shown the leaching and subsequent gasification to produce hydrogen as a viable process for olive oil waste [156–171].

Pacific Northwest National Laboratory studied the gasification of biomass to produce a variety of gaseous fuels using appropriate catalysts. The earlier studies used a catalytic steam gasification of biomass with concurrent separation of hydrogen in a membrane reactor that employed a permselective membrane to separate the hydrogen as it is produced [156–171]. The process was particularly well suited for wet biomass and may be conducted at temperatures as low as 300°C. One experiment was conducted at 4000 psi pressure and 450°C, although most others were at 15–30 psi. The process was named SepRx. Optimal gasification conditions were found to be at about 500°C, an atmospheric pressure, and a steam/biomass ratio of 10/1. In the presence of a nickel catalyst, the product hydrogen concentration of 65 vol% was generated under these optimal conditions. Rapagna [168] examined steam gasification of almond shell in the temperature range of 500°C–800°C. Smaller particle size yielded more hydrogen. Rapagna and Foscolo [169] examined catalytic steam gasification in a fluidized bed reactor followed by a fixed-bed catalytic reactor. Over a temperature range of 660°C–830°C, the catalytic converter using different steam reforming nickel catalysts and dolomite gave as high as 60% hydrogen yield.

Steam gasification and steam reforming can be coupled processes. Mckinley et al. [165] examined various biomass gasification processes for the production of hydrogen. Turn et al. [166] showed that for a noncatalytic gasification of sawdust, the highest hydrogen yield was obtained at 825°C and for a steam/biomass ratio of 1.7. Zhou et al. [167], however, showed that for the production of hydrogen, adding steam to the gasification process was not as effective as adding steam to downstream nickel-catalyzed steam reforming process.

4.5.2.6 Mixed Feedstock

In the recent years, significant efforts have been made to gasify and steam reform mixed feedstock of coal and waste, coal and biomass, and various types of biomass. These studies are described in a recent publication by Lee and Shah [2] and others [5,29–32,172]. Gasification and steam reforming of mixed feedstock has a very bright future. De Ruyck et al. [172] examined the co-utilization of biomass and natural gas in a combined cycle through primary steam reforming of natural gas.

The study proposed a method in which external firing is combined with the potential high efficiency of combined cycles through co-utilization of natural gas with biomass. Biomass is burned to provide heat for partial reforming of the natural gas feed. In this way, biomass energy is converted into chemical energy contained in the produced syngas. Waste heats from reformer and biomass combustor are recovered through a waste heat recovery system. This way, biomass can replace up to 5% of the energy in the natural gas feed. It also shows that in the case of combined cycles, this alternate path allows for external firing of biomass without an important drop in cycle efficiency.

4.5.2.7 Carbon and Carbon Monoxide

These are perhaps the most basic steam reforming reactions leading to the production of a mixture of hydrogen and carbon monoxide and the subsequent reaction to produce hydrogen. The reactions are as follows [18]:

$$C + H_2O \rightleftarrows CO + H_2 \qquad \Delta H = 131.2 \text{ kJ/mol} \tag{4.68}$$

and

$$CO + H_2O \rightleftarrows CO_2 + H_2 \qquad \Delta H = -41.1 \text{ kJ/mol} \tag{4.69}$$

The first reaction is the basis of all different types of steam reforming reactions outlined earlier. The second reaction is called "water–gas shift reaction," and in this section, we mainly focus on this reaction. While the water–gas shift reaction was first reported in 1888 [173], it became the most popular for producing hydrogen in the Haber process for manufacturing ammonia. In the early stages of ammonia process, the hydrogen was obtained by burning coal, coke, and carbon according to reaction at a temperature about 1000°C [1]. At lower temperature, another reaction

$$C + 2H_2O \rightleftarrows CO_2 + 2H_2 \qquad \Delta H = 90 \text{ kJ/mol} \tag{4.70}$$

produced needed hydrogen. Pure hydrogen can be obtained by separating CO_2 using absorption, adsorption, or membrane separation technique. CO can be separated by liquefaction or copper liquor scrubbing. Later, Bosch and Wild [174] discovered that a mixture of carbon monoxide and steam can be converted to hydrogen and carbon dioxide at 400°C–500°C by iron and chromium oxides, thereby generating additional hydrogen for the Haber process. Thus, the use of water–gas shift reaction became a very important part of hydrogen generation from carbonaceous materials.

In the recent years, water–gas shift reaction has been extensively studied and new catalysts for this reaction have been developed. These catalysts are analyzed in a recent excellent review by Ratnaswamy and Wagner [18]. According to these authors, there are basically four types of catalysts for water–gas shift reaction. At moderately high temperature (350°C–450°C), promoted iron oxide catalysts are used and these catalysts are called HTS catalysts. At low temperatures

(190°C–250°C), copper–zinc oxide catalysts are used and called low-temperature shift (LTS) catalysts. The third type of catalysts employs cobalt and molybdenum sulfides as active ingredients, and they are sulfur tolerant and used to treat "sour gas"-containing sulfur. They are therefore called sour gas catalysts. Finally, medium-temperature shift catalysts operate between 275°C and 350°C, and they are copper–zinc catalysts modified with iron oxide. Besides these four types of catalysts, Pt and Gold catalysts have been intensely investigated and promoters such as Cu and Al_2O_3 have been added to the conventional iron and chromium oxide HTS catalysts.

As discussed earlier, the water–gas shift reaction is moderately exothermic and equilibrium controlled. The equilibrium constant first sharply decreases with an increase in the temperature above around 190°C and the levels of around 480°C. The rate expression is as follows [18]:

$$K_p = \exp\left(\frac{4477.8}{T} - 4.33\right) \tag{4.71}$$

where:

T is expressed in kelvin

Thus, high forward conversion of water–gas shift reaction is favored at low temperature, and it is essentially unaffected by the total pressure. At high temperature, reverse water–gas shift reaction dominates. The reaction is reversible and the forward reaction rate is strongly inhibited by the reaction products: CO_2 and H_2. Low CO level can be obtained by maintaining the reactor temperature at around 200°C. At low temperature, however, condensation of water and its contact with the catalyst should be avoided. The equilibrium carbon monoxide concentration is also affected by the steam-to-gas ratio. Higher steam-to-gas ratio lowers the product CO concentration and increases the hydrogen and carbon dioxide production rates. Since the water–gas shift reaction is always present and equilibrium controlled, in any steam reforming process a substantial amount of carbon dioxide is present in the reaction mixture. The presence of carbon dioxide also forces the "dry reforming" reaction between hydrocarbons and carbon dioxide. Thus, in any autothermal reactors, it is more than likely that steam reforming reaction, partial oxidation, water–gas shift reaction, and dry reforming reaction all occur simultaneously.

4.5.2.8 Bio-Oil

Catalytic steam reforming of bio-oil at 750°C–850°C over a nickel-based catalyst is a two-step process that includes the shift reaction [24–26]:

$$\text{Bio-oil} + H_2O \rightarrow CO + H_2 \tag{4.72}$$

$$CO + H_2O \rightarrow CO_2 + H_2 \tag{4.73}$$

The overall stoichiometry gives a maximum yield of 11.2% based on wood. The overall reaction is

$$CH_{1.9}O_{0.7} + 1.26\ H_2O \rightarrow CO_2 + 2.21\ H_2 \tag{4.74}$$

For this process, bio-oil from regional facility is generally transported to central reforming facility. The process is compatible with other organic waste streams such as aqueous steam fractionation processes used for ethanol production and trap grease. Methanol and ethanol can also be produced from biomass by a variety of technologies and used for on-board reforming for transportation. Methane from anaerobic digestion could be reformed along with natural gas. A system analysis has shown that biomass gasification/shift conversion is economically unfavorable compared to natural gas reforming except for very low-cost biomass and potential environmental incentives.

4.6 STEAM GASIFICATION AND REFORMING REACTORS

4.6.1 Steam Gasification Reactors

Fundamentally, three types of gasifiers are used in the commercial processes: fixed bed, fluidized bed and/or CFB, and entrained bed [5,8–10,33,75–83,125,127,156, 175–191]. In some specific applications, plasma and free radical gasifiers as well as molten salt gasification reactors are also used. Although in most conventional applications the first two types are most commonly used, all types of gasification reactors are briefly described below.

4.6.1.1 Fixed-Bed Gasifiers

There are two major types of fixed-bed gasifiers: countercurrent or "updraft" and co-current or "downdraft" gasifiers. In countercurrent gasifier, the carbonaceous materials (coal, biomass, waste, etc.) flow downward, whereas steam, oxygen, and/or air flow upward in the reactor. The ash is removed either dry or as slag. The slagging gasifiers have a lower ratio of steam to carbon achieving a temperature higher than the ash fusion temperature. The fuel must be permeable and noncaking. The throughput for this type of gasifier is relatively low. In this type of reactor, while thermal efficiency is high, both tar and methane productions are high and the product gases need to be extensively cleaned. The tar can be recycled. In gasification of rice hulls, the gas gets very hot (up to 1000°C) and has to be forced (by fan) into the reactor.

In both updraft and downdraft gasifiers, drying and devolatilization occur at the top of the reactor [56,191]. In the updraft reactor, this is followed by reduction and combustion. But in the downdraft gasifier, combustion precedes reduction. In an updraft reactor, in the devolatilization zone volatile species are released and considerable quantities of tars are formed. In the reduction zone, permanent gases are formed and finally char and remaining solids are combusted in the final combustion zone. The updraft reactor produces low tar content and the temperature in the gasification zone can also be controlled by co-feeding steam and air or humidifying air.

The product gases are cooled down to 200°C–300°C before leaving the gasifier. The overall energy efficiency of the updraft gasifier is high.

In a "down draft" reactor, both raw materials and gasification agent flow co-currently downward. Heat needs to be added in the upper part of the bed either by combusting small portion of fuel or by some external sources. The produced gas leaves the reactor at high temperature, and most of the heat is transferred to the entering gasification agent. This creates an energy efficiency similar to that for "updraft" reactors. Since tar passes through the hot zones within the reactor, its level in the product gas is much lower than that in the "updraft" reactor [56,191].

The downdraft gasifier has four distinct zones [56,191]: (1) upper drying zone, (2) upper medium pyrolysis section, (3) lower medium oxidation zone, and (4) lower reduction zone. The temperature in the oxidation zone is 1000°C–1400°C and the tar produced is exclusively tertiary tar. The downdraft gasifier produces clean gas but has low thermal efficiency, and it is not suitable for handling biomass with high moisture and ash content.

Besides updraft and downdraft gasifiers, sometimes cross-flow gasifiers are used where raw materials (coal, biomass, etc.) flow downward and air or steam is introduced from the side. The product gases at about 800°C–900°C are withdrawn from the top of the gasifier. A hot combustion/gasification zone forms around the air entrance, with both pyrolysis and drying zones being formed higher up in the vessel. Ash is removed from the bottom of the reactor. The gasifier gives low-energy efficiency and produces high tar content.

The fixed-bed reactors are easy to design; however, they produce gas with low heating value and high tar content. The use of oxygen along with steam improves the product gas heating value. The heating value also significantly depends on the nature of the feedstock.

A novel HTAG unit was used to study biomass waste such as bark, charcoal, and wood pellets with diameters ranging from 6 to 12 mm as well as densified and not densified plastic wastes [56,75–84,178–181]. The facility consisted of a batch-type, countercurrent (updraft), fixed-bed vertical column gasifier. The reactor had three sections: a wind box, the feedstock section, and gas reaction section. The afterburning combustion chamber was coupled to a gasifier to burn completely produced fuel gas. After burner had an inlet for fuel gas, an outlet for flue gas, and a set of air nozzles to ensure complete combustion. Flue gas outlet channel was connected with the afterburning combustion chamber that was equipped with the cooling system. The air was preheated to 600°C with a capability to go up to 1300°C. The unit has worked very successfully.

4.6.1.2 Suspended Bed Reactor

There are three types of suspended bed for steam gasification used in commercial practice: fluidized bed, CFB, and entrained bed.

4.6.1.2.1 Fluidized Bed Reactor

In this type of reactor, fuel is fluidized by air (or oxygen) and steam. The ash is removed dry or as heavy agglomerates that defluidized. In dry ash gasifier, the temperature is relatively low, thus generating high methane content gases. The agglomerating gasifiers have higher temperatures and are more suitable for high-rank coals. The flow rate in fluidized bed reactor is higher than that in fixed-bed reactor. The conversion per pass is usually low due to elutriation of carbonaceous materials. The mixing in the reactor is high giving more uniform temperature. Fluidized bed is most useful for raw materials such as biomass which form highly corrosive ash that can damage the walls of slagging gasifiers. The fluidized bed reactor is generally operated under "bubbling fluidized bed" conditions. A modeling and experimental validation of biomass–steam gasification in the bubbling fluidized bed reactor is given by Gopalakrishnan [82]. His analysis indicated that for steam gasification of biomass, an increase in temperature in such a reactor increases the production of hydrogen and carbon monoxide and decreases the production of carbon dioxide and methane. An increase in steam-to-biomass ratio increases the production of carbon dioxide, decreases the production of hydrogen and carbon monoxide, and has no effect on the production of methane.

4.6.1.2.2 CFB Reactor

One way to improve conversion in fluidized bed reactor is to recycle solids back into the reactor. In this type of reactor, the solids coming out of reactor are separated from gas and recycled back into the reactor. This reactor thus provides more flexibility on the solids residence time within the reactor. The solids recycling also provides better solids mixing and uniform temperature distribution.

A variation of single CFB was examined by Matsuoka et al. [83] who examined a circulating dual bubbling fluidized bed system. In this system, two bubbling fluidized beds were used as a gasifier and a combustor. The gasifier and combustor had identical inner diameters (80 mm), and the static bed heights of the bed material in the gasifier and combustor were 270 and 150 mm, respectively. The inner diameter of the riser was 18 mm and its height from the top of the gasifier to the cyclone was about 1800 mm. Porous γ-alumina particles with a diameter of 75–150 μm were used as refractory materials. The system was used to treat sawdust at temperatures ranging from 773 to 1073 K. The data were obtained at different steam-to-biomass ratios and different set of residence times. The system was found to be very efficient, and higher carbon conversion and hydrogen yield were achieved in this system compared to those obtained in conventional CFB.

Another variation of the dual fluidized bed steam gasification process was developed by Pfeifer et al. [81] at the Vienna University of Technology. This system is graphically depicted in Figure 4.5 later in the chapter. In this process, heat for the gasification reactor is provided by circulating bed material. This system was a further development of the so-called fast internally CFB (FICFB) technology [75–83,178–181,187]. In this technology, biomass enters a bubbling fluidized bed gasifier in which drying, thermal degasification, and partially heterogeneous char gasification take place at temperatures of about 850°C–900°C. Residual biomass char leaves the gasifier together with the bed material through an inclined, steam fluidized chute toward the combustion reactor. The combustion reactor serves for heating up the bed material and is designed as highly expanded fluidized bed (riser). Air is used as the fluidization agent in the riser. The circulating rate can be adjusted easily by changing the amount of primary and secondary air in the combustion chamber. After particle separation from the flue gas in a cyclone, the hot bed material flows back to the gasifier via a loop seal. The solids in both the loop seal and the chute are fluidized with steam, which effectively prevent gas leakage between gasification and combustion zone, and also allow high solid throughput. The temperature difference between the combustion and the gasification reactor is determined by the energy needed for gasification as well as bed material circulation rate. The system is inherently auto-stabilizing since a decrease in the gasification temperature leads to a higher amount of residual char, which results in more fuel for the combustion reactor. This, in turn, transports more energy into the gasification zone and therefore stabilizes the temperature. Both the gasifier and the combustor operate at atmospheric pressure. The process yields two separate gas streams: a high-quality producer gas and a conventional flue gas at high temperatures. The high-quality producer gas contains low amounts of tars and nitrogen, and high concentration of hydrogen. For practical use, olivine, a natural mineral, has proven to be a suitable bed material with enough resistance to attrition and moderate tar-cracking activity [72–84]. This concept was proven to be more efficient for converting the primary fuel energy into producer gas than conventional dual fluidized bed steam gasification because of the lower operating temperature.

4.6.1.2.3 Entrained Bed Reactor

This type of reactor uses fine solids and high rate of gas flow to provide uniform temperature distribution and low residence time within the reactor [5]. The reactor is generally operated at high temperatures so that the tar and methane concentrations in the product gases are very low. The oxygen requirement in this type of gasifier is higher than those in other types of gasifiers. All entrained bed gasifiers remove the major part of the ash as a slag as the operating temperature is well above the ash fusion temperature. These types of gasifiers do not suffer from corrosive slags and can better handle biomass that can generate corrosive slag.

For processing fuels with very high ash fusion temperatures, some limestone is mixed with fuel, which lowers the ash fusion temperature. The need for fine solids requires the fuel pulverization process before gasification. The reactor needs more energy due to fuel pulverization and the production of oxygen that is used for gasification.

4.6.1.3 Plasma and Free Radical Gasifiers

Both of these types of gasifiers use either thermolytic, photolytic, or high-voltage torch to supply heat for the gasification process. These are high energy intensive reactors and mostly produce clean syngas. They are not often used for steam gasification.

4.6.1.4 Molten Salt Steam Gasification Reactors

There are at least four different designs that use molten salt media to gasify coal in the presence of steam [1,3]. Two of these four designs, namely, Rockwell molten salt gasifier and Rummel–Otto single-shaft gasifier, are graphically illustrated in Figure 4.3a and b, respectively. Here, we briefly describe the remaining two, namely, Kellogg–Pullman molten salt process and Atgas molten iron coal gasification process.

4.6.1.4.1 Kellogg–Pullman Molten Salt Process

In this process, the coal is gasified in a bath of molten sodium carbonate through which steam is passed. The process offers the following advantages [1,3]:

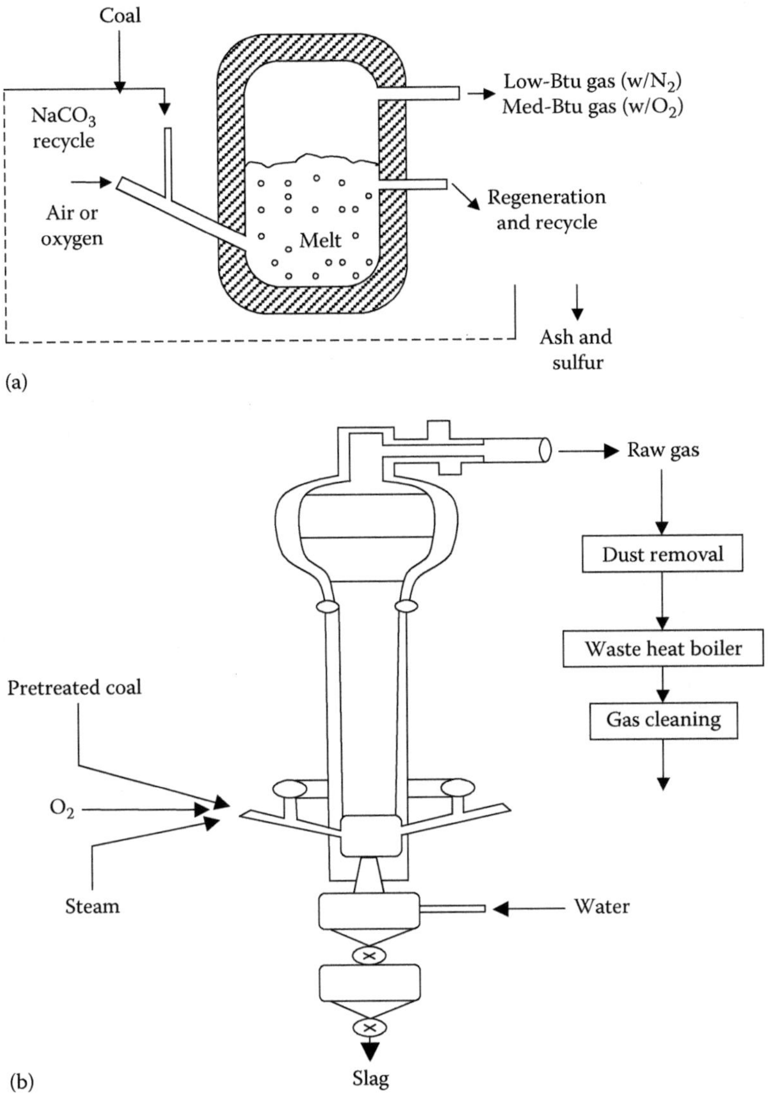

FIGURE 4.3 Schematics of two typical molten salt gasifiers: (a) Rockwell molten salt gasifier; (b) Rummel–Otto single-shaft gasifier. (Adapted from Lee, S., Speight, J.G., and Loyalka, S.K., *Handbook of Alternative Fuel Technologies*. Taylor & Francis, Boca Raton, FL, 2007; Lee, S., "Gasification of coal," in Lee, S., Speight, J.G., and Loyalka, S.K., eds., *Handbook of Alternative Fuel Technologies*. Taylor & Francis, Boca Raton, FL, 26–78, 2007.)

1. Salt bath supplies the necessary heat for gasification. Due to high and uniform temperature in the bath, products are free of impurities such as tars and tar acids.
2. Gasification of caking coal without carbonization is possible due to uniform distribution of coal and steam and good contacting between the two reactants.
3. Complete gasification at a lower temperature is possible due to catalytic effect of sodium carbonate for coal steam gasification reactions.

In this process, the preheated oxygen and steam transport coal and unreacted recycled coal (after ash removal) in the molten salt gasifier. A significant portion of oxygen and steam is also admitted into the bottom of the reactor to provide the necessary gases for the complete gasification reactions. Sulfur in the coal is accumulated as sodium sulfide at equilibrium level and it reacts with molten salt as

$$Na_2CO_3 + H_2S \rightarrow Na_2S + CO_2 + H_2O \tag{4.75}$$

Ash accumulates in the melt and leaves with a bleed stream of salt where it is separated and the clean salt is recycled back into the reactor. The bleed salt is quenched in the water to dissolve sodium carbonate and the ash is separated by filtration. Sodium carbonate is further carbonated to make sodium bicarbonate ($NaHCO_3$), which is then separated and heated to regenerate sodium carbonate for reuse in the reactor. The entrained salt and heat in the product gas are recovered and the purified gas stream is further processed to make synthesis gas, pipeline gas, or synthetic natural gas (SNG).

4.6.1.4.2 Atgas Molten Iron Coal Gasification

In this process, coal is injected with steam in the molten iron bath [1,3]. Thermal cracking of coal along with steam dissociation generates a mixture of carbon monoxide and hydrogen. The sulfur in coal is captured by iron and transferred to lime slag from which elemental sulfur is recovered. The Atgas process produces gases with a heating value of about 900 Btu/scf. The Atgas molten iron process has the following advantages over conventional fixed- and fluidized bed steam gasification processes:

1. Sulfur in coal is recovered as elemental sulfur, which can be sold, and this helps process economics. The product gas is essentially free of sulfur.
2. Gasification is carried at low pressure; hence, the coal feeding problem in pressurized operation is eliminated. Coking properties, ash fusion temperature and generation of coal fines, are not problematic.
3. Tar formation is minimal due to high-temperature operation.
4. The system is very flexible and does not cause any environmental problems. Relatively large coal particles can be handled without any pretreatment.
5. Reactor start-up and shutdown procedures are much simpler compared to those for fixed and fluidized bed reactors.

The coal and limestone are injected into the molten iron through tubes using steam as a carrier gas. Coal gasifies and produces carbon monoxide, and sulfur (both inorganic and organic) migrates to slag and reacts with lime to produce CaS. The product gas at 1425°C is cooled and compressed, and passes through a shift converter to convert CO into water gas with a H_2-to-CO ratio of 3–1. The carbon dioxide is removed from the final product, and the gas is again cooled and passed through a methanator to produce methane by the reaction: $CO + 3H_2 \rightarrow CH_4 + H_2O$. Excess water is removed from the methane-rich product.

4.6.2 Steam Reforming Reactors

While hydrogen production can be achieved by a number of commercially proven technologies such as gasification of coal, biomass, and residue (waste); methanol decomposition; and steam reforming of methane, renewable materials, and liquid hydrocarbons, it is the last technology that produces the largest portion of hydrogen production [5,33,126,128,175–190]. With the considerable advances in unconventional production of natural gas that includes shale gas, deep gas, tight gas, coal bed methane, gas from geopressurized zones, and gas hydrates, the steam reforming of natural gas is likely to become even more important. An increase in natural gas production is likely to make the steam reforming of methane the choice of significant hydrogen production.

The design of a steam reforming plant requires the considerations that

1. The economics of the process is very scale dependent [33]. For example, for 5×10^6 Nm^3/day plant, the operating cost can be as low as \$80/kW of H_2, whereas for 2300 Nm^3/day plant, the same cost would be \$4000/kW for hydrogen.
2. The capital cost can be large due to large size of the plant (i.e., in large plants, reformer tubes can be as long as 12 m) and the need for expensive alloy materials for high-temperature and high-pressure operations.
3. The small-scale operation, while expensive, is often used for niche application such as FC technology and hydrogen refueling station, and this requires small and compact reformers at low cost.

Due to these considerations, both large- and small-scale reformers have been developed.

In normal commercial reformers, the steam-to-hydrocarbon ratio is kept high enough to prevent coking but to avoid overloading the reformer duty. Generally, the ratio of 3 is used. The inlet temperature of 760°C is used, and because reforming reaction is endothermic, additional heat is added as mixture flows down the catalyst-filled reformer tubes. A critical factor in the reformer heater design is keeping the tube wall temperature uniform and hot enough to promote reforming reaction. For this purpose, two types of heater design, side-fired reforming furnace and roof-fired heater design, have been employed [33].

In side-firing furnace, two parallel rectangular boxes are connected at the top with horizontal ductwork into the vertical convection stack. Several rows (typically four) are used to directly fire the tubes. A typical reformer furnace has 300 burners. Reformer tubes are 5 inch in diameter with a wall thickness of 0.5 inch and about 34 ft of wall is exposed to the burners. The tubes are generally 25% chrome, 20% nickel, or a high nickel steel such as HL40 [33].

The top-fired reformer is a rectangular box, the tubes are still vertical, and the inlet and outlet are pigtails to the pigtail inlet header and the outlet transfer line. The burners have a pencil-shaped flame design. All burners are located above the inlet manifold. Hydrogen plants with single reformer heaters and a capacity of up to 100,000 ft^3/day are used in the vertical down-firing approach. The outlet transfer line from the reformer is used to generate high-pressure (650 psig) steam. The reformer effluent gas exits through the transfer line at about 760°C [33].

While the large commercial reformers are designed as described above, more compact and economical designs are used in the smaller scale reformers [120–134, 188–190]: (1) annular bed reformers, (2) plate-type reformers, (3) membrane reactors, (4) auto-thermal reactors, (5) ITM reformers, (6) sorbent-enhanced reformers, (7) plasma reformers, and (8) micro-channel reformers. These different designs consider the ways to improve the heat transfer rate, the area and the efficiency since reforming requires a large supply of heat due to an endothermic nature of the reaction. Improvement of material cost is another important consideration. Finally, a process that carries out simultaneous reaction and separation of hydrogen is important to improve the conditions for equilibrium and purity of hydrogen product. The following paragraphs briefly summarizes the descriptions of these eight reformers given in References 120–134.

The annular bed reformer is used for FCs and low hydrogen production (on the order of 2 kW) needs. In the latter case, it is generally operated at a low temperature of about 700°C and a low pressure of about 3 atm. These mild conditions reduce the cost of materials and produce an energy efficiency of about 70%–80%. This type of reformer is used by industries such as Haldor Topsoe, Ballard Power Systems, Sanyo Electric, and International Fuel Cells. The technology produces more compact reformers at a lower cost than conventional reformers.

Plate-type reformers are more compact than annular or conventional long tube reformers and are often used for proton exchange membrane (PEM) FC or residential-type FC (20 kW) applications. It has the same energy efficiency as that of annular reformer. The plates are arranged in a stack in which one side of the plate is coated with the catalyst and on the other side (anode) exhaust gas from FC undergoes catalytic combustion to supply heat for the endothermic steam reforming reaction. The unit is compact and low cost, and has good heat transfer and small heat-up period. For PEM FC applications, Osaka Gas Co., Japan, is developing a low cost reformer with an integrated plate design that carries out sulfur removal, steam reforming, water–gas shift reaction, and CO removal steps all in one unit making the final device more compact and economical. GASTEC is applying the technology for residential-type FCs and minimizing the cost by testing the variables such as combustion catalysts, coatings, and substrate materials.

In the membrane reactor, reforming, water–gas shift reaction, and further CO cleaning step all occur in the same unit. The reaction and separation functions are thus combined. The reactor operates under high pressure and uses the Pd membrane on one side through which H_2 permeates with high selectivity. The constant removal of hydrogen on the downstream side allows equilibrium to be shifted to achieve better conversion by reforming at a lower temperature. The reactor also produces high-purity hydrogen.

In the auto-thermal reformer, endothermic reforming reaction is accompanied by partial oxidation reaction that generates enough heat to supply the heat needed for reforming reaction. Thus, the reactor does not need any external source for the heat. Arthur D. Little, Nuvera, Epyx, and a consortium of McDermott Technology/Catalytica, among others, have developed a 50 kW FC reformer of this type. Small-scale (10–50 kW) auto-thermal reactors have been developed for PEM FC by Honeywell, DaimlerChrysler, Analytical Power, and IdaTech, among others [120]. Generally, auto-thermal reactors use gasoline, diesel, and logistic fuels along with natural gas. The use of diesel and logistic fuels makes them specially useful for FC applications on ships [120].

In the ITM technology (being developed by a consortium of industries and universities headed by Air Product), one side of the membrane separates oxygen from air at around the room temperature and 0.03–0.20 atm pressure; on the other side, methane and steam react at high pressure (3–20 atm) to produce syngas. The membrane is made up of nonporous multicomponent oxides that operate at a temperature higher than 725°C and has high permeability and selectivity for oxygen transfer. Partial oxidation provides the heat for reforming reaction. The syngas can either be reformed to produce hydrogen or converted to produce fuels and chemicals. The ITM technology generally uses flat plate system.

In the sorbent-enhanced reforming (SER) technology, the steam reforming is accompanied by simultaneous removal of carbon dioxide and carbon monoxide by calcium oxide. The removal of carbon dioxide allows the reforming reaction to occur at 400°C–500°C as opposed to the normal reforming temperature of 800°C–1000°C. The reaction also produces reasonably pure hydrogen (90% H_2, 9.5% CH_4, 0.5% CO_2, and <50 ppm CO), and this alleviates the downstream expensive purification processes such as water–gas shift reaction, preferential oxidation, and membrane separation.

Thermal plasma technology is a high-temperature (2,700°C to about 10,000°C) process to generate hydrogen and hydrogen-rich gas from a variety of feedstock. High temperature accelerates the rate of reforming process. The products generally contain ethylene and acetylene along with hydrogen, carbon monoxide, and carbon dioxide. The process can handle various reaction volumes, interelectrode gap, sulfur impurities, and carbon deposit. The process can be operated in auto-thermal mode. The process can generate a large range of fuel power (10–40 kW) and can give up to 90% conversion of methane.

One attractive method to improve the transport limitations in the reforming reactor is to use micro-channel reactor that can operate at 10 ms or lower residence time compared to conventional reactor that operates at the residence time of 1 s. Since intrinsic reforming reaction is very fast, at a high residence time, a significant portion of the catalyst volume in the steam reformer is wasted. The micro-channel reactor allows a reduction of plant volume by about a factor of 30, and thereby reducing both capital and operating costs for steam reforming of methane. Also, the micro-channel reactors allow high reaction rates by increasing the heat transfer rates. For highly active catalyst, equilibrium can be reached in less than 0.5–1 ms residence time. This indicates that further lowering of transport resistances can further reduce residence time to reach the desired equilibrium.

Besides the eight different types of reforming reactors that are being developed (particularly for small-scale applications), solar reforming reactors that use solar energy to carry out steam reforming reactions are gaining more popularity. These reactors are described in the following sections.

4.7 NOVEL STEAM GASIFICATION AND REFORMING PROCESSES

4.7.1 Solar Gasification Technology

Solar thermochemistry refers to a number of process technologies such as thermal or thermochemical splitting of water, solar electrolysis, solar gasification, and reforming or cracking of water and other carbonaceous materials [192–214]. Many of these endothermic reactions are carried out by energy harnessed by concentrated solar beams. Solar gasification generally deals with upgrading and decaronization of fossil fuels. Such gasification is often carried out in the presence of steam. Successful solar gasification of carbonaceous materials was first reported in the 1980s in which coal, activated carbon, coke, and coal/biomass mixtures were employed in a fixed-bed windowed reactor. Charcoal, wood, and paper were gasified with steam in a fixed-bed reactor. More recently, steam gasification of oil shale and coal, biomass, waste tires and plastics, and coal in a fluidized bed reactor as well as petroleum coke and vacuum residue in fixed, fluidized, and entrained bed reactors were examined [192,193,195–199,202–210] (Piatkowski 2012, pers. comm.; Yeheskel et al., 2012, pers. comm.). In the last type of reactor, dry coke particles, coal–water slurries, and vacuum residues were tested for the steam gasification.

In a conceptual solar gasification process using steam, biomass is heated rapidly in a solar furnace to achieve flash pyrolysis at temperatures of about 900°C [192]. Some steam is added to the pyrolyzer to increase the gas yield relative to char. The char constituting about 10%–20% of the biomass by weight is steam gasified with external heating at temperatures of 900°C–1000°C; all of the volatile hydrocarbons are then steam reformed in a solar reformer. Steam for the process is generated from heat recovered from the product gas. The composition of the syngas is adjusted to the user's needs utilizing conventional operation involving the water–gas shift reaction and CO_2 stripping. This conceptual process can be modified in a number of different ways depending on the specific needs.

A number of gasification experiments were carried out using small quantities of biomass, coal, oil shale, and residual oil with external heat supplied by the Sun [192–207]. These experiments included cellulose gasification and oil shale gasification with carbon recovery approaching nearly 100% at a temperature of 950°C and short residence times [196,208,209]. While these experiments confirmed the applicability of the flash pyrolysis approach, they did not provide the data for design and scale-up of a solar gasification process. More work is being pursued to improve the design and scale-up capabilities of solar gasification process [192–207].

4.7.2 Solar Gasification Reactors and Processes

A number of different types of solar steam gasification reactors have also been examined in the literature [192–206,209]. The reactor configuration examined by Z'Graggen [202] at Swiss Federal Institute of Technology (ETH) at Zurich consisted of a cylindrical cavity receiver of 21 cm in length and 12 cm in inside diameter, and an aperture of 5 cm in diameter for solar beams. The cavity-type geometry was designed to effectively capture the incident solar radiation and its apparent absorption is estimated to exceed 0.95. The cavity was made of Inconel 601 lined with Al_2O_3 and insulated with an Al_2O_3/ZrO_2 ceramic foam. The aperture was closed by 0.3-cm-thick clear fused quartz window mounted in a water-cooled aluminum ring that also served as a shield for spilled radiation. The window was actively cooled and kept away from particles and condensable gases.

Steam and particles were injected separately into the reactor cavity, permitting the separate control of mass flow rates and stoichiometry. Steam was introduced through several ports. The carbonaceous material feed unit was positioned on the top of the reactor vessel with its inlet port located at the same plane as the primary steam injection system, allowing for the immediate entrainment of particles by the steam flow. Reactor temperature was measured at 12 separate locations by thermocouples inserted in the Inconel walls. Both inlet and exit temperatures were also measured by the thermocouples. The dry, slurry, and liquid feeding of raw materials were carried out by different devices.

Piatkowski et al. [197], Piatkowski (2012, pers. comm.), and Piatkowski and Steinfeld [195] used a packed bed solar steam gasification reactor as shown in Figure 4.4. This reactor was specially designed for beam-down incident solar radiation,

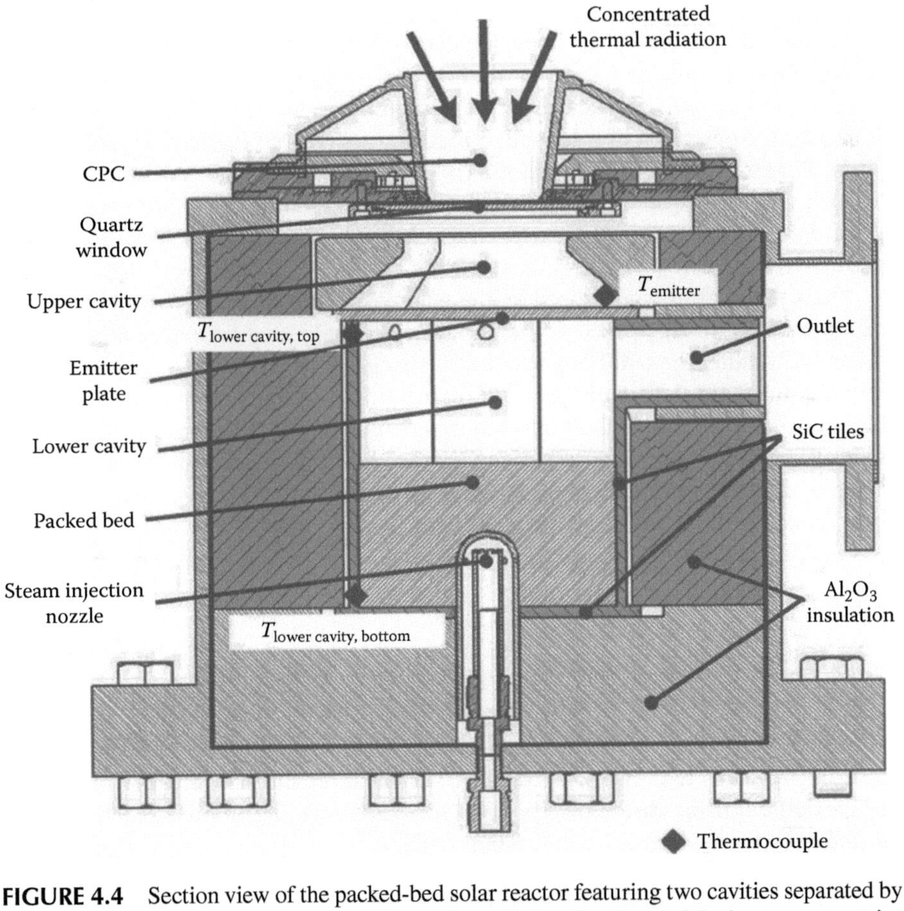

FIGURE 4.4 Section view of the packed-bed solar reactor featuring two cavities separated by an emitter plate, with the upper one serving as the radiative absorber and the lower one containing the reacting packed bed that shrinks as the reaction progresses. CPC, compound parabolic concentrator. (Reprinted from *Fuel Processing Technology*, 90, Piatkowski, N., Wieckert, C., and Steinfeld, A., Experimental investigation of a packed bed solar reactor for the steam gasification of carbonaceous feedstocks, 360–366, Copyright 2009, with permission from Elsevier.)

which was obtained through a Cassegrain optical configuration that made use of a hyperbolic reflector at the top of the solar tower to redirect the sunlight collected by a heliostat field to a receiver located at the ground level. The reactor had two cavities in series. The upper one absorbed the solar radiation and contained a small aperture to gather concentrated solar radiation. The lower cavity contained carbonaceous materials on the top of a steam injector. An emitter plate separated the two cavities.

A 3D compound parabolic concentrator (CPC) was incorporated in the aperture of the reactor, further augmenting the incident solar flux before passing it through a quartz window in the upper cavity. The emitter plate acted as a transmitter of the radiation to the lower cavity, thus avoiding the direct contact between the quartz window and the reactants and products. This set-up also provided uniform temperature in the lower cavity and a constant supply of radiant heat through the upper cavity that can act as energy storage, which was needed due to intermittent supply of radiant heat. This type of batch, two-cavity solar reactor, has been successfully used for the carbothermal reduction of ZnO and the detoxification of solid waste. The reactor can be operated with a wide variety of particle sizes, and as the reaction proceeds, both the particle size and the packed bed reactor volume decrease. The detailed dimensions and the operation of this type of reactor are given by Piatkowski and Steinfeld [195]. Piatkowski et al. [197] and Perkins et al. [196] also showed an effective use of such a reactor to produce syngas from coal, biomass, and other carbonaceous feedstock. Z'Graggen [198] and Z'Graggen et al. [202] produced hydrogen from petroleum coke using solar gasification process.

The solar energy is also used as the heat carrier for the pressurized coal gasification process. In this process, finely powdered coal is fed by a specially designed injection system. The oxidizing and fluidizing agent is a superheated steam. The heat required for the endothermic gasification reaction is introduced by means of a tubular heat exchanger assembly immersed in the fluidized bed. The technical feasibility of a solar power tower and pressurized gasifier integration has been demonstrated in a small pilot plant [194,199,202]. Solar energy has also been used to gasify biomass in different types of reactors [196,203,204,206,208,209].

4.7.3 Solar Reforming

The high temperatures required for solar reforming effectively limit the nature of solar energy collector [192,194,199–201,207–214] (Yeheskel et al., 2012, pers. comm.). The bulk energy production, whether in closed-loop or open-loop configurations, probably must be carried out on a large scale to compete with fossil fuels and probably requires the tower (central receiver) solar technology. Solar reforming can be carried out using different processes such as direct and indirect, each requiring different type of reformer configuration [192].

4.7.3.1 ASTERIX: Solar Steam Reforming of Methane

Advanced Steam Reforming of Methane in Heat Exchange (ASTERIX) experiment, an earlier joint Spanish–German project, examined steam reforming of methane using solar-generated high-temperature process heat by an indirectly heated reformer [192,194,199]. The specific objectives of the ASTERIX experiments were to collect and store an amount of solar energy to obtain the maximum conversion of methane and to produce consistently high-quality synthesis gas. The experiment used gas-cooled solar tower (GAST) system to produce hot air (up to 0.36 kg/s at 1000°C and 9 bars) to drive separate steam reformer. This air was then fed back into the GAST cycle. The GAST technology program is described by Becker and Bohmer [194].

During normal operation, the heating medium, air, is taken from the GAST circuit (receiver) at a temperature of 1000°C over a suitable bench line and fed through the electric heater to the reforming reactor inlet. In this solar-only operating mode, air flows through the heater passively without any additional electric heating. Methane reforming is initiated at the process gas end of the reformer. A liquid natural gas storage tank directly provides the reforming unit with natural gas at the required pressure via the Liquid natural gas evaporator. The process gas mixture is heated by air from 500°C to about 850°C as it passes through the catalyst bed. The endothermic reforming reaction results in the production of hydrogen and carbon monoxide with a ratio of 3/1. More details of the ASTERIX experiment are given in References 192–194 and 199.

4.7.3.2 The Weizmann Institute Tubular Reformer/Receiver

The WIS (Weizmann Institute of Science) operated a solar central receiver for the development of high-temperature technology including the storage and transport of solar energy via methane reforming [192,199,200] (Yeheskel et al., 2012, pers. comm.). The WIS had a designed facility for testing reformers up to about 480 kW absorbed energy. The facility was designed for either steam or carbon dioxide reforming and can accommodate the reformer that operates between 1 and 18 bars. The reformer systems were operated in coordination with a matching methanator system that recovered the energy from the reverse reaction [192].

A cavity receiver containing eight vertical reformer tubes (2 inch schedule 80 and 4.5 m long) was designed. The overall dimension of the device was about 5 m high, 4.5 m wide, and 3 m deep. The reactor was designed to produce syngas at 800°C. It resembled commercial reformers except that a solar cavity receiver had replaced the conventional gas-fueled radiant furnace [192].

4.7.3.3 Soltox Process

In the Soltox process, a parabolic dish is used to concentrate sunlight through a quartz window into an internally insulated aluminum reactor vessel in which it is absorbed on a rhodium-coated reticulated ceramic foam absorber [192,199,201,207,210, 212–214]. Concentrated organic waste and steam are mixed and flow through the hot (>1000°C) catalyst bed, in which they react completely in fractions of a second to produce hydrogen, carbon dioxide, carbon monoxide, and halogen acids (which are easily neutralized to simple salts). The extremely good heat and mass transfer within the reactor result in a compact, highly efficient system [192–194].

When a vaporized organic waste is mixed with steam and passed through the reactor, highly specific, irreversible, endothermic reforming reactions take place on the catalyst-coated surface of the radiantly heated absorber to quantitatively destroy the waste. For example, trichloroethylene (TCE) reacts with steam to produce hydrogen, carbon monoxide, and hydrogen chloride. Because reforming is not a combustion process, neither fuel, nor air, nor oxygen needs to be supplied to the reactor. Thus, unlike incineration, solar-driven, high-temperature catalytic reforming produces neither NO_x nor products of incomplete combustion (PICs). Furthermore, variable absorber thickness and adjustable gas flow rates mean that residence times within the absorber and thus reaction times and destruction efficiency can be controlled [192–194].

4.7.3.4 Open-Loop Solar Syngas Production

The applications of open-loop solar syngas production include the following [188,189,192,199,214]:

1. Natural gas reforming for power plants—A number of European countries have imported natural gas via pipelines from North Africa and have reformed this gas to either syngas or hydrogen, increasing its calorific value by about 25% before combustion in gas turbine or FC power plants [192].
2. Syngas production from municipal, agricultural, and organic industrial waste—In sunbelt countries, concentrated waste streams can be gasified to syngas with solar energy at potentially acceptable costs and with essentially no emissions to the atmosphere [192].
3. Soltox type processing—It provides an option for environmentally acceptable disposal of a number of toxic organic materials [192].

Open-loop syngas production can also be used for the generation of synthesis gas that is being supplied worldwide for the production of hydrogen, methanol, ammonia, and oxyalcohols.

4.7.3.5 Other Solar Reforming Processes

A number of studies have focused on the production of hydrogen by steam reforming of methane and other hydrocarbons using solar reactor [189,192,193,199–201,212,213] (Yeheskel et al., 2012, pers. comm.). A schematic of the solar reactor used by Seinfeld and coworkers is depicted in Figure 4.4. Yeheskel et al. (2012, pers. comm.) studied the chemical kinetics of high-temperature hydrocarbon reforming using a solar reactor. Watanuki et al. [189] examined methane steam reforming using a molten salt membrane reforming reactor. In this type of the reactor, the reforming reaction takes place in tubular reactors that consist of selective membranes, generally palladium, which separates hydrogen as it is produced. The principal advantages of a solar membrane reforming process compared to the conventional reforming process are as follows:

1. The reforming is carried out at a lower temperature (550°C). This means a significant reduction in the energetic consumption. Low-temperature reactors also use less costing materials for the reforming reactor tubes.
2. Hydrogen is obtained with a higher purity due to highly efficient membrane separation process.
3. Methane conversions up to 90% can be reached due to high hydrogen extraction through the membrane.

4. A big part of CO–CO_2 conversion is produced inside the reactor itself.
5. Emissions are reduced by about 34%–53% due to the use of concentrated solar energy to obtain the process heat.

In this study, steam reforming of methane proceeded with the original module having palladium membrane below the decomposition temperature of molten salt (around 870 K). The SOLREF (solar reforming) process [211] and its various options for solar reforming of natural gas by steam are also described by Moller [201]. A review of hydrogen production technologies from solar energy is also given by Suarez-Gonzalez et al. [212].

The above-described process can be easily adapted to solar gasification, but for this case, heavy hydrocarbons are used as feedstock. These are transformed into cleaner fuels for a combined cycle or in the process that can produce hydrogen. As mentioned earlier, a solar gasification plant using petroleum coke has been tested in the solar platform of Almería, Spain. The reactor has reached the hydrogen production efficiency of 60% working at 1500 K [192,199,202,211–213].

4.7.4 Microwave-Assisted Reforming

In the recent years, a significant interest in the use of microwave to carry out high-temperature operations such as steam reforming, pyrolysis, dry reforming, and cracking has been reported [215–217]. Microwave heating is very different from conventional heating in that it heats the materials from inside out unlike outside in heating that normally takes place in conventional heating. This means that all heat is generated and absorbed by the materials and not the surroundings (like microwave cooking at home). The microwave heating, however, requires materials with good dielectric properties such that it not only absorbs microwave but also converts microwave energy into thermal energy. The use of porous, activated carbon for this purpose has been successfully demonstrated [215–217]. Oxides of various materials can also be useful for this purpose.

Menendez et al. [215–217] have shown that microwave-assisted reforming can give better results than the reforming carried out by conventional heating, particularly at lower temperatures. They studied both activated carbon and numerous catalysts deposited on the activated carbon. They also showed that microwave heating is more energy efficient than conventional heating. This approach has a significant potential. More research and development in this area is needed.

4.7.5 Underground Coal Gasification

When coal is imbedded underground in steep seams, it is very difficult to mine. Often the energy from such steep coal seams is recovered by *in situ* underground gasification. Just like conventional gasification, underground gasification is often carried out with oxygen or with oxygen and steam mixture. Yang et al. [190,218] studied the product distribution from underground coal in China. They found that with pure oxygen gasification hydrogen volume percentage in product gas varied from 23.63% to 30.24% and carbon monoxide volume percentage varied from 35.22% to 46.32%. When oxygen–steam mixture was used for the gasification, the gas compositions virtually remained stable and $CO + H_2$ were basically between

61.66% and 71.29%. Moving-point gasification improved the changes in the cavity in the coal seams or the effect of roof in-break (i.e., hole in roof for solar energy) on gas quality. For steep seams, during oxygen–steam mixture gasification, the composition of $CO + H_2$ remained within 58% and 72%. The average oxidation zone temperature reached 1200°C, and it was higher for forward gasification than for backward gasification. In general, for both types of seams, hydrogen concentration increased and carbon monoxide concentration decreased with an increase in steam-to-oxygen ratio. The hydrogen concentration reached about 60% at the steam/oxygen ratio of about 3.

4.7.5.1 Underground Gasification Reactors

A typical underground gasification reactor is illustrated in Figure 4.5 [1,3]. In this type of reactor, the combustion process can be handled in either forward or reverse mode. The forward combustion involves the movement of the combustion front and injected air in the same direction. In the reverse combustion, the combustion front moves in the opposite direction to the injected air. The process involves drilling and subsequent linking of the two boreholes to enable gas flow between the two. Combustion is initiated at the bottom of one borehole (called injection well) and is maintained by the continuous injection of air and steam. A typical underground reaction system involves linking of a series of such a unit reactor system.

There are two principal methods for underground steam gasification which have been tried successfully: shaft methods and shaftless methods (and a combination of two). Selection of a specific method depends on the parameters such as natural permeability of coal seam; the geochemistry of coal deposit; the seam thickness, depth, width, and inclination; closeness to the metropolitan areas; and the amount of mining desired. Shaft methods involve driving of shafts and drilling of other large diameter openings which require the underground labor and shaftless methods use boreholes for gaining access to the coal seam which do not require any underground labor.

The shaft method can be further divided into three subdivisions: (1) chamber or warehouse method in which underground galleries are prepared and the coal panels are isolated with brick wall, (2) borehole producer method in which parallel underground galleries are created about 500 ft apart within the coal bed, and (3) stream method in which inclined galleries following the dip of the coal seam of steeply pitched coal beds are constructed parallel to each other.

The shaftless method carries out gasification through a series of boreholes drilled from the surface to the coal seam. The coal beds are made more permeable between the inlet and outlet boreholes by a chosen linking method, ignite the coal seam, and gasify it by passing air and steam from the inlet to the outlet borehole. In percolation or filtration method, multiple boreholes, at a distance that depends on the seam permeability, are used to gasify the underground coal.

The potential problems in all of these methods include (1) high and constant quality of product gas; (2) high-percentage recovery of coal energy; (3) control of groundwater contamination; (4) combustion control; (5) roof structure control; (6) product gas leakage control; (7) proper control of permeability, linking, and fracturing; and (8) proper monitoring of underground processes. An ideal underground steam gasification system must be the following: (1) it is operable on large scale; (2) no large deposit of coal remains ungassified; (3) the process is controllable and the quantity

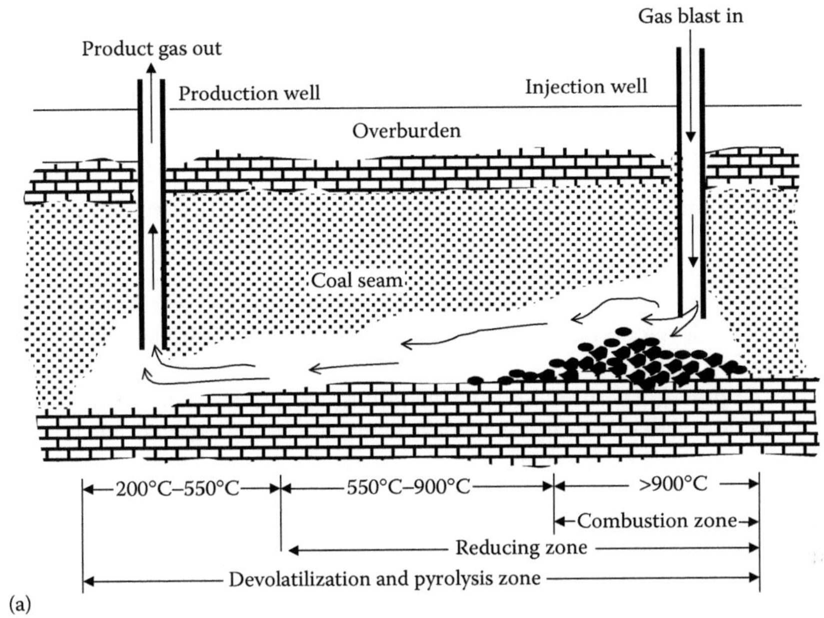

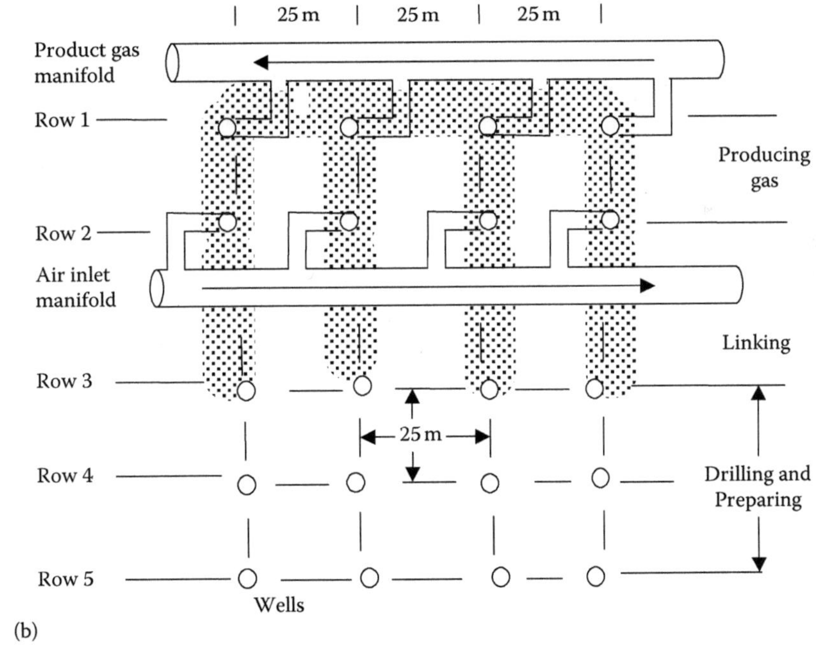

FIGURE 4.5 (a) Schematic of *in situ* underground gasification process. (b) Plane view of linked vertical well underground gasification plant operated near Moscow. (Adapted from Lee, S., Speight, J.G., and Loyalka, S.K., *Handbook of Alternative Fuel Technologies*. Taylor & Francis, Boca Raton, FL, 2007; Lee, S., "Gasification of coal," in Lee, S., Speight, J.G., and Loyalka, S.K., eds., *Handbook of Alternative Fuel Technologies*. Taylor & Francis, Boca Raton, FL, 26–78, 2007.)

and quality of product gases are constant and uniform; (4) it is mechanically stable and removed from any leakages to the groundwater; and (5) the process requires a minimal or no underground work.

4.7.6 Other Novel Processes

Sato and White [219] showed that using a physical mixture of powdered Texas lignite and platinized titania, in the presence of water vapor and ultraviolet (UV) light, a catalytic reaction to produce H_2 and CO_2 at 23°C can be achieved. Quantum yields were very low, but improvements were thought to be possible. Belghit and El Issami [220] developed a theoretical model of a moving bed chemical reactor for gasifying coal with steam. The heat was supplied by a high-temperature nuclear reactor. Cypres [221] discussed the metallurgical process for hydrogen production from coal and other carbonaceous materials, including coal gasification in a molten iron bath. An argument was made to place such a gasifier in the vicinity of steel manufacturing plant.

A steam–iron process is one of the oldest commercial methods for the production of hydrogen from syngas [222–230]. Various types of oxides of iron were examined. Neither chemical composition nor porosity of the ores was found to govern the efficiency. Potassium salts enhanced the activity of both natural and synthetic oxides. A number of recent studies have examined the classical steam–iron (sponge iron) process for upgrading synthesis gas (mainly CO and H_2) to pure hydrogen for use in FCs and other energy devices. Friedrich et al. [226] looked at this purification of nitrogen containing "reduction" gas from biomass gasifier using wood and wood wastes. The process involved two steps: (1) cleaning of gas from solid biomass, coal, or methane, and (2) energy storage in sponge iron. This study investigated woody biomass and commercially available sponge iron. The reactions are as follows:

$$Fe_3O_4 + 4CO \rightarrow 3Fe + 4CO_2 \quad \text{(coal, biomass, or natural gas)} \tag{4.76}$$

$$3Fe + 4H_2O \rightarrow Fe_3O_4 + 4H_2 \tag{4.77}$$

This process was stated to have little risk. Jannach et al. [230] extended the sponge iron process to FeO, as well as Fe as the oxidant. The sponge iron reaction was further studied by Hacker et al. [228,229] and Jannach et al. [230] in TGA (thermogravimetric analysis) and tube furnace devices. Other types of reactors were also examined by Fankhauser et al. [225] and Hacker et al. [227–229]. Biollaz et al. [223] explored the iron redox process to produce clean hydrogen from biomass. In the first step, iron oxide in the form of Fe_3O_4 reacted with the reducing components of wood gas to produce FeO, CO_2, and H_2. The kinetics of the second step, $3FeO + H_2O \rightarrow H_2 + Fe_3O_4$, could be improved by adding other transitional metal oxides. The reduction of iron oxide with biosyngas to sponge iron and later oxidation of sponge iron with steam offers the potential of shifting and purifying biosyngas, and storing and transporting its energy. Bijetima and Tarman [222] described the steam–iron process for hydrogen production operated in a large-scale pilot facility. Economic advantages of the process were also presented.

Another novel process is steam combustion to recover oil from reservoir. This enhanced oil recovery method is briefly examined in Chapter 2.

REFERENCES

1. Lee, S., Speight, J.G., and Loyalka, S.K., *Handbook of Alternative Fuel Technologies.* Taylor & Francis, Boca Raton, FL (2007).
2. Lee, S. and Shah, Y.T., *Biofuels and Bioenergy: Processes and Technologies.* Taylor & Francis, Boca Raton, FL (2012).
3. Lee, S., "Gasification of coal," in Lee, S., Speight, J.G., and Loyalka, S.K. (eds.), *Handbook of Alternative Fuel Technologies.* Taylor & Francis, Boca Raton, FL, 26–78 (2007).
4. "Coal gasification," Wikipedia, the free encyclopedia, 1–7 (2012).
5. Ratafia-Brown, J., Haslbeck, J., Skone, T., and Rutkowski, M., "Assessment of technologies for co-converting coal and biomass to a clean syngas—Task 2 report (RDS)," DOE/NETL-403.01.08 Activity 2 Report (May 10, 2007).
6. Clayton, S.J., Stiegel, G.J., and Wimer, J.G., *Gasification Technologies: Gasification Markets and Technologies—Present and Future: An Industry Perspective.* US Department of Energy, Washington, DC (2002).
7. Lang, R., "Exxon 1997 Coal liquefaction residue steam gasification," US Patent No. 4060478, Exxon Research and Engineering Company, Linden, NJ (1997).
8. Kamka, F., Jochmann, A., and Picard, L. (eds.), "Development status of BGL gasification," International Freiberg Conference on IGCC & XtL Technologies, June 16–18, Freiberg, Germany (2005).
9. Higman, C. and van der Burgt, M., *Gasification*, 2nd edn. Gulf Professional Publishing, Oxford (2008).
10. Basu, P., *Biomass Gasification and Pyrolysis: Practical Design and Theory.* Academic Press, Burlington, MA (2010).
11. Linares-Solano, A., Mahajan, O.P., and Walker, P.L, "Reactivity of heat-treated coals in steam," *Fuel*, 58, 327–332 (1979).
12. Demirbas, A., "Hydrogen production from carbonaceous solid wastes by steam reforming," *Energy Sources, Part A*, 30, 924–931 (2008).
13. Navarro, R., Pena, M., and Fierro, J., "Hydrogen production reactions from carbon feedstocks: Fossil fuels and biomass," *Chemical Reviews*, 107, 3952–3991 (2007).
14. Mühlen, H.-J., van Heek, K.H., and Jüntgen, H., "Kinetic studies of steam gasification of char in the presence of H_2, CO_2 and CO," *Fuel*, 64, 944–949 (1985).
15. Hüttinger, K.J. and Merdes, W.F., "The carbon-steam reaction at elevated pressure: Formations of product gases and hydrogen inhibitions," *Carbon*, 30 (6), 883–894 (1992).
16. Weeda, M., Abcouwer, H.H., Kapteijn, F., and Moulijn, J.A., "Steam gasification kinetics and burn-off behaviour for a bituminous coal derived char in the presence of H_2," *Fuel Processing Technology*, 36, 235–242 (1993).
17. Moilanen, A., Saviharju, K., and Harju, T., "Steam gasification reactivities of various fuel chars," in Bridgwater, A.V. (eds.), *Advances in Thermochemical Biomass Conversion.* Blackie Academic & Professional, Glasgow, 131–141 (1993).
18. Ratnaswamy, C. and Wagner, J.P., "Water gas shift catalysis," *Catalysis Reviews*, 51, 325–440 (2009).
19. Wang, D., Yuan, W., and Ji, W., "Effective syngas cleanup and reforming using Ni/y-Al_2O_3 catalyst," *International Journal of Agricultural and Biological Engineering*, 3 (2), 1–7 (2010).
20. Lee, W.S., Kim, T., and Woo, S., "High throughput screening for the promoters of alumina supported Ni catalysts in autothermal reforming of methane," *Top Catalysis*, 53, 123–128 (2010).

21. Watanabe, M., Yamashita, H., Chen, X., Yamanaka, J., Katobuki, M., Suzuki, H., and Uchida, H., "Nano-sized Ni particles on hollow alumina ball: Catalysts for hydrogen production," *Applied Catalysis B: Environmental*, 71, 237–245 (2007).
22. Bradford, M.C.J. and Vannice, M.A., "Catalytic reforming of methane with carbon dioxide over nickel catalysts. II. Reaction kinetics," *Applied Catalysis A*, 142 (1), 97–122 (1996).
23. Bradford, M.C.J. and Vannice, M.A., "CO_2 reforming of CH_4," *Catalysis Reviews—Science and Engineering*, 41 (1), 1–42 (1999).
24. Kechagiopoulos, P., Voutetakis, S., Lemonidou, A., and Vassalos, I., "Hydrogen production via steam reforming of the aqueous phase of bio-oil in a fixed bed reactor," *Energy & Fuels*, 20, 2155–2163 (2006).
25. Lucas, C., Szewczyk, D., Blasiak, W., and Mochida, S., "High temperature air and steam gasification of densified biofuels," *Biomass & Bioenergy*, 27, 563–575 (2004).
26. Ye, T., Yuan, L., Chen, Y., Kan, T., Tu, J., Zhu, X., Torimoto, Y., Yamamoto, M., and Li, Q., "High efficient production of hydrogen from bio-oil using low-temperature electrochemical catalytic reforming approach over NiCuZn-Al2O3 catalyst," *Catalysis Letters*, 127, 323–333 (2009).
27. Kumabe, K., Hanaoka, T., Fujimoto, S., Minowa, T., and Sakanishi, K., "Co-gasification of woody biomass and coal with air and steam," *Fuel*, 86, 684–689 (2007).
28. Weiland, N., Means, N., and Morreale, B., "Kinetics of coal/biomass co-gasification," NETL 2010 Workshop on Multiphase Flow Science, May 4–6, Pittsburgh, PA (2010).
29. Sami, M., Annamalai, K., and Wooldridge, M., "Co-firing of coal and biomass fuel blends," *Progress in Energy and Combustion Science*, 27, 171–214 (2001).
30. Seo, M., Goo, J., Kim, S., Lee, S., and Choi, Y., "Gasification characteristics of coal/biomass blend in a dual circulating fluidized bed reactor," *Energy & Fuels*, 24 (5), 3108–3118 (2010).
31. Yamada, T., Akano, M., Hashimoto, H., Suzuki, T., Maruyama, T., Wang, Q., and Kamide, M., "Steam gasification of coal biomass briquettes," *Nippon Enerugi Gakkai Sekitan Kagaku Kaigi Happyo Ronbunshu*, 39, 185–186 (2002).
32. Ji, K., Song, B., Kim, Y., Kim, B., Yang, W., Choi, Y., and Kim, S., "Steam gasification of low rank fuel-biomass, coal and sludge mixture in a small scale fluidized bed," *Proceedings of the European Combustion Meeting*, April 14–17, Vienna, Austria (2009).
33. Elshout, R., "Hydrogen production by steam reforming," *Chemical Engineering*, 117, 34–38 (2010).
34. Meijer, R., Kapteijn, F., and Moulijn, J.A., "Kinetics of the alkali-carbonate catalysed gasification of carbon: H_2O gasification," *Fuel*, 73 (5), 723–730 (1994).
35. Sørensen, L.H., Henriksen, U., Risnes, H., Poulsen, K.T., Hansen, L.K., Olsen, A., and Rathman, O., "Straw—H_2O gasification kinetics," Determination and discussion, Nordic Seminar on Thermochemical Conversion of Solid Fuels, December 3, Chalmers University of Technology, Sweden (1997).
36. Wei, J. and Iglesia, E., "Isotopic and kinetic assessment of the mechanism of reactions of CH_4 with CO_2 or H_2O to form synthesis gas and carbon on nickel catalysts," *Journal of Catalysis*, 224 (2), 370–383 (2004).
37. Rostrup-Nielsen, J.R., Hansen, J., and Bak, H. "CO_2-reforming of methane over transition metals," *Journal of Catalysis*, 144 (1), 38–49 (1993).
38. Li, J. and van Heiningen, A.R.P., "Kinetics of gasification of black liquor char by steam," *Industrial & Engineering Chemistry Research*, 30 (7), 1594–1601 (1991).
39. Jackson, S.D., Thomson, S.J., and Webb, G., "Carbonaceous deposition associated with catalytic steam reforming of hydrocarbons over nickel alumina catalysts," *Journal of Catalysis*, 70, 249–263 (1981).
40. Van Heek K., "General aspects and engineering principles for technical application of coal gasification," in Figuieiredo, J.L. and Moulijn, J.A. (eds.), *Carbon and Coal Gasification: Science and Technology*. Springer, Alvor, Portugal, 383–402 (1986).

41. Dalai, A.K., Sasaoka, E., Hikita, H., and Ferdous, D., "Catalytic gasification of sawdust derived from various biomass," *Energy & Fuels*, 17 (6), 1456–1463 (2003).
42. Hu, G., Xu, S., Li, S., Xiao, C., and Liu, S., "Steam gasification of apricot stones with olivine and dolomite as downstream catalysts," *Fuel Processing Technology*, 87 (5), 375–382 (2006).
43. Sutton, D., Kelleher, B., and Ross, J., "Review of literature on catalysts for biomass gasification," *Fuel Processing Technology*, 73, 155–173 (2001).
44. Aznar, M.P., Delgado, J., Corella, J., and Aragues, J.L., "Fuel and useful gas by steam gasification of biomass in fluidized bed with downstream methane and steam reforming: New results," in Grassi, G., Collina, A., and Zibetta, H. (eds.), Biomass for Industry, Energy and Environment: 6th EC Conference, Elsevier Applied Science, New York, 707–713 (1992).
45. Aznar, M.P., Corella, J., Delgado, J., and Lahoz, J., "Improved steam gasification of lignocellulosic residues in a fluidized bed with commercial steam reforming catalysts," *Industrial & Engineering Chemistry Research*, 32 (1), 1–10 (1993).
46. Aznar, M.P., Caballero, M.A., Gil, J., Marte, J.A., and Corella, J., "Commercial steam reforming catalysts to improve biomass gasification with steam/oxygen mixtures. 2. Catalytic tar removal," *Industrial & Engineering Chemistry Research*, 37 (97), 2668–2680 (1998).
47. Xu, G., Murakami, T., Suda, T., Kusama, S., and Fujimori, T., "Distinctive effects of CaO additive on atmospheric gasification of biomass at different temperatures," *Industrial & Engineering Chemistry Research*, 44 (15), 5864–5868 (2005).
48. Lee, W.J., "Catalytic activity of alkali and transition metal salt mixtures for steam-char gasification," *Fuel*, 74 (9), 1387–1393 (1995).
49. Lee W., Nam, S., Kim, S., Lee, K., and Choi, C., "The effect of Na_2CO_3 on the catalytic gasification of rice straw over nickel catalysts supported on Kieselguhr Seung," *Korean Journal of Chemical Engineering*, 17 (2), 174–178 (2000).
50. Simell, P., Stahlberg, P., Solantausta, Y., Hepola, J., and Kurkela, E., "Gasification gas cleaning with nickel monolith catalysts," in Bridgewater, A. and Boocock, D.G. (eds.), *Development of Thermochemical Biomass Conversion*. Blackie Academic & Professional, Glasgow, 1103–1116 (1997).
51. Simell, P., Leppalahti, J., and Bredenberg, J., "Catalytic purification of tarry fuel gas with carbonate rocks and ferrous materials," *Fuel*, 71, 211–218 (1992).
52. Simell, P., Leppalahti, J., and Kurkela, E., "Tar decomposition activity of carbonate rocks under high CO_2 pressure," *Fuel*, 74 (6), 938–945 (1995).
53. Olivares, A., Aznar, M.P., Caballero, M.A., Gill, J., Franes, E., and Corella, J., "Biomass gasification: Produced gas upgrading by in-bed use of dolomite," *Industrial & Engineering Chemistry Research*, 36, 5220–5226 (1997).
54. Rapagna, S., Jand, N., Kiennemann, A., and Foscolo, P., "Steam gasification of biomass in a fluidized bed of olivine particles," *Biomass & Bioenergy*, 19, 187–197 (2000).
55. Ciferno, J. and Marano, J., "Benchmarking biomass gasification technologies for fuels, chemicals and hydrogen production," DOE Report by National Energy Technology Laboratory, Washington, DC (2002).
56. Kumar, A., Jones, D., and Hanna, M., "Thermochemical biomass gasification: A review of the current status of the technology," *Energies*, 2, 556–581 (2009).
57. Shah, Y.T. and Gardner, T., "Dry reforming of hydrocarbons," *Catalysis Reviews Science and Engineering* (in press).
58. York, P.E., Xiao, T.C., Green, M.L.H., and Claridge, J.B., "Methane oxyforming for synthesis gas production," *Catalysis Reviews*, 49 (4), 511–560 (1995).
59. Dibbern, H.C., Olesen, P., Rostrup-Nielsen, J.R., Tottrup, P.B., and Udengaard, N.R., "Make low H_2/CO syngas using sulfur passivated reforming," *Hydrocarbon Processing*, 65 (1), 71–74 (1986).

60. Osaki, T., Horiuchi, T., Suzuki, K., and Mori, T., "Suppression of carbon deposition in CO_2-reforming of methane on metal sulfide catalysts," *Catalysis Letters*, 35 (1/2), 39–43 (1995).
61. Rozovskii, Y. and Lin, G., *Theoretical Bases of Methanol Synthesis*. Khimia, Moscow, Russia, 190.
    Also as "Fundamentals of methanol synthesis and decomposition," *Topics in Catalysis,* 22 (3/4), 137 (2003).
62. Gao, J., Hou, Z.Y., Shen, K., Lou, H., Fei, J.H., and Zheng, X.M. "Autothermal reforming and partial oxidation of methane in fluidized reactor over highly dispersed Ni catalyst prepared from Ni complex," *Chinese Journal of Chemistry*, 24 (6), 721–723 (2006).
63. Halmann, M. and Steinfeld, A., "Thermoneutral tri-reforming of flue gases from coal and gas fired power stations," *Catalysis Today*, 115, 170–178 (2006).
64. Kang, J.S., Kim, D., Lee, S., Hong, S., and Moon, D., "Nickel based tri-reforming catalyst for the production of synthesis gas," *Applied Catalysis A: General*, 332, 153–158 (2007).
65. Puolakka, J., "CO_2 reforming," Thesis for the Degree of Licentiate of Science and Technology, Helsinki University of Technology, Finland (2007).
66. Corella, J., Toledo, J., and Molina, G., "Steam gasification of coal at low-medium (600-800°C) temperature with simultaneous CO2 capture in fluidized bed at atmospheric pressure: The effect of inorganic species. 1. Literature review and comments," *Industrial & Engineering Chemistry Research*, 45, 6137–6146 (2006).
67. Sharma, D.K., "Enhancing the steam gasification reactivity of coal by boosting the factors affecting the gasification reactions in the stepwise coal conversion," *Energy Sources, Part A*, 32, 1727–1736 (2010).
68. Sharma, D.K., "Modelling of steam gasification reactions for reactor design," *Energy Sources, Part A*, 33, 57–71 (2011).
69. Inayat, A., Ahmad, M., Yusup, S., and Abdul Mutalib, M., "Biomass steam gasification with in-situ CO_2 capture for enriched hydrogen gas production: A reaction kinetics modeling approach," *Energies*, 3, 1472–1484 (2010).
70. Demirbas, A., "Hydrogen from mosses and algae via pyrolysis and steam gasification," *Energy Sources, Part A*, 32, 172–179 (2010).
71. Li, J., Liu, J., Liao, S., Zhou, X., and Yan, R., "Syn-gas production from catalytic steam gasification of municipal solid wastes in a combined fixed bed reactor," *Proceedings of the International Conference on Intelligent System Design and Engineering Application*, November 29–December 1, IEE Computer Society, 530–534 (2010).
72. Demirbas, M., "Hydrogen from various biomass species via pyrolysis and steam gasification processes," *Energy Sources, Part A*, 28, 245–252 (2006).
73. Demirbas, A., "Hydrogen rich gases from biomass via pyrolysis and air-steam gasification," *Energy Sources, Part A*, 31, 1728–1736 (2009).
74. Salaices, E., de Lasa, H., and Serrano, B., "Steam gasification of a cellulose surrogate over a fluidizable Ni/alpha-alumina catalyst: A kinetic model," *AIChE Journal* (2011), doi:10.1002/aic.12696.
75. Hofbauer, H., Stoiber, H., and Veronik, G., "Gasification of organic material in a novel fluidization bed system," *Proceedings of the 1st SCEJ Symposium on Fluidization*, December 13–14, Kagaku Kogaku Kyokai, Tokyo, Japan (1995).
76. Hofbauer, H., Veronik, G., Fleck, T., and Rauch, R., "The FICFB gasification process," in Bridgwater, A.V. and Boocock, D. (eds.), *Developments in Thermochemical Biomass Conversion*, Vol. 2. Blackie Academic & Professional, Glasgow, 1016–1025 (1997).
77. Hofbauer, H., Rauch, R., Löffler, G., Kaiser, S., Fercher, E., and Tremmel, H., "Six years experience with the FICFB-gasification process," in Palz, W. et al. (eds.), *Proceedings of the 12th European Biomass Conference*, ETA-Florence, Florence, Italy, Elsevier, 982–985 (2002).

78. Hofbauer, H. and Rauch, R., "Stoichiometric water consumption of steam gasification by the FICFB-gasification process," Progress in Thermochemical Biomass Conversion, September, Innsbruck, Austria (2000), http://www.ficfb.at.
79. Hofbauer, H., Rauch, R., Bosch, K., Koch, R., and Aichernig, C., "Biomass CHP plant Guessing—A success story," in Bridgwater, A.V. (ed.), *Pyrolysis and Gasification of Biomass and Waste*, CPL Press, Newbury, 527–536 (2003).
80. Herguido, J., Corella, J., and Gonzalez-Saiz, J., "Steam gasification of lignocellulosic residues in a fluidized bed at a small pilot scale. Effect of the type of feedstock," *Industrial & Engineering Chemistry Research*, 31, 1274–1282 (1992).
81. Pfeifer, C., Proll, T., Puchner, B., and Hofbauer, H., "H_2-rich syngas from renewable sources by dual fluidized bed steam gasification of solid biomass," The 12th International Conference on Fluidization New Horizons in Fluidization Engineering, The Berkeley Electronic Press, Berkeley, CA, 889–895 (2011).
82. Gopalakrishnan, P., "Modelling and experimental validation of biomass-steam gasification in bubbling fluidized bed reactor," PhD thesis, University of Canterbury, Canterbury, New Zealand (2007).
83. Matsuoka, K., Kuramoto, K., Murakami, T., and Suzuki, Y., "Steam gasification of woody biomass in a circulating dual bubbling fluidized bed system," *Energy & Fuels*, 22 (3), 1980–1985 (2008).
84. Donaj, P., Yang, W., and Blasiak, W., "High-temperature steam gasification of straw pellets," Report of KTH-ITM Division of Energy and Furnace Technology, Royal Institute of Technology, Stockholm, Sweden, 1–25 (2011).
85. Gupta, A.K. and Cichonski, W., "Ultra high temperature steam gasification of biomass and solid wastes," *Environmental Engineering Science*, 24 (8), 1179–1189 (2007).
86. Butterman, H. and Castaldi, M., "CO_2 enhanced steam gasification of biomass fuels," *Proceedings of the 16th Annual North American Waste-to-Energy Conference*, May 19–21, Philadelphia, PA (2008).
87. Aznar, M., Caballero, M., Molina, G., and Toledo, J., "Hydrogen production by biomass gasification with steam-O_2 mixtures followed by a catalytic steam reformer and a CO shift system," *Energy & Fuels*, 20, 1305–1309 (2006).
88. Seo, J., Youn, M., Nam, I., Hwang, S., Chung, J., and Song, I., "Hydrogen production by steam reforming of liquefied natural gas over mesoporous Ni-Al_2O_3 catalysts prepared by a co-precipitation method: Effect of Ni/Al atomic ratio," *Catalysis Letters*, 130, 410–416 (2009).
89. Seo, J., Youn, M., Park, D., Jung, J., and Song, I., "Hydrogen production by steam reforming of liquefied natural gas over mesoporous Ni-Al_2O_3 composite catalysts prepared by a single step non-ionic surfactant—Templating method," *Catalysis Letters*, 130, 395–401 (2009).
90. Demirbas, A., Karshoglu, S., and Ayas, A., "Hydrogen resources conversion of black liquor to hydrogen rich gaseous products," *Fuel Science and Technology International* 14 (3), 451–463 (1996).
91. Demirbas, A. and Caglar, A., "Catalytic steam reforming of biomass and heavy oil residues to hydrogen," *Energy, Education, Science & Technology*, 1, 45–52 (1998).
92. Demirbas, A., "Yields of hydrogen-rich gaseous products via pyrolysis from selected biomass samples," *Fuel*, 80, 1885–1891 (2002).
93. Demirbas, A., "Sustainable cofiring of biomass with coal," *Energy Conversion and Management*, 44 (9), 1465–1479 (2003).
94. Demirbas, A., "Co-firing coal and municipal solid waste," *Energy Sources, Part A*, 30, 361–369 (2008).
95. Chmielniak, T. and Sciazko, M., "Co-gasification of biomass and coal for methanol synthesis," *Applied Energy*, 74, 393–403 (2003).
96. Pan, Y.G., Roca, V., Manya, J., and Puigianer, L., "Fluidized bed co gasification of residual biomass/poor coal blends for fuel gas production," *Fuel*, 79, 1317–1326 (2000).

97. Satrio, J., Shanks, B., and Wheelock, T., "A combined catalyst and sorbent for enhancing hydrogen production from coal or biomass," *Energy & Fuels*, 21, 322–326 (2007).
98. Indrawati, V., Manaf, A., and Purwadi, G., "Partial replacement of non renewable fossil fuels energy by the use of waste materials as alternative fuels," in Handoko, L. and Siregar, M. (eds.), *Proceedings of the International Workshop on Advanced Materials for New and Renewable Energy*. American Institute of Physics, College Park, MD, 179–184 (2009).
99. Baker, E.G., Brown, M.D., Elliott, D.C., and Mudge, L.K., *Characterization and Treatment of Tars from Biomass Gasifiers*. Pacific Northwest National Laboratory, Richland, WA, 1–11 (1988).
100. Milne, T.A., Evans, R.J., and Abatzoglou, N., *Biomass Gasification "Tars": Their Nature, Formation, and Conversion*. National Renewable Energy Laboratory, Golden, CO (1998).
101. Gopal, G., Adhikari, S., Thangalazhy-Gopakumar, S., Christian, B.C., Bhavnani, S., and Taylor, S., "Tar analysis in syngas derived from pelletized biomass in a commercial stratified downdraft gasifier," *Bioresources*, 6 (4), 4652–4661 (2011).
102. Devi, L., Ptasinski, K.J., and Janssen, F.J.J.G., "A review of the primary measures for tar elimination in biomass gasification processes," *Biomass & Bioenergy*, 24 (2), 125–140 (2003).
103. El-Rub, Z., Bramer, E., and Brem, G., "Review of catalysts for tar elimination in biomass gasification processes," *Industrial & Engineering Chemistry*, 43, 6911–6919 (2004).
104. Huang, H. and Ramaswamy, S., "Thermodynamic analysis of black liquor steam gasification," *Bioresources*, 6 (3), 3210–3230 (2011).
105. Whitty, K.J., "Gasification of black liquor char with H_2O under pressurized conditions," Report No. 93-4, Department of Chemical Engineering, Combustion Chemistry Research Group, Åbo Akademi, Turku, Finland (1993).
106. Ferdous, D., Dalal, A., Bej, S., and Thring, R., "Production of H_2 and medium heating value gas via steam gasification of lignins in fixed bed reactor," *The Canadian Journal of Chemical Engineering*, 79 (6), 913–922 (2001).
107. Kumar, V., "Pyrolysis and gasification of lignin and effect of alkali addition," Ph.D. thesis, Georgia Institute of Technology, Atlanta, GA (Auguest 2009).
108. Kumar, V., Ilsa, K., Benerjee, S., and Frederick, W., "Characterization of lignin gasification and pyrolysis," *Proceedings of the Pulping and Environment Conference*, August 24–27, TAPPI Engineering, Portland, OR (2008).
109. Mas, V., Bergamini, M., Baronetti, G., Amadeo, N., and Laborde, M., "A kinetic study of ethanol steam reforming using a nickel based catalyst," *Top Catalysis*, 51, 39–48 (2008).
110. Biswas, P. and Kunzru, D., "Steam reforming of ethanol on Ni-CeO_2-ZrO_2 catalysts: Effect of doping with copper, cobalt and calcium," *Catalysis Letters*, 118 (1/2), 36–49 (2007).
111. Akdim, O., Cai, W., Fierro, V., Provendier, H., van Deen, A., Shen, W., and Mirodatos, C., "Oxidative steam reforming of ethanol over Ni-Cu/SiO_2, Rh/Al_2O_3 and Ir/CeO_2: Effect of metal and support on reaction mechanism," *Top Catalysis*, 51, 22–38 (2008).
112. Fajardo, H., Longo, E., Mezalira, D., Nurenberg, G., Almerindo, G., Collasiol, A., Probst, L., Garcia, I., and Carreno, N., "Influence of support on catalytic behavior of nickel catalysts in the steam reforming of ethanol for hydrogen production," *Environmental Chemistry Letters*, 8, 79–85 (2010).
113. Sekine, Y., Kazama, A., Izutsu, Y., Matsukata, M., and Kikuchi, E., "Steam reforming of ethanol over cobalt catalyst modified with small amount of iron," *Catalysis Letters*, 132, 329–334 (2009).
114. Dong, T., Wang, Z., Yuam, L., Torimoto, Y., Sadakata, M., and Li, Q., "Hydrogen production by steam reforming of ethanol on potassium doped 12CaO-7Al_2O_3 catalyst," *Catalysis Letters*, 119, 29–39 (2007).
115. Song, H., Tan, B., and Ozkan, U., "Novel synthesis techniques for preparation of Co/CeO_2 as ethanol steam reforming catalysts," *Catalysis Letters*, 132, 422–429 (2009).

116. He, L., Barntsen, H., Ochoa-Fernandez, E., Walmsley, J., Blekkan, E., and Chen, D., "Co-Ni catalysts derived from hydrotalcite-like materials for hydrogen production by ethanol steam reforming," *Top Catalysis*, 52, 206–217 (2009).
117. Henpraserttae, S. and Toochinda, P., "Effects of preparation of Cu/Zn over Al_2O_3 catalysts for hydrogen production from methanol reforming," *Suranaree Journal of Science and Technology*, 16 (2), 103–112 (2009).
118. Lee, S., *Methanol Synthesis Technology*. CRC Press, New York (1989).
119. Purnama, H., Ressler, T., Jentoft, R., Soerijanto, H., Schlogl, R., and Schomacker, R., "Steam reforming of methanol on Cu/ZrO2 catalysts," *Applied Catalysis A*, 259, 83 (2004).
120. Lindström, B., "Development of methanol reformer for fuel cell vehicles," Thesis, Department of Chemical Engineering and Technology, KTH, Stockholm, Sweden (2003).
121. Lindström, B. and Pettersson, L.J., "Deactivation of copper-basedcatalysts for fuel cell applications," *Catalysis Letters*, 74, 27–30 (2001).
122. Lindström, B. and Pettersson, L.J., "Steam reforming of methanol for fuel cell applications," *Proceedings of the 9th Nordic Symposium on Catalysis*, June 4–6, Lidingö, Sweden (2000).
123. Lindström, B. and Pettersson, L.J., "A study of ethanol and methanol as a fuel for onboard hydrogen generation by steam reforming on copper-based catalysts," *Proceedings of the 13th International Symposium on Alcohol Fuels*, July 3–6, Stockholm, Sweden (2000).
124. Lindström, B. and Pettersson, L.J., "Steam reforming of methanol for automotive applications," *Proceedings of the 2000 Fuel Cell Seminar*, October, Portland, OR (2000).
125. Lindström, B., Agrell, J., and Pettersson, L.J., "Combinatorial reforming of methanol for hydrogen generation over monolithic catalysts," *Proceedings of the 17th North American Catalysis Society Meeting*, June 3–8, Toronto, ON (2001).
126. Lindström, B. and Pettersson, L.J., "Catalytic steam reforming of methanol for automotive fuel cell applications," *Proceedings of the 5th European Congress on Catalysis*, September 2–7, Limerick, Ireland (2001).
127. Lindström, B. and Pettersson, L.J., "Steam reforming of methanol over copper-based monoliths: The effects of zirconia doping," *Proceedings of the 7th Grove Fuel Cell Symposium*, September 11–13, London (2001).
128. Kolb, G., Keller, S., Tiemann, D., Schelhaas, K., Schürer, J., and Wiborg, O., "Design and operation of a compact microchannel 5 kWel, net methanol steam reformer with novel Pt/In_2O_3 catalyst for fuel cell applications," *Chemical Engineering Journal*, 207–208, 388–402 (2012).
129. Lindström, B., Pettersson, L.J., and Menon, P.G., "Influence of the operating conditions on the performance of a methanol reformer," *Proceedings of the 10th Nordic Symposium on Catalysis*, June 2–4, Helsingør, Denmark (2002).
130. Zhang, L., Pan, L.-W., Ni, C.-J., Zhao, S.-S., Wang, S.-D., Hu, Y.-K., Wang, A.-J., and Jiang, K., "Optimization of methanol steam reforming for hydrogen production," *Journal of Fuel Chemistry and Technology*, 41 (1), 116–122 (2013).
131. Pan, L. and Wang, S., "Modeling of a compact plate-fin reformer for methanol steam reforming in fuel cell systems," *Chemical Engineering Journal*, 108, 51–58 (2005).
132. Pan, L. and Wang, S., "Methanol steam reforming in a compact plate-fin reformer for fuel cell systems," *International Journal of Hydrogen Energy*, 30, 973–979 (2005).
133. Pan, L., Ni, C., Zhang, X., Yuan, Z., Zhang, C., and Wang, S., "Study on a compact methanol reformer for a miniature fuel cell," *International Journal of Hydrogen Energy*, 36, 319–325 (2011).
134. Li, Z., Gao, D., Chang, L., Liu, P., and Pistikopoulos, E., "Coal-derived methanol for hydrogen vehicles in china: Energy, environment and economic analysis for distributed reforming," *Chemical Engineering Research and Design*, 88, 73–80 (2010).

135. Olafadehan, O., Susu, A., and Jaiyeola, A., "Mechanistic kinetic models for n-heptane reforming on platinum/alumina catalysts," *Petroleum Science and Technology*, 26, 1459–1480 (2008).
136. Salaun, M., Capela, S., De Costa, S., Gagnepain, L., and De Costa, P., "Enhancement of 3-way CNG catalyst performance at high temperature due to the presence of water in the feed: On the role of steam reforming of methane and on the influence of ageing," *Top Catalysis*, 52, 1972–1976 (2009).
137. Futamura, S. and Kabashima, H., "Steam reforming of aliphatic hydrocarbons with nonthermal plasma," *IEEE Transactions of Industrial Applications*, 40 (6), 1476–1481 (2004).
138. Mota, N., Alvarez-Galvan, A.C., Villoria, J.A., Rosa, F., Fierro, J., and Navarro, R., "Reforming of diesel fuel for hydrogen production over catalysts derived from $LaCo_{1-x}M_xO_3$ (M = Ru, Fe)," *Top Catalysis*, 52, 1995–2000 (2009).
139. Ambroise, E., Courson, C., Keinnemann, A., Roger, A., Pajot, O., Samson, E., and Blanchard, G., "On-board hydrogen production through catalytic exhaust-gas reforming of isooctane: Efficiency of mixed oxide catalysts $Ce_2Zr_{1.5}Me_{0.5}O_8$ (Me = Co, Rh or Co–noble metal)," *Top Catalysis*, 52, 2101–2107 (2009).
140. Krauppila, R., "Concurrent production of syngas and steel from coal, biomass and iron ore," U.P. Steel, DE-PS02-07ER07-36, Topic 57a (November, 2007).
141. Kaila, R., Gutierrez, A., Korhonen, S., and Krause, A., "Autothermal reforming of n-dodecane, toluene and their mixture on mono- and bimetallic noble metal zirconia catalysts," *Catalysis Letters*, 15 (1/2), 70–78 (2007).
142. Kiryanov, D., Smolikov, M., Pashkov, V., Proskura, A., Zatolokina, E., Udras, I., and Belyi, A., "Modern state of catalytic reforming of petrol fractions. Experience of manufacture and industrial exploitation of reforming catalysts of the PR series," *Russian Journal of General Chemistry*, 77 (12), 2255–2264 (2007).
143. Encinar, J., Gonzalez, J., Martinez, G., Sanchez, N., and Sanguino, J., "Hydrogen production by means pyrolysis and steam gasification of glycerol," *International Conference on Renewable Energies and Power Quality*, April 13–15, Las Palmas de Gran Canaria, Spain (2010).
144. Maciel, C. and Ishikura, S., "Steam reforming of renewable feedstock's for the production of hydrogen," Universidad de Zaragoza, Zaragoza, Spain (2007).
145. Zhang, B., Tang, X., Li, Y., Xu, Y., and Shen, W., "Hydrogen production from steam reforming of ethanol and glycerol over ceria supported metal catalysts," *International Journal of Hydrogen Energy*, 32 (13), 2367–2373 (2007).
146. Sadanandam, G., Sreelatha, N., Sharma, P.M., Reddy, S., Srinivas, B., Venkateswarlu, K., Krishnudu, T., Subrahmanyam, M., and Kumar, V., "Steam reforming of glycerol for hydrogen production over Ni/Al_2O_3 catalyst," *ISRN Chemical Engineering*, 2012, 10 (2012), http://dx.doi.org/10.5402/2012/591587.
147. Hakim, L., Yaakob, Z., Ismail, M., Daud, W., and Sari, R., "Hydrogen production by steam reforming of glycerol over Ni/Ce/Cu hydroxyapatite-supported catalysts," *Chemical Papers*, 67 (7), 703–712 (2013).
148. Simonetti, D., Kunkes, E., and Dumesic, J., "Gas phase conversion of glycerol to synthesis gas over carbon supported platinum and platinum-rhenium catalysts," *Journal of Catalysis*, 247 (2), 298–306 (2007).
149. Reddy, R., Patel, S., Nair, S., and Suvikram, Y., "Preparation of hydrogen from glycerol via steam reforming process," *International Conference on Current Trends in Technology*, NUICONE, December 8–10, Nirma University, Ahmedabad, India (2011).
150. Wang, H., Wang, X., Li, M., Li, S., Wang, S., and Ma, X., "Thermodynamic analysis of hydrogen production from glycerol autothermal reforming," *International Journal of Hydrogen Energy*, 34, 5683–5690 (2009).

151. Adhikari, S., Fernando, S., Gwaltney, S., Filipito, S., Mark Bricka, R., Steele P., and Haryanto, A., "A thermodynamic analysis of hydrogen production by steam reforming of glycerol," *International Journal of Hydrogen Energy*, 32, 2875–2880 (2007).
152. Douette, A., Turn, S., Wang, W., and Keffer, V., "Experimental investigation of hydrogen production from glycerin reforming," *Energy & Fuels*, 21, 3499–3504 (2007).
153. Adhikari, S., Fernando, S., and Haryanto, A., "Hydrogen production from glycerol: An update," *Energy Conversion and Management*, 50, 2600–2604 (2009).
154. Slinn, M., Kendall, K., Mallon, C., and Andrews, J., "Steam reforming of biodiesel by-product to make renewable hydrogen," *Bioresource Technology*, 99, 5851–5858 (2008).
155. Chen, H., Ding, Y., Cong, N., Dou, B., Dupont, V., Ghadiri V., and Williams, P., "A comparative study on hydrogen production from steam-glycerol reforming: Thermodynamics and experimental," *Renewable Energy*, 36, 779–788 (2011).
156. Encinar, J., González, J., Rodriguez J., and Ramiro, M., "Catalysed and uncatalysed steam gasification of eucalyptus char: Influence of variables and kinetic study," *Fuel*, 80, 2025–2036 (2001).
157. Encinar, J., González, J.F., and González, J., "Steam gasification of *Cynara cardunculus* L.: Influence of variables," *Fuel Processing Technology*, 75, 27–43 (2002).
158. Timpe, R.C. and Hauserman, W.B., "The catalytic gasification of hybrid poplar and common cattail plant chars," in Klass, D.L. (ed.), *Energy from Biomass and Wastes XVI*, Orlando, FL. Institute of Gas Technology, Chicago, IL, 903–919 (1992).
159. Kojima, T., Assavadakorn, P., and Furusawa, T., "Measurement and evaluation of gasification kinetics of sawdust char with steam in an experimental fluidized bed," *Fuel Processing Technology*, 36, 201–207 (1993).
160. Capart, R. and Gélus, M., "A volumetric mathematical model for steam gasification of wood char at atmospheric pressure, Energy from Biomass 4," *Proceedings of the 3rd Contractors' Meeting*, May 25–27, Paestum, Italy (1988).
161. Hemati, M. and Laguerie, C., "Determination of the kinetics of the wood sawdust steam-gasification of charcoal in a thermobalance," *Entropie*, 142, 29–40 (1988).
162. Moilanen, A. and Saviharju, K., "Gasification reactivities of biomass fuels in pressurised conditions and product gas mixtures," in Bridgewater, A. and Boocock, D. (eds.), *Developments in Thermochemical Biomass Conversion*. Blackie Academic & Professional, Glasgow, 828–837 (1997).
163. Milne, T.A., Elam, C., and Evans, R., "Hydrogen from biomass-state of the art and research challenges," Report for the International Energy Agency, Agreement on the Production and Utilization of Hydrogen, Task 16, Hydrogen from carbon containing materials, IEA/H2/TR-02/001, 1–78 (2001).
164. Hansen, L.K., Rathmann, O., Olsen, A., and Poulsen, K., "Steam gasification of wheat straw, barley straw, willow and giganteus," Project No. ENS-1323/95-0010. Optics and Fluid Dynamics Department, Risø National Laboratory, Roskilde, Denmark (1997).
165. McKinley, K.R., Browne, S.H., Neill, D.R., Seki, A., and Takahashi, P.K., "Hydrogen fuel from renewable resources," *Energy Sources*, 12, 105–110 (1990).
166. Turn, S., Kinoshita, C., Zhang, Z., Ishimura, D., and Zhou, J., "An experimental investigation of hydrogen production from biomass gasification," *International Journal of Hydrogen Energy*, 23 (8), 641–648 (1998).
167. Zhou, J., Ishimura, D.M., and Kinoshita, C.M., "Effect of injecting steam on catalytic reforming of gasified biomass," *Proceedings of the 4th Biomass Conference of the Americas*, August 29–September 2, Oakland, CA (1999).
168. Rapagna, S., and Foscolo, P.U., "Hydrogen from biomass by steam gasification," *Proceedings of the 11th World Hydrogen Energy Conference*, June 23–28, Stuttgart, Germany, Vol. 1 (1996).

169. Rapagna, S. and Foscolo, P.U., "Cataly tic gasification of biomass to produce hydrogen rich gas," *International Journal of Hydrogen Energy*, 23 (7), 551–557 (1998).
170. Cox, J.L., Tonkovich, A.Y., Elliott, D.C., Baker, E.G., and Hoffman, E.J., "Hydrogen from biomass: A fresh approach," *Proceedings of the 2nd Biomass Conference of the Americas* (NREL/CP-200-8098; CONF-9508104) August, Portland, OR (1995).
171. Weber, S.L., Sealock, L.J., Mudge, L.K., Rboertus, R.J., and Mitchell, D.H., "Gasification of biomass in the presence of multiple catalysts for the direct production of specific products," AIAA Paper Symposium on Energy from Biomass and Wastes 4, January 21–25, Lake Buena, Vista, FL, 351–367 (1980).
172. De Ruyck, J., Delattin, F., and Bram, S., "Co-utilization of biomass and natural gas in combine cycles through primary steam reforming of natural gas," *Energy*, 32, 371–377 (2007).
173. Mond, L. and Langer, C., British Patent No. 12608 (1888).
174. Bosch, C. and Wild, W., Canadian Patent No. 153379 (1914).
175. Proll, T., Rauch, R., Aichernig, C., and Hofbauer, H., "Performance characteristics of an 8 MW (th) combined heat and power plant based on dual fluidized bed steam gasification of solid biomass," *Proceedings of the 12th International Conference on Fluidization—New Horizons in Fluidization Engineering*, Vienna, Austria (2007).
176. Gallucci, F., Annaland, M., and Kuipers, J., "Autothermal reforming of methane with integrated CO_2 capture in a novel fluidized bed membrane reactor. Part 1. Experimental demonstration," *Top Catalysis*, 51, 133–145 (2008); "Part 2. Comparison of reactor configuration," *Top Catalysis*, 51, 146–157 (2008).
177. Fleck, T., Hofbauer, H., Rauch, R., and Veronik, G., "The FICFB gasification process," *Proceedings of the IEA Bioenergy Meeting*, May, Banff, AL (1996).
178. Zschetzsche, A., Hofbauer, H., and Schmidt, A., "Biomass gasification in an internally circulating fluidized bed," *Proceedings of the 8th European Conference on Biomass for Agriculture and Industry*, October 3–5, Vienna, Austria, Pergamon Press, Oxford (1994).
179. Fercher, E., Hofbauer, H., Fleck, T., Rauch, R., and Veronik, G., "Two years experience with the FICFB—Gasification process," *Proceedings of the 10th European Conference and Technology Exhibition*, June, Würzburg, Germany (1998).
180. Karl, J., Schmitz, W., and Hein, D., "Allothermal gasification in fluidized bed gasifiers," *Proceedings of the 10th Engineering Foundation Conference on Fluidization*, May 20–25, Beijing, China (2001).
181. Paisley, M.A., Farris, M.C., Black, J.W., Irving, J.M., and Overend, R.P., "Preliminary operating results from the battelle/ferco gasification demonstration plant in Burlington, Vermont, USA," in Kyrisits, S. et al. (eds.), *Proceedings of the 1st World Conference on Biomass for Energy and Industry*, June 5–9, Sevilla, Spain, James & James, London (2000).
182. Pröll, T., Rauch, R., Aichernig, C., and Hofbauer, H., "Fluidized bed steam gasification of solid biomass: Analysis and optimization of plant operation using process simulation," ASME Paper FBC2005-78129, *Proceedings of the 18th International Conference on Fluidized Bed Combustion*, May 23–25, Toronto, ON (2005).
183. Gil, J., Aznar, M.P., Caballero, M.A., Francés, E., and Corella, J., "Biomass gasification in fluidized bed at pilot scale with steam-oxygen mixtures. Product distribution for very different operating conditions," *Energy & Fuels*, 11, 1109–1118 (1997).
184. Bilbao, R., García, L., Salvador, M.L., and Arauzo, J., "Steam gasification of biomass in a fluidized bed, effect of a Ni-Al catalyst," *Proceedings of the 10th European Conference and Technology Exhibition on Biomass for Energy and Industry*, June 8–11, Würzburg, Germany (1998).
185. Evans, R., Knight, R.A., Onischak, M., and Babu, S.P., "Process performance and environmental assessment of the Renugas process," in Klass, D.L. (ed.), *Energy from Biomass and Wastes X*, Institute of Gas Technology, Chicago, IL, Elsevier Applied Science Publishers, 677–696 (1987).

186. Groeneveld, M.J., "The co-current moving bed gasifier," Ph.D. thesis, Twente University of Technology, Enschede, the Netherlands (1980).
187. Phillips, J., "Different types of gasifiers and their integration with gas turbines," Report from EPRI, Palo Alto, CA (2003).
188. Olsson, D., "Comparison of reforming process between different types of biogas reforming reactors," Project Report, 2008 MVK 160, Heat and Mass Transport, Lund, Sweden (2008).
189. Watanuki, K., Nakajima, H., Hasegawa, N., Kaneko, H., and Tamaura, Y., "Methane-steam reforming by molten salt membrane reactor using concentrated solar thermal energy," *Proceedings of the 16th World Hydrogen Energy Conference*, June, Lyon, France (2006).
190. Yang, L., Liu, S., Yu, L., and Zhang, W., "Underground coal gasification field experiment in the high dipping coal seams," *Energy Sources, Part A*, 31, 854–862 (2009).
191. McKendry, P., "Energy production from biomass (part 3): Gasification technologies," *Bioresource Technology*, 83, 55–63 (2002).
192. Spiewak, I., Tyner, C., and Langnickel, U., "Applications of solar reforming technology," SANDIA Report No. SAND93-1959, UC-237, Sandia National Laboratories, Albuquerque, NM (1993).
193. Muller, W.D., "Solar reforming of methane utilizing solar heat," in Baker, M. (ed.), *Solar Thermal Utilization, German Studies on Technology and Applications*, Vol. 3. Springer, Berlin, Germany, 1–179 (1987).
194. Becker, M. and Bohmer, M. (eds.), "GAST, the gas cooled solar tower technology program," *Proceedings of the Final Presentation*, May 30–31, Lahnstein, FRG. Springer, Berlin, Germany (1989).
195. Piatkowski, N. and Steinfeld, A., "Solar driven coal gasification in a thermally irradiated packed-bed reactor," *Energy & Fuels*, 22, 2043–2052 (2008).
196. Perkins, C.M., Woodruff, B., Andrews, L., Lichty, P., Lancaster, B., Bringham, C., and Weimer, A., "Synthesis gas production by rapid solar thermal gasification of corn stover," Report of Midwest Research Institute Under Contract No. DE-AC36-99GO10337, Department of Energy, Washington, DC (2009).
197. Piatkowski, N., Wieckert, C., and Steinfeld, A., "Experimental investigation of a packed bed solar reactor for the steam gasification of carbonaceous feedstocks," *Fuel Processing Technology*, 90, 360–366 (2009).
198. Z'Graggen, A., "Solar gasification of carbonaceous materials—Reactor design, modeling and experimentation," Doctorate thesis, ETH No. 17741, ETH, Zurich, Switzerland (2008).
199. Meier, A. and Sattler, C., "Solar fuels from concentrated sunlight," SolarPACES, IEA SolarPACES implementing agreement, Platforma Solar de Almería, Almería, Spain (August 2009).
200. Petrasch, J. and Steinfeld, A., "Dynamics of a solar thermochemical reactor for steam reforming of methane," *Chemical Engineering Science*, 62, 4214–4228 (2007).
201. Moller, S., "Solar reforming of natural gas," Report by Deutsches Zentrum, DLR fur Luft-und Raumfahrt e.V., Mgheb-Europ project, June 14, Lyon, France (2006).
202. Z'Graggen, A., Haueter, P., Trommer, D., Romero, M., De Jesus, J., and Steinfeld, A., "Hydrogen production by steam-gasification of petroleum coke using concentrated solar power—II. Reactor design, testing, and modeling," *International Journal of Hydrogen Energy*, 31 (6), 797–811 (2006).
203. Gordillo, E. and Belghit, A., "A bubbling fluidized bed solar reactor model of biomass char high temperature steam-only gasification," *Fuel Processing Technology*, 92 (3), 314–321 (2010).
204. Gordillo, E. and Belghit, A., "A downdraft high temperature steam-only solar gasifier of biomass char: A modelling study," *Biomass & Bioenergy*, 35 (5), 2034–2043 (2011).

205. Hathaway, B.J., Davidson, J.H., and Kittelson, D.B., "Solar gasification of biomass: Kinetics of pyrolysis and steam gasification in molten salt," *Journal of Solar Energy Engineering*, 133 (2), 021011 (2011).
206. Lichty, P., Perkins, C., Woodruff, B., Bingham, C., and Weimer, A., "Rapid high temperature solar thermal biomass gasification in a prototype cavity reactor," *Journal of Solar Energy Engineering*, 132 (1), 011012 (2010).
207. Klein, H.H., Karni, J., and Rubin, R., "Dry methane reforming without a metal catalyst in a directly irradiated solar particle reactor," *Journal of Solar Energy Engineering*, 131 (2), 021001 (2009).
208. Flechsenhar, M. and Sasse, C., "Solar gasification of biomass using oil shale and coal as candidate materials," *Energy*, 20 (8), 803–810 (1995).
209. Weimer, A., Perkins, C., Mejic, D., Lichty, P., and inventors; WO Patent WO/2008/027,980, as signee. Rapid Solar-Thermal Conversion of Biomass to Syngas. United States Patent No. WO2008027980 (June 3, 2008).
210. Zedtwitz, P. and Steinfeld, A., "The solar thermal gasification of coal—Energy conversion efficiency and CO_2 mitigation potential," *Energy*, 28 (5), 441–456 (2003).
211. Sattler, C. and Raeder, C., "SOLREF—Solar steam reforming of methane rich gas for synthesis gas production," Final Activity Report, DLR, Cologne, Germany (2010).
212. Suarez-Gonzalez, M., Blanco-Marigorta, A., and Peria-Quintana, A., "Review on hydrogen production technologies from solar energy," International Conference on Renewable Energies and Power Quality, April 13–15, Los Palmas de Gran Canaria, Spain (2011).
213. Ogden, J.M., "Review of small stationary reformers for hydrogen production," Report for IEA, Agreement on the production and utilization of hydrogen, Task 16, Hydrogen from carbon containing materials, IEA/H2/TR-02/002 (2002).
214. Padban, N. and Becher, V., "Clean hydrogen rich synthesis gas," Literature and state of art review (Re: Methane Steam Reforming), Report No. CHRISGAS, WP11 D89 (October 2005).
215. Fernández, Y., Arenillas, A., Bermúdez, J., and Menéndez, J., "Comparative study of conventional and microwave-assisted pyrolysis, steam and dry reforming of glycerol for syngas production, using a carbonaceous catalyst," *Journal of Analytical and Applied Pyrolysis*, 88 (2), 155–159 (2010).
216. Fidalgo, B. and Menéndez, J., "Syngas production by CO_2 reforming of CH_4 under microwave heating—Challenges and opportunities," in Indarto, A. and Palgunadi, J. (eds.), *Syngas: Production, Applications and Environmental Impact*. Nova Science Publishers, Inc., Hauppauge, NY, 121–149 (2011).
217. Fidalgo, B., Domínguez, A., Pis, J., and Menéndez, J., "Microwave-assisted dry reforming of methane," *International Journal of Hydrogen Energy*, 33 (16), 4337–4344 (2008).
218. Yang, L., Liang, J., and Yu, L., "Clean coal technology—Study on the pilot project experiment of underground coal gasification," *Energy*, 14, 1445–1460 (2003).
219. Sato, S. and White, J.M., "Photocatalytic production of hydrogen from water and Texas lignite by use of a platinized titania catalyst," *Industrial & Engineering Chemistry Product Research and Development*, 19, 542–544 (1980).
220. Belghit, A. and El Issami, S., "Hydrogen production by steam gasification of coal in gas–solid moving bed using nuclear heat," *Energy Conversion and Management*, 42 (1), 81 (2001).
221. Cypres, R., "Modern carbochemical processes for hydrogen production from coal," *International Journal of Hydrogen Energy*, 12 (7), 451–460 (1987).
222. Bijetima, R. and Tarman, P.B., "Development states of the steam-iron process for hydrogen production," *Alternative Energy Sources*, 2, 3335–3347 (1981).
223. Biollaz, S., Sturzenegger, M., and Stucki, S., "Redox process for the production of clean hydrogen from biomass," Progress in Thermochemical Biomass Conversion Abstracts, September 17–22, Tyrol, Austria (2000).

224. Das, A., Chatterjae, D.S., and Mukherjee, P.N., "Steam-iron process for production of hydrogen," *Indian Journal of Technology*, 15, 339–341 (1977).
225. Fankhauser, R., Hacker, V., Spreitz, B., Kleindienst, K., Fuchs, H., Rabel, A., Friedrich, K., and Faleschini, G., "First operation results of a small scale sponge iron reactor (SIR) for hydrogen production," *Proceedings of the 12th World Hydrogen Energy Conference*, Buenos Aires, Argentina, Pergamon Press, Oxford (1998).
226. Friedrich, K., Kordesch, K., Simader, G., and Selan, M., "The process cycle sponge iron/hydrogen/iron oxide used for fuel conditioning in fuel cells," *Proceedings of the International Symposium on New Materials for Fuel Cell Systems*, July 9–13, Montreal, QC, Pennington, NJ (1995).
227. Hacker, V., Fuchs, H., Fankhauser, R., Spreitz, B., Friedrich, K., and Faleschini G., "Hydrogen production from gasified biomass by sponge iron reactor (SIR)," *Proceedings of the 12th World Hydrogen Energy Conference*, Buenos Aires, Argentina, Pergamon Press, Oxford (1999).
228. Hacker, V., Faleschini, G., Fuchs, H., Fankhauser, R., Simader, G., Ghaemi, M., Spreitz, B., and Friedrich, J., "Usage of biomass gas for fuel cells by the SIR process," *Journal of Power Sources*, 71, 226–230 (1998).
229. Hacker, V., Fankhauser, R., Faleschini, G., Fuchs, H., Friedrich, J., Muhr, M., and Kordesch, K., "Hydrogen production by steam-iron process," *Journal of Power Sources*, 86, 531–535 (2000).
230. Jannach, G., Krammer, G., Staudinger, G., and Friedrich, K., "Hydrogen from biomass gasification for fuel cell systems by application of the sponge iron/iron oxide process cycle," *Proceedings of the 2nd International Symposium on New Materials for Fuel Cell and Modern Battery Systems*, July 6–10, Montreal, QC (1997).

5 Hydrothermal Processes in Subcritical Water

5.1 INTRODUCTION

As discussed in Chapter 4, gaseous water, steam, at high temperature and pressure is a powerful reactant to produce gaseous synthetic fuels from a variety of carbonaceous feedstock. While the products formed depend on the operating conditions, the catalyst, and the nature of feedstock, steam plays a very powerful role in the gasification and reforming processes. In recent years, significant efforts have been placed to explore the role of liquid water as a reactant and/or a reaction medium at high temperature and pressure for the production of a variety of synthetic gaseous, liquid, and solid fuels [1–10].

The properties of water at high temperature and pressure are significantly different from those at room temperature and pressure [5,11–13]. For many different types of carbohydrate feedstock, water provides an environment such that at a temperature of 180°C–250°C, a residence time of 1–12 h, and a pressure above saturation pressure at this temperature range, hydrothermal carbonization (HTC) occurs producing mainly (50%–80%) solid char with about 5%–20% liquid dissolved in water and a small amount of gas (2%–5%). As the temperature increases to about 270°C–390°C, the residence time of few hours, and the pressure below the critical pressure of about 213 atm, carbohydrates are converted to liquids by the process called hydrothermal liquefaction (HTL). The amount and nature of liquid produced depend on the nature of feedstock, the operating conditions, and the nature of the catalyst (if present). At much higher temperatures (>300°C), hydrothermal gasification (HTG) occurs.

The chapter addresses this hydrothermal biomass–water conversion chemistry under subcritical conditions. The issues of purification, upgrading, and utilization of the products obtained from the three processes—HTC, HTL, and HTG—are also examined. In particular, a hydrothermal upgrading (HTU) process to upgrade the products from the HTL is briefly described.

The chapter also addresses various aspects of coal–water chemistry under high-temperature and high-pressure conditions. While water does not have as much affinity for coal as it has for biomass, water can also play an important role in coal liquefaction. The weathered coal created by pretreatment with water can have a significant negative effect on the yield and products of coal liquefaction. Water can also act as a hydrogen donor for the coal liquefaction process under high-temperature and high-pressure conditions. Finally, coal–water slurry, if prepared properly, can be a good feedstock for combustion in boilers, diesel engine, or gas turbines. The chapter briefly examines these three roles of water in coal–water interactions under subcritical conditions.

5.1.1 Properties of Water at High Temperature and Pressure

Water at room temperature and pressure is a rather benign, polar substance with low diffusivity, high dielectric constant (i.e., low permittivity), and low dissociation constant. These properties do not allow any meaningful reactions with organic molecules. As the temperature and pressure increase, the dielectric constant quickly decreases, the shared electron by oxygen and hydrogen atoms tends to circulate more evenly, and electronegativity of the oxygen molecule is reduced (i.e., less polar). When the temperature of the water increases from 25°C to 300°C, the dielectric constant decreases from 78.85 to 19.66, resulting in water molecules to become fairly nonpolar. This nonpolarity increases the affinity of water for organic hydrocarbons.

As the temperature and pressure increase, the dissociation constant also significantly increases. The dissociation constant for water at 300°C is about 500 times higher than that at room temperature. An increase in pressure also increases the ionization of water. At room temperature and pressure, low dissociation constant allows H^+ and OH^- ions in hydrolysis or dissociation in equilibrium balance and the rate of acid- or base-catalyzed reaction rates is low. High dissociation constant at higher temperature and pressure facilitates more acid–base-catalyzed reactions [5].

For the above two reasons, water becomes a good solvent for typically nonpolar and hydrophobic hydrocarbons at high temperature and pressure. Water at 300°C possesses the properties of acetone at 25°C. The increased solubility of organics in water at high temperature enhances the possibilities of ionic reactions and the contacts of dissociated H^+ with hydrocarbons, thereby accelerating the activities of hydrolysis. These dramatic changes in physical and chemical properties thus allow various organic reactions to take place in the water [5,14]. In addition, water has the ability to carry out condensation, cleavage, and hydrolysis reactions and to affect the selective ionic chemistry, which are more compatible with the organic reactions. Thus, water becomes a medium similar to organic hydrocarbons in which different types of organic reactions can freely occur. As the temperature increases and crosses the critical temperature, water and organic hydrocarbons become more homogeneous in carrying out the various types of organic reactions. The role of water under supercritical conditions is discussed in Chapter 10.

High-temperature water, due to its properties, can act as a reactant and a catalyst for a second pathway to cascade organic molecular transformation of biomass (and to some extent coal) that leads to refined biomass, oil, or synthetic gaseous fuel. Water can cause organic material from biomass to disintegrate and reform by the addition of H^+ to open carbon bond into fragments that can then be converted to different types of hydrocarbons. Thus, hot water can be a catalyst for a series of ionic reactions. Hydrothermal operation differentiates itself from dry pyrolysis in that degradation of biomass in dry pyrolysis is caused by thermal forces, whereas the disintegration of biomass in water can occur due to acid- or base-catalyzed reactions. Water can act as a base to nibble certain organic molecules, and once the reaction conditions are changed, it can act as an acid promoting different sets of reactions. Thus, basic mechanisms for changes in hydrothermal operations (i.e., wet pyrolysis) are different from those occurring in the dry pyrolysis, and this difference is largely caused by the changes in the physical and chemical properties of water as its temperature increases [15].

To understand hydrothermal operation in sub- and supercritical regions, it is important to illustrate the behavior of various properties of water under these conditions. We will mainly focus on the water properties in the subcritical conditions. The properties of water under supercritical conditions are illustrated in Chapter 10. Interphase transport resistances can be considerably reduced at higher water temperatures. The water properties will vary considerably with temperature to facilitate various types of organic reactions and also the separations of products from byproducts. When feedstock contain inorganics such as sulfates, nitrates, and phosphates, hydrothermal operation can facilitate the recovery and recycling of these chemicals in their ionic forms for eventual use as fertilizers. Also, in hydrothermal operations, product streams are completely sterilized with respect to any possible pathogens including biotoxins, bacteria, and viruses. For temperatures greater than about 250°C and the contact time of few seconds, proteins are destructively hydrolyzed so that even prions would be destroyed [5,16].

Figure 5.1 illustrates the variations in dissociation constant, dielectric constant, and density as a function of temperature at ~30 MPa pressure. The figure shows that the density, the dissociation constant, and the static dielectric constant all vary significantly between the room temperature and the critical temperature. These changes cause enormous changes in the solvation behavior of water; it is changed from the polar, highly hydrogen-bonded solvent to the behavior of nonpolar solvent such as hexane. The dielectric constant changes from 80 to <2 in the temperature range of 25°C–450°C.

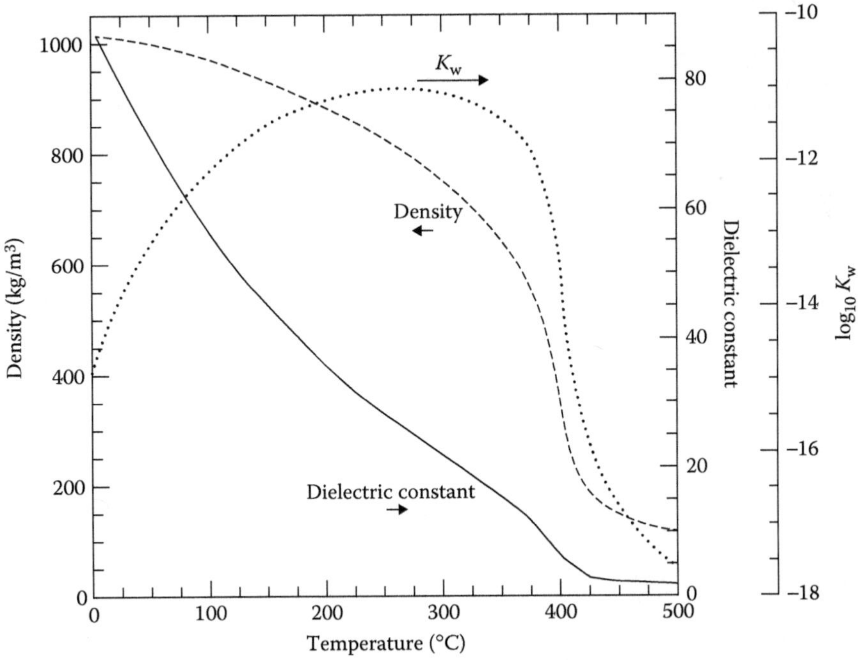

FIGURE 5.1 Variations of water density, static dielectric constant, and ion dissociation constant (K_w) as a function of temperature at ~30 MPa. (From Peterson, A., Vogel, F., Lachance, R., Frolling, M., Antal, M., and Tester, J., *Energy & Environmental Science*, 1, 32–65, 2008. With permission.)

While the dissociation constant goes through a maximum with changes in temperature, it is also changed by about 5 orders of magnitude in nearly the same temperature range.

One of the indicators that water becomes polar to nonpolar solvent as temperature increases is the solubility of various inorganic salts and organic acids in water. Table 5.1 shows the solubility of various inorganic salts at 25 MPa pressure in water at different temperatures [5]. These data clearly indicate that as temperature increases, water becomes more nonpolar and the solubility of inorganic salts rapidly decreases. In fact, at very high temperature (in supercritical conditions), inorganic salts precipitate out of the water phase. Tables 5.2 and 5.3 show that the solubility of various fatty

TABLE 5.1
Solubility Limits of Various Salts in Water at 25 MPa

	Salt Concentration (ppm)				
Temperature (°C)	**$CaCl_2$**	**KCl**	**NaCl**	**Na_2SO_4**	**$CaSO_4$**
350	6000[a]	–	–	60,000	0.6
400	18	1000[a]	1200[a]	30	0.07
450	6	200	250	0.7	0.004
500	3	100	110	–	0.0017
550	2	100[a]	100	–	0.008
Decrease[b]	9	10	12	–	8.75

Source: Peterson, A., Vogel, F., Lachance, R., Frolling, M., Antal, M., and Tester, J., *Energy & Environmental Science*, 1, 32–65, 2008. With permission.

[a] The best extrapolated estimates from the graphical data.

[b] Decrease ratio indicates solubility at 400°C/solubility at 550°C.

TABLE 5.2
Fatty Acid Concentration in Water at Different Temperatures at 15 MPa

	Solubility (kg/kg)					
Temperature (°C)	**$n = 8$**	**$n = 10$**	**$n = 12$**	**$n = 14$**	**$n = 16$**	**$n = 18$**
75	4×10^{-3}	7×10^{-4}	10^{-4}	1.5×10^{-5}	–	–
100	9×10^{-3}	10^{-3}	1.8×10^{-4}	5×10^{-5}	–	–
150	4×10^{-2}	7×10^{-2}	10^{-3}	4×10^{-4}	9×10^{-6}	–
200	1.5×10^{-1}	2×10^{-2}	10^{-2}	1.5×10^{-3}	8×10^{-4}	5×10^{-5}
225	1.2	10^{-1} [a]	5×10^{-2}	5×10^{-3}	5×10^{-3}	6×10^{-4}
Increase[b]	3000	142.8[a]	500	333.3	–	–

Source: Peterson, A., Vogel, F., Lachance, R., Frolling, M., Antal, M., and Tester, J., *Energy & Environmental Science*, 1, 32–65, 2008. With permission.

[a] The best extrapolated estimates from the graphical data.

[b] Increase numbers denote the ratio of fatty acid concentration at 225°C/fatty acid concentration at 75°C.

n, number of carbon atoms in fatty acid.

TABLE 5.3
Water Concentration in Two Fatty Acids at Different Temperatures and at the Vapor Pressure of the System

	Solubility (kg/kg)	
Temperature (°C)	**Coconut Fatty Acids**	**Tallow Fatty Acids**
100	0.05	0.01
150	0.075	0.02
200	0.1	0.05
250	0.28	0.1
300[a]	0.70	0.23
Increase[b]	14	23

Source: Peterson, A., Vogel, F., Lachance, R., Frolling, M., Antal, M., and Tester, J., *Energy & Environmental Science*, 1, 32–65, 2008. With permission.

[a] The best extrapolated estimates from the graphical data.

[b] Increase ratio indicates solubility at 300°C/solubility at 100°C.

acids in water and vice versa increases with temperature, meaning that fatty acids dissolve well in high-temperature water allowing organic reactions to occur in the water phase. This is another indication of the fact that at high temperatures, water behaves more like an organic nonpolar solvent than a polar solvent.

In hydrothermal operations, water can act as a reactant as well as a solvent. Recently, Savage [14] outlined a number of organic reactions that can occur in the supercritical or near supercritical conditions. These reactions can be complete oxidation, decomposition of organic materials and compounds, and a variety of chemical syntheses. A review of chemical oxidation reactions near the critical region was also recently reported by Ding et al. [17]. The water thus provides a suitable medium for many hydrocarbon refinery operations involving a host of organic syntheses.

5.2 HYDROTHERMAL CARBONIZATION (WET PYROLYSIS)

HTC is a thermochemical conversion process to convert biomass into a solid, coal-like product in the presence of liquid water. This process is often called a wet or hydropyrolysis process and results in the production of "hydrochar" that has high carbon content and low oxygen content compared to original biomass. The main advantage of the HTC process over conventional pyrolysis process is that it can convert wet feedstock into carbonaceous material without having to remove water with an energy-intensive and energy-expensive drying process. The potential feedstock that can be used for this process are wet animal manures, human waste, sewage sludges, municipal solid waste (MSW), aquaculture and algal residues, and many other wet energy crops. The process is of course most beneficial when biomass is accompanied by a large amount of water.

The solids produced from this process have been given many names such as char, biocoal or biochar, or more accurately hydrochar to differentiate it from the char or coal produced by the conventional dry pyrolysis. Significant reviews of hydrochar have been published recently particularly on its production processes [18–28]. The interest in hydrochar has increased very rapidly due to its connection to understand natural coal formation [18–20,27], its use in creating new innovative materials [21], and its application in soil quality improvement [22–28].

HTC can be an exothermic process, which lowers both oxygen and hydrogen content of the original feedstock mainly by dehydration and decarboxylation. The overall reaction identifying the heating value of the process can be expressed as [18–28]

$$C_6H_{12}O_5 \rightarrow C_{5.25}H_4O_{0.5} + 0.75CO_2 + 3H_2O \tag{5.1}$$

The initial phase of this overall reaction, that is, hydrolysis of cellulose, is an endothermic reaction [29]. As shown in Figure 5.2, this process is not as harsh as dry pyrolysis in the reduction of hydrogen/carbon (H/C) and oxygen/carbon (O/C) ratios, and it produces "coal-like" material, which can be similar to bituminous or sub-bituminous coals. Typical effects of residence time and temperature on selected feedstock such as cellulose, peat bog, and wood are illustrated by Libra et al. [15]. Generally, the process occurs in the temperature range of 180°C–220°C at saturated pressure and for the reaction conditions that last for several hours. The process is accompanied by numerous reaction mechanisms such as hydrolysis, dehydration, decarboxylation, condensation polymerization, and aromatization, which are further discussed in Sections 5.2.1 through 5.2.4. These are not consecutive but parallel reaction paths, and the detailed

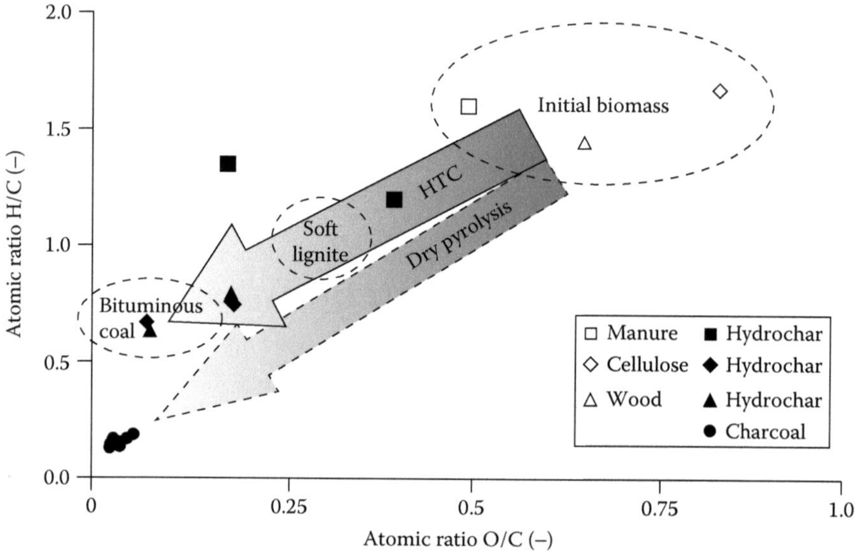

FIGURE 5.2 Comparison of H/C and O/C ratio variations for dry and wet pyrolysis processes (HTC) for various feedstock. (From Libra, J., Ro, K., Kammann, C., Funke, A., Berge, N., Neubauer, Y., Titirici, M., Fuhner, C., Bens, O., and Emmerich, K., *Biofuels*, 2, 89–124, 2011. With permission.)

nature of these mechanisms and their relative significance during the course of reaction primarily depend on the nature of the feedstock.

During the HTC process, biomass components are hydrolyzed to produce a large amount of monomers and oligomers [15]. Simultaneously, water-soluble extractables are also produced. These monomers, oligomers, and extractables then further undergo dehydration, decarboxylation, and condensation reactions. Many intermediates such as 5-hydroxymethylfurfural (HMF) are very reactive and some of them have high chemical values. These intermediates further undergo polymerization to produce humic acids, bitumen, and insoluble solids, some of which precipitate as HTC coal or hydrochar. Some components of biomass (e.g., crystalline cellulose) or oligomer cellulose do not hydrolyze under these reaction conditions. More details on the reaction mechanisms during the HTC process are given in the published literature [15,18,22–27,30–32].

Carbonization of biomass has a number of advantages over biological treatment. First, it takes only few hours as opposed to days and months taken by the biological process allowing more compact reactor design. While toxic feedstock cannot be converted biochemically, high temperature and chemical reactivity of hydrothermal environment can destroy pathogens and potential organic contaminants such as pharmaceutically active compounds [33–40]. HTC also produces useful gas, liquid, and solid products that can be further utilized, which contribute to greenhouse gas (GHG) and climate change mitigations, odor reduction, and soil amelioration [33–40]. The discussion in Sections 5.2.1 through 5.2.5 closely follows an excellent review by Libra et al. [15] and others [2,18–40].

5.2.1 Reaction Mechanisms

As mentioned earlier, HTC is accompanied by a series of chemical reactions such as hydrolysis, dehydration, decarboxylation, polymerization, and aromatization [14,15,18,21]. These reaction mechanisms are briefly described below.

Hydrolysis reactions during the HTC process mainly break down ether and ester bonds resulting in a wide range of products that include saccharides of cellulose and phenolic fragments of lignin. Along with other degradation mechanisms mentioned later, the intermediate products are also further hydrolyzed such as HMF converted to levulinic acid and formic acid. This transformation is further discussed in Chapter 7 on biofine process. Hemicellulose is hydrolyzed around 180°C and cellulose is hydrolyzed above ~200°C. The detailed mechanism of cellulose hydrolysis is given by Peterson et al. [5]. While initial hydrolysis reactions are favored by the alkaline conditions, further degradation of glucose is accelerated by the acidic conditions. In a pH range of 3–7, the rate of reaction is largely independent of H^+ and OH^- concentrations. The hydrolysis of lignin occurs around 200°C and produces highly active low-molecular-weight substances. Some of these substances go through condensation reactions and precipitate from the solutions. In general, hydrolysis reactions are fast and transport limited. The structures produced from hemicellulose and lignin interact with each other resulting in high solubility of aromatic structures. At high temperatures, condensation reactions are likely to occur.

Chemical dehydration of biomass generally results in the elimination of hydroxyl group and the production of water. For example, the dehydration of cellulose is as follows [40]:

$$4(C_6H_{10}O_5)_n \rightleftarrows 2(C_{12}H_{10}O_5)_n + 10H_2O \tag{5.2}$$

The rate of decarboxylation versus dehydration is generally measured by the factor F = mole of CO_2/mole of H_2O that varies from 0.2 for cellulose and 1 for lignite.

Condensation of fragments can also regenerate water during the HTC process, which results in a partial elimination of carboxyl groups producing CO_2 and CO above 150°C [2,15,18,21]. Generally, CO_2 is produced from carboxyl groups and CO is produced from carbonyl groups. One likely source for CO_2 is formic acid that is formed in a significant amount by degradation of cellulose. CO_2 can also be produced by condensation reactions, cleavage of intramolecular bonds, and destruction of oxidized molecules at high temperatures.

In an HTC process, intermediates that are created by dehydration and decarboxylation reactions are highly active and can polymerize to produce larger molecules. Condensation reactions are also accounted for the production of CO_2. The rate of carbonization is increasingly determined by stearic influences with a higher condensation degree of aromatics [15,18,19,21]. Thus, condensation polymerization is the main reason for the formation of biocoal in the HTC process. The condensation polymerization is most likely governed by the step-growth polymerization [15,18,19,21].

Cellulosic structures are capable of forming aromatic structures under hydrothermal conditions [2,15,18,19,21]. Aromatic structures show high stability under hydrothermal reaction conditions and may be considered as a basic building block of HTC coal. Alkaline conditions favor aromatization. Cross-linking condensation of aromatic rings also makes up the major constituents of HTC coal. A large number of aromatic bonds reduce the effects of HTC process on the carbon content. High temperature and residence time favor aromatization. Cellulose aromatizes most in the temperature range of 200°C–300°C.

Besides the reaction mechanisms mentioned earlier, certain transformation reactions for crystalline structures in cellulose or certain oligomers are also possible. Their contribution at temperatures below 200°C appears to be small [15,18,19,21]. Demethylation has been used to explain the conversion of phenolic structure to catechol-like structure in HTC coal. The production of a small amount of methane substantiates this hypothesis. At temperatures higher than 200°C, pyrolytic reactions may also compete with the mechanisms mentioned earlier, although a significant amount of tar and CO (major products of pyrolysis) has not been found. Finally, Fischer–Tropsch (FT)-type reactions may also occur during the HTC process.

In general, wet pyrolysis is more effective on cellulose than on lignin. The literature data show that during HTC of lignin, the decline in the H/C ratio is not as severe as that of cellulose or even wood [15,18,19]. The decline in the H/C ratio of wood is in between that of lignin and cellulose.

5.2.2 Effects of Operating Conditions

A number of operating conditions such as (1) the water and solid concentrations, (2) the feed slurry pH value, (3) the reaction temperature and pressure, (4) the reaction residence time, and (5) the nature of feedstock affect the product distributions. These effects are well described in the literature [2,15,18–40]. Here we briefly summarize these literature results.

HTC process and formation of hydrochar requires the presence of water. It is known that the biomass above the water surface does not carbonize, although only small amount of water is necessary. The process of carbonization is accelerated by water because of its active role as a solvent, a reactant, and a catalyst during various steps of biomass degradation and subsequent condensation and aromatization processes. Water helps thermally driven pyrolysis. The initial step of hydrolysis is very important and the role of water for this step increases with an increase in temperature. Water facilitates the condensation polymerization of active intermediate species and also dissolves numerous compounds formed during the HTC process. The amount of water can also affect the transport of fragments from the influence of reactive centers. Generally, very low concentration of biomass in water may result in very low production of precipitated carbonized solids since most biomass may be dissolved. However, excessive biomass may result in some unreacted organic materials. Generally, increase in feed solid concentration increases the monomer concentration in the liquid phase. The key is to optimize the effect of residence time–solids concentration interplay on the extent of polymerization reaction. For each feedstock, there will be an optimum solids concentration to achieve the highest yield of carbonized solids.

It is known that during HTC, pH drops due to the formation of acetic, formic, lactic, and levulinic acids [15,18,19,21]. It is also known that natural coalification requires a neutral-to-weak acidic environment [15,18,19,21]. The effects of the nature and quantity of acids and bases on the product characteristics are well described in many of the studies mentioned earlier. In general, high pH values result in the product with a high H/C ratio, hydrolysis reactions are favored by the acidic conditions, and a weak acidic condition improves the overall rate of HTC [2,15,18–40].

Depending on the nature of the feedstock, an exothermal effect during hydrothermal operation can occur at as low as 100°C temperature (e.g., for peat). Temperature is the most important parameter in the HTC process. It is known that hydrolysis of glucose with subsequent dehydration may take several seconds at 270°C, but it will take up to several hours at 150°C. The rate of polymerization is also temperature dependent. The temperature has a definite influence on the nature of biomass that can be hydrolyzed. Hemicellulose hydrolyzes at 180°C and lignin at 200°C, but cellulose hydrolysis requires 220°C [2,15,18,19,21]. Pyrolytic reactions also become more important at higher temperatures.

The temperature also affects all the physical and chemical properties that are important in the HTC process. The viscosity of water is also decreased by twofold when the temperature is increased from the room temperature to around 350°C. Lower viscosity helps the water penetrate in porous biomass media. While pressure is a more expensive process variable, it does not affect the HTC process as significantly as the temperature. While both dehydration and decarboxylation reactions are suppressed with an increase in pressure, this, however, does not significantly affect the overall HTC process. An increase in pressure facilitates (1) the removal of extractable, (2) solubilization of compressed gases and physical compacting of bio-coal, and (3) hydrogen ion transfer and condensation polymerization between solids allowing the use of higher biomass/water ratio.

Although initial extraction and hydrolysis are rapid reactions, overall HTC process is a slow-limiting process. The diffusion-controlled transport mechanisms during biomass degradation and condensation polymerization govern the overall rate of reaction. Due to its slow nature, HTC coal yield increases with an increase in residence time.

Finally, feedstock characteristics such as chemical composition, volatile and non-combustible fractions, moisture content, particle size, and energy content significantly affect conversion efficiency and char characteristics. These effects are well described in the studies mentioned earlier [2,15,18–40].

5.2.3 Comparison of HTC and Dry Pyrolysis Process

HTC process produce a high amount of solids, more water-soluble organics, and fewer gases that mainly contain CO_2 [2,15,18,19,22]. The chemical structure of hydrochar more closely resembles coal than charcoal in terms of elemental composition and types of chemical bonds and their relative quantities. As discussed earlier, HTC hydrochar has higher H/C and O/C ratios than the char coming from dry pyrolysis (Figure 5.2). Thus, the ratio of decarboxylation to dehydration reaction is higher in HTC than in dry pyrolysis. Figure 5.2 shows that even though feed composition of animal-derived biomass is different from that of plant materials, the final hydrochar products coming from these materials have similar elemental compositions. The aromatic structure of hydrochar product is substantially different from that of char from dry pyrolysis. These and other structural differences indicate that different reaction mechanisms govern these two processes. Radical mechanism pathways taken in dry pyrolysis are completely suppressed in hot water in favor of ionic reactions.

HTC primarily starts out with hydrolysis reactions of biomacromolecules resulting in the production of oligosaccharides, hexoses, pentoses, and fragments of lignin [2,15,18,19,22]. These intermediates in the aqueous phase follow completely different reaction pathways than those in thermally driven dry pyrolysis. For example, in dry pyrolysis of glucose, the major product is anhydrous glucose, which appears only in a very small amount in the HTC process. In the HTC process, HMF is a crucial intermediate, which provides a standard platform for many subsequent reactions. Hydrolysis in the HTC process completely disintegrates the physical structure of biomass. This is not the case for dry pyrolysis. The degree of hydrolysis, however, depends on the temperature and process design [2,15,18,19,22]. Final H/C ratios produced by HTC of lignin and cellulose are different. This is a result of different reaction paths of these two processes.

The literature has shown that HTC is more energy efficient as a pretreatment process than dry pyrolysis for wood combustion when the water content is >50% [2,15,18–40]. Generally, when the water content of feed slurry is >50%–70%, wet pyrolysis is preferred over dry pyrolysis. For such a slurry, dry pyrolysis will only be capable of producing charred material. However, HTC process can be used for a slurry containing 70%–90% water or even higher. The amount of external heat required will depend on the process design. Generally, HTC process with a slurry containing >40% water will have an energetic advantage over dry pyrolysis. HTC process needs to manage the required pressure and its effect on the safety and material issues. The energy from hot water also needs to be recovered. In general, energetic requirements to run the process and its auxiliary equipment are higher for a hydrothermal system than for dry pyrolysis.

5.2.4 Product Characteristics and Usages

HTC process largely produces hydrochar (solid) and liquid with dissolved nutrients. The fate of heavy metals and organic chemicals (present in the original feedstock or created by the reactions) is generally not known and must be traced during the process. Wet pyrolysis cannot destroy heavy metals. Since they have a toxic risk potential, their fate needs to be followed. If they accumulate in the solid char, which is subsequently used as soil nutrient, they can affect the food chain. Generally, except for zinc, heavy metal concentrations in char do not exceed the allowable limits [15,18,19,21,30–40]. Also, heavy metal contents of hydrochar obtained by wet pyrolysis are less than those obtained in biochar from dry pyrolysis [15,18,19,21,30–40]. Just like for heavy metals, a systematic knowledge of the fate of organic compounds during the HTC process is not well understood. Unlike heavy metals, new organic compounds can be formed during condensation, polymerization, and aromatization reactions. The fate of compounds such as polychlorinated biphenyl (PCB) and hexachlorobenzene (HCB) needs to be particularly followed along with all other organic chemicals during the HTC process [15,18,19,21,30–40].

For both animal manures and sewage sludge, hydrochar retains a significant level of calcium, potassium, and phosphorus. pH affects the mobility and sorption capability of the nutrients, particularly for the case of phosphorus. In the HTC process, dissolution of water-soluble minerals can be significant [15,18,19,21,30–40]; however, the nutrient content will also depend on the technique for dewatering the solid conversion product. The ratio between evaporation and dewatering governs the amount of plant nutrients that will be adsorbed or retained at the hydrochar interface. Nutrient retention should be an important parameter in the detailed process design.

Generally, an increase in temperature decreases the hydrochar yield and increases the yield of liquids and gases such as CO_2, CO, and H_2. An increase in temperature also decreases the H/C and O/C ratios in the hydrochar. The maximum allowable yields (which may be obtained at very large residence time) for various feedstock are illustrated by Libra et al. [15] and others [18,19,21]. A lower biomass solid concentration (i.e., high water concentration) generally gives a lower hydrochar yield.

In general, 60%–84% of the biomass carbon remains in the hydrochar. The char composition is mainly affected by the nature of the feedstock, the temperature, and the reaction time. For all feedstock, an increase in temperature increases the carbon content and decreases the oxygen content of the hydrochar.

Hydrochar is often used for soil application as a fertilizer or a carbon sequester. Some discussion in this regard is given by Libra et al. [15] and others [30–40]. Experimental evidences indicate that depending on the nature of char, it can remain in soil for a long period. While less aromatic hydrochar stays longer in soils than uncarbonized carbon, it decomposes faster than the char from dry pyrolysis. The oxidation and degradation of char can be affected by the swelling–shrinking of clay materials by the weather. The degradation of hydrochar is also accelerated by labile carbon substrates and white-rot fungi, which are dominant char decomposers. In general, the degradation and stability of hydrochar in top soils of Earth's surface can be affected by surface erosion and dissolution as dissolved organic carbon or transported to subsoil as small particles with rain water. The char particles that are imbedded in subsoil surface are less susceptible to erosion and degradation.

Hydrochar promotes the fungal growth and soil aggregation [15,30–40]. It will very likely reduce the tensile strength, increase the hydraulic conductivity, and enhance the soil water holding capacity (WHC). Hydrochar does not have a very large internal surface area that may affect the penetration by water and nutrients and the resulting microbial activity. While the WHC of hydrochar is generally higher than that of mineral soils, it can be considerably reduced after it is fully dried. Hydrochars are more acidic than biochars and they do undergo aging process that can change the functional groups and therefore its effectiveness as nutrient. The hydrochar can also be used as (1) activated carbon adsorbents, (2) raw materials for the generation of nanostructured materials, (3) catalyst supports or as catalysts, (4) CO_2 sorption materials, and (5) energy production and storage materials. These applications of hydrochar are described in more detail in an excellent review by Libra et al. [15].

The HTC process is accompanied by a large number of intermediate products due to complex reaction mechanism. The solids coming out of the HTC process represent the agglomerates of chemical substances. An elemental analysis of the hydrochar shows that it may approach lignite or even sub-bituminous coal depending on the reaction severity (Figure 5.2). An exception is resin whose H/C ratio remains unaffected by the HTC process. As mentioned earlier, HTC coal from lignin tends to have a lower hydrogen content, whereas coal from cellulose tends to achieve higher carbon content [15,18,19,21].

HTC coal (hydrochar) is soluble in benzol–alcohol mixtures, alkaline solutions, and ammonia [15]. The skeletal of HTC coal is very similar to that of natural coal, although it exhibits a higher amount of functional groups compared to natural bituminous coal. The removal of hydroxyl and carboxyl groups during the HTC process makes HTC coal with a lower hydrophobicity than the original materials [15,18,19,21]. While the inorganics largely remain in HTC coal, their relation with process conditions is not well known. While HTC coal has a small surface area, this area can be increased significantly (by 2 orders of magnitude) by removing extractables or by thermal treatment [15]. Observations of the nanostructure of HTC coal reveal its potential technical applications as functionalized carbonaceous materials. Libra et al. [15] and others [30–40] show that with the use of proper starting materials and appropriate catalyst/template-assisted treatment methods, various types of nanospheres, nanocables, nanofibers, microcables, submicrotubes, and porous structures can be created from HTC coal.

5.2.5 Process Considerations

The HTC process generates more water than carbon dioxide. Water acts as a solvent and a reactant, and therefore carries a significant amount of inorganics and organics, many of which can be valuable chemicals. The solids in water can cause problems upon precipitation due to condensation or polymerization reactions. The wastewater of the HTC process can be processed with aerobic or anaerobic treatment to lower its total organic content. Some inorganics in water may be good nutrients for soil. The gases coming out of the HTC process mainly contain carbon dioxide with minor CO, CH_4, and H_2 as well as traces of C_mH_n. The dissolution of carbon dioxide in the liquid phase may affect the critical temperature condition in the process. An increase in temperature generally decreases CO and increases H_2 and CH_4. The progress of the HTC process can be monitored by following the production of carbon dioxide. The decarboxylation of feedstock is a major reaction during the HTC process [15,18,19,21]. While its high rate produces high heating value HTC coal, it also reduces carbon efficiency. Thus, the process should be optimized based on the end use of the final product.

In the HTC process, the ratio of biomass to water should be kept as high as possible to enhance polymerization. Less water will also give less energy loss and less pumping costs for the total throughput. The feedstock can be submerged in the water by mechanical compacting device to take advantage of the best reaction conditions. The residence time should be as large as possible to get the complete reaction to occur and minimize the loss or organics in the wastewater. A recirculation of water is one way to achieve this objective.

The reaction conditions should take advantages of the possible effects of organic acids; they may give faster polymerization and higher ash content of the produced HTC coal. While higher temperature accelerates the process and gives higher carbon content in the HTC product, high pressure required to achieve high temperature may be expensive. Pressure–temperature relationship should be optimized based on the intended use of the end product. Since hydrolysis is a diffusion-controlled reaction, small particle size of the feedstock may be beneficial. This, however, increases the energy demand and the investment cost [15,18,19,21].

5.3 HYDROTHERMAL LIQUEFACTION

Just like the HTC process, HTL is a wet pyrolysis process in which complex organic (particularly carbohydrate base) molecules from biowaste (manure and food processing waste), lignocellulose (crop residue), algae, and others are converted to crude oil type and other liquid fuel products as well as chemicals. To some extent, it mimics the natural geological process which is thought to be involved in the production of fossil fuels. HTL involves a direct liquefaction of biomass in the presence of water (and may be a catalyst) to liquid fuels in subcritical conditions. The process conditions for HTL thus differ from the HTC process described earlier in that generally HTL process requires the temperature range of about 250°C–400°C, a range that is higher than that used in the HTC process and that does not use the conditions of supercritical water. It is analogous to intermediate-temperature dry pyrolysis in that the objective of both processes is to produce liquid with minimum amounts of solids and gases.

The HTL process is not a selective catalytic process such as aqueous-phase reforming (APR) and its derivative technologies described in Chapter 6, in which hydrogen, alkanes, and monofunctional groups are produced at low temperature (range of 215°C–265°C) from a selective group of oxygenated biomass by a set of selective catalysts or biofine process described in Chapter 7. When high-temperature HTL reaction conditions are used with a selective group of catalysts, HTG process evolves. This process largely generates methane and carbon dioxide with some hydrogen. The HTG process is described in Section 5.4. Numerous excellent reviews on HTL are available in the literature [41–47] (Elliott, 2012, pers. comm.).

5.3.1 Reaction Mechanisms

HTL is a chemical transformation process of biomass in a heated and pressurized water environment where long-chain organic compounds break into short-chain hydrocarbons. All fossil fuels found underground, petroleum, natural gas and coal, and so on based on biogenic hypothesis, are formed through the HTL process from biomass buried beneath the ground and subjected to high pressure and temperature. In the recent years, it has been found that kerogens (which are a large part of oil shale) break down much easily in the presence of water than without it [41–48] (Elliott, 2012, pers. comm.). Gas hydrates and several carbon isotope studies have shown the involvement of water (and hydrogen from it) in the creation of natural gas. HTL process reaction paths depend on the temperature, the pressure, the reaction time, the water pH, the solids particle size, and the nature of the catalysts (if present).

While the exact reaction pathway for the HTL process is as yet not known, the study of Appell et al. [49–52] at Pittsburgh Energy Technology Center made some important points for the process. They studied liquefaction of wood particles with hydrogen and carbon monoxide at 370°C and 27 MPa pressure in the presence of sodium carbonate catalyst. Alkali salts such as sodium carbonate and potassium carbonate can initiate the hydrolysis of cellulose and hemicellulose into smaller fragments. The degradation of biomass into smaller products mainly proceeds by depolymerization and deoxygenation reactions. The amount of solids residue remained depends on the lignin content. Lignin contains alkyl phenols and free phenoxyl radicals formed by its thermal decomposition above 250°C, and it is likely to recombine and form the solids residue through condensation or repolymerization reaction.

Appell et al. suggested that during the conversion of carbohydrates to oil, sodium carbonate reacts with carbon monoxide and water to form sodium formate as

$$Na_2CO_3 + 2CO + H_2O \rightarrow 2HCO_2Na + CO_2 \quad (5.3)$$

which in turn reacts with cellulose in the wood wastes to form oil and regenerate sodium carbonate as

$$2C_6H_{10}O_5 + 2HCO_2Na \rightarrow 2C_2H_{10}O_4 + H_2O + CO_2 + Na_2CO_3 \qquad (5.4)$$

Vicinal hydroxyl groups in the carbohydrates undergo dehydration to form an enol followed by its isomerization to ketone. The newly formed carbonyl group is reduced to the corresponding alcohol with formate ion and water. The hydroxyl ion then reacts with additional carbon monoxide to regenerate the formate ion.

The above set of basic reactions is accompanied by a multitude of side reactions producing a whole host of intermediates. Some of the beneficial side reactions are facilitated by the alkaline conditions. When two carbonyl groups become vicinal, a benzylic type of rearrangement occurs, which results in a hydroxyl acid. The hydroxyl acid readily decarboxylates causing a net effect of reducing the remainder of the carbohydrate-derived molecule [49–52].

For the HTL process, the segments produced by hydrolysis are further degraded by dehydration, dehydrogenation, decarboxylation, and deoxygenation. These types of reactions result in the formation of paraffin-type structures that have less oxygen than the original compounds.

There are several other features in this reaction mechanism. Aldol condensation may be a part of the reaction. This can occur between a carbonyl group on one molecule and two hydrogens on another molecule, resulting in the elimination of water. In the absence of a reducing agent such as CO or H_2, condensation reactions dominate, which lead to polymerization and the formation of solid-like products. The reducing agents keep the concentration of carbonyl groups low enough to produce liquid products instead of solid products.

Appell et al. [49–52] also pointed out that hydrogen radicals formed by the addition of CO and the presence of water–gas shift reaction can react with various carbonyl and hydroxyl groups to form paraffins and water, thus avoiding various condensation reactions. The addition of CO is thus more useful than that of molecular hydrogen (Table 5.4).

TABLE 5.4
Effect of Reducing Gas on Maximum Liquid Oil Yield as a Function of Liquefaction Temperature

Reducing Gas	Maximum Yield of Liquid Oil (wt%)
Air	25
Nitrogen	36
Hydrogen	42
Carbon monoxide	50

Source: Akhtar, A. and Amin, N., *Renewable & Sustainable Energy Reviews*, 15, 1615–1624, 2011. With permission.

Note: These are best estimates from the graphical data.

In the HTL process, other sets of complex reactions such as cracking and reduction of polymers such as lignin and lipids, hydrolysis of cellulose and hemicellulose to glucose and other simple sugars, hydrogenolysis in the presence of hydrogen, reduction of amino acids, dehydration, decarboxylation, C–O and C–C bond ruptures, and hydrogenation of various functional groups result in the production of liquids from biomass rather than solids as they occur in the HTC process [41–52] (Elliott, 2012, pers. comm.). More fundamental work in this area is still needed.

5.3.2 Effects of Operating Conditions on HTL Process

Unlike the HTC process, the main objective of the HTL process is to generate bio-oil of high quality. The process is designed to minimize the production of solids and gas. Numerous process parameters affect the performance of the HTL process and these are well examined in the literature [41–48,53–66] (Elliott, 2012, pers. comm.). The major process operating parameters are (1) temperature, (2) residence time, (3) solids concentration, (4) pressure, (5) biomass heating rate, (6) biomass particle size, (7) presence of hydrogen donor solvent and reducing gas environment, and (8) pH of slurry. Since the most important variable is the nature of feedstock, its effect on process performance is discussed in Section 5.3.3. The literature information on the effects of various operating parameters on the product distribution was well summarized in an excellent review by Akhtar and Amin [41]. Sections 5.3.2.1 through 5.3.2.3 briefly summarize their assessments.

5.3.2.1 Pressure, Temperature, and Residence Time

In any HTL process, pressure must be at least equal or above the saturation pressure to maintain the process in a single-phase operation. High pressure allows a better manipulation of hydrolysis reaction and the reaction pathways are thermodynamically favorable to produce liquids and gases. High pressure also increases the solvent density, resulting in better extraction capability of the solvent. For a catalytic operation, however, high solvent density can block the active catalyst sites and therefore reduce C–C bond breakage and the resulting degradation rate.

Generally, high temperature increases both the concentration of free radicals and the probability of repolymerization of fragmented species. The hydrolysis and fragmentation of free radicals dominate in the early stages of the reactions, whereas repolymerization occurs in the later stages of the reaction, which in turn forms char. Generally, at very high-temperature bio-oil production is reduced due to (1) the secondary decompositions and Boudouard gas reactions that become active at high temperature leading to high gas formation or (2) the recombination of free radicals to form char. The overall process conditions and the presence of a catalyst generally dictate the dominant reaction mechanism. For most feedstock, however, the maximum bio-oil is obtained at temperatures around 300°C–350°C [41]. Also, the literature results show that the largest shift in the optimum temperature for bio-oil occurs for algae [67–79]. Both softwood and grass are generally more difficult to liquefy because of their higher lignin content and less reactive cellulose content.

TABLE 5.5
Yield of HTL Products as a Function of Temperature

Temperature (°C)	Residue (%)	Oil (%)	WS/W (%)	Gas (%)	O/S	O/G
150	72	6	21	1	0.083	6
200	56	13	26	5	0.23	2.6
250	32	33	29	6	1.03	5.5
300	16	38	30	16	2.37	2.37
350	16	27	33	24	1.69	1.37

Source: Akhtar, A. and Amin, N., *Renewable & Sustainable Energy Reviews*, 15, 1615–1624, 2011. With permission.

Note: These are the best calculations/estimations from the graphical data.

O/G, oil/gas ratio; O/S, oil/solid ratio; W, water; WS, water solubles.

Akhtar and Amin [41] compared the bio-oil productions from various studies and showed that bio-oil production showed an optimum in each study, although the exact location of the maximum oil production depended on the nature of the feedstock. At temperatures above 300°C, the gas production can also increase particularly when a suitable catalyst is used. The typical temperature dependence of the product distribution of solids residue, oil, water solubles plus water, and gas are illustrated in Table 5.5 [41]. Again, the exact phase composition will depend on the nature of the feedstock. The table also shows that both oil/solid and oil/gas ratios show maxima at a temperature around 300°C. Akhtar and Amin [41] suggested that a variation in the solids residue can be set as a reference point to measure the optimum liquefaction temperature for bio-oil yield for a given feedstock. As shown in the table, an increase in temperature increases gas yield and decreases residue yield. The oil yield shows an optimum with respect to the temperature at around 300°C.

The effect of the residence time on the HTL process has been examined by numerous investigators and this is well summarized by Akhtar and Amin [41]. Both biomass conversion and the nature of product distribution depend on the residence time. Since the initial hydrolysis process is fast, normally short residence is preferred in HTL. Boocock and Sherman [67] showed that the bio-oil production was suppressed at high residence time except when biomass concentration in the feed was very high. The effect of residence time on the bio-oil yield also depended on the temperature. At low temperatures, an increase in residence gave higher bio-oil yield [41], whereas at high temperatures (250°C–280°C), high residence time gave poorer bio-oil yields due to an increase in gas yields. In general, higher residence time gave higher biomass conversion. Qu et al. [69] found a decrease in heavy oil production at high residence time.

The effect of the residence time on the product distribution in the HTL process is complex once the biomass conversion is leveled off. The intermediate products formed during this time can form gas, liquid, or solid products by the secondary and tertiary reactions depending on the nature of intermediates and the prevailing local reaction environment. Karagoz et al. [70] showed that the decomposition products were not the same at low and high residence times for both low-temperature (180°C) and high-temperature (250°C) operations. In general, the composition and yield of bio-oil can be optimized by suitably adjusting the temperature and the residence time such that heavy residues containing asphaltenes and pre-asphaltenes are converted selectively to oil (and not gas). Often an addition of a reducing agent such as CO or H_2 prevents polymerization reactions and stabilizes the active free radicals.

5.3.2.2 Biomass Particle Size, Heating Rate, and Concentration

While a reduction in a particle size of biomass (particularly for biomass such as wheat straw, barley straw, and switchgrass) is energy intensive, in general, a smaller particle size results in higher degree of hydrolysis and fragmentation. However, the literature data show that the particle size has a secondary effect on biomass conversion and product distribution in an HTL process because of high solvation and extraction powers of water at high temperature and pressure. Zhang et al. [64] found no effect of the particle size variation from 0.5 to 2 mm of grass perennials on the yield of bio-oils. Akhtar and Amin [41] recommended that the particle size between 4 and 10 mm should be suitable to overcome the heat and mass transfer limitations at a reasonable grinding cost.

While the studies mentioned earlier indicated the marginal effects of particle size on the herbaceous biomass liquefaction process, Kobayashi et al. [68] showed a significant effect of particle size of woody biomass on the HTL process. Wood powder pulverized by the vibration mill, cutter mill, and grinder was used as a liquefaction material. The wood powder was sieved between 212 and 500 μm. Based on the results of water solubles and specific surface areas for three different milling processes obtained in this study, it was concluded that an increase in specific surface area increased the production of water soluble (saccharine); however, the difference in water solubles between the grinder and the cutter mill was only marginal. This indicated that the crystallinity of the wood powder also affects the water solubles yield.

Bio-oil production during the HTL process generally occurs at moderate heating rates. Slow heating rates usually lead to the formation of char residue due to secondary condensation and polymerization reactions. Very high heating rates also promote secondary reactions that generally result in more gas production. The heating rate is important for both dry and wet pyrolyses, although it is less important in the HTL process due to the better dissolution and stabilization of fragmented species in hot-compressed water medium. Zhang [45] and Zhang et al. [64] observed that for a heating rate range of 5°C–140°C/min during the HTL process for grassland perennials, bio-oil yield increased from 63% to 76% with an increase in the heating rate.

The solids concentration also affects the bio-oil production [41]. In general, high amount of water favors the production of liquids largely due to enhanced extraction and higher degree of solvation of biomass. The solvent enhances the stability and solubility of fragmented components, thereby reducing the production of solids residues and gases. At high biomass concentration, the interactions between the fragmented biomass components and the water decrease, the reactions among various fragments increase, and thus the influence of water on the product distributions diminishes. At high biomass concentration, the HTL process behaves much like the intermediate-temperature dry pyrolysis resulting in less bio-oil production.

5.3.2.3 Gas and Liquid Properties

Generally, a reducing gas or a hydrogen donor stabilizes the fragmented products of liquefaction. Reducing the environment inhibits condensation, cyclization, and repolymerization of free radicals, thereby reducing the char formation [14,41]. The stabilization occurs by the following reactions [14,41]:

$$Ar* + H_2 \rightarrow ArH + H* \tag{5.5}$$

$$Ar* + H* \rightarrow ArH \tag{5.6}$$

While H_2 is an effective reducing agent, it is also an expensive one. Often syngas (CO and H_2), steam, N_2 and Argon, and so on are also used to provide a reducing environment [41]. The effects of various reducing agents on the maximum bio-oil production for cattle manure by the HTL process are illustrated in Table 5.4. The results showed that CO was the most effective reducing agent and provided the maximum bio-oil yield of 50% at 310°C [41]. This is in line with the assertion of Appell et al. [49–52]. The temperature (310°C) at which the maximum occurred was independent of the nature of the gas. Air was ineffective because it led to combustion of biomass. The results also indicated that reactive gases gave better maximum oil yield than an inert gas such as nitrogen.

The use of hydrogen donor solvents such as tetralin and phenanthrene is also an effective way to stabilize free radicals and improve the bio-oil yield. This has been extensively examined to improve the yield in direct coal liquefaction processes. In general, this is, however, expensive for its commercial applications. A suitable catalyst can also induce or accelerate hydrogen transfer reactions to improve the bio-oil yield. However, catalyst stability and cost can be the limiting factors. In general, an adsorption of reducing gas (such as H_2) on the catalyst surface can increase the probability of hydrogen transfer reaction for the free radicals.

While most of the literature studies have focused on the neutral and alkaline water conditions for the HTL process, recently Yin et al. [59] have examined the effect of water pH on reaction mechanism and product distribution of HTL of cellulose. The study examined the water pH of 3, 7, and 14 in the temperature range of 275°C–320°C, and the residence time of 0–30 min. The results showed that the composition of the products from HTL varied with pH. In acidic and neutral conditions, the main liquid product was HMF. Under alkaline conditions, the main compound was C_{2-5} carboxylic acid. At all pH levels, high temperature and long residence times had negative effects on the bio-oil yields. The reaction mechanisms also depended on the pH level. Under acidic conditions, polymerization of HMF to solids reduced the bio-oil production. Under neutral conditions, HMF was converted to both solids and gases. Under alkaline conditions, bio-oil was converted to gases by the formation of short-chain acids and aldehydes. Different reaction mechanisms and product characteristics mean different strategies required to improve the quality and quantity of bio-oil under different pH conditions.

5.3.3 Role of Feedstock

The nature of feedstock is the most important variable affecting the quantity and quality of bio-oil by the HTL process. HTL produces high-density liquid fuels and operates in the presence of water eliminating the need for feedstock drying that is important in dry pyrolysis process. The HTL process uses agricultural biomass and biowaste including crop residues and wood, food processing waste, animal and human manure, and algae. The HTL process can also be used as a pretreatment process for the subsequent fermentation of the feedstock that are difficult to convert.

While animal and food processing waste contain lipids, proteins, and small amounts of lignocellulose, crop residues and wood primarily contain lignocellulose. The primary basic compounds in these feedstock are various isomers of glucose (such as D-glucose and L-glucose), hemicellulose, cellulose, lignin, amino acids, proteins, lipids including fatty acids such as stearic acid and palimitic acid, and so on. Over the years, more and more efforts have been made to examine the effectiveness of the HTL process for a variety of lignocellulosic wastes containing different amounts of lignins and crystalline cellulose. Some of the materials examined are swine manure, garbage, Indonesian biomass residue, birch wood, sawdust, rice husk, phytomass, and chlorella [41–47] (Elliott, 2012, pers. comm.). These feedstock have been tested in the temperature range of 280°C–375°C, the pressure range of 5–50 MPa, and the residence time of 5–180 min depending on the feedstock. In some cases, sodium carbonate was used as a catalyst. The final yield varied from 21% (in case of garbage) to as high as 61% (in case of swine manure) [41–47] (Elliott, 2012, pers. comm.). More recently, HTL has been applied to algae for making biofuel. HTL can thus be applied to a variety of biomass with a varying degree of success. In the following discussion, feedstock is broken into three categories: biowastes, lignocellulosic wastes, and algae. These three categories cover the range of feedstock property variation.

5.3.3.1 Biowastes

The best raw materials for the HTL process are perhaps biowastes such as various types of manures and wastewaters because they are mostly cellulosic (with very little lignin) and can be easily converted into bio-oil under hydrothermal conditions. In general, presence of high cellulose and hemicellulose content in biomass yields more bio-oil. Appell et al. [49–52] were the first to examine various waste streams such as urban refuse, cellulosic wastes, and sewage sludge, and found that at a temperature around 380°C, a pressure of 1500 psig, and a residence time of 20 min, even in the absence of a catalyst, an oil yield of about 24.5% was obtained. At low temperatures, oil was largely paraffinic and cycloparaffinic in the presence of carboxyl and carbonyl groups, whereas at high temperatures some aromatics were present. Following this pioneering study, a significant number of additional studies were published [41–48,53–68] (Elliott, 2012, pers. comm.), indicating that the applicability of the HTL process to cellulosic wastes is easy. These studies are also extensively described in a number of recent reviews [41–48] (Elliott, 2012, pers. comm.). The studies included swine manure, bovine manure, cellulose, activated sludge, sewage sludge, artificial garbage, protein-containing biomass (such as food wastes), various types of wastewaters (including wastewater from paper and pulp industries), glucose, glycine, dairy manure, poultry litter, and so on. Some of these studies are briefly described below.

A study by Minowa et al. [71–74] using glucose and glycine as model compounds of carbohydrates and proteins indicated that a significant oil production started at temperatures >250°C and increased with the temperature. This and other studies have shown that fatty acids and lipid are the main reactants in HTL process. Below 300°C, aliphatic compounds are the major source of bio-oil. Protein is widely involved in HTL reaction possibly by peptide bond splitting and amino acid conversion dehydration. Within the range of 300°C–450°C, the protein conversion reaction intensifies and the peptide bond begins to react. Saccharide reaction mainly belongs to the splitting of branched chains and the group transfer while considerable dehydration and cyclization of the main chain still appear to be dominant. The decomposition of an individual cellulosic biomass differs based on its structure. Decomposition is easier in hemicelluloses due to amorphous structure. Cellulose is little crystalline to decompose due to beta(1–4)-glycosidic linkages and relatively intermediate degree of polymerization (500–10,000). The major products of holocellulose degradation include cellohexaose, cellopentaose, cellotriose, cellobiose, fructose, glucose, erythrose, glycolaldehyde, glyceraldehyde, pyruvaldehyde, and furfurals [71–74].

5.3.3.2 Lignocellulose

The presence of liquid water as solvent is essential for HTL of lignocellulose feedstock. Water in this case acts as a solvent and reactant along with its role as a vehicle for biomass and a carrier for the catalyst. Furthermore, water is simple to use, inexpensive, and environmentally benign.

Lignocellulose is the largest segment of the total biomass and contains a significant amount of lignin along with cellulose. It is the lignin component along with crystalline cellulose that is difficult to convert to bio-oil in the HTL process. In the absence of a catalyst, lignin produces very little bio-oil and ends up as a solids residue in the HTL process. While water is an excellent medium for the intermediate hydrolysis of cellulose and other higher molecular-weight carbohydrates to water-soluble sugars, it is not as effective for hydrolysis of heavily aromatic and multiring aromatic structures. The breakdown of lignin requires high temperature or the presence of a catalyst. Within lignocellulosic substances, softwood gives much lower yield than hardwood because of the difference in their lignin contents. Zhang [45] and Akhtar and Amin [41], among others, have given an extensive review of HTL of a variety of lignocellulosic biomass such as various energy crops, herbaceous products, forestry and other agricultural wastes, and various crop oil wastes. Midgett [42] examined the HTL process for materials such as tallow seed, switchgrass, and pine dust. Zhong and Wei [80] studied the effect of temperature on four different types of woods and concluded that the bio-oil yield was affected by both the temperature and the lignin content of the wood.

An interesting study was carried out by Sugano et al. [48] in which they examined the effectiveness of black liquor, paper regeneration wastewater, and the water on HTL of herbaceous eucalyptus biomass. Like softwood, this material contains lignin. The study showed that black liquor gave very low oil yield and high yield of water-soluble components. The liquefaction of eucalyptus by water resulted in high residue production due to dehydration and polymerization, such as the formation of aliphatic ester bonds. The paper regeneration wastewater gave low residue and high yield of bio-oil compared to water. It appears that condensation reaction observed during the liquefaction in water was inhibited because carboxylic acid formed during the liquefaction of eucalyptus was neutralized with the cations in the wastewater. In the temperature range of 150°C–350°C, process wastewater gave the best oil yield compared to other two solvents. The optimum temperature for bio-oil yield was 300°C. This study indicated that solvent pH and other additives can affect the bio-oil production of lignocellulosic materials.

Finally, high degree of polymerization (>10,000) and complex branching make lignin difficult to decompose even at high temperatures. The studies described earlier show that the conversion of lignin containing lignocellulosic materials requires catalysts to produce bio-oil instead of solids and gases. An extensive literature review of various catalyst studies for the HTL process applied to lignocellulosic biomass is given by Zhang [45].

5.3.3.3 Algae

Ever since Glen Meier of Renewable Energy Group introduced the concept of producing fuel from algae, it has caught everybody's attention [75–79,81–88]. Like biowaste, algae do not compete with food materials for fuel. The HTL of algae has been given some attention in the recent years. Minowa et al. [89] converted *Dunaliella tertiolecta* with a moisture content of 78.4% directly into 37% oil by the HTL process operated at 300°C and 10 MPa. The oil had a viscosity of 150–330 MPa-s and a calorific value of 36 kJ/g, numbers comparable to that of fuel oil. Dote et al. [90] hydrothermally converted the artificially cultivated *Botryococcus braunii* Kützing Berkeley strain. The strain contained about 50% hexane solubles. The HTL process of this strain resulted in the production of 57% petroleum-like bio-oil at 300°C. Similar work was carried out for *Microcystis viridis* harvested from a lake.

Brown et al. [91] reported hydrothermal conversion of marine microalgae *Nannochloropsis* sp. into bio-oil in the temperature range of 200°C–500°C and for the residence time of 60 min. The highest bio-oil of 43 wt% was obtained at 350°C with a heating value of 39 MJ/kg, a number comparable to petroleum crude oil. The H/C and O/C ratios changed from 1.73 and 0.12 at 200°C to 1.04 and 0.05 at 500°C, respectively. The major components of bio-oil were phenol and its alkylated derivatives, heterocyclic N-containing compounds, long-chain fatty acids, alkanes, alkenes, derivatives of phytol, and cholesterol. Gases largely contained CO_2 and H_2.

Metal catalysts had been used in microalgae liquefaction. Matsui et al. [92] investigated the liquefaction of *Spirulina*, a high protein algae in water at 300°C–425°C using $Fe(CO)_5$–S catalyst. Other metal catalysts used were $Ru_3(CO)_{12}$ and $Mo(CO)_6$. Continuous culturing of the *B. braunii* Berkeley strain in the secondary treated sewage was conducted and then liquefied by Sawayama et al. [93]. The liquefaction was carried out at 200°C, 300°C, and 340°C. The yield of the hexane-soluble fraction was 97% compared to that in the feedstock algal cells. The heating value of the liquefied oil obtained from this reaction was 49 MJ/kg and the viscosity was 64 MPa-s at 50°C.

Different microalgae are not the same in producing oil through liquefaction. Recent literature have shown a wide variety of performances in the production rate of bio-oil and its quality [75–79,81–94]. The performance of HTL of microalgae has been improved by the use of a catalyst such as Na_2CO_3 among others [75–79,81–94]. Zhou et al. [79] examined the HTL of marine macroalgae *Enteromorpha prolifera* in the temperature range of 220°C–320°C for 30 min and in the presence of 5 wt% Na_2CO_3 catalyst. The highest bio-oil yield of 23 wt% with higher heating value of 28–30 MJ/kg at 300°C was obtained. These numbers are smaller than what are reported for microalgae [81]. The bio-oil contained ketones, aldehydes, phenols, alkenes, fatty acids, esters, aromatics, and nitrogen-containing heterocyclic compounds. Acetic acid was the main component of water-soluble components.

Vardon et al. [77,78] studied the HTL of *Scenedesmus* (raw and defatted) and *Spirulina* algal biomass at 300°C and 10–12 MPa pressure and compared the performance with that of Illinois shale oil and bio-oil produced by dry pyrolysis (at 450°C). Both wet and dry pyrolyses gave energy-dense bio-oil (35–37 MJ/kg) that approached shale oil (41 MJ/kg). Bio-oil yields (24%–45%) and physicochemical characteristics were highly influenced by the conversion route and feedstock selection. Sharp differences were observed for the mean bio-oil molecular weight (dry pyrolysis: 280–360 Da; HTL: 700–1330 Da) and the percentage of low-boiling compounds (bp < 400°C) (dry pyrolysis: 62%–66%; HTL: 45%–54%). For wet algal biomass containing 80% moisture, the energy consumption ratio (ECR) for HTL (0.44–0.63) was more favorable than that for dry pyrolysis (0.92–1.24). In another study, Vardon et al. [77,78] showed that *Spirulina* algal biomass gave 32.6% biocrude as opposed to 9.4% for digested sludge under the same reaction conditions as mentioned earlier and for 30 min residence time. While swine manure, digested sludge, and *Spirulina* algae gave biocrudes of similar heating value (32–34.7 MJ/kg), they differ substantially in their detailed chemistry. The molecular weights tracked with obdurate carbohydrate content followed the order: *Spirulina* < swine manure < digested sludge.

Duan and Savage [81] were the first to evaluate the effects of various hydroprocessing catalysts on HTL of microalgae *Nannochloropsis* sp. The experiments were performed at 350°C with Pd/C, Pt/C, Ru/C, Ni/SiO_2–alpha-Al_2O_3, CoMo/l_3-Al_2O_3 (sulfide), and zeolite catalysts. In the absence of hydrogen, all catalysts gave higher yields of bio-oil, but the elemental compositions and heating value of bio-oil (about 38 MJ/kg) were insensitive to the nature of the catalyst used. Gases contained H_2, CO_2, CH_4, and lesser amounts of C_2H_4 and C_2H_6. Ru and Ni catalysts produced nitrogen. The H/C and O/C ratios of the products were about 1.7 and 0.09, respectively. While the presence of hydrogen and higher pressure suppressed the gas formation, the bio-oil yield and its characteristics did not significantly change.

Generally, high lipid content of algal mass limits its conversion to bio-oil by the HTL process. Yu et al. [83] examined the HTL process for low-lipid microalgae *Chlorella pyrenoidosa* and found that at 280°C and 120 min reaction time, the bio-oil yield of 39.4% was obtained. The bio-oil yield, water solubles, and gases strongly depended on the temperature and reaction time. Biller and Ross [84] correlated the performances of various types of algal biomass in the HTL process by correlating the bio-oil yield with the biochemical content of the biomass. They examined microalgae *Chlorella vulgaris*, *Nannochloropsis oculata*, and *Porphyridium cruentum*, and cyanobacteria *Spirulina* and found that the yields of biocrudes from these species were 5–25 wt% higher than the lipid content of the algae depending on the biochemical composition. The yields of biocrudes follow the order: Lipids > proteins > carbohydrates.

Ross et al. [94] examined the effects of alkalis and organic acids on HTL of low lipid content *C. vulgaris* and *Spirulina* algae at 300°C and 350°C, respectively. The effects of the temperature and the catalyst types on the product yields and composition were examined. The catalysts used were alkali, potassium hydroxide, sodium carbonate, and the organic acids, acetic acid and formic acid. The yields of biocrudes were higher using an organic acid catalyst and these crudes had a lower boiling point and improved the flow properties. The higher heating value ranged from 33.4 to 39.9 MJ/kg. The biocrude contained 70%–75% carbon, 10%–16% oxygen, and 4%–6% nitrogen.

Biller et al. [87] examined a range of microalgae and lipids extracted from the terrestrial oil seed for the HTL process at 350°C and 150–200 atm pressure in the presence of a variety of heterogeneous catalysts. The results showed that the HTL process converted triglycerides to fatty acids and alkanes in the presence of certain heterogeneous catalysts. While heterogeneous catalysts increased biocrudes only slightly with the use of heterogeneous catalysts, higher heating value and deoxygenation of the products increased by up to 10% due to the presence of the heterogeneous catalysts.

5.3.4 HTU Process

The HTU process is a successful pilot-scale HTL process in which biomass reacts with liquid water at an elevated temperature and pressure but under subcritical conditions. The reactor is operated under complex phase equilibria due to the simultaneous presence of water, supercritical carbon dioxide, and various alcohols along with biocrude that contains 10%–13% oxygen. In subsequent upgrading, a large portion of oxygen is removed as carbon dioxide.

In the HTU process, biomass chips (or other organic materials) are first digested by water under pressure at about 200°C–250°C. The digested slurry is then passed into a reactor that is generally operated at 300°C–350°C, 12–18 MPa, and a residence time of 5–20 min. The feed slurry contains about 25% of biomass such as wood or other organic wastes. Once biomass is converted, the product biocrude that is a mixture of light (oil) and heavy (solid) materials is separated. Light biocrude is dehydrogenated and upgraded to premium diesel fuel or kerosene, or used as a refinery feedstock. Heavy biocrude is combusted along with coal to generate electricity.

The product biocrude has higher energy density than the feed biomass and it contains alcohols, acids, and numerous other water-soluble components. The typical data of a pilot plant and the typical feedstock, the reaction conditions, and the products of HTU process are given by Demirbas [3,4]. The process is simple and of high efficiency.

5.4 HYDROTHERMAL GASIFICATION

The fast hydrolysis of organic molecules such as biomass at high temperature leads to a rapid degradation of the polymeric structure of biomass [6–11,91]. A series of consecutive reactions lead to the formation of gas whose composition depends on the temperature and pressure of water, the contact time, and the catalyst if it is present. High solubility of intermediates in water, particularly at high temperature and pressure, allows further organic reactions to occur in aqueous media and prevents the formation of tar and coke. The reactive species originating from biomass (or other species) are diluted by solvation in water, thereby preventing polymerization to unwanted products. These conditions also lead to the formation of high gas yield at relatively low temperatures. The HTG process is thus the process of gaseous fuel generation in an aqueous medium, which differs from "steam gasification" in which solids react with gaseous steam to produce a set of gaseous products.

The goal of HTG under subcritical conditions is to obtain high quality and yield of fuel gas. Two most important components of fuel gas are hydrogen and methane. As discussed earlier, steam gasification and reforming generates gas with high hydrogen concentration. Thermochemical formation of methane is possible only by low-temperature hydrothermal route since in conventional steam or oxygen gasification process, temperatures are generally too high for the methane production from biomass. The HTG under subcritical conditions can be divided into two parts: (1) low-temperature APR and its derivative technologies, and (2) high-temperature catalytic gasification. The APR and its derivative technologies are discussed in Chapter 6. Here we address the subject of high-temperature catalytic gasification.

At higher temperatures up to supercritical temperature, in the presence of a catalyst, biomass or organic compounds are gasified mainly to methane and carbon dioxide. In the absence of a catalyst, this region of temperature (250°C to critical temperature, 374°C) is also called HTL region wherein carbohydrates are liquefied to various organic products. In the catalytic HTG process, the heat recovery is important for an efficient operation. The catalytic HTG process converts biomass/water slurry into fuel gas and water that are subsequently separated. The gaseous fuel can be used for heat, power, or the generation of various chemicals. The role of catalysts on HTG is described in Section 5.4.1.

5.4.1 Catalysts for HTG

The HTG can be divided into three regions depending on the range of temperature [6–11,95–105]. Osada et al. [98–101] identified region 1 as the one with a temperature range of 500°C–700°C; supercritical water in which biomass decomposes and the activated carbon can be used to avoid char formation or alkali catalyst to facilitate water–gas shift reaction. In this region, very little solids are remained and the main product of the gasification is hydrogen. In region 2, where the temperature range is 374°C–500°C that is again in the supercritical region, biomass hydrolyzes and metal catalyst facilitates gasification. In this region, the main product is hydrogen with some carbon dioxide, carbon monoxide, and methane. Both regions 1 and 2 producing fuel in supercritical water are discussed in Chapter 10.

In this section, we focus on region 3 where temperature is below the critical temperature of 374°C. In this case, biomass hydrolysis is slow and catalysts are required for gas formation. In the subcritical region, the gas product distribution will be dictated by the thermodynamic equilibrium at a given temperature and pressure. In general, in the subcritical region, more methane is produced compared to hydrogen. The partial pressure of water can also affect the gas composition. Higher partial pressure and lower biomass concentration can result in more steam reforming producing more hydrogen. An appropriate catalyst (such as nickel) can also reform methane to produce more hydrogen. The catalyst can also help to reduce the gasification temperature while maintaining useful kinetics. Lower temperature and pressure help in lowering the capital costs for the equipment as well as lowering the possible corrosion effect on the reactor walls, thus allowing the use of less costly alloys for the reactor vessel.

The catalysts for biomass gasification under subcritical conditions are discussed in an excellent review by Elliot [9]. His analysis is briefly described below.

Elliot et al. [102–105] examined the subcritical gasification of biomass feedstock that included cellulose, lignin, hollocellulose (cellulose and hemicellulose), and a Douglas fir wood flour using nickel catalyst and added sodium carbonate cocatalyst. The results showed that at 350°C, the catalyst gave 42% of carbon fed compared to 15% of carbon fed in the absence of catalyst. Both hydrogen and methane concentrations were higher for the catalytic operations compared to those without catalyst. The carbon monoxide concentration was close to zero in the presence of catalyst. With regard to the activity of alkali additions, the activity follows the order: Cs > K > Na. The study by Elliot et al. [102–105] also indicated that conventional support for nickel, namely, alumina (other than alpha-alumina), silica, various ceramic supports, minerals such as kieselguhr and other silica-alumina, were unstable in a hot liquid water environment due to mechanisms such as dissolution, phase transition, and hydrolysis. They reported useful supports such as carbon, monoclinic zirconia or titania, and alpha-alumina.

Elliot evaluated the base metal catalysis, noble metal catalysis, and activated carbon catalysis for HTG. His important conclusions are summarized as follows:

1. Of all the base metal catalysts examined [102–105], such as nickel, magnesium, tungsten, molybdenum, zinc, chromium, cobalt, rhenium, tin, and lead, nickel was found to be the most active and stable catalyst. Various supports such as kieselguhr, silica-alumina, alpha-alumina, alumina-magnesia in spinel form, and carbon examined in the literature [6–11,91,102–105] gave a varying degree of success. The most useful promoters were ruthenium, copper, silver, and tin impregnated at 1 wt%.
2. For noble metal catalysis, while some conflicting results are reported by various investigators [6–11,91,95–105], in general, platinum, palladium, and silver showed minor activities to HTG at 350°C; iridium had some activities but the best activities were shown by ruthenium and rhodium. Rutile form of titania and carbon supports was found to be effective. Vogel et al. [96] and Vogal and Hildebrand [97] found ruthenium doping on nickel catalyst on carbon to be effective for HTG.
3. While activated carbon and charcoal were found to be the effective catalysts by some investigators [6–11,91,95–105], these results were mostly obtained under supercritical conditions.

The study by Minowa and Ogi [71] indicated that the cellulose gasification depends on the nature of support and the size of metal particles on the support. They presented the following mechanism for the cellulose gasification:

$$\text{Cellulose} \xrightarrow{\text{Decompose}} \text{Water soluble products} \xrightarrow{\text{Gasification/Ni}}$$
$$\text{Gases } (H_2 + CO_2) \xrightarrow{\text{Methanation/Ni}} \text{Gases } (CH_4 + CO_2) \quad (5.7)$$

Vogel group [96,97] indicated that Raney nickel was more effective than alpha-alumina-supported nickel. They also studied the nickel catalysts with ruthenium, copper, and molybdenum doping. The most effective results were obtained from ruthenium doping on nickel catalysts. Elliot [9] and Elliot et al. [102–105] reported that at 350°C, bimetallic Ru/Ni, Ru/C, and Cu/Ni gave favorable gas production by HTG of a variety of biomass. Favorable yields were obtained for lignin gasification by Ru/TiO_2, Ru/Al_2O_3, Ru/C, and Rh/C catalysts.

Favorable results for HTG of various biomass have been obtained for both batch and continuous systems. Ro et al. [95] showed that the subcritical HTG of hog manure feedstock can be the net energy producer for the solids concentration >0.8 wt%. While the costs for gasification are higher than those for anaerobic digestion lagoon system, the land requirement for the gasification process and the cost of transportation and tipping fees are lower. In addition, the catalytic gasification process would destroy pathogens and bioactive organic compounds, and will produce relatively clean water for reuse. The ammonia and phosphate byproducts generated in gasification have also the potential value in the fertilizer market.

5.5 COAL–WATER CHEMISTRY

While the affinity of coal with water is not as pronounced as that of biomass, evidences have shown that chemical interactions between coal and water can be significant. Here, we examine these interactions for three different cases: (1) the effect of pretreatment of coal by water on the coal conversion during coal liquefaction, (2) the effect of water on coal liquefaction at high temperatures and pressures, and (3) the effectiveness of coal–water slurry as a fuel for combustion in boilers, diesel engines, and gas turbines.

5.5.1 Effect of Water Pretreatment of Coal on Coal Liquefaction

The effect of water pretreatment of coal on coal liquefaction was studied by Serio et al. [106,107] (Serio et al., 2012, pers. comm.) and Ross and Hirschon [108]. Serio et al. examined four different types of coals (Zap lignite, Wyodak subbituminous, Illinois No. 6 bituminous, and Pittsburgh bituminous) pretreated by water at 4000 psig and 350°C and for the treatment times from 5 to 1200 min in a batch reactor. For each experiment, the yields of gases, water-soluble materials, and residues were determined. The residues were subjected to an analysis by a variety of techniques such as thermogravimetry coupled with Fourier transform infrared spectroscopy (TG-FTIR), solvent extraction, donor solvent liquefaction, and FTIR. The study resulted in the following conclusions:

1. At short pretreatment times, the process loosened up the coal structure resulting in the increase of extractables and the yield. The oxygen content also decreased when coal was subjected to an accelerated aging process. However, the liquefaction yields appear to decrease relative to the raw coal.
2. At longer pretreatment times, the process partly recombines the structure resulting in a decline of extractable and tar yields. Oxygen continues to be removed, but ether groups go through a maximum. The liquefaction yields were closer to values for the raw coal.
3. The solvent adduction may be the reason for the decline in liquefaction yields for coals with short pretreatment times.
4. For Illinois coal, the yields were very sensitive to the amount of oxygen exposure. The participation by the oxidized form of pyrite in the liquefaction pretreatment chemistry appeared possible.

Bienkowski et al. [109,110] evaluated the effect of steam pretreatment on coal liquefaction. For a Wyodak coal stored under water (to avoid weathering), they pretreated the coal using 750 psig steam for 30 min at 200°C. Pretreatment of suction dried coal at 200°C increased the production of extractables at 400°C from 30.5% to 38.5%. While an increase in the pretreatment temperature to 240°C increased the yield to 40.3% an increase in the pretreatment temperature to 320°C reduced the conversion to 33.8%. Bienkowski et al. [109,110] argued that an increase in the pretreatment temperature increased coal matrix loosening and stabilization of some reactive components of the coal resulting in higher conversion. A further increase in temperature set up higher rate of retrogressive reaction, which in turn decreased the conversion. Bienkowski et al. [110] also found that an addition of ammonia in both the pretreatment and subsequent liquefaction stages gave even higher conversion due to the reactions between hydrogen and oxygen functional groups.

Graff and Brandes [111,112] (Graff and Brandes, 2012, pers. comm.) and Brandes et al. [113] observed higher yields of liquid products from pyrolysis and solvent extraction of Illinois No. 6 coal that was pretreated by steam at 320°C–360°C and 50 atm pressure. A similar pretreatment with helium had no effect and the exposure to air of steam-pretreated coal lost the increase in yields. The study concluded that the pretreatment disrupts the hydrogen bonds, reduces the number of covalent cross-links, and increases the hydroxyl groups in the coal [111,112] (Graff and Brandes, 2012, pers. comm.). The exposure to air weathers the coal with a negative effect on liquefaction yield. Khan et al. [114] showed that the steam pretreatment at a pressure of 1100–1300 psig and a temperature of 300°C–320°C for five coals of different ranks did not increase the tar yields when pyrolyzed at a slow heating rate. The steam treatment reduced the concentration of oxygen functional groups for the low-rank coals and increased the tar yields when pyrolyzed at a rapid heating rate.

Ross et al. [115–119] evaluated the effects of water pretreatment on Illinois No. 6 and Wyodak subbituminous coals and found no effects on toluene solubles in a subsequent donor solvent liquefaction process. Significant changes were, however, observed in the composition and molecular weight distributions of the liquid products of the liquefaction process due to water pretreatment. The pretreatments were carried out at 250°C and 38 atm pressure. The coal liquefaction was carried out at 400°C and 500 psi H_2 pressure for 20 min in the tetralin solvent.

5.5.2 Coal Liquefaction in High-Pressure and High-Temperature Water

A number of studies [52,120–124] have examined the coal liquefaction in water at high temperature and pressure. Mikita et al. [52] and Blaustein et al. (Blaustein et al., 2012, pers. comm.) found tetrahydrofuran (THF) conversion of Illinois No. 6 coal in water to be about 67%, in water and solvent-refined coal (SRC) II solvent about 87%, and in water and 1000 ppm of Mo about 90% for reactions at 385°C, 1200 psig H_2 pressure, and 30 min residence time. A synergism was observed at low ratios (≤0.5) of donor solvent to coal upon combination of SRC II distillate and water. A similar effect was not observed when cyclododecane replaced water. The addition of Mo catalyst precursors to the water allowed a complete elimination of donor solvent without loss in conversion.

Yoneyyama et al. [120] examined noncatalytic hydrogenation of several bituminous and subbituminous coals with or without water addition at 400°C. By comparison, similar experiments in nitrogen or undecane (n-C_{11}) were also carried out. In nitrogen or hydrogen atmosphere, water promoted coal conversion, but the addition of undecane neither changed nor decreased the conversions. For higher rank coal, undecane inhibited coal conversion in nitrogen. The conversion of coals using nitrogen and water increased with increasing carbon content of coals. However, when hydrogen and water were used, there existed no clear relationship between the coal conversion and the carbon content of coals. Under pressurized hydrogen, coals containing pyrites gave significantly larger conversions implying their catalytic role in the conversion process. A synergistic effect existed between hydrogen and water on the conversion of coals, and the effect was more obvious for the coals containing larger amount of pyrite.

Ross and Blessing [121] and Ross et al. [122,123] found that for Illinois No. 6 coal in the CO/H_2O system at 4000–5000 psig pressure and 400°C (under supercritical conditions), better toluene solubles were achieved than for tetralin under the same conditions. The CO/H_2O system was more effective than the H_2/H_2O system and the latter system was not very effective for demineralized coal. The results were explained in terms of an ionic mechanism involving the initial formation of formate ion by which hydrogen is donated to the coal.

Recently, Anderson [124] examined hydrothermal dissolution of coal and found that at high temperature and pressure, coal dissolution is rapid and can be taken to completion. Breaking cross-linking structures will convert high-molecular-weight structures into low-molecular-weight products that can be processed and used as high-value chemical feedstock. Product is a pumpable liquid that can be further processed. Up to 90% of the original carbon is recoverable as water-soluble product. Finally, inorganic components (pyrites, calcite) are readily converted to soluble products that can be recovered and/or treated in the liquid phase.

The studies described earlier clearly indicate that water plays an active role as a reactant for the coal liquefaction under high-temperature and high-pressure conditions. The reactive role of water is further increased near and above the supercritical conditions. Thus, water should be evaluated as a possible solvent for the coal liquefaction process.

5.5.3 Coal–Water Mixture as Fuel

A slurry of finely powdered coal and water (coal–water mixture as fuel [CWF]) has been found to be an effective fuel for combustion purposes. Presence of water in CWF reduces harmful emissions into the atmosphere, makes the coal explosion proof, and also makes the coal equivalent to liquid fuel [125–129] (Penn State's coal-water slurry fuel program, 2012, pers. comm.). CWF can be used in place of oil and gas in any size of heating and power station. It can be used in oil and coal boilers. It can also be used in the diesel engine power plants and the combined cycle gas turbines. While the energy efficiency of CWF may be somewhat lower (by about 3%) compared to natural gas and oil, depending on the geographical area, the price per unit energy of CWF can be 30%–70% lower than the equivalent oil or gas. Low emissions and low BTU cost make CWF a very cost-effective and environment-friendly fuel for heat and power generation. Another advantage of CWF production process is the separation of noncarbon material that reduces the ash content by about 2% in CWF, making it a viable alternative to diesel fuel #2 for use in large stationary engines or diesel electric locomotives [125–129] (Penn State's coal-water slurry fuel program, 2012, pers. comm.).

While the first patent on coal–water fuels was granted in 1891 [125], the real development of coal–water fuels from high-quality coal commenced in earnest in the United States, Germany, and the former Soviet Union in the 1960s. In the United States, the research was accelerated in 1970 and early 1980 following oil embargo and subsequent increase in oil price. While the original objectives of the research were to produce a cheap substitute for heavy oils in boilers, in the subsequent years fuels that met environmental regulations and that can also be used for diesel engines and turbines became important. The use of CWF in boilers, diesel engines, and turbines required different level of chemical and physical properties and specialized equipment for handling and transporting slurries [125–129] (Penn State's coal-water slurry fuel program, 2012, pers. comm.).

While the thermal efficiency of CWF in boilers is around 2%–3% lower than that of coal, the intense environmental regulations in 1990 for clean coal technology forced more research for cleaner and the one with better physical and chemical properties of CWF. The R&D leads to processes that can produce ultralow-ash feed coals, especially for high-value metallurgical applications, such as the production of electrode carbons [125–129]. Processes include advanced physical processing to produce "super coal" that has a very low residual ash and very fine particle size so 60%–70% coal in CWF can be burned cleanly and possess the physical and chemical properties that are acceptable to boilers, diesel engines, and turbines. The use of CWF in diesel engines and turbines particularly required very fine coal particles in the slurry. The research led to the development of chemical processes that either remove fine residual ash (e.g., ultra clean coal [UCC] and CENfuel), from coal or coal dissolution processes that produce ash free synthetic coal-like material (e.g., Hypercoal). Although none of these processes reached immediate commercial development, these advanced coal beneficiation techniques gave a significant impetus to the use of CWF for the following reasons [125–129] (Penn State's coal-water slurry fuel program, 2012, pers. comm.):

1. CWF produced from ultra clean coal can replace fuel oils used in high-efficiency gas turbines and low–medium speed diesel engines. The cycle efficiency for gas turbines or diesel engines is not negatively impacted by the water content in CWF.
2. CWF facilitates pipeline transportation and storage, and gives additional reductions in greenhouse gases (GHGs). The convenience of easy transport of CWF is a benefit for many countries with overloaded transport infrastructure.
3. Large and fuel-efficient diesel engines for stationary power generation are especially suitable for retrofit to burn CWF. Small gas turbines and diesel engines can also utilize CWF with high efficiency.

The preparation of CWF involves crushing the coal particles to 10–65 μm particle size, although the particle size of 10–25 μm is more desirable. This can be achieved by the standard grinding or crushing processes. This is generally followed by the wet milling and homogenization process using standard milling processes. Some additives may be used to facilitate the process. CWF has been prepared with a number of coals such as lignite, flame and gas flame coals, anthracite, and brown coals [125–129] (Penn State's coal-water slurry fuel program, 2012, pers. comm.). If CWF is to be used for gas/oil boilers, the ash content should be <10%. For coal boilers, no limit on ash content is necessary. According to the literature [125–129] (Penn State's coal-water slurry fuel program, 2012, pers. comm.), CWF for brown coal (lignite) has been successfully tested. For flame coal with 40–45 vol% slurry and gas flame coal and gas coal with 28–40 vol% slurry, the systems are well developed. For anthracite with 7–28 vol% slurry, CWF is possible and has been successfully tested [125–129] (Penn State's coal-water slurry fuel program, 2012, pers. comm.).

CWF can be used in several different applications [125–129] (Penn State's coal-water slurry fuel program, 2012, pers. comm.) such as a possible substitute for heavy-grade fuel oils such as diesel #6, bunker C, and bunker D residual fuel oils. When a particle size is ≤80 μm, it can be used as co-fuel and substitute fuel in diesel engines [125–129]. Low-speed marine and modular power plant diesels can operate on pure CWF. Medium-speed diesels such as locomotives sometimes need coinjection of CWF and diesel #2 fuel that acts as an ignition source for the CWF. For the use of CWF in gas turbines, fine particles such as 5–10 μm of coal are needed to substitute petroleum and natural gas in these usages. The particle size of coal is an important factor in making homogeneous CWF that can be easily atomized in various types of engines [125–129] (Penn State's coal-water slurry fuel program, 2012, pers. comm.).

Coal impurities prohibit the use of CWF in turbines and diesel engines. Although CWF with significantly higher specifications than coals used in previous turbine and diesel engine tests is now possible with the development of the UCC product described in Section 5.5.3.1.1. Several programs initiated by the Department of Energy in the late 1970s help the development better injection systems for the use of CWF in diesel engines and turbines. The major considerations for CWF use are ignition timing, plugging, and sticking issues [125–129]. The use of CWF in diesel engines and gas turbines are now possible. Direct firing of coal requires micronizing to <20–30 μm for diesel engines and <10 μm for gas turbines and producing a CWF containing around 50 wt% coal. In the past, CWF was largely used for the compression ignition (diesel) engine [125–129]. The characteristics of injection and combustion of CWF in diesel engines are significantly different to those for diesel fuels due to the combined effects of poorer atomization and the time required to evaporate the slurry water. However, combustion and thermal efficiencies matching diesel fuel have been achieved for CWF at up to 1900 rpm [125–129].

The most researched area has been the design of the injectors, which gives the optimum atomization of fuel for the best combustion and thermal efficiencies. Coal particle size and rheology of coal–water mixture play a very important role in efficient atomization. The engine modifications such as purged shuttle fuel pump plunger, electronically timed injection, diamond compact injector tip nozzles, tungsten carbide-sprayed cylinder liner and top ring set, and pilot injection of diesel are some of the engine modifications considered for the successful direct injection of CWF [125–129]. The fate of mineral matter and its effect on the engine wear and how to minimize coal agglomeration during the evaporation of individual CWF droplets are also required further investigations [125–129] (Penn State's coal-water slurry fuel program, 2012, pers. comm.).

5.5.3.1 Production of CWF

The production of CWF requires sophisticated treatment of coal to remove mineral matters and sulfur, and prepare a very fine particle size such that CWF can be used as a replacement of heavy fuel oil not only in boilers but also in diesel engines and gas turbines. The preparation of such coal follows multiple steps: physical cleaning, advanced coal processing projects, advanced coal milling, and chemical cleaning.

Physical cleaning of coal is carried out by a wide array of solid–liquid and solid–solid separation processes. Floatation technologies and various dewatering systems are some of the processes used for this purpose. The objective is to remove ash and other mineral matters without losing coal. Most of the current milling, separation, and dewashing techniques allowed coal particles to get down to 30–40-μm size. Further removal of impurities and mineral matters required the applications of chemical methods.

Advanced physical coal cleaning was developed by Bechtel and AMAX [130]. The primary objective was to produce UCCs suitable for conversion to stable and highly loaded CWF. The main specification was an ash content of <1%–2%.The separation technologies were advanced column froth floatation and selective agglomeration. A more novel process for preparation of ultraclean micronized coal was researched in China, based around high-pressure water jet milling [131,132]. It was found that hydraulically milled coal significantly increased the liberation of minerals (97% vs. 90% for ball milling) and led to an improved overall mineral separation.

Ultrafine coal milling is an essential part of firing coal into gas turbines or diesel engines. For gas turbines, the particle size of ≤10 μm is desirable. For diesel engine, the top size of 20 μm has been specified, although this depends on the size and speed of the engine. Milling energy depends on both the type of mill and the material especially at small particle size required for CWF in turbines and engines. A number of advanced mills that are now available include ball mills, centrificial or planetary mills, nutating mills, opposed flow jet mills, impact jet mills, spiral jet mills, and high-pressure water jet mill [125–129] (Penn State's coal-water slurry fuel program, 2012, pers. comm.). While nutating mill appears to offer the lowest energy consumption, the final choice of mill may depend on the interaction of milling with the deashing process. Another factor affecting the choice may be the particle size distribution required for the deashing technology to be employed for each CWF product.

Two types of processes for chemical cleaning of coal are (1) those that attempt to dissolve the mineral components of coal (e.g., UCC, CENfuel) and (2) those that dissolve the coal leaving a mineral-rich insoluble coal byproduct (e.g., Hypercoal). We briefly examine these three important processes in Sections 5.5.3.1.1 through 5.5.3.1.3.

5.5.3.1.1 Ultra Clean Coal

The UCC production process [116] involves two main steps: a caustic pressure leach to convert silicates and clays to dissolved sodium silicates and sodalite-type minerals. The sodalite material is then dissolved in acid so that it can be removed with the filtrate in a simple filtering operation [125–129]. The key features of the technology are as follows:

1. Since coal pulverization is not required, the solid–liquid separation easy.
2. Digestion removes both extraneous and a large portion of minerals within the coal particles. The process also removes most of alkalis, all of the inorganic sulfur, and some of the organic sulfur. The UCC product contains about 30% moisture.
3. The process is capable of treating most bituminous coals.

The process is capable of meeting gas turbine specifications for all bituminous coals [125–129,133] (Penn State's coal-water slurry fuel program, 2012, pers. comm.).

5.5.3.1.2 CENfuel

CENfuel produces ultralow-ash coal by an acid regeneration and with the removal of other deleterious elements from coal. In the process, the main ash components such as SiO_2, Al_2O_3, and TiO_2 are removed by leaching granular coal (2 mm) with an aqueous solution of hydrofluoric acid and fluosilicic acids. Sulfoides such as iron pyrites are not affected by leach. The rich liquor contains soluble fluosilicates and undissolved FeS particles. The liquor is passed to a distillation unit where metal fluorides are recovered and removed from the system. The residue is dried and stored. The spent liquor is dried and sent to the gas absorber where hydrogen fluoride (HF) and H_2SiF_6 are recovered and excess H_2SiF_6 is passed to a hydrolyzer for conversion to silica and HF is returned to the dissolution circuit [134,135]. More improved process contains two dissolution steps [125–129] involving hydrofluoric acid and fluosilicic acids (Penn State's coal-water slurry fuel program, 2012, pers. comm.).

5.5.3.1.3 Hypercoal

Hypercoal is a low-ash, low-alkali coal product produced by dissolving the coal matter into an organic solvent, then flashing off the solvent for recycling to the dissolution step of the process [125–129,136] (Penn State's coal-water slurry fuel program, 2012, pers. comm.). The insolubles are retained in the high-ash byproduct coal. Hypercoal process is very different from UCC process in that this process aims to separate solvent-soluble coaly matter from the ash and insoluble coal, thereby producing a high-ash coproduct. The process involves five steps: slurry preparation, extraction at 360°C temperature, separation of the extracts, removal of alkali from the liquid component and removal of ultrafine particles by filtration, and finally drying of the final product. A number of solvents such as tetralin, 1-methyl-naphthalene, dimethylnaphthalene, and light cycle oil have been used as solvents for extraction. The key features of the technology are as follows [125–129]:

1. On a dry coal basis, the yield can be as high as 80%. The process removes most alkalis from raw coal.
2. The higher ash (coproduct) is suitable for domestic power generation.
3. The process can be applied to all subbituminous and bituminous coals. Yields are lower for subbituminous coals than for bituminous coals. Yields for subbituminous coals can be increased by pretreating the coal with HCl or weaker acids.

5.5.3.2 Fuel Preparation and Transportation

The science and technology behind the production of UCC CWF has been well published [125–130] (Penn State's coal-water slurry fuel program, 2012, pers. comm.). Most systems involve the preparation of coal–water slurries containing 60%–70% coal, together with additives to provide slurry stabilization and to lower the viscosity. The energy density of such slurry is about 18 GJ/m^3. Additives consist of dispersants and stabilizers. The dispersants such as sodium sulfonate of naphthalene, polystyrene, polymethacrylate, and polyolefin maintain the separation of coal particles within the slurry [125–130]. Stabilizers include additives such as cellulose or xanthum gums. For gas turbines and diesel engines, the water penalty for CWF is much smaller and probably negligible when the overall power cycle is considered, and therefore, CWF slurry transportation is a preferred form. Final preparation of CWF requires that the coal is either premilled dry before slurry preparation or milled (micronized) wet as either a part of slurry preparation process or immediately prior to combustion. In general, wet milling has lower cost and lower energy consumption [125–130].

In China and Japan, CWF has been produced in large plants for more than 30 years. In a typical CWF production plant, the coal is mixed with water and some additives (to help forming good mixture) and passed through one or more pulverizers and multiple milling (high-load and low-load) processes. For boiler application of CWF, high slurry concentration (65–70 wt%) and better stability requires a wide particle size distribution (often bimodal) with mass mean particle size between 5 and 10 μm. This is often achieved by using several mills and/or recycle streams.

The efficient atomization of CWF slurry in a combustor governs the required particle size and slurry concentration. For gas turbine and diesel engine applications of CWF, generally lower slurry concentration and mass mean particle size of coal are required. For diesel engines, the preferable slurry concentration is 50–55 wt% with a mass mean particle size of 5–15 microns. For turbines, the preferable slurry concentration is 55–60 wt% with a mass mean particle size of 4–6 microns. Since the cost of milling and pulverizing rises exponentially with a decrease in particle size below 30–50 microns, low particle sizes required in diesel engines and turbines will necessitate the use of special milling and pulverizing processes.

CWF exhibits the rheological properties different from fuel oils. Fuel oils tend to be more Bingham fluids. CWF is pourable and pumpable, but its viscosity decreases significantly with the shear caused by agitation and pumping. Also unlike fuel oils, the viscosity of CWF is unaffected by the temperature. Slurry viscosities are strongly affected by the coal characteristics, concentration, and flow conditions. Currently, CWF containing 65%–70% coal has an apparent viscosity around 1000 MPa-s at room temperature. This is too high for atomization of slurry in the combustion chamber, and it is normally reduced either adding water (10%–20%) or heating before the injection, which can promote flashing.

The use of CWF in diesel engines and gas turbines besides its use as a substitute for heavy oil in boilers requires the following considerations and additional research:

1. CWF is more difficult to atomize than diesel fuel due to its much higher viscosity. The effective atomization is more critical to combustion due to the effect of droplet size on ignition delay (which is caused by the time required for water evaporation) and burnout. Pressure atomization can be improved by increasing the liquid velocity through the nozzles; however, this greatly increases the nozzle wear.
2. CWF causes chronic wear of injection nozzles, with wear being exacerbated by cavitation effects.

More research has been continuing to address these issues. However, CWF combustion has a very promising future.

5.5.3.3 Combustion of CWF

Fu et al. (2012, pers. comm.) carried out the earlier combustion experiments for CWF in a boiler using oxygen-enriched air. Their study for 700 hp watertube boiler with bituminous coal indicated that the use of oxygen-enriched air resulted in the required lower air preheating and the improvement in the carbon burnout.

The reduction in the volume of flue gas lowered the heat losses and increased the boiler efficiency. The air-preheating temperature was reduced by 192°F by enriching combustion air by 22.8% volume of oxygen. The boiler performance was significantly improved even with a small addition of oxygen (2%–3%) in air. With the use of staged air admissions, the NO_x emission was also reduced by one-third. This was, however, accompanied by some decrease in combustion efficiency. The use of oxygen-enriched air in the primary combustion stage increased flame stability, reduced carbon burnout, and moderately reduced the overall NO_x emissions. The carbon conversion (96%–98%) and the boiler efficiencies (81%–83%) remained high in their entire study.

The issues of CWF atomization and its effect on the optimization of combustion efficiencies in boilers, diesel engines, and gas turbines are continued to be investigated. Coal–water slurry has, however, proven to be an important synthetic fuel. It has a significant economic potential [137] and its application to generate heat and power will continue to grow [138–140].

REFERENCES

1. Toor, S., Rosendahl, L., and Rudolf, A., "Hydrothermal liquefaction of biomass: A review of subcritical water technologies," *Energy*, 36, 2328–2342 (2001).
2. Gullon, P., Conde, E., Moure, A., Dominguez, H., and Parajo, J., "Selected process alternatives for biomass refining: A review," *The Open Agriculture Journal*, 4, 135–144 (2010).
3. Demirbas, A., "Biorefineries: Current activities and future developments," *Energy Conversion and Management*, 50, 2782–2801 (2009).
4. Demirbas, A., "Sub and supercritical water depolymerization of biomass," *Energy Sources, Part A*, 32, 1100–1110 (2010).
5. Peterson, A., Vogel, F., Lachance, R., Frolling, M., Antal, M., and Tester, J., "Thermochemical biofuel production in hydrothermal media: A review of sub and supercritical technologies," *Energy & Environmental Science*, 1, 32–65 (2008).
6. Azadi, P. and Farnood, R., "Review of heterogeneous catalysts for sub and supercritical water gasification of biomass and wastes," *International Journal of Hydrogen Energy*, 36, 9529–9541 (2011).
7. Tanksale, A., Beltramini, J., and Lu, G., "A review of catalytic hydrogen production processes from biomass," *Renewable & Sustainable Energy Reviews*, 14, 166–182 (2010).
8. Kruse, A., "Hydrothermal biomass gasification," *The Journal of Supercritical Fluids*, 47, 391–399 (2009).
9. Elliott, D.C., "Catalytic hydrothermal gasification of biomass," *Biofuels, Bioproducts and Biorefining*, 2, 254–265 (2008).
10. Kong, L., Li, G., Zhang, B., He, W., and Wang, H., "Hydrogen production from biomass wastes by hydrothermal gasification," *Energy Sources, Part A*, 30, 1166–1178 (2008).
11. Akiya, N. and Savage, P.E., "Roles of water for chemical reactions in high-temperature water," *Chemical Reviews*, 102 (8), 2725–2750 (2002).
12. Siskin, M. and Katritzky, A.R., "Reactivity of organic compounds in superheated water: General background," *Chemical Reviews*, 101 (4), 825–835 (2001); *Environmental Science & Technology*, 45, 5696–5703 (2011).
13. Wantanabe, M., Sato, T., Inomata, H., Smith, R., Arai, K., Kruse, A., and Dinjus, E., "Chemical reactions of C1 compounds in near critical and supercritical water," *Chemical Reviews*, 104, 5803–5821 (2004).
14. Savage, P., "Organic chemical reactions in supercritical water," *Chemical Reviews*, 99, 603–621 (1999).

15. Libra, J., Ro, K., Kammann, C., Funke, A., Berge, N., Neubauer, Y., Titirici, M., Fuhner, C., Bens, O., and Emmerich, K., "Hydrothermal carbonization of biomass residuals: A comparative review of the chemistry, processes and applications of wet and dry pyrolysis," *Biofuels*, 2 (1), 89–124 (2011).
16. Kumar, S., Lognathan, V., Gupta, R., and Barnett, M., "An assessment of U (VI) removal from groundwater using biochar produced from hydrothermal carbonization," *Journal of Environmental Management*, 92, 2504–2512 (2011).
17. Ding, Z., Frisch, M., Li, L., and Gloyna, E., "Catalytic oxidation in supercritical water," *Industrial & Engineering Chemistry Research*, 35, 3257–3279 (1996).
18. Funke, A. and Ziegler, F., "Hydrothermal carbonization of biomass: A literature survey focusing on its technical application and prospects," *Proceedings of the 17th European Biomass Conference and Exhibition*, June 29–July 3, Hamburg, Germany (2009).
19. Funke, A. and Ziegler, F. "Hydrothermal carbonization of biomass: A summary and discussion of chemical mechanisms for process engineering," *Biofuels, Bioproducts and Biorefining*, 4 (2), 160–177 (2010).
20. Titirici, M., White, R., Falco, C., and Sevilla, M., "Black perspectives for a green future: Hydrothermal carbons for environment protection and energy storage," *Energy & Environmental Science*, 5, 6796–6822 (2012).
21. Hu, B., Wang, K., Wu, L., Yu, S., Antonietti, M., and Titiricia, M., "Engineering carbon materials from the hydrothermal carbonization process of biomass," *Advanced Materials*, 22, 813–828 (2010).
22. Erlach, B. and Tsatsaronis, G., "In upgrading of biomass by hydrothermal carbonisation: Analysis of an industrial-scale plant design," *Proceedings of ECOS 23rd International Conference on Efficiency, Cost, Optimization, Simulation and Environmental Impact of Energy Systems*, June 14–17 (2010).
23. Titirici, M.M., Thomas, A., Yu, S.H., Muller, J.O., and Antonietti, M., "A direct synthesis of mesoporous carbons with bicontinuous pore morphology from crude plant material by hydrothermal carbonization," *Chemistry of Materials*, 19 (17), 4205–4212 (2007).
24. Sevilla, M. and Fuertes, A.B., "The production of carbon materials by hydrothermal carbonization of cellulose," *Carbon*, 47 (9), 2281–2289 (2009).
25. Sevilla, M. and Fuertes, A.B., "Chemical and structural properties of carbonaceous products obtained by hydrothermal carbonizatioon of saccharides," *Chemistry—A European Journal*, 15, 4195–4203 (2009).
26. White, R.J., Budarin, V., Luque, R., Clark, J.H., and Macquarrie, D.J., "Tuneable porous carbonaceous materials from renewable resources," *Chemical Society Reviews*, 38, 3401–3418 (2009).
27. Antal, M.J. and Gronli, M., "The art, science, and technology of charcoal production," *Industrial & Engineering Chemistry Research*, 42 (8), 1619–1640 (2003).
28. Regmi, P., Moscosq, J., Kumar, S., Cao, X., Mao, J., and Schafran, G., "Removal of copper and cadmium from aqueous solution using switchgrass biochar produced via hydrothermal carbonization process," *Journal of Environmental Management*, 93, 1–9 (2012).
29. Ball, R., McIntosh, A.C., and Brindley, J., "The role of char-forming process in the thermal decomposition of cellulose," *Physical Chemistry Chemical Physics*, 1, 5035–5043 (1999).
30. Yu, S., Cui, X., Li, L., Li, K., Yu, B., Antonietti, M., and Colfen, H., "From starch to metal/carbon hybrid nanostructures: Hydrothermal metal-catalyzed carbonization," *Advanced Materials*, 18, 1636–1640 (2004).
31. Titirici, M. and Antonietti, M., "Chemistry and materials options of sustainable carbon materials made by hydrothermal carbonization," *Chemical Society Reviews*, 39, 103–116 (2010).
32. Liu, Z., Zhang, F.-S., and Wu, J., "Characterization and application of chars produced from pinewood pyrolysis and hydrothermal treatment," *Fuel*, 89, 510–514 (2010).

33. Mao, J.-D., Fang, X., Lan, Y., Schimmelmann, A., Mastalerz, M., Xu, L., and Schmidt-Rohr, K., "Chemical and nanometer-scale structures of kerogen and their changes during thermal maturation investigated by advanced solid-state NMR spectroscopy," *Geochimica et Cosmochimica Acta*, 74, 2110–2127 (2010).
34. Mao, J.-D., Schimmelmann, A., Mastalerz, M., Hatcher, P.G., and Li, Y. "Structural features of a bituminous coal and their changes during low-temperature oxidation and loss of volatiles investigated by advanced solid-state NMR spectroscopy," *Energy & Fuels*, 24, 2536–2544 (2010).
35. Cao, X., Ro, K.S., Chappell, M., Li, Y., and Mao, J.-D., "Chemical structures of swine-manure chars produced under different carbonization conditions investigated by advanced solid-state 13C NMR spectroscopy," *Energy & Fuels*, 25, 388–397 (2011).
36. Mao, J.-D., Ajakaiye, A., Lan, Y., Olk, D.C., Ceballos, M., Zhang, T., Fan, M.Z., and Forsberg, C.W., "Chemical structures of manure from conventional and phytase transgenic pigs investigated by advanced solid-state NMR spectroscopy," *Journal of Agricultural and Food Chemistry*, 56, 2131–2138 (2008).
37. Ramke, H.G., Blohse, D., Lehmann, H.J., and Fettig, J., "Hydrothermal carbonization of organic waste," in Cossu, R., Diaz, L.F., and Stegman, R. (eds.), *Proceedings of the 12th International Waste Management and Landfill Symposium*, January 13, CISA, S. Margherita di Pula, Sardinia, Italy (2009).
38. Spokas, K.A. and Reicosky, D.C., "Impacts of sixteen different biochars on soil greenhouse gas production," *Annals of Environmental Science*, 3, 179–193 (2009).
39. Mursito, A.T., Hirajima, T., and Sasaki, K., "Upgrading and dewatering of raw tropical peat by hydrothermal treatment," *Fuel*, 89 (3), 635–641 (2010).
40. Russell, J.A., Miller, R.K., and Molton, P.M., "Formation of aromatic compounds from condensation reactions of cellulose degradation products," *Biomass*, 3 (1), 43–57 (1983).
41. Akhtar, A. and Amin, N., "A review on process conditions for optimum bio oil yield in hydrothermal liquefaction of biomass," *Renewable & Sustainable Energy Reviews*, 15, 1615–1624 (2011).
42. Midgett, J., "Assessing hydrothermal liquefaction process using biomass feedstocks," MS thesis, Department of Biological and Agricultural Engineering, Louisiana State University, Baton Rouge, LA (May 2008).
43. Huber, G., Iborra, S., and Corma, A., "Synthesis of transportation fuels from biomass: Chemistry, catalysts, and engineering," *Chemical Reviews*, 106, 1–51 (2006).
44. Demirbas, A., "Progress and recent trends in biofuels," *Progress in Energy and Combustion Science*, 33, 1–18 (2007).
45. Zhang, Y., "Hydrothermal liquefaction to convert biomass into crude oil," in Blaschek, H., Ezeji, T., and Scheffran, J. (eds.), *Biofuels from Agricultural Wastes and Byproducts*. Blackwell Publishing, New York, 201–232 (2010).
46. Behrendt, F., Neubauer, Y., Oevermann, M., Wilmes, B., and Zobel, N., "Direct liquefaction of biomass," *Chemical Engineering Technology*, 31 (5), 667–677 (2008).
47. Chen, P., Min, M., Chen, Y., Wang, L., Li, Y., Chen, Q., Wang, C. et al., "Review of the biological and engineering aspects of algae to fuels approach," *International Journal of Agricultural and Biological Engineering*, 2 (4), 1–30 (2009).
48. Sugano, M., Takagi, H., Hirano, K., and Mashimo, K., "Hydrothermal liquefaction of plantation biomass with two kinds of wastewater from paper industry," *Journal of Materials Science*, 43, 2476–2486 (2008).
49. Appell, H.R., Wender, I., and Miller, R.D., "Solubilization of Low Rank Coal with Carbon Monoxide and Water," *Chemistry & Industry*, 47, 1703 (1969).
50. Appell, H.R., *Energy*, 1, 24 (1976).
51. Blaustein, B.C., Bockrath, B.C., Davis, H.M., Friedman, S., Illig, E.C., and Mikita, M.A., *American Chemical Society, Division of Fuel Chemistry*, preprints, 30 (2), 359 (1985).

52. Mikita, M., Bockrath, B., Davis, H., Friedman, S., and Illig, E., "Water and nondonor-vehicle-assisted liquefaction of Illinois bituminous coal," *Energy & Fuels*, 2, 534–538 (1988).
53. Burkhard, K., Werner, H., and Friedhelm, B., "Catalytic hydroliquefaction of biomass with red mud and cobalt monoxide molybdenum trioxide catalysts," *Fuel*, 69 (4), 448–455 (1990).
54. Sudong, Y. and Zhongchao, T., "Hydrothermal liquefaction of cellulose to bio-oil under acidic, neutral and alkaline conditions," *Applied Energy*, 92, 234–239 (2012).
55. Theegala, C. and Midgett, J., "Hydrothermal liquefaction of separated dairy manure for production of bio-oils with simultaneous waste treatment," *Bioresource Technology*, 107, 456–463 (2012).
56. Liu, H., Xie, X., Li, M., and Sun, R., "Hydrothermal liquefaction of cypress: Effects of reaction conditions on 5-lump distribution and composition," *Journal of Analytical and Applied Pyrolysis*, 94, 177–183 (2012).
57. Kang, S., Li, B., Chang, J., and Fan, J., "Antioxidant abilities comparison of lignins with their hydrothermal liquefaction products," *BioResources*, 6 (1), 243–252 (2011).
58. Liu, Z. and Zhang, F., "Removal of copper (II) and phenol from aqueous solution using porous carbons derived from hydrothermal chars," *Desalination*, 267 (1), 101–106 (2011).
59. Yin, S., Dolan, R., Harris, M., and Tan, Z., "Subcritical hydrothermal liquefaction of cattle manure to bio-oil: Effects of conversion parameters on bio-oil yield and characterization of bio oil," *Bioresource Technology*, 101 (10), 3657–3664 (2010).
60. Xiu, S., Shahbazi, A., Shirley, V., and Cheng, D., "Hydrothermal pyrolysis of swine manure to bio-oil: Effects of operating parameters on products yield and characterization of bio-oil," *Journal of Analytical and Applied Pyrolysis*, 88 (1), 73–79 (2010).
61. Liu, Z. and Zhang, F., "Removal of lead from water using biochars prepared from hydrothermal liquefaction of biomass," *Journal of Hazardous Materials*, 167 (1–3), 933–939 (2009).
62. Kruse, A., Maniam, P., and Spieler, F., "Influence of proteins on the hydrothermal gasification and liquefaction of biomass. 2. Model compounds," *Industrial & Engineering Chemistry Research*, 46 (1), 87–96 (2007).
63. Yanagida, T., Fujimoto, S., and Minowa, T., "Application of the severity parameter for predicting viscosity during hydrothermal processing of dewatered sewage sludge for a commercial PFBC plant," *Bioresource Technology*, 101 (6), 2043–2045 (2010).
64. Zhang, B., von Keitz, M., and valentas, K., "Thermochemical liquefaction of high diversity grassland perennials," *Journal of Analytical and Applied Pyrolysis*, 84 (1), 18–24 (2009).
65. Balan, V., Kumar, S., Bals, B., Chundawat, S., Jin, M., and Dale, B., "Biochemical and thermochemical conversion of switchgrass to biofuels," in Monti, A. (ed.), *Switchgrass: A Valuable Biomass Crop for Energy*. Springer, New York, 153–185 (2012).
66. Karagoz, S., Bhaskar, T., Muto, A., Sakata, Y., Oshiki, T., and Kishimoto, T., "Low-temperature catalytic hydrothermal treatment of wood biomass: Analysis of liquid products," *Chemical Engineering Journal*, 108 (12), 127–137 (2005).
67. Boocock, D.G.B. and Sherman, K.M., "Further aspects of powdered poplar wood liquefaction by aqueous pyrolysis," *Canadian Journal of Chemical Engineering*, 3, 627–633 (2009).
68. Kobayashi, N., Okada, N., Hirakawa, A., Sato, T., Kobayashi, J., Hatano, S., Itaya, Y., and Mori, S., "Characteristics of solid residue obtained from hot compressed water treatment of woody biomass," *Industrial & Engineering Chemistry Research*, 48, 373–379 (2009).
69. Qu, Y., Wei, X., and Zhong, C., "Experimental study on the direct liquefaction of *Cunninghamia lanceolata* in water," *Energy*, 28, 597–606 (2003).
70. Karagoz, S., Bhaskar, T., Muto, A., Sakata, Y., and Azhar Uddin, Md., "Low temperature hydrothermal treatment of biomass: Effects of reaction parameters on products and boiling point distributions," *Energy & Fuels*, 18, 234–241 (2004).

71. Minowa, T. and Ogi, T., "Hydrogen production from cellulose using a reduced nickel catalyst," *Catalysis Today*, 45, 411–416 (1998).
72. Minowa, T. and Inoue, S., "Hydrogen production from biomass by catalytic gasification in compressed water," *Renewable Energy*, 16, 1114–1117 (1999).
73. Minowa, T., Murakami, M., Dote, Y., Ogi, T., and Yokoyama, S., "Oil production from garbage by thermochemical liquefaction," *Biomass & Bioenergy*, 8, 117–120 (1995).
74. Minowa, T., Kondo, T., and Sudirjo, S., "Thermochemical liquefaction of Indonesian biomass residues," *Biomass & Bioenergy*, 14, 517–524 (1998).
75. Patil, V., Tran, K., and Giselrod, H., "Towards sustainable production of biofuels from microalgae," *International Journal of Molecular Science*, 9, 1188–1195 (2008).
76. Zhou, D., Zhang, L., Zhang, S., Fu, H., and Chen, J., "Hydrothermal liquefaction of macroalgae *Enteromorpha prolifera* to bio-oil," *Energy & Fuels*, 24, 4054–4061 (2010).
77. Vardon, D., Sharma, B., Blazina, G., Rajagopalan, K., and Strathmann, T., "Thermochemical conversion of raw and defatted algal biomass via hydrothermal liquefaction and slow pyrolysis," *Bioresource Technology*, 109, 178–187 (2012).
78. Vardon, D., Sharma, B., Scott, J., Yu, G., Wang, Z., Schideman, L., Zhang, Y., and Strathmann, T., "Chemical properties of biocrude oil from the hydrothermal liquefaction of Spirulina algae, swine manure and digested anaerobic sludge," *Journal of Bioresource Technology*, 102 (17), 8295–8303 (2011).
79. Zhou, S., Wu, Y., Yang, M., Imdad, K., Li, C., and Junmao, T., "Production and characterization of bio oil from hydrothermal liquefaction of microalgae *Dunaliella tertiolecta* cake," *Energy*, 35 (12), 5406–5411 (2010).
80. Zhong, C.L. and Wei, X.M., "A comparative experimental study on the liquefaction of wood," *Energy*, 29 (11), 1731–1741 (2004).
81. Duan, P. and Savage, P., "Hydrothermal liquefaction of a microalga with heterogeneous catalysts," *Industrial & Engineering Chemistry Research*, 50 (1), 52–61 (2011).
82. Anastasakis, K. and Ross, A., "Hydrothermal liquefaction of the brown macro-alga *Laminaria saccharina*: Effect of reaction conditions on product distribution and composition," *Bioresource Technology*, 102(7), 4876–4883 (2011).
83. Yu, G., Zhang, Y., Schideman, L., Funk, T., and Wang, Z., "Hydrothermal liquefaction of low lipid content microalgae into bio crude oil," *Transactions of the ASABE*, 54 (1), 239–246 (2011).
84. Biller, P. and Ross, A., "Potential yields and properties of oil from the hydrothermal liquefaction of microalgae with different biochemical content," *Bioresource Technology*, 102 (1), 215–225 (2011).
85. Ross, A., Biller, P., Kubacki, M., Li, H., Lea-Langton, A., and Jones, J., "Hydrothermal processing of microalgae using alkali and organic acids," *Fuel*, 89 (9), 2234–2243 (2010).
86. Chakraborty, M., Miao, C., McDonald, A., and Chen, S., "Concurrent extraction of bio oil and value added polysaccharides from *Chlorella sorokiniana* using a unique sequential hydrothermal extraction technology," *Fuel*, 95, 63–70 (2012).
87. Biller, P., Riley, R., and Ross, A., "Catalytic hydrothermal processing of microalgae: Decomposition and upgrading of lipids," *Bioresource Technology*, 102 (7), 4841–4848 (2011).
88. Li, D., Chen, L., Xu, D., Zhang, X., Ye, N., Chen, F., and Chen, S., "Preparation and characteristics of bio-oil from marine brown alga *Sargassum patens* C. Agardh," *Bioresource Technology*, 104, 737–742 (2012).
89. Minowa, T., Yokoyama, S., Kishimoto, M., and Okakurat, T., "Oil production from algae cells of *Dunaliella tertiolecta* by direct thermochemical liquefaction," *Fuel*, 74, 1735–1738 (1995).
90. Dote, Y., Sawayama, S., Inoue, S., Minowa, T., and Yokoyama, S., "Recovery of liquid fuel from hydrocarbon-rich microalgae by thermochemical liquefaction," *Fuel*, 73, 1855–1857 (1994).

91. Brown, T., Duan, P., and Savage, P., "Hydrothermal liquefaction and gasification of *Nannochloropsis* sp.," *Energy & Fuels*, 24, 3639–3648 (2010).
92. Matsui, T., Nishihara, A., Ueda, C., Ohtsuki, M., Ikenaga, N., and Suzuki, T., *Fuel*, 76, 1043–1048 (1997).
93. Sawayama, S., Minowa, T., and Yakoyama, S., "Possibility of renewable energy production and CO_2 mitigation by thermochemical liquefaction of microalgae," *Biomass & Bioenergy*, 17, 33–39 (1999).
94. Ross, A.B., Jones, J., Kubacki, M., and Bridgeman, T., "Classification of microalgae as fuel and its thermochemical behavior," *Bioresource Technology*, 99, 6494–6504 (2008).
95. Ro, K., Cantrell, K., Elliot, D., and Hunt, G., "Catalytic wet gasification of municipal and animal wastes," *Industrial & Engineering Chemistry Research*, 46, 8839–8845 (2007).
96. Vogel, F., Waldner, M., Rouff, A., and Rabe, S., "Synthetic natural gas from biomass by catalytic conversion in supercritical water," *Green Chemistry*, 9, 616–619 (2007).
97. Vogel, F. and Hildebrand, F., "Catalytic hydrothermal gasification of woody biomass at high feed concentrations," *Chemical Engineering Transactions*, 2, 771–777 (2002).
98. Osada, M., Sato, T., Watanabe, M., Shirai, M., and Arai, K., "Catalytic gasification of wood biomass in subcritical and supercritical water," *Combustion Science and Technology*, 178, 537–552 (2006).
99. Osada, M., Hiyoshi, N., Sato, O., Arai, K., and Shirai, M., "Subcritical water regeneration of supported ruthenium catalyst poisoned by sulfur," *Energy & Fuels*, 22, 845–849 (2008).
100. Osada, M., Hiyoshi, N., Sato, O., Arai, K., and Shirai, M. "Reaction pathways for catalytic gasification of lignin in the presence of sulfur in supercritical water," *Energy & Fuels*, 21, 1854–1858 (2007).
101. Osada, M., Sato, T., Watanabe, M., Adschiri, T., and Arai, K., "Low temperature catalytic gasification of lignin and cellulose with a ruthenium catalyst in supercritical water," *Energy & Fuels*, 18, 327–333 (2004).
102. Elliot, D., Sealock, L., and Baker, E., "Chemical processing in high pressure aqueous environment. 3. Batch reactor process development experiments for organic destructions," *Industrial & Engineering Chemistry Research*, 33, 558–565 (1994).
103. Elliot D., Peterson, K., Muzatko, D., Alderson, E., Hart, T., and Neuenschwander, G., "Effects of trace contaminants on catalytic processing of biomass derived feedstocks," *Applied Biochemistry and Biotechnology*, 113–116, 807–825 (2004).
104. Elliot, D., Neuenschwander, G., Phelps, M., Hart, T., Zacher, A., and Silva, L., "Chemical processing in high pressure aqueous environment. 6. Demonstration of catalytic gasification for chemical manufacturing wastewater cleanup in industrial plants," *Industrial & Engineering Chemistry Research*, 38, 879–883 (1999).
105. Elliot, D., Neuenschwander, G., Hart, T., Butner, R., Zacher, A., and Englelhard, M., "Chemical processing in high pressure aqueous environments. 7. Process development for catalytic gasification of wet biomass feedstocks," *Industrial & Engineering Chemistry Research*, 43, 1999–2004 (2004).
106. Serio, M., Solomon, P., Kroo, E., and Charpenay, S., "Fundamental studies of water pretreatment of coal," DOE Report under Contract No. DE-ACR22-89PC89878, September 5, 1989–March 5, 1992, Advanced Fuel Research, East Hartford, CT (1993).
107. Serio, M.A., Solomon, P.R., Bassilakis, R., Woo, E., Malhotra, R., and McMillen, D., "Fundamental studies of retrograde reactions in direct liquefaction," Quarterly Reports under DOE/PETC Contract No. DE-AC22-88PC88814 (1989).
108. Ross, D. and Hirschon, A., "The effects of hydrothermal pretreatment on the liquefaction of coal," *American Chemical Society, Division of Fuel Chemistry*, preprint, 35 (1), 37–45 (1990).
109. Bienkowski, P.R., Narayan, R., Greenkorn, R.A., and Choa, K.W., *Industrial & Engineering Chemistry Research*, 26, 202 (1987).

110. Bienkowski, P.R., Narayan, R., Greenkorn, R.A., and Choa, K.C., *Industrial & Engineering Chemistry Research*, 26, 206 (1987).
111. Graff, R.A. and Brandes, S.D., *Proceedings of the New Fuel Forms Workshop*, US DOE Fossil Energy, Washington, DC, 35 (1986).
112. Graff, R.A. and Brandes, S.D., *Energy & Fuels*, 1, 84 (1987).
113. Brandes, S.D., Graff, R.A., Gorbaty, M.L., and Siskin, M., *Energy & Fuels*, 3, 494 (1989).
114. Khan, M.R., Chen, W.-Y., and Suuberg, E.M., *Energy & Fuels*, 3, 223 (1989).
115. Ross, D.S., Green, T.K., Mansani, R., and Hum, G.P., *Energy & Fuels*, 1, 287 (1987).
116. Ross, D.S., Green, T.K., Mansani, R., and Hum, G.P., *Energy & Fuels*, 1, 292 (1987).
117. Ross, D.S. and Hirschon, A., *American Chemical Society, Division of Fuel Chemistry*, preprint, 35 (1), 37 (1990).
118. Ross, D.S., Hirschon, A., Tse, D.S., and Loo, B.H., *American Chemical Society, Division of Fuel Chemistry*, preprint, 35 (2), 352 (1990).
119. Ross, D.S., McMillen, D., Ogier, W, Fleming, R., and Hum, G., "Exploratory study of coal conversion chemistry," Quarterly Report No. 4, February 19–May 18, DOE/PC/40785-4, 26ff (1982).
120. Yoneyyama, Y., Okamura, M., Morinaga, K., and Tsubaki, N., "Role of water in hydrogenation of coal without catalyst addition," *Energy & Fuels*, 11, 1–8 (2001).
121. Ross, D.S. and Blessing, J.D., *Fuel*, 57, 379 (1978).
122. Ross, D.S., Blessing, J.E., Nguyen, Q.C., and Hum, G.P., *Fuel*, 63, 1206 (1984).
123. Ross, D.S., Nguyen, Q.C., and Hum, G.P., *Fuel*, 3, 1211 (1984).
124. Anderson, K., "Hydrothermal dissolution of coal and other organic solids," SIUC Technology Expo, Carbondale, IL (October 9, 2009).
125. Wibberley, L., Palfreyman, D., and Scaife, P., "Efficient use of coal water fuels," Technology Assessment Report No. 74, CSIRO Energy Technology, New South Wales (April 2008).
126. Wibberley, L., "Alternative pathways to low emission electricity," COAL21 Conference, September 18–19, Crown Plaza, Hunter Valley, CA (2007).
127. Schwartz, E., "Process for the production of low-polluting CWF," in Sens, P. and Wilkinson, J. (ed.), *Coal–Water Mixtures*, Elsevier, New York (1987).
128. "Available and emerging technologies for reducing greenhouse gas emissions from coal-fired electric generating units," prepared by sector policies and programs division, Office of Air quality planning and standards, EPA, Research Triangle, NC (October 2010).
129. "Clean coal technologies in Japan—Technology innovation in the coal industry," Japan Coal Energy Center Report, NEDO, Tokyo, Japan (January 2007).
130. Mahesh, J. and Smit, F., "Engineering development of advanced physical fine coal cleaning for premium fuel applications," AMAX Research Centre, DOE Contract No. DE-AC22-92PC92208 (1993).
131. Cui, L., An, L., Gong, W., and Jiang, H., "A novel process for preparation of ultra-clean micronised coal by high pressure water jet communication technique," *Journal of Fuel*, 86 (5–6), 750 (2007).
132. Mazurkiewicz, D., "Investigation of ultra-fine coal disintegration effect by high pressure water jet," Rock Mechanics and Explosives Research Center, University of Missouri, Rolla, MI (2002), http://www.wjta.org/Book%203/4_1_Mazurkiewicz.pdf.
133. Brooks, P., Clark, K., Langley, J., Lothringer, G., and Waugh, B., "UCC as a gas turbine fuel" (1999), http://www.australiancoal.csiro.au/pdfs/ucc.pdf.
134. Steel, K. and Patrick, J., "The production of ultra clean coal by sequential leaching with HF followed by HNO_3," *Fuel*, 82, 1917 (2003).
135. Steel, K. and Patrick, J., "Regeneration of HF and selective separation of Si(IV) in a process for producing ultra clean coal," *Fuel Processing Technology*, 86 (2), 179 (2004).

136. Kashimura, N., Masaki, M., Matsumura, A., Saito, I., Sato, S., and Takanohashi, T., "Effect of pretreatment with carbonic acid on 'Hypercoal' (ash-free coal) production from low-rank coals," *Energy & Fuels*, 19, 2021–2025 (2005).
137. Siemon, J., "Economic potential of coal-water mixtures," ICEASIE8, September, London, IEA Coal Research (1985).
138. Thambimuthu, K., "Developments in coal-liquid mixtures," IEA Report (1994).
139. Khodakov, G., Gorlov, E., and Golovin, G., "Coal-slurry fuel," *Solid Fuel Chemistry*, 39 (6) (2005).
140. Khodakov, G., "Coal-water suspensions in power engineering," *Thermal Engineering*, 54 (1), 36–47 (2007).

6 Aqueous-Phase Reforming and BioForming Process

6.1 INTRODUCTION

In Chapter 5, we examined the conversion of biomass to biochar, bio-oil, and fuel gases such as methane and hydrogen in water at high temperature and pressure but under subcritical conditions. We showed that water under high-temperature and high-pressure conditions possesses unique physical and chemical properties that allow its strong interactions with biomass to generate solid, liquid, and gaseous fuels. Water, in this case, not only acts as a solvent but also as a reactant and a catalyst to carry out wet pyrolysis reactions. The quality of products depends on the reaction temperature, pressure, reaction time as well as the presence of any catalyst. The chapter also showed that water at high temperature possesses the properties very similar to several organic chemicals and is capable of carrying out various types of organic chemical reactions. While the level of the conversion by the hydrothermal processes (hydrothermal carbonization [HTC], hydrothermal liquefaction [HTL], or hydrothermal gasification [HTG]) can be improved with the use of a suitable catalyst, these processes are basically nonselective.

Biomass can produce hydrogen and liquid fuels in a number of different ways. These production methods can be thermochemical, biochemical, or catalytic. Current processes to convert biomass to liquid fuels include (1) fermentation of glucose to ethanol, (2) pyrolysis or high-pressure liquefaction of biomass to bio-oils, (3) gasification of biomass to syngas followed by Fischer–Tropsch (FT) synthesis to alkanes, (4) anaerobic digestion of cellulosic waste to produce hydrogen and methane, (5) Mobil process of conversion of carbohydrates to aromatic hydrocarbons and coke with Zeolite Socony Mobil (ZSM)-5 catalyst, and (6) supercritical water extraction or gasification of biomass to hydrogen or liquid fuels. In this chapter, we describe yet another selective process to generate hydrogen, syngas, alkanes, and monofunctional groups using low-pressure catalytic process in an aqueous environment. Monofunctional groups can also be further upgraded to various liquid fuels (such as diesel, gasoline, and jet fuel) using a selective catalytic process known as "bioforming process." Numerous excellent reviews on both aqueous-phase reforming (APR) and bioforming process are available in the literature [1–14].

We briefly examine in this chapter a set of catalytic reactions that can be carried out for a select group of oxygenated compounds such as sugar, glucose, sorbitol, glycerol, ethylene glycol, and methanol (with carbon/oxygen [C/O] ratio close to one) to produce hydrogen, syngas, lower alkanes (C_1–C_6), and various monofunctional groups. The monofunctional groups can be subsequently converted to a variety of liquid fuels such as gasoline, diesel, and jet fuels (or their additives) with another set of catalysts. This overall process is called "APR and its derivative technologies" or bioforming process [1]. The selective APR reactions generally occur in the temperature range of 215°C–265°C. The upgrading of monofunctional groups generally requires a somewhat higher temperature. The nature and quality of products strongly depend on the feedstock and the nature of the catalyst, support, and promoter along with other operating conditions such as temperature, pressure, acidity of slurry and catalyst, and solid concentration in the feed slurry. Unlike the process of hydrothermal conversion described in Chapter 5, this is a very selective catalytic process targeted to only certain types of compounds, producing targeted fuels and chemicals [1–6].

6.2 AQUEOUS-PHASE REFORMING

The pioneering work carried out by Dumesic et al. [1–6] showed that carbohydrates such as sugars (e.g., glucose) and polyols such as methanol, ethylene glycol, glycerol, and sorbitol can be efficiently converted to hydrogen and carbon dioxide at 500 K by reforming under aqueous conditions. The process can be applied to all carbohydrates found in wastewater from biomass processing of cheese whey, beer brewery, sugar processing as well carbohydrate streams from agricultural products, such as corn and sugar beets and hemicellulose from any biomass [4,15]. Typical feedstock that can be used for APR and bioforming process are listed in Table 6.1 [4,15]. The secondary feedstock mentioned in the table are first converted to primary feedstock (by hydrolysis and/or hydrogenation processes depending on the feedstock), before using them for APR process. The produced hydrogen can be used to hydrogenate many components of lignocellulosic biomass to produce glycols and other polyols, thus enlarging the feedstock possibilities for APR. The hydrogen can also be used to produce ammonia and fertilizer, an additive to gasification products to produce liquid fuels via FT synthesis and fuel source for polymer electrolyte membrane (PEM) fuel cells.

Besides hydrogen, APR can also produce syngas (CO and H_2), alkanes, and monofunctional groups depending on the nature of the catalyst and the operating conditions. As will be discussed later, the production of hydrogen and syngas requires the breakage of C–C bonds within oxygenated compounds, whereas the production of alkanes and monofunctional groups requires the breakage of C–O bonds within the oxygenated compounds. With most feedstock examined so far, the alkane production is limited to six carbon atoms. More feedstock, catalysts, and reactor designs are needed to produce C_8–C_{15} alkanes from the biomass-derived reactants. The alkanes and monofunctional groups can be further upgraded catalytically by creating new C–C bondages (through condensation reactions) to produce higher alkanes and liquid fuels. The light fuel additives such as pentane and hexane have limited values due to their high volatility. Various reaction paths that can be produced by APR process are schematically illustrated in Figure 6.1 [4].

TABLE 6.1
Typical Feedstock for Aqueous-Phase Reforming

Primary Feedstock
Water-soluble oxygenated hydrocarbons such as sugars, sugar alcohols, saccharides, and other polyhydric alcohols
Secondary Feedstock[a]
Sugar crops
Grain crops
Agricultural waste (cornstalks, straw, seed hulls, sugarcane leavings)
Bagasse, nutshells, manure (from cattle, poultry, and hogs)
Wood materials (wood or bark, sawdust, timber slash, mill scrap)
Municipal waste (waste paper, yard clippings)
Energy crops (poplars, willows, alfalfa, switchgrass, prairie bluestem, corn, soybean)

Source: Huber, G.W., Cortright, R.D., and Dumesic J.A., *Angewandte Chemie International Edition*, 43, 1549–1551, 2004. With permission; Davda, R. and Dumesic, J., *Angewandte Chemie International Edition*, 42, 4068, 2003. With permission; Tao, J., Shishi, C., and Fahai, C., *Chemical Industry and Engineering Progress*, 31, 1010–1017, 2012. With permission; Alonso, D.M., Bond, J.Q., and Dumesic, J.A., *Green Chemistry*, 12, 1493–1513, 2010. With permission; Cortright, R., Davda, R., and Dumesic, J., *Nature*, 418, 964–967, 2002. With permission; Huber, G. and Dumesic, J., *Catalysis Today*, 111, 119–132, 2006. With permission; Blommel, P.G. and Cortright, R.D., "Production of conventional liquid fuels from sugars," A White Paper for European Platform on Biofuels, 2012. With permission.

[a] These are used to generate primary feedstock.

6.3 APR VERSUS STEAM REFORMING

The low-temperature APR to produce hydrogen has significant advantages over conventional steam reforming mentioned in Chapter 4 in that

1. The process occurs in one liquid phase eliminating energy requirement to vaporize water and carbohydrates. Steam reforming requires high temperature and is accompanied by a phase change.
2. The raw materials for APR are nonflammable and nontoxic allowing them to store and handle safely and conveniently. We have established technologies for the storage of sugar, starch, and carbohydrates.
3. The temperature and pressure used in APR favors the thermodynamics of water–gas shift reaction allowing high conversion of CO in one reactor. This allows the production of nearly pure hydrogen stream (with very low CO concentration).
4. The conventional PSA, cryogenic separation, and membrane technologies are easily applicable to the product stream to separate carbon dioxide from hydrogen since pressures used in APR vary from 15 to 50 atm. Steam reforming is often carried out at low pressure, thus requiring pressurization

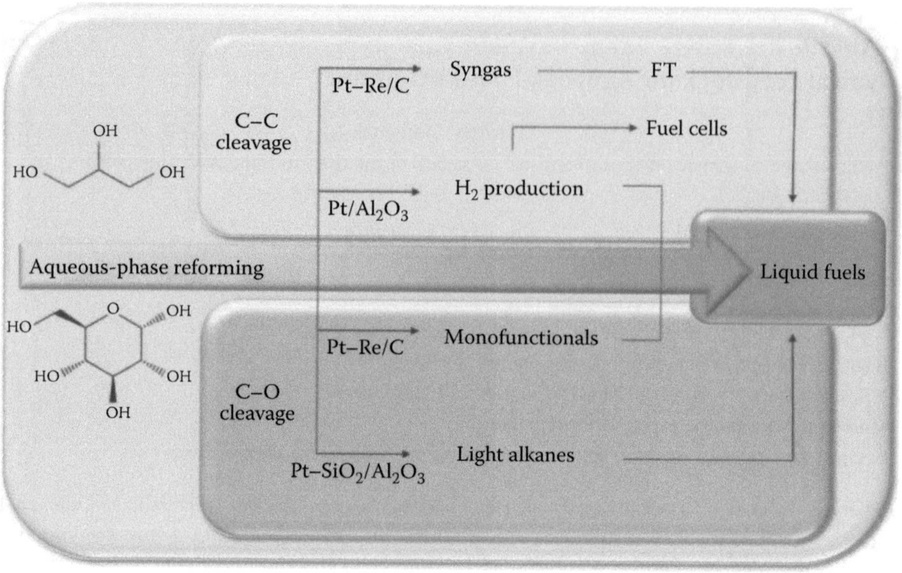

FIGURE 6.1 **(See color insert.)** Possible reaction paths for APR for water-soluble oxygenated hydrocarbons. (Reprinted from *Green Chemistry*, 12, Alonso, D.M., Bond, J.Q., and Dumesic, J.A., Catalytic conversion of biomass to biofuels, 1493–1513, Copyright 2010, with permission from Elsevier.)

of the product to carry out effective separation. Pure hydrogen can thus be produced more easily by APR process.

5. Low temperatures used in APR minimize the decomposition reactions for carbohydrates and resulting coking of the catalysts. Coking of the catalyst is a significant issue in the conventional steam reforming.
6. APR can produce hydrogen in a single reactor as opposed to conventional steam reforming process that will generally require a multistage process.
7. Since APR produces hydrogen, syngas, lower alkanes, and monofunctional groups (which can be further processed to generate different types of liquid fuels), the operating conditions and catalysts can be manipulated to obtain the desired selectivity among various products. This process thus offers more product possibilities than conventional steam reforming process. It should, however, be reemphasized that APR is a selective process that can only be used for a certain type of feedstock. However, steam reforming can be used for all carbonaceous feedstock.

6.4 THERMODYNAMICS OF APR

The discussion in this section closely follows excellent reviews by Dumesic et al. [1–6,16] on the subject. The prevailing thermodynamic forces for the steam reforming of alkanes and oxygenated compounds along with the water–gas shift reaction are illustrated in Figure 6.2 in the form of a plot of Gibbs free energy versus

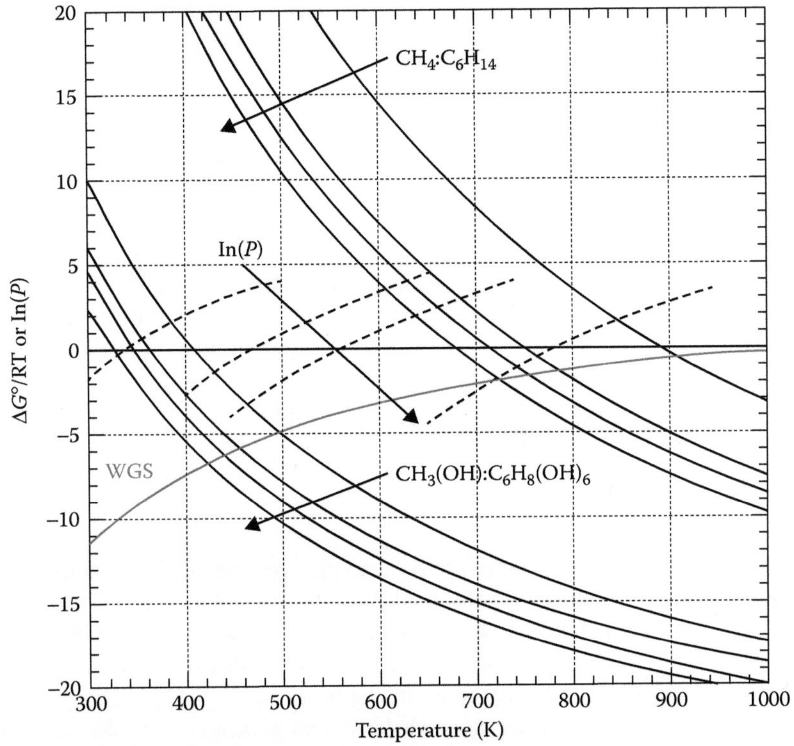

FIGURE 6.2 DG/RT vs. temperature for production of CO and H_2 from vapor-phase reforming of CH_4, C_2H_6, C_3H_8, and C_6H_{14}; $CH_3(OH)$, $C_2H_4(OH)_2$, $C_3H_5(OH)_3$, and $C_6H_8(OH)_6$; and water–gas shift reaction. Dotted lines show the values of ln(P) for the vapor pressures vs. the temperature of $CH_3(OH)$, $C_2H_4(OH)_2$, $C_3H_5(OH)_3$, and $C_6H_8(OH)_6$ (pressure in units of atmosphere). (Reprinted from *Applied Catalysis B: Environmental*, 56, Davda, R., Shabaker, J., Huber, G., Cortright, R., and Dumesic, J., A review of catalytic issues and process conditions for renewable hydrogen and alkanes by aqueous-phase reforming of oxygenated hydrocarbons over supported metalcatalysts, 171–186, Copyright 2005, with permission from Elsevier.)

temperature [1–6,16]. The favorable thermodynamic forces for these reactions require negative Gibbs free energy. Based on this condition, the figure shows that both oxygenate reforming (of methanol, ethylene glycol, glycerol, sorbitol, and glucose) and water–gas shift reactions are favorable at low temperatures. Also, methanation reaction is favorable at reasonably low temperatures. However, steam reforming reactions for methane and other alkanes are only favorable at higher temperatures.

The concept of APR is based on the fact that at moderate temperature and pressure, oxygenated carbohydrates react with water to produce either alkanes or hydrogen and carbon monoxide by the following reforming reaction [1–6,16]:

$$C_nH_{2n+2} + nH_2O \rightleftarrows nCO + (2n+1)H_2 \tag{6.1}$$

Also, at these temperatures and pressures, the following water–gas shift reaction is favored:

$$CO + H_2O \rightleftarrows CO_2 + H_2 \quad (6.2)$$

Figure 6.2 presents the Gibbs free energy ($\Delta G°/RT$) associated with the steam reforming of a series of alkanes such as CH_4, C_2H_6, C_3H_8, and C_6H_{14} normalized per mole of CO produced along with that for the water–gas shift reaction. For a reaction to occur, negative free energy value in Figure 6.2 is needed. These results show that while water–gas shift reaction is favorable at low temperature, the thermodynamics of steam reforming of alkanes is only favorable at higher temperatures (e.g., $T > 675$ K for C_6H_{14} and $T > 900$ K for CH_4). Thus, at lower temperatures, lower alkanes cannot be reformed to syngas.

The oxygenated hydrocarbons having a C/O ratio of 1:1 form carbon monoxide and hydrogen according to the following reaction [1–7,16]:

$$C_zH_{2y}O_z \rightleftarrows zCO + yH_2 \quad (6.3)$$

The Gibbs free energy diagrams for some typical oxygenated compounds such as methanol (CH_3OH), ethylene glycol [$C_2H_4(OH)_2$], glycerol [$C_3H_5(OH)_3$], and sorbitol [$C_6H_8(OH)_6$] are also shown in Figure 6.2. These results indicate that the APR of these compounds at low temperatures are thermodynamically favorable. Sorbitol is generally obtained by the hydrogenation of glucose [$C_6H_6(OH)_6$]. Thus, oxygenated hydrocarbons can be reformed at much lower temperatures than the alkanes with similar carbon number. A combination of aqueous (or steam) reforming of oxygenated carbohydrates and water–gas shift reaction will allow the production of hydrogen at low temperatures.

Figure 6.2 also illustrates the logarithms of vapor pressure as a function of temperature for methanol, ethylene glycol, glycerol, and sorbitol. For the first three substances, steam reforming (in the gas phase) can be carried out at temperatures of ≥550 K, while for sorbitol, vapor-phase steam reforming requires a temperature of at least 750 K. Thus, at low temperatures (<750 K), reforming of sorbitol (and glucose) can be carried out in the aqueous phase producing hydrogen and syngas. The favorable thermodynamics for APR of oxygenated compounds illustrated in this figure prompted a significant research to evaluate favorable kinetic conditions to produce hydrogen, syngas, and alkanes via the APR process [1–7,16].

Since the thermodynamics of steam reforming of alkanes at low temperatures are not favorable, hydrogen and carbon dioxide formed from oxygenates at lower temperatures are not stable and alkanes can be formed by the methanation and FT reactions between hydrogen and carbon monoxide and carbon dioxide. For example, at 500 K, the equilibrium constant for methanation reaction is favorable [1–7,16]:

$$CO_2 + 4H_2 \rightleftarrows CH_4 + 2H_2O \quad (6.4)$$

Thus, forming hydrogen selectively and inhibiting the formation of alkanes would require a catalyst that promotes C–C scission followed by the water–gas shift reaction and inhibits C–O scission followed by the hydrogenation.

6.5 KINETICS AND CATALYSIS OF APR PROCESS

As shown in Figure 6.1, the APR process can be tailored toward the four distinct kinetic steps depending on the desired product [1–6,11–33] (Tanksale et al., 2008, pers. comm.). APR can produce hydrogen, syngas, alkanes, or monofunctional groups depending on the catalyst and support system, promoters, and other operating conditions.

The original purpose of APR was to generate either hydrogen or alkanes by an APR of sugar, other oxygenated compounds, and polyols (with a ratio of 1:1). The kinetics of APR depends on the temperature, the pressure, the nature of the catalyst and its support, the presence of promoters, the pH of the slurry, the acidity of catalyst active sites, and the nature of the feedstock. Here, we briefly examine the effects of these operating variables on the APR process. While the literature has shown numerous ways to generate hydrogen from biomass under high-temperature conditions, APR is unique in that it is the only process that can be carried out in liquid water [5,34–57]. While APR can only be used for selective feedstock without their pretreatments, it is faster than anaerobic digestion process for generating hydrogen from cellulosic waste [5,34–57].

6.5.1 Effects of Temperature, Carbon Number, and Pressure

As shown in Figure 6.1, APR provides multiple options of reaction paths depending on the operating conditions. Figure 6.3 illustrates that hydrogen selectivity decreases with an increase in carbon number of oxygenated compounds and an increase in temperature. The temperature effect shown in this figure is valid for all oxygenated feedstock. The literature has shown that compounds such as furanone and acetic acid are not amenable to the production of hydrogen by APR [1–7,16]. The hydrogen selectivity depends on the nature of the bond breaking in oxygenated compounds; the breakage of C–C bond favors the hydrogen formation and the breakage of C–O bond favors the formation of alkanes. Following the preferred pathway is the key to the hydrogen formation. Dumesic et al. [1–6,16] also showed that an increase in pressure reduced the hydrogen selectivity. For example, for the reaction of 5 wt% sorbitol over Pt–SiAl at 498 K, the hydrogen selectivity at 25.8 atm pressure was 21, whereas the same selectivity at a pressure between 33.1 and 52.1 atm was <2 [1–6,16].

6.5.2 Effects of Catalysts and Supports

As shown in Table 6.1, the major primary feedstock for APR are glucose (and sorbitol), alcohols, ethylene glycol, and glycerol. For all of these feedstock, Dumesic et al. [1–6,16], among others [5,15,17–26,27–57] (Tanksale et al., 2008, pers. comm.), have clearly shown that the nature of metal and support has an important influence on the reaction paths and the rates of reactions in the APR process. The product selectivity can be tuned depending on the metal and support. For example, Pt-black and Pt supported on Al_2O_3, TiO_2, and ZrO_2 have been demonstrated to be active and selective for the APR of methanol and ethylene glycol to produce hydrogen. Catalysts based on Pd have shown similar activity compared to Pt analogs. Ru, Rh, and Ni, however, showed lower activity for hydrogen.

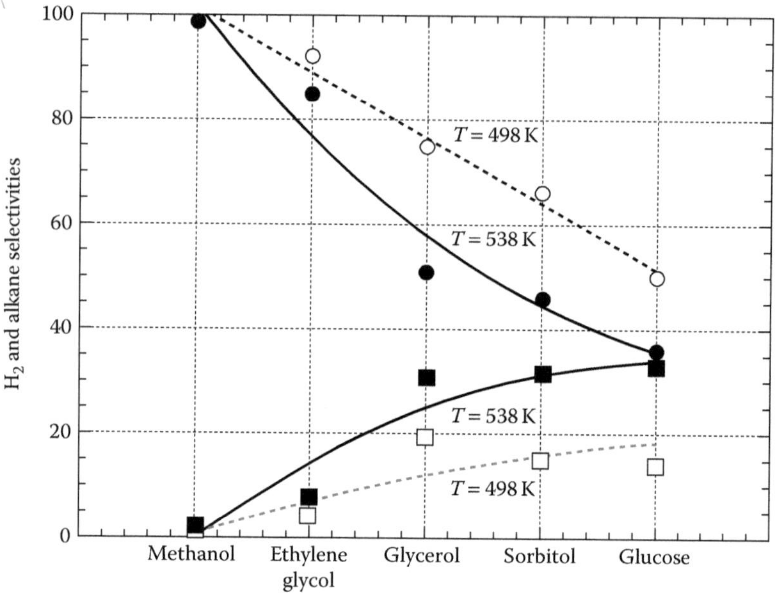

FIGURE 6.3 Selectivities vs. oxygenated hydrocarbon. H_2 selectivity (circles) and alkane selectivity (squares) from APR of 1 wt% oxygenated hydrocarbons over 3 wt% Pt/Al_2O_3 at 498 K (open symbols and dashed curves) and 538 K (filled symbols and solid curves). (Reprinted from *Applied Catalysis B: Environmental*, 56, Davda, R., Shabaker, J., Huber, G., Cortright, R., and Dumesic, J., A review of catalytic issues and process conditions for renewable hydrogen and alkanes by aqueous-phase reforming of oxygenated hydrocarbons over supported metalcatalysts, 171–186, Copyright 2005, with permission from Elsevier.)

Davda et al. [16] and others [15–33] (Tanksale et al., 2008, pers. comm.) have examined the effectiveness of various group VIII metal catalysts, such as Ru, Rd, Pt, Ir, Pd, and Ni, for APR. The studies compare the selectivity for hydrogen, alkanes, and carbon dioxide by Pt, Pd, Ru, Rh, and Ni catalysts for various oxygenated compounds and at various temperatures. The results show that CO_2 selectivity was the highest for Pt and Ni catalysts, and the lowest for Rh and Pd catalysts. The alkanes selectivity was the highest for Ru and Rh catalysts followed by Pt and Ni. Very little alkanes were produced by Pd catalysts. Finally, Pt and Pd (followed by Ni) showed good reforming activity and high hydrogen production rates. Good catalysts for hydrogen production by APR should show high activity for water–gas shift reaction and for cleavage of C–C bonds. Both Pd and Pt catalysts gave poor activity for C–O scission and subsequent methanation and FT reactions [1–6].

In the final analysis, since Pt catalysts gave good product distributions for all three (hydrogen, carbon dioxide, and alkanes) components, Pt was considered to be the best catalyst. Ni catalyst, although cheap, gave preference to alkanes. Park et al. [29] studied the production of biohydrogen by APR of polyols over Pt catalysts supported on three-dimensionally bimodal mesoporous carbon (3D-BMC). The 3D-BMCs with mesopores of tunable size (controlled through the polymerization of the carbon precursor) were synthesized. After loading with platinum, the catalysts were used in APR of polyols, and superior performance was shown in terms of carbon conversion, hydrogen yield, selectivity, and hydrogen production rate compared to Pt catalysts supported on activated carbon or two-dimensional CMK-3 (ordered mesoporous carbon synthesized by silica hard template).

The study by Davda et al. [2,16,41,58,59] also indicated that the best support for Pt was Al_2O_3 for hydrogen production and the effect of support on reforming activity and selectivity is greater than that of metal dispersion. They also analyzed bimetallic catalysts and concluded that Ni–Sn catalysts show potential for APR. The selectivity for hydrogen and alkanes for different oxygenates at 225°C and 265°C using Pt/Al_2O_3 catalyst is illustrated in Figure 6.3 [16].

6.5.3 Effects of Promoters and Acidity of Liquid and Solids

The addition of a promoter can also have some effect on the catalyst performance. Re was found to be an effective promoter for Pt/C catalyst. The selectivity of Pt–Re/C was found to be different from that of Pt/C. Hydrogen selectivity with promoter was lower, although hydrogen productivity was higher. Following reduction, Pt–Re/C catalyst was significantly more active for APR of glycerol than Pt/C catalyst. The presence of Re created surface acidity that favored a pathway of C–O bond breaking (dehydration), resulting in lower hydrogen and $CO(CO_2)$ selectivity and higher alkanes selectivity [1–6,16]. The literature [1–6,16] also showed that an addition of KOH (base) affected APR selectivity of glycerol for 3%Pt3%Re/C catalyst.

The effects of liquid and solid acidities on carbon selectivity for sorbitol at 538 K and 57.6 bar with Pt/Al catalysts were also examined by Dumesic et al. [1–6,16]. The results indicated that lower pH of both liquids and solids produce higher carbon number alkanes. In general, an increase in acidity by either the use of acid catalyst support (i.e., SiO_2/Al_2O_3) or the addition of the mineral acid such as HCl to increased the feed alkanes selectivity due to the increased rate of dehydration and hydrogenation pathways compared to hydrogenolysis and reforming reactions. The nickel supported on SiO_2 or AI_2O_3 was found to have low selectivity for hydrogen and favored the formation of alkanes. However, an addition of an Sn promoter to Raney R–Ni-based catalysts enhanced the production of hydrogen from sorbitol, glycerol, and ethylene glycol [1–6,16]. While the promoters and acidity can be used to produce alkanes, some C–C bond needs to be broken to produce hydrogen needed for the production of alkanes. For example, the hydrogenation and complete deoxygenation of sorbitol results in the following set of reactions [1–6,16]:

$$C_6H_{14}O_6 + 6H_2 \rightarrow C_6H_{14} + 6H_2O \quad (6.5)$$

However, complete deoxygenation occurs as [1–6,16]

$$C_6H_{14}O_6 \rightarrow 13/19\,C_6H_{14} + 36/19\,CO_2 + 42/19\,H_2O \quad (6.6)$$

These reactions indicate the need for hydrogen for the production of alkanes.

6.5.4 Effects of Feedstock

As shown in Table 6.1, while APR can be applied to both the primary and secondary feedstock, so far, most of the work has been focused on the primary feedstock such as sugar, glucose, sorbitol, alcohols, ethylene glycol, and glycerol. As Virent's BioForming process based on APR develops, it is intended to apply to the secondary feedstock as well. This will require some acid and/or enzyme hydrolysis pretreatments to the feedstock. Here we briefly review some of the reported studies on APR for both the primary and secondary feedstock. For the primary feedstock, the discussion is further broken into two parts: (1) individual compounds having high vapor pressure such that APR is carried out in both gas and liquid phases, and (2) individual compounds having low vapor pressure such that APR occurs largely in the aqueous environment. Since the literature for APR of biomass-derived products is extensive [1–7,12,14,16,17,56,60–64], here we focus only on few recent studies on individual compounds and materials [1–7,12,14,16,17,56,60–64].

6.5.4.1 APR of Ethylene Glycol, Alcohols, and Glycerol (Primary Feedstock with High Vapor Pressure)

APR of these compounds can occur in both the liquid and gas phases due to their high vapor pressure under the reaction conditions. Dumesic et al. [4,16,48,49] have extensively studied APR of ethylene glycol, particularly for alumina-supported Pt catalysts. Their results are well reviewed by Alonso et al. [4] and Davda et al. [16]. Two recent novel studies are by D'Angelo et al. [31] who examined APR of ethylene glycol in a novel microchannel reactor and Chu et al. [38] who examined APR of ethylene glycol on Co/ZnO catalysts prepared by the coprecipitation method. APRs of various alcohols (methanol, ethanol, and other polyols) are well examined by Dumesic et al. [1–6,16,48], Park et al. [29], Cruz et al. [50], and Zhang et al. [54].

The Department of Energy has identified glycerol as one of the 12 important platform chemicals from biomass (see Chapter 7). In Chapter 4, we examined a significant work reported on the steam reforming of glycerol. The reforming reaction with glycerol results in [1–7,16]

$$C_3H_8O_3 \rightarrow 3CO + 4H_2 \tag{6.7}$$

APR of glycerol has also been widely studied, as indicated by the large amount of literature on the subject [34–36,44,47,51–57,65]. The subject is also extensively covered in a number of reviews by Dumesic et al. [1–6], Davda et al. [16], and Vaidya and Rodrigues [47]. Here we examine only few recent studies on the subject in brief detail.

Tuza et al. [65] examined the production of renewable hydrogen by APR of glycerol over Ni–Cu catalysts derived from hydrotalcite precursors. The reforming was carried out in a batch reactor at 250°C and 270°C. The catalyst with 5% of Cu showed high H_2 selectivity at 250°C. At 270°C, there was consumption of H_2 with time due to hydrogenolysis of glycerol. The study proposed the main reaction routes, which considered liquid byproducts. At 250°C, hydrogen selectivity was always higher than 80% and the formation of CO was very low (<3%). The addition of Cu decreased the formation of methane.

Wen et al. [39] examined the activities and stabilities of Pt, Ni, Co, and Cu catalysts and supports for H_2 production by APR of glycerol. The experimental data were taken in a continuous flow fixed-bed reactor. It was found that the activity of the metal catalysts increased in the order of Co, Ni, Cu, and Pt. Additionally, Pt was highly stable, whereas Ni and Co showed a significant deactivation with time on stream. It was also found that the activity of Pt catalysts on various supports follows the order: SAPO-11 < active carbon (AC) < HUSY < SiO_2 < MgO < Al_2O_3. Moreover, the basic support resulted in high activity and higher hydrogen molar concentration, whereas acidic support and neutral Al_2O_3 support tended to increase alkanes formation. It was shown by X-ray diffraction (XRD) that Pt was caused to sinter on all of the supported Pt catalysts during the reaction. In addition, a trace amount of carbon deposition was found on all of the supported Pt catalysts. However, no remarkable deactivation was observed over Pt/Al_2O_3, Pt/SiO_2, Pt/AC, and Pt/HUSY catalysts. Two zeolite-supported catalysts showed low activities as well as the collapse of the support. In addition, little influence of the collapse of the support on the stability of Pt/HUSY was observed. Pt/SAPO-11 catalyst exhibited very high deactivation.

Cho et al. [57] examined APR of glycerol over Ni-based catalysts for hydrogen production. The reforming was carried out at 225°C, 23 bar, and liquid hourly space velocity (LHSV) = 4 h^{-1}. The Ni-based catalyst was prepared by an incipient wetness impregnation method. It was found that Ni (20 wt%)–Co (3 wt%)/γ-Al_2O_3 catalyst showed higher glycerol conversion and hydrogen selectivity than Ni (20 wt%)/γ-Al_2O_3 catalyst. There were no major changes in Ni particles after the reaction over Ni–Co/γ-Al_2O_3 catalyst. The results suggest that the Ni–Co/γ-Al_2O_3 catalyst can be applied to the hydrogen production system using APR of glycerol.

6.5.4.2 APR of Sugar and Glucose (Primary Feedstock with Low Vapor Pressure)

Tanksale et al. [42] examined the hydrogen production by APR of sugar solutions using metal-supported catalysts. The aim of this study was to examine the influence of several reaction parameters on hydrogen production using liquid-phase reforming of sugar solution over Pt, Pd, and Ni supported on nanostructured supports. It was found that the desired catalytic pathway for H_2 production involves cleavage of C–C, C–H, and O–H bonds that adsorb on the catalyst surface. Thus, a good catalyst for the production of H_2 by liquid-phase reforming must facilitate the C–C bond cleavage and promote the removal of adsorbed CO species by the water–gas shift reaction, but the catalyst must not facilitate the C–O bond cleavage and hydrogenation of CO or CO_2. Apart from studying various catalysts, a commercial Pt/γ-alumina catalyst was also examined at three different temperatures: 458, 473, and 493 K. On the surface of some of the spent catalysts, the amorphous and organized form of coke was found. APR of sugar solution was also studied by Blommel and Cortright [15], Cortright [8,66], and Held [67].

APR of cellulose can form hydrogen by the following reaction [1–6,16]:

$$C_6O_6H_{12} + 6H_2O \rightarrow 6CO_2 + 12H_2 \tag{6.8}$$

Also, dehydration/hydrogenation results in the formation of alkanes as

$$C_6O_6H_{12} + 7H_2 \rightarrow C_6H_{14} + 6H_2O \tag{6.9}$$

which gives the combined reaction as

$$1.6C_6O_6H_{12} \rightarrow C_6H_{14} + 3.5CO_2 + 2.5H_2O \tag{6.10}$$

Alkanes contained 95% of the heating value and only 30% of the mass of the biomass-derived reactant.

Davda et al. [16] proposed that a way to increase hydrogen selectivity from glucose is to operate in two stages: (1) to carry out the low-temperature hydrogenation step followed by the high-temperature reforming process and (2) to co-feed hydrogen with liquid reactant stream to the reforming reactor. This co-feeding argument leads them to propose a reactor scheme shown in Figure 6.4 to obtain the product of desired specification using APR [16].

For biomass application, APR of glucose is very important because it is the basic sugar component of all starch and carbohydrates [1–6,16,48]. The hydrogenation of glucose leads to the formation of sorbitol, and both glucose and sorbitol can be reformed to form carbon dioxide and hydrogen. As the glucose concentration in the feed increases, the hydrogen selectivity decreases. Also, these reactions are favored at low temperatures. The reforming of both glucose and sorbitol can occur on Pt and Ni–Sn alloy by cleavages of C–C bonds followed by the water–gas shift reaction [1–6,16]. The alkanes are produced on the acidic sites of metals from both glucose and sorbitol. Glucose also produces acids, aldehydes, and so on through homogeneous side reactions. Since undesirable side reactions are first order with respect to glucose and the desirable reactions have a fraction-order dependence on glucose, an increase in glucose concentration reduces hydrogen selectivity [1–6,16]. The hydrogenation of glucose to sorbitol also occurs at a higher rate at low temperature (400 K) and high hydrogen pressure. Recently, Wen et al. [39] examined the catalytic properties of Ni on alumina for the APR of glucose.

6.5.4.3 APR of Biomass and Cellulose (Secondary Feedstock)

APR has also been tested on biomass and cellulose waste paper [37,45]. Valenzuela et al. [37] studied APR of woody biomass in a batch reactor. In this study, APR was used to produce hydrogen from actual biomass. The experiments were carried out in a 100 mL Parr micro reactor heated to 225°C. Both acid hydrolysis of woody biomass and subsequent APR of soluble molecules by a Pt/Al_2O_3 catalyst were carried out in a single reactor. The experiments showed that increasing the acid concentration from 1% to 5% resulted in more than a twelve-fold increase in H_2 concentration in the product gas. However, hydrogen accounted for only 18% of

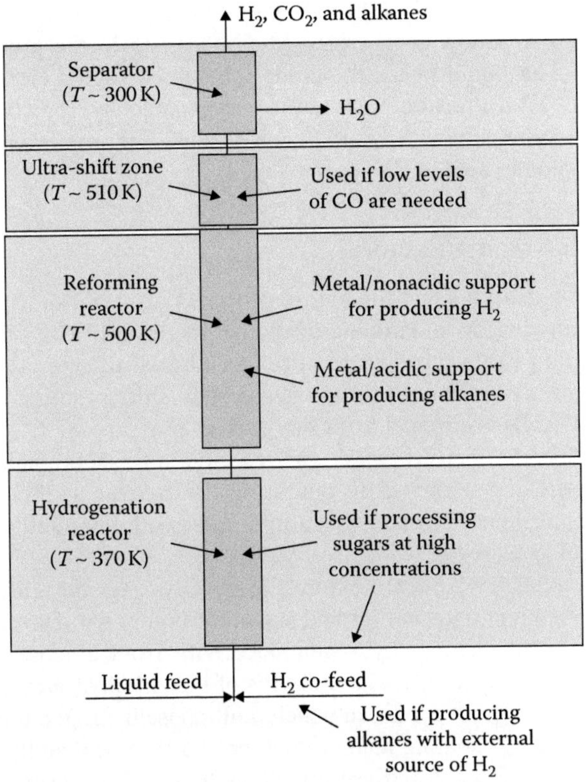

FIGURE 6.4 Summary of the process conditions employed to obtain a product of the desired specifications using the APR process. (Reprinted from *Applied Catalysis B: Environmental*, 56, Davda, R., Shabaker, J., Huber, G., Cortright, R., and Dumesic, J., A review of catalytic issues and process conditions for renewable hydrogen and alkanes by aqueous-phase reforming of oxygenated hydrocarbons over supported metalcatalysts, 171–186, Copyright 2005, with permission from Elsevier.)

the noncondensable gas phase with CO_2 as a major product. The presence of the Pt/Al_2O_3 reforming catalyst enhanced both the selectivity and the yield of hydrogen in the gas phase. This was also accompanied by a noticeable decrease in carbon monoxide production due to a faster water–gas shift reaction catalyzed by platinum. In comparison with other feeds such as glucose, wastepaper, and ethylene glycol, the amount of hydrogen produced from biomass was of a comparable magnitude per gram of feed, although biomass yielded more hydrogen per gram of carbohydrate than either glucose or wastepaper. Baseline experiments were carried out to confirm that the observed hydrogen production was originated from the biomass.

Tungal and Shende [45] reported APR of wastepaper in the presence of a homogeneous $Ni(NO_3)_2$ catalyst for biocrude and H_2 production. In this study, reforming of aqueous wastepaper slurry (0.1 g/cc) was performed using 5 wt% catalyst at

200°C–275°C. At 250°C, about 44 wt% biocrude and 3.8 mol% H_2 were observed after 120 min of reaction time. Other gases observed in the products were CO_2, CO, and CH_4. The liquid phase (biocrude) contained sugars (7.5 wt%), hydroxymethylfurfural (HMF)/furfural (~1 wt%), oxygenated hydrocarbons (42.4 wt%), and monocarboxylic acids (49.1 wt%) such as acetic, formic, propionic, and lactic (2-hydroxypropionic) acids.

6.5.5 Novel Reactor Designs

The operating conditions and the nature of the catalyst not only affect the selectivity between hydrogen and alkanes, but also the level of CO production. A low CO concentration in the product requires an ultra shift operation in which the reaction conditions are such that the water–gas shift reaction is favored. The lowest level of CO requires the lowest partial pressure of CO_2 and H_2 in the gas phase so that the forward water gas reaction is thermodynamically favored. These conditions are achieved by operating the reactor at the saturation pressure for water (at the reaction temperature) and using low feed concentration of oxygenates [16].

Very few studies have been done on the novel reactor design to carry out the APR process. As mentioned earlier and shown in Figure 6.4, Davda et al. [16] proposed that a way to increase hydrogen selectivity from glucose is to operate in two stages. D'Angelo et al. [31] studied APR of biocarbohydrates in a catalytically stable wash-coated micro reactor, in which multiphase hydrogen removal enhanced hydrogen efficiency. A coating method to deposit a Pt-based catalyst on the microchannel walls was selected and optimized. APR reactivity tests were performed using ethylene glycol as the model compound. Optimum results were achieved with a static wash coating technique in which a highly uniform and well-adhered 5 μm layer was deposited on the walls of a 320 μm internal diameter (ID) microchannel in one single step. During APR of ethylene glycol, the catalyst layer exhibited high stability over 10 days after limited initial deactivation. The microchannel presented higher conversion and selectivity to hydrogen than a fixed-bed reactor. They concluded that the benefits of using a micro reactor for APR can be further enhanced by utilizing the increased Pt loadings, higher reaction temperatures, and larger carbohydrates (e.g., glucose). The use of micro technology for APR can allow a significant reduction in the reformer size, thus rendering it promising for distributed hydrogen production.

Subsequently, D'Angelo et al. [9] used the 1.7 m long, 320 μm ID microchannel reactor with a 5 μm Pt-based wash-coated catalyst layer described earlier to study APR of sorbitol. The performance of this microchannel reactor was correlated to the mass transfer properties, reaction kinetics, hydrogen selectivity, and product distribution. While mass transfer did not affect kinetically controlled sorbitol consumption, it did affect hydrogen selectivity and the product distribution. Compared to a fixed-bed reactor, the hydrogen selectivity in the microchannel reactor was higher by a factor of 2. The yield of side products (mainly C_3 and heavier hydrodeoxygenated species) was suppressed, whereas the yield of hydrogen was increased from 1.4 to 4.0 moles per mole of sorbitol fed.

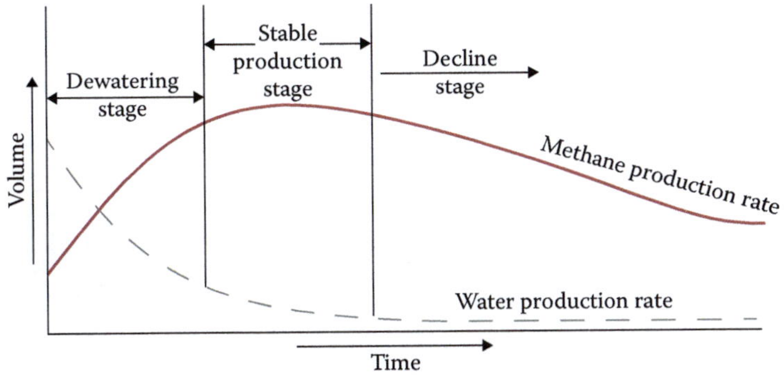

FIGURE 2.1 Typical production curve for a coal bed methane well showing relative methane and water production. (Adapted from Rice, D., "Coal bed methane—An untapped energy resource and environment concern," US Geological Survey, Energy Resource Surveys Program, USGS Fact Sheet FS-019-97, 1997.)

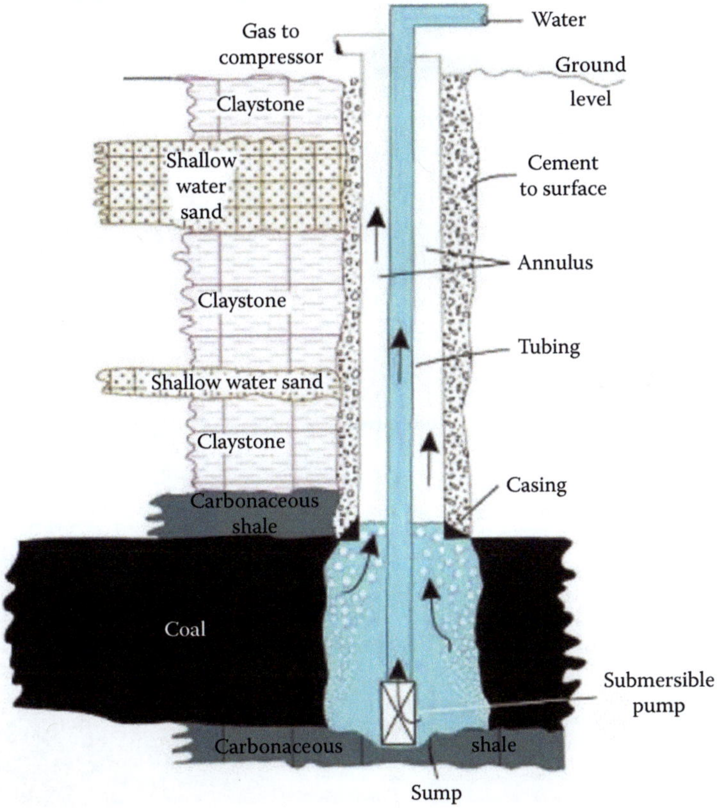

FIGURE 2.2 Simplified illustration of a coal bed methane production well. (From Huth, E., Sule, M., Todman, L., Brant, J., and Templeton, M., "Treatment and reuse of coalbed methane produced water using pervaporation irrigation," 22nd Annual Produced Water Society Conference, January 17–19, 2012. With permission.)

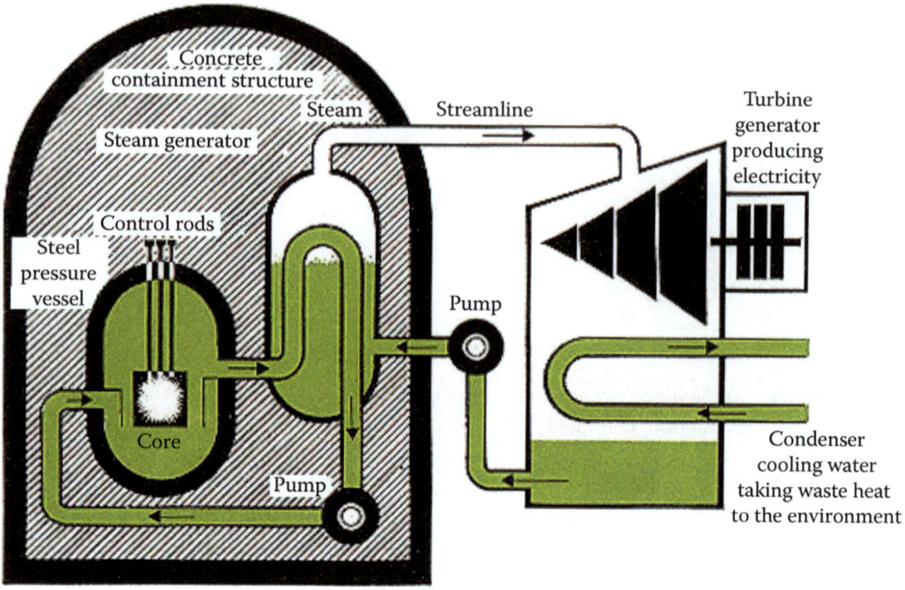

FIGURE 3.2 PWR—A common type of LWR. (Adapted from "Nuclear power reactors," a document of World Nuclear Association, July 2013.)

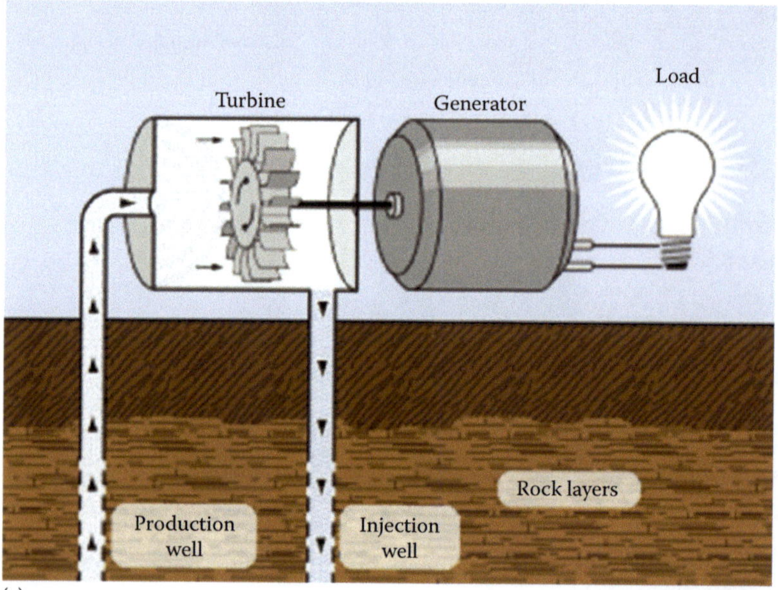

(a)

FIGURE 3.3 Three methods of recovering geothermal energy: (a) dry steam. (Adapted from Union of Concerned Scientists, *How Geothermal Energy Works*, Union of Concerned Scientists, Cambridge, MA, 2012; National Renewable Energy Laboratory. Planta Solar 20. http://www.nrel.gov/csp/solarpaces/project_detail.cfm/projectID=39.)

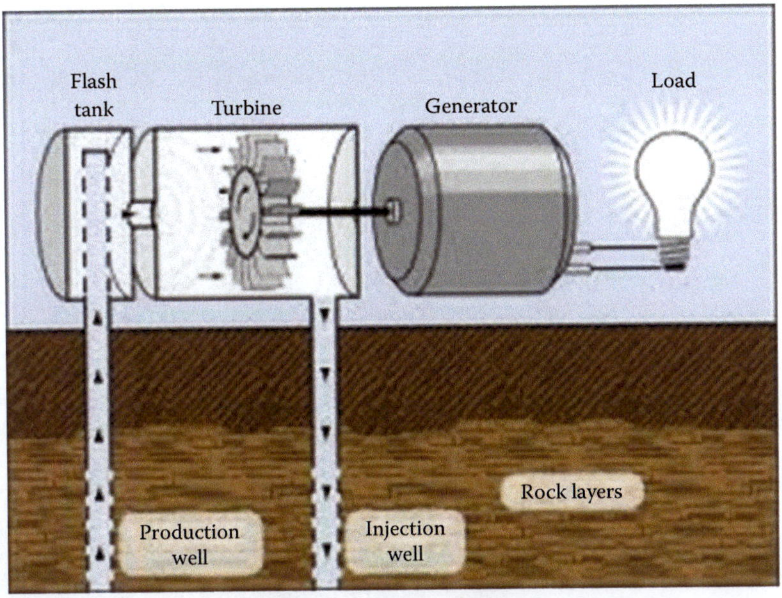

(b)

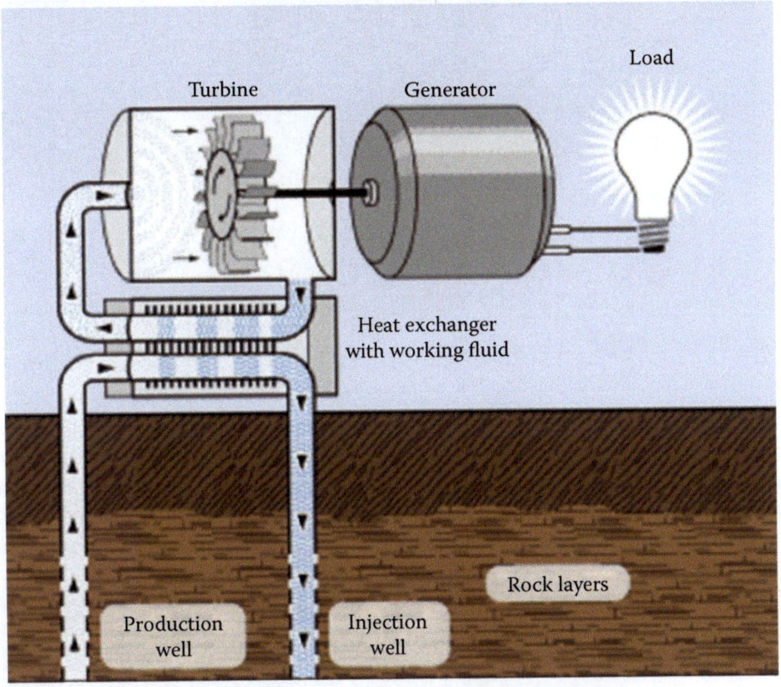

(c)

FIGURE 3.3 (Continued) Three methods of recovering geothermal energy: (b) flash steam and (c) binary cycle. (Adapted from Union of Concerned Scientists, *How Geothermal Energy Works*, Union of Concerned Scientists, Cambridge, MA, 2012; National Renewable Energy Laboratory. Planta Solar 20. http://www.nrel.gov/csp/solarpaces/project_detail.cfm/projectID=39.)

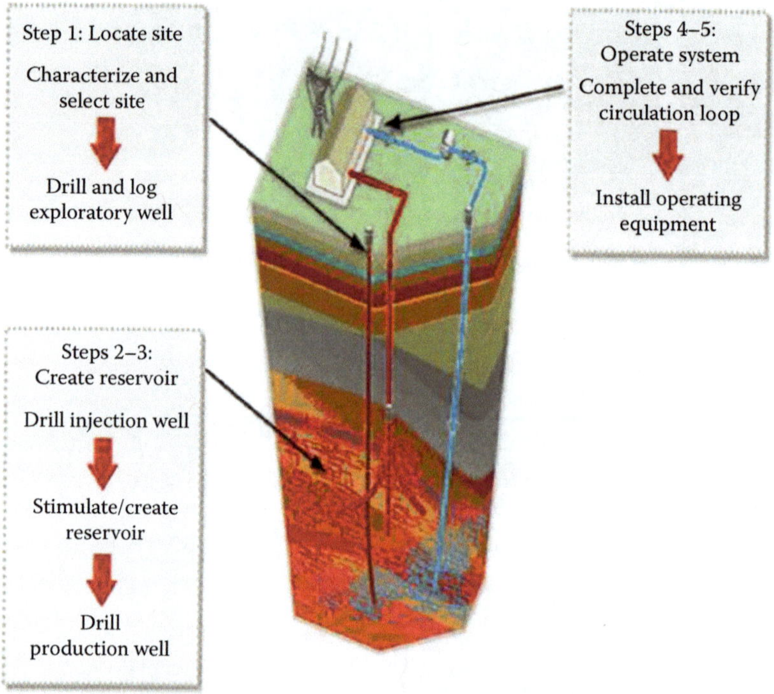

FIGURE 3.4 Steps taken to recover geothermal energy via the EGS. (Adapted from Union of Concerned Scientists, *How Geothermal Energy Works*, Union of Concerned Scientists, Cambridge, MA, 2012; Office of Energy Efficiency and Renewable Energy, *An Evaluation of Enhanced Geothermal Systems Technology*, US Department of Energy, Washington, DC, 2008.)

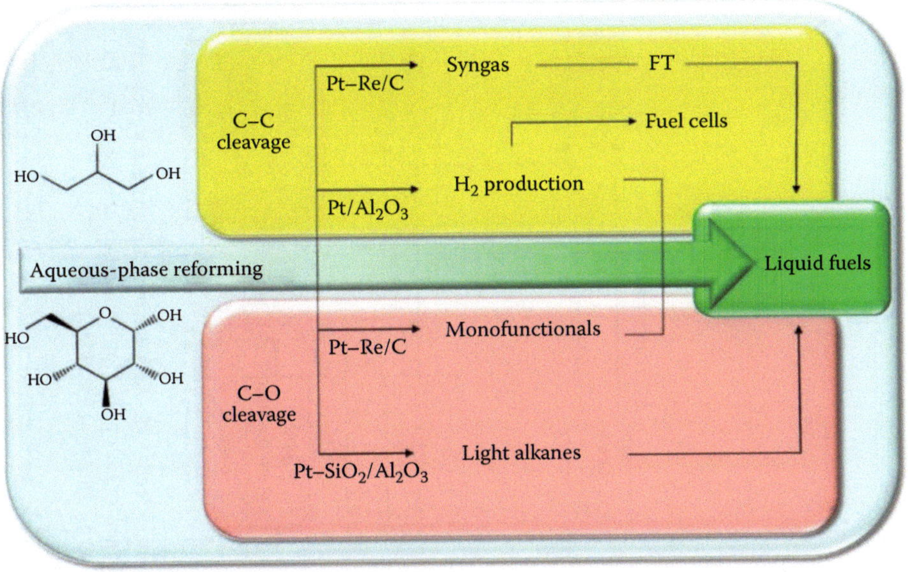

FIGURE 6.1 Possible reaction paths for APR for water-soluble oxygenated hydrocarbons. (Reprinted from *Green Chemistry*, 12, Alonso, D.M., Bond, J.Q., and Dumesic, J.A., Catalytic conversion of biomass to biofuels, 1493–1513, Copyright 2010, with permission from Elsevier.)

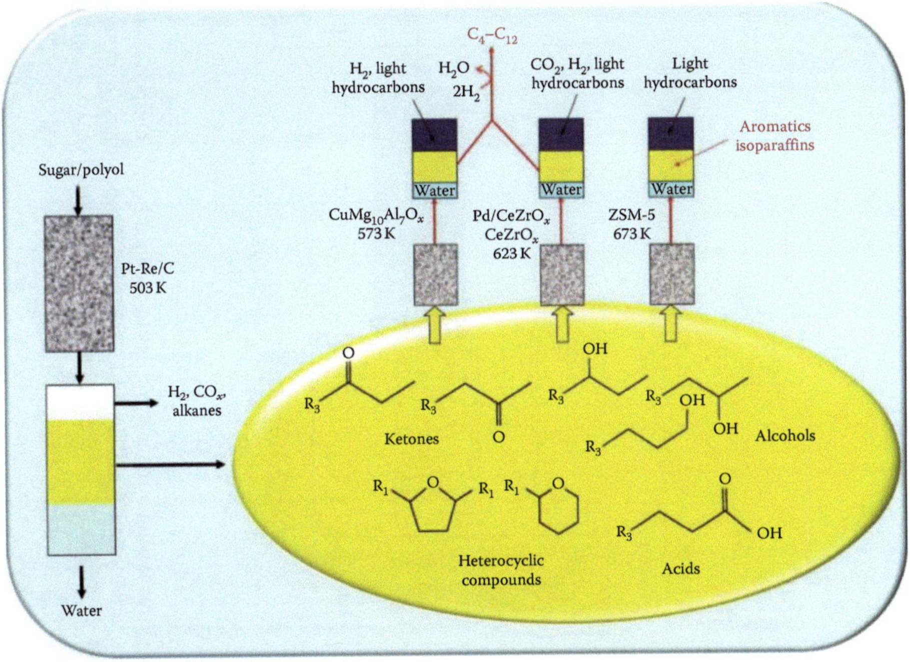

FIGURE 6.5 Schematic pathways to convert sugar and polyols to biofuel through production of monofunctional intermediates. (Reprinted from *Green Chemistry*, 12, Alonso, D.M., Bond, J.Q., and Dumesic, J.A., Catalytic conversion of biomass to biofuels, 1493–1513, Copyright 2010, with permission from Elsevier.)

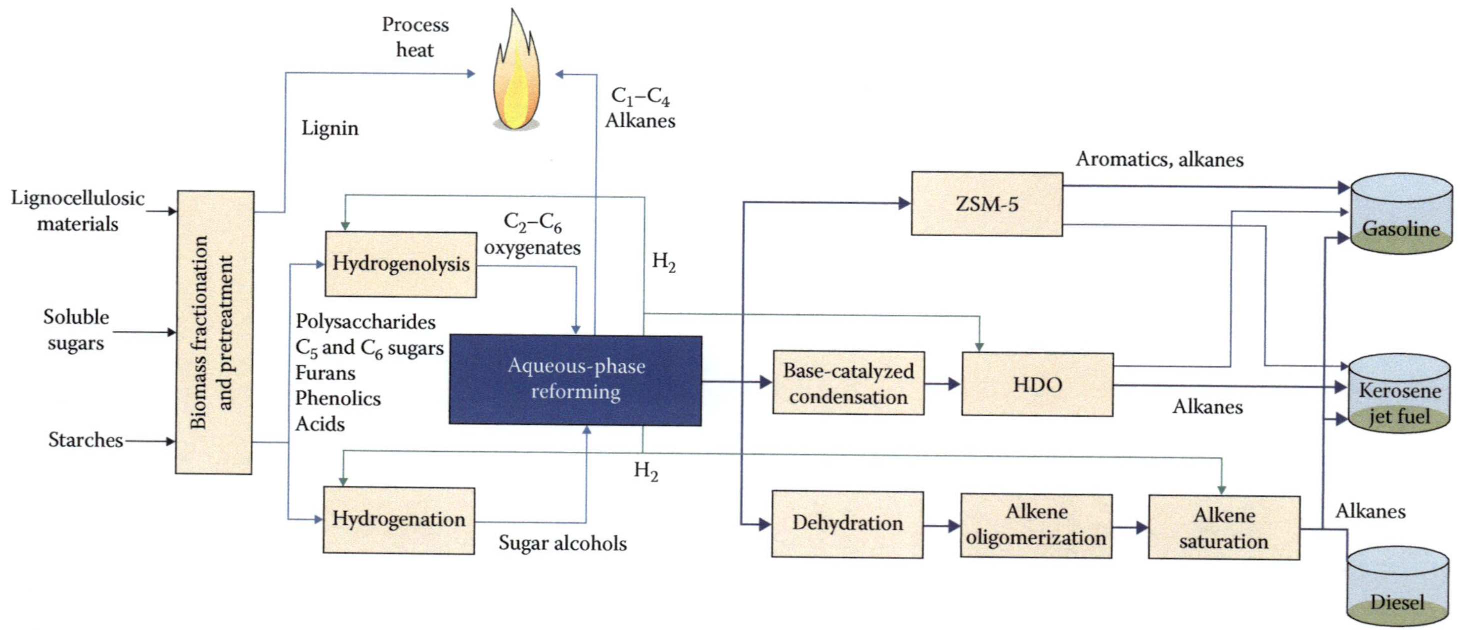

FIGURE 6.6 Virent's BioForming® process to produce liquid transportation fuels from biomass feedstock. APR enables the process to partially defunctionalize carbohydrate feedstock for further catalytic upgrading. (Reprinted from White Paper for European Platform on Biofuels, Blommel, P.G. and Cortright, R.D., Production of conventional liquid fuels from sugars, Copyright 2012, with permission from Elsevier.)

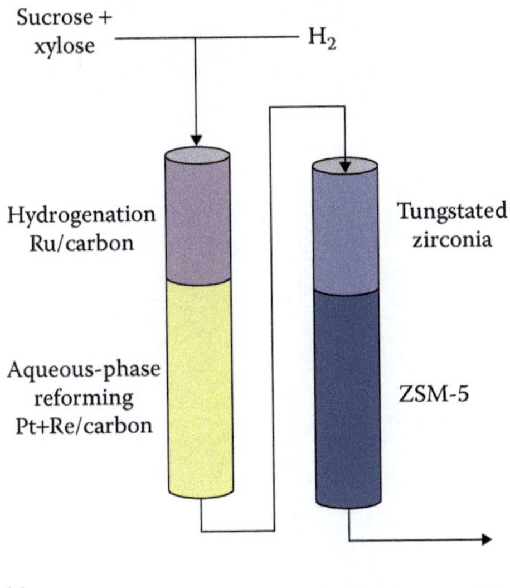

(a)

	Carbon (%)	LHV (%)
	Feeds	
Sucrose + xylose	100	87
Hydrogen	–	12
	Products	
CO_2	25	–
H_2	–	3
C_1–C_4	26	31
C_5–C_9	43	53
C_{10+}	5	6

(b)

FIGURE 6.7 Detailed two-stage reactor setup for BioForming process: Panel (a) illustrates the catalytic steps used to convert glucose and xylose to gasoline-range hydrocarbons; panel (b) summarizes the molar carbon and heating value yields of the resulting products. (Reprinted from White Paper for European Platform on Biofuels, Blommel, P.G. and Cortright, R.D., Production of conventional liquid fuels from sugars, Copyright 2012, with permission from Elsevier.)

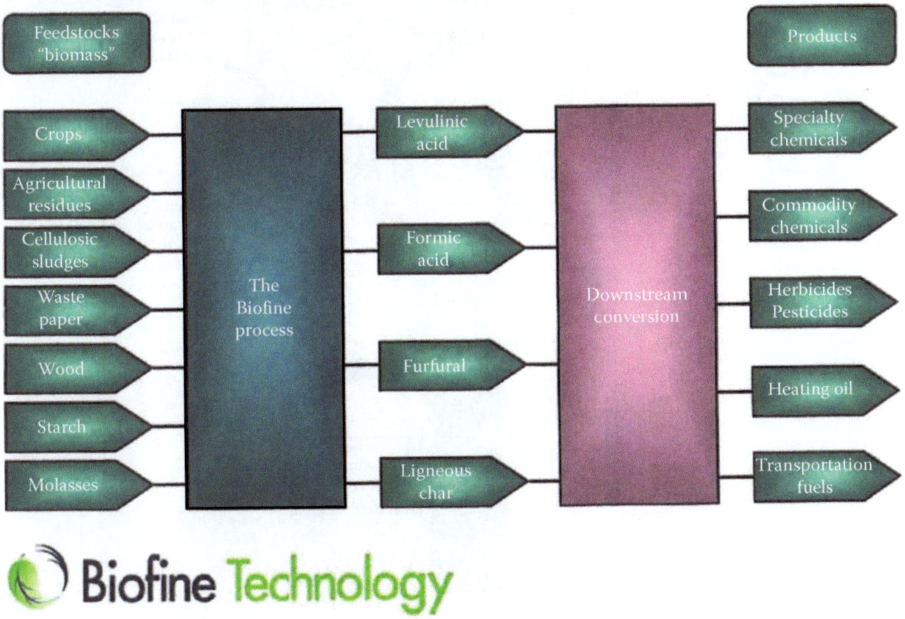

FIGURE 7.1 A schematic of the overall Biofine process that includes product upgrading. (From Fitzpatrick, S. and Nace, P., "Biofine Technology, LLC: Renewable chemicals and biofuels," Paper presented for Sustainable Bioplastics Council of Maine, 2012. With permission.)

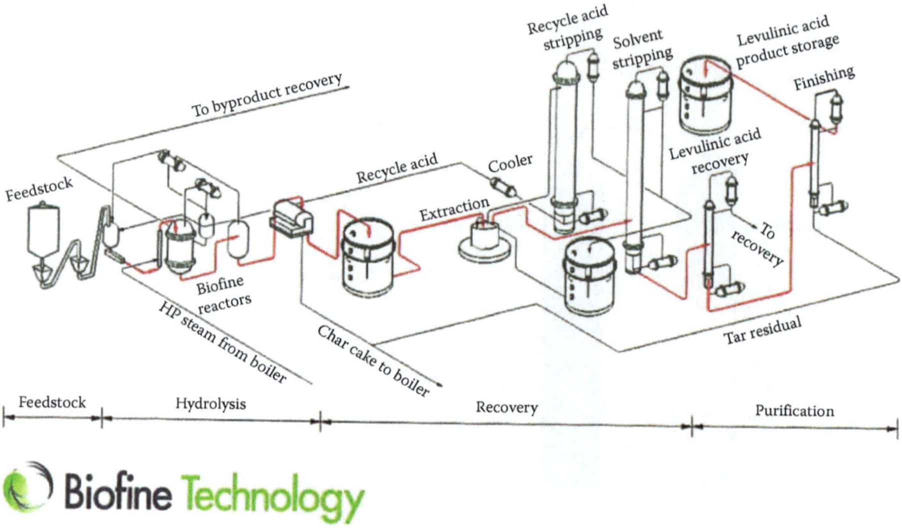

FIGURE 7.4 A complete process flow diagram of the Biofine process. (From Fitzpatrick, S. and Nace, P., "Biofine Technology, LLC: Renewable chemicals and biofuels," Paper presented for Sustainable Bioplastics Council of Maine, 2012. With permission.)

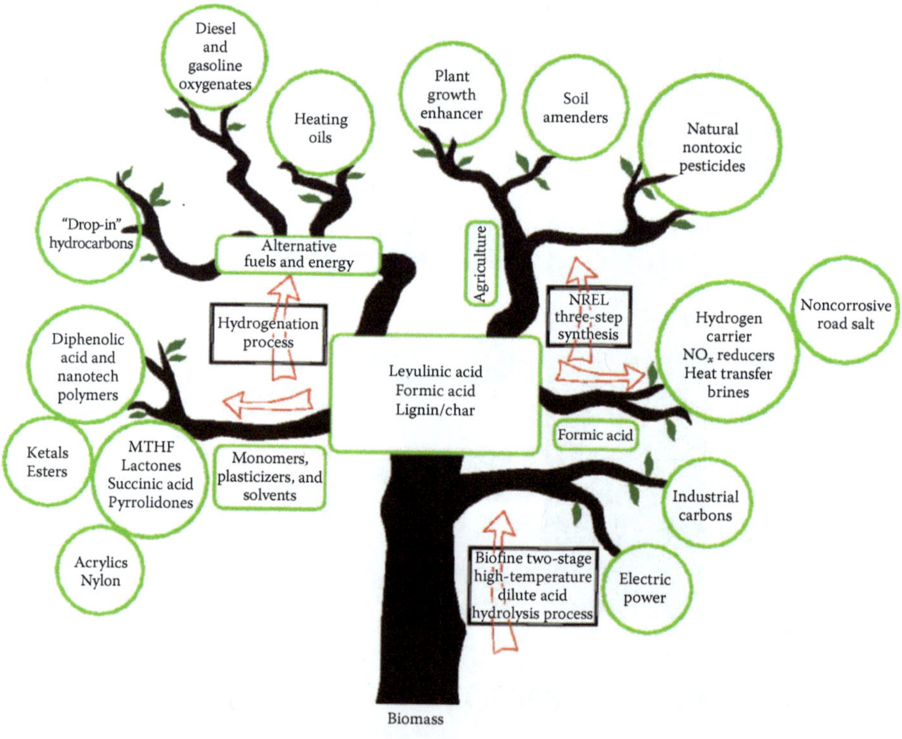

FIGURE 7.5 Biorefinery products "family tree." (From Fitzpatrick, S. and Nace, P., "Biofine Technology, LLC: Renewable chemicals and biofuels," Paper presented for Sustainable Bioplastics Council of Maine, 2012. With permission.)

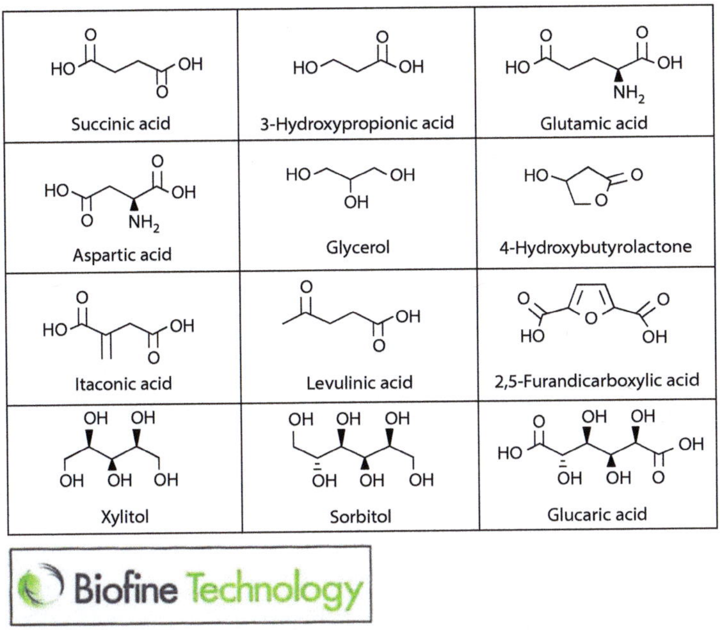

FIGURE 7.6 "Select 12" platform chemicals from biomass as identified by the Department of Energy. (From Fitzpatrick, S. and Nace, P., "Biofine Technology, LLC: Renewable chemicals and biofuels," Paper presented for Sustainable Bioplastics Council of Maine, 2012. With permission.)

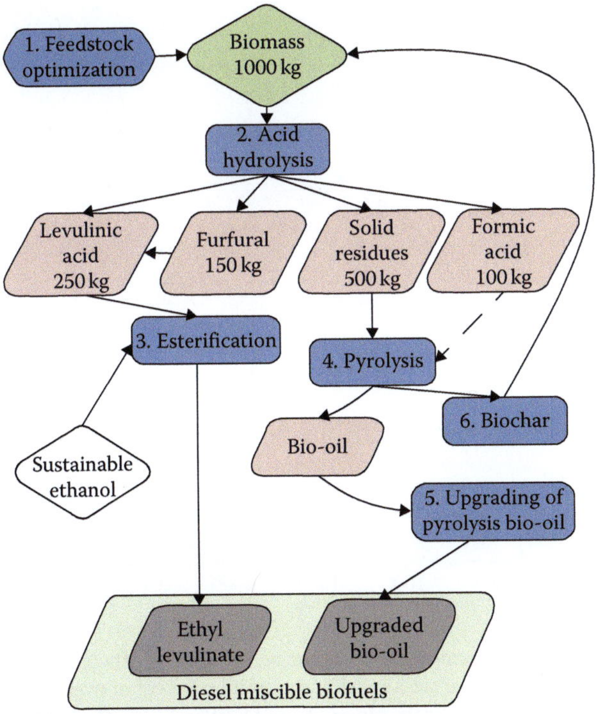

FIGURE 7.7 Process chain to produce maximum yields of DMB from organic waste and residue. (From Hayes, D., DiBANET project, 2013. With permission.)

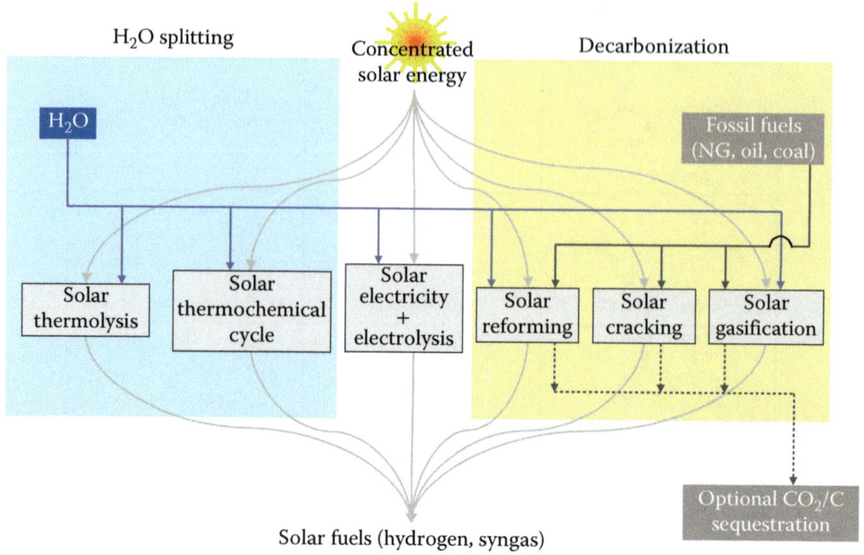

FIGURE 11.1 Thermochemical routes for solar hydrogen production. (From Meier, A. and Sattler, C., "Solar fuels from concentrated sunlight," SolarPACES, Solar Power and Chemical Energy Systems, IEA report, 2009. With permission.)

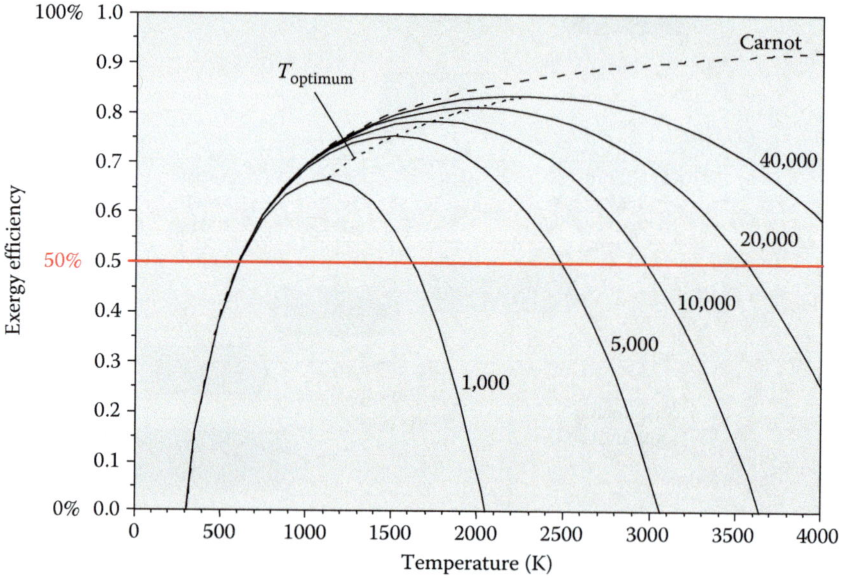

FIGURE 11.2 Exergy efficiency—Variation of the exergy efficiency as a function of the process operating temperature for a blackbody cavity receiver converting concentrated solar energy into chemical energy. (From Meier, A. and Sattler, C., "Solar fuels from concentrated sunlight," SolarPACES, Solar Power and Chemical Energy Systems, IEA report, 2009. With permission; Fletcher, E.A. and Moen, R.L., *Science*, 197, 1050–1056, 1977. With permission.)

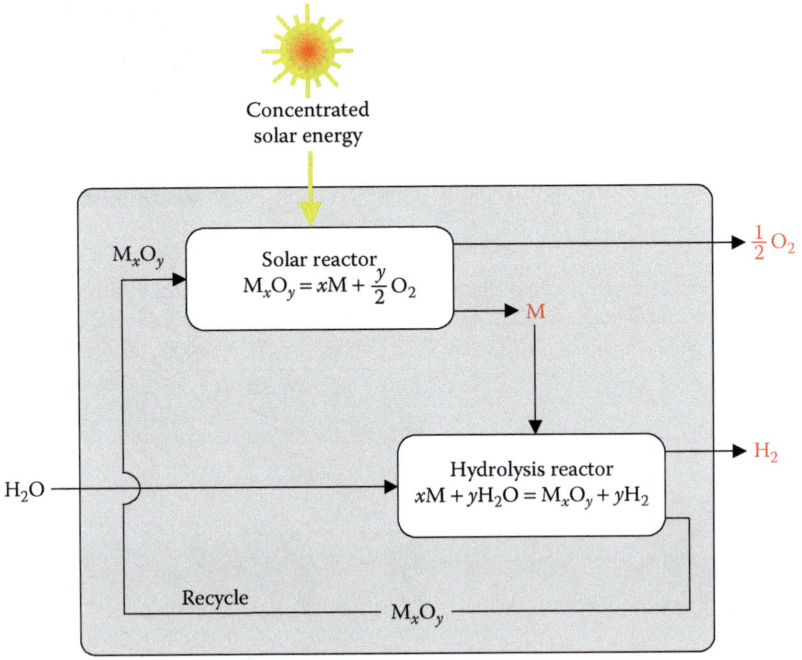

FIGURE 11.3 Thermochemical route based on metal oxide–redox reactions. (From Meier, A. and Sattler, C., "Solar fuels from concentrated sunlight," SolarPACES, Solar Power and Chemical Energy Systems, IEA report, 2009.)

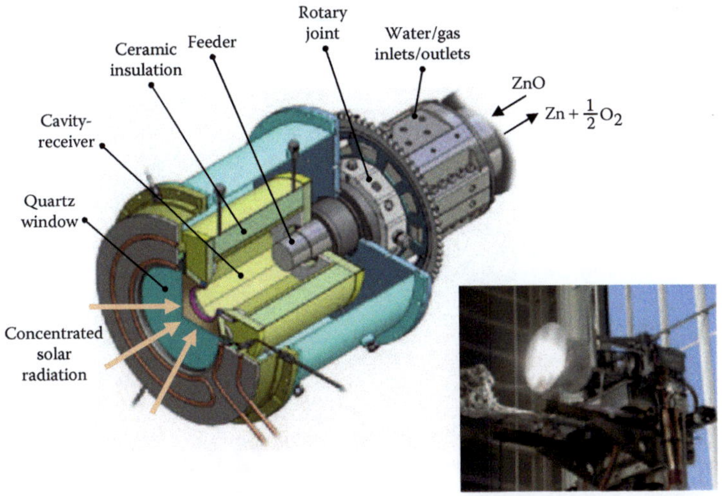

FIGURE 11.4 Rotary solar reactor for the thermal dissociation of zinc oxide to zinc and oxygen at above 1700°C. (From Meier, A. and Sattler, C., "Solar fuels from concentrated sunlight," SolarPACES, Solar Power and Chemical Energy Systems, IEA report, 2009.)

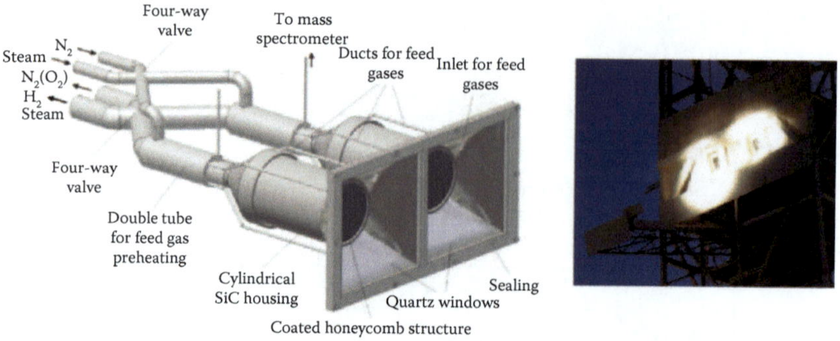

FIGURE 11.5 Monolithic dual-chamber solar receiver reactor for continuous hydrogen production. (From Meier, A. and Sattler, C., "Solar fuels from concentrated sunlight," SolarPACES, Solar Power and Chemical Energy Systems, IEA report, 2009.)

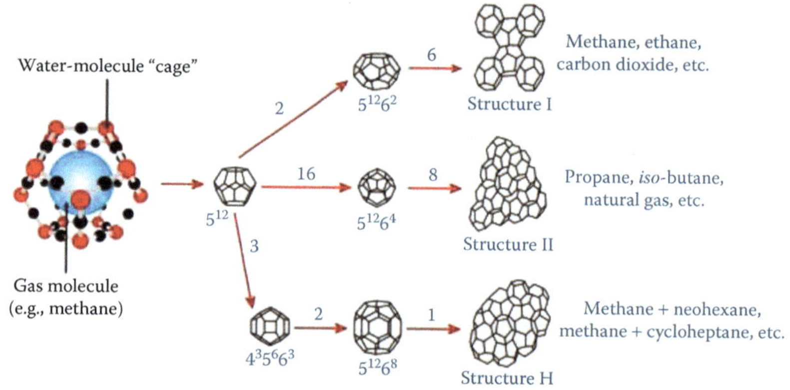

FIGURE 12.1 Various molecular structures of gas hydrate and clathrate depending on guest molecules. (From "Methane hydrates," A communication by Center for Gas Hydrate Research, Heriot-Watt University, Edinburgh, The Hydrate forum Org., 2012. With permission.)

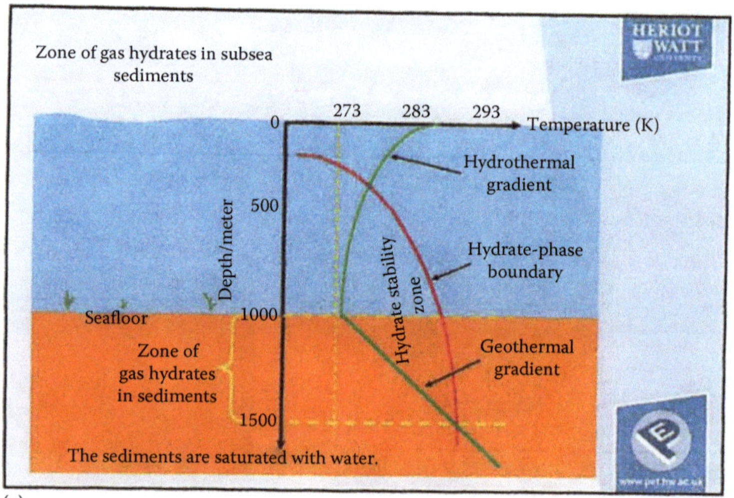

(a)

FIGURE 12.2 Gas hydrate stability fields for (a) nominal marine settings. (From Tohidi, 2013, pers. comm. With permission.)

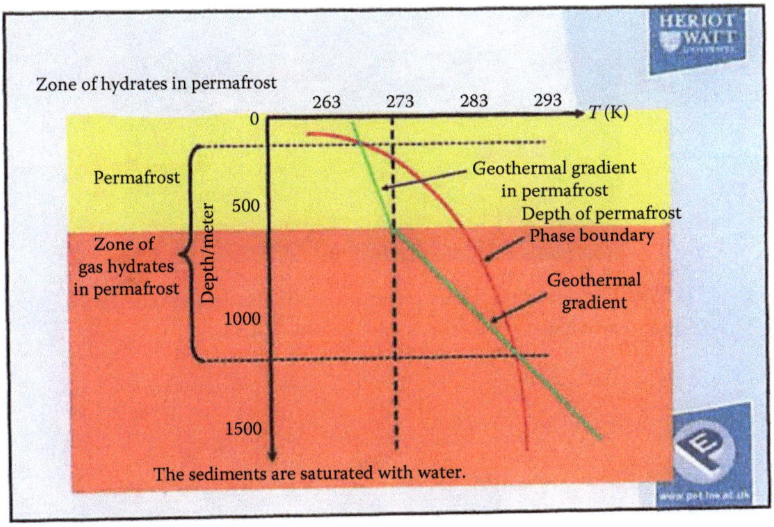

(b)

FIGURE 12.2 (Continued) Gas hydrate stability fields for (b) permafrost settings. (From Tohidi, 2013, pers. comm. With permission.)

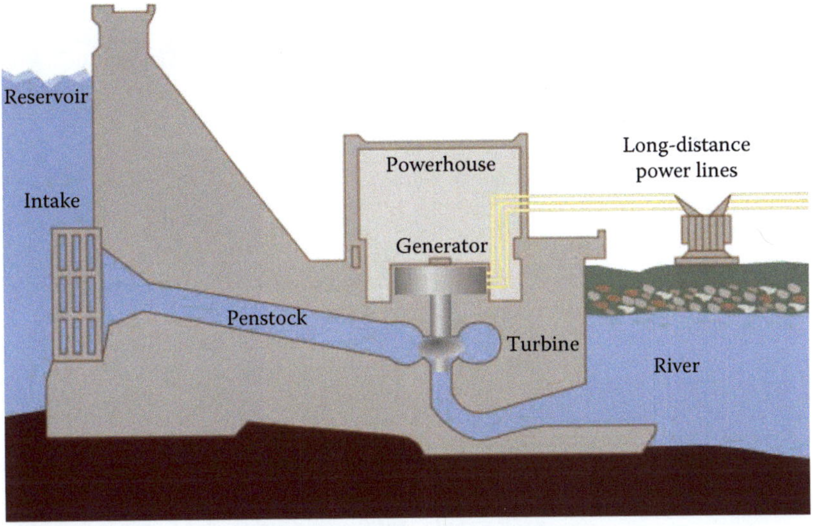

FIGURE 13.1 Cross section of a conventional hydroelectric dam. (Adapted from "Hydroelectricity," Wikipedia, the free encyclopedia, 1–7, 2012.)

FIGURE 13.3 WaveRoller wave energy farm installation in Peniche, Portugal. (Adapted from "Wave power," Wikipedia, the free encyclopedia, 2013.)

FIGURE 13.4 Wave Dragon seen from reflector. (Adapted from "Wave power," Wikipedia, the free encyclopedia, 2013.)

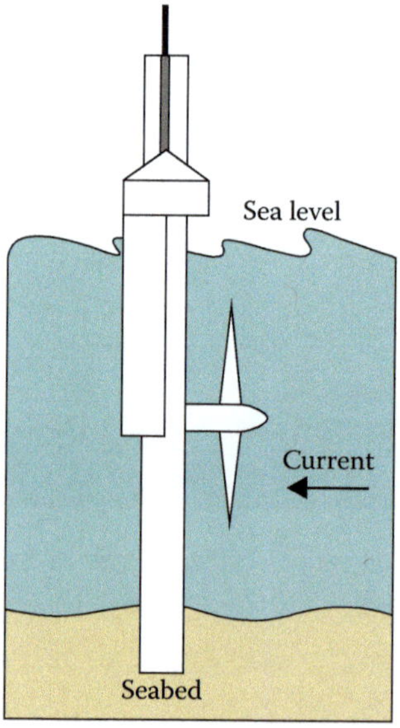

FIGURE 13.5 A horizontal-axis hydrokinetic rotating device, tidal turbine. (Adapted from "How hydrokinetic energy works?" Union of Concerned Scientists, 1–5, 2012.)

FIGURE 13.6 Cross-flow turbine used in Alaska Rivers: ORPC's TidGen^{TM} power system. [Adapted from "Hydrokinetic energy (in river, tidal, and ocean current)," *Alaska Energy Wiki*, Alaska Center for Energy and Power, 1–4, 2012.]

FIGURE 13.7 The world's first commercial scale and grid-connected tidal stream generator—SeaGen—in Strangford Lough. (Adapted from "Tidal power," Wikipedia, the free encyclopedia, 2013.)

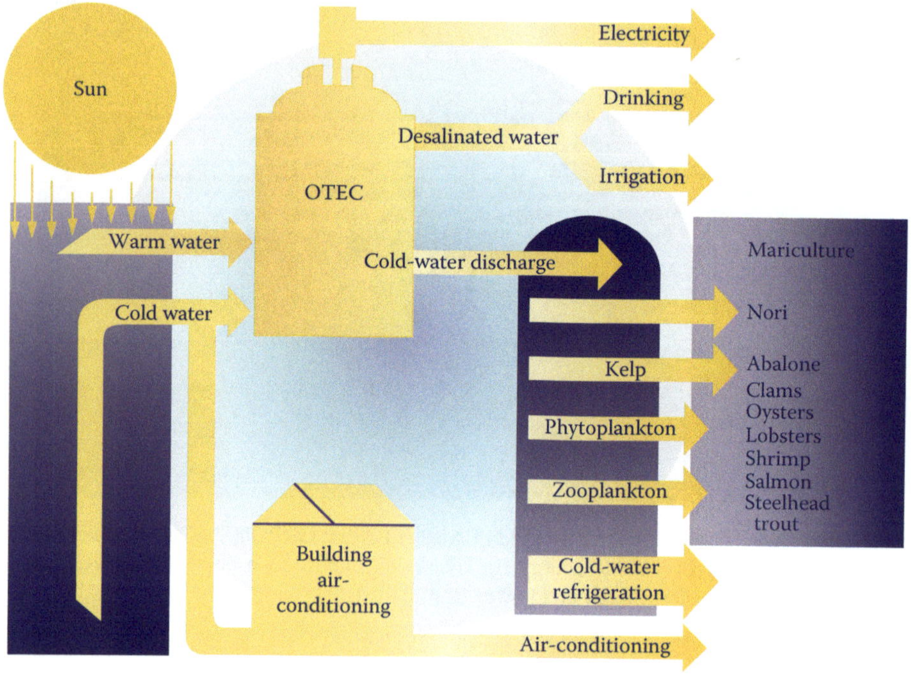

FIGURE 13.9 A schematic of OTEC process with applications. (Adapted from "Ocean thermal energy conversion," Wikipedia, the free encyclopedia, 2013.)

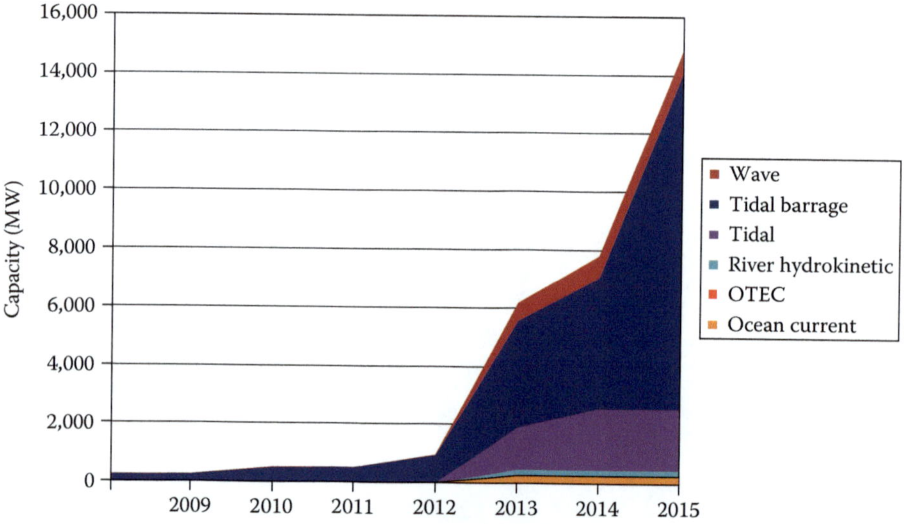

FIGURE 13.10 Cumulative marine and hydrokinetic energy installed capacity by technology, world market: 2008–2017. (Adapted from Gauntlett, D. and Asmus, P., "Executive summary: Hydrokinetic and Ocean Energy; Renewable power generation from ocean wave, tidal stream, river hydrokinetic, ocean current, and ocean thermal technologies," Research report by Pike Research, Cleantech Market Intelligence, Boulder, CO, 2012.)

6.5.6 Summary

The above-described discussion on thermodynamics, kinetics, and catalysis of the APR process gives the following conclusions about the APR process [1–6,16]:

1. The basis for the APR process is that while alkanes reforming is only favorable at high temperatures, the reforming of oxygenated carbon (with a C/O ratio of 1:1) and the water–gas shift reaction are possible at low temperatures. This allows APR to be carried out in the liquid phase.
2. The activation energy required to break up the C–C bond in oxygenated compounds is lower than that required in alkanes. Thus, H_2 and CO_2 from oxygenated compounds can be obtained in a single reactor. This can be accomplished in liquid phase only for high boiling compounds such as glucose and sorbitol, whereas for low boiling compounds such as glycerol, ethylene glycol, and methanol, the reactions can occur in both the gas and liquid phases.
3. The choice of a catalyst can affect the products. Pt, Pd, and Ni–Sn alloys show high selectivity for hydrogen, whereas Ni catalysts tend to make more alkanes. Ru and Rh catalysts also make alkanes with very little hydrogen. More acidic support favors alkanes production, whereas more basic/neutral support such as alumina favors hydrogen production. The acidic aqueous solution similarly promotes alkanes production due to acid-catalyzed dehydrogenation reactions (followed by the hydrogenation on the metal). The basic aqueous solution favors hydrogen production. The promoters such as Re add acidity to the catalyst, thereby reducing hydrogen formation.
4. The type of feed and its concentration affect the product distribution. Sorbitol gives higher selectivity for hydrogen than glucose. Within polyols, hydrogen selectivity decreases with an increase in carbon number of the feed and an increase in feed concentration due to an increase in side reactions. APR of platform chemical glycerol has been very widely studied. APR can also be applied to the secondary feedstock as long as they are properly pretreated by hydrolysis (either acid or enzymatic) and/or hydrogenation depending on the nature of the feedstock.
5. Davda et al. [2,16,41,58,59] outlined a number of different pathways that can occur in the APR reactor depending on the nature of catalyst, its acidity and acidity level of aqueous solution, the temperature, and the pressure to obtain the desired product distributions. Generally, higher carbon number in the feed and more acidity on the catalyst or aqueous solution favor C–O scission and more alkanes production. The reverse conditions promote C–C bond cleavages to form hydrogen and CO_2. The latter compounds can, however, undergo undesirable methanation and FT reactions to produce more alkanes. Some metals such as Ru and Rh favor C–O scission and form more alkanes. Pt and Pd, however, favor C–C scission. More bifunctional catalysis can occur by the combination of metal, support, and solutions. In general, high hydrogen selectivity requires high

C–C scission, low rates of C–H scission, and low rates of methanation and F–T reactions. Low CO level can be obtained by operating the reactor with low partial pressures of hydrogen and carbon dioxide. In the recent years, the use of microchannel reactors has been found to have positive effects on the APR process [9,31].

6.6 PRODUCTION OF SYNGAS AND MONOFUNCTIONAL GROUPS AND THEIR UPGRADING

6.6.1 Syngas

Besides hydrogen and alkanes, reforming has also been used to produce syngas from glycerol feedstock [34–36,44,47,51–57,65]. This once again requires the selective breakage of C–C bonds. This can be achieved with Pt catalyst in the temperature range of 498–548 K but at lower pressure. Under these conditions, Pt surface is covered by CO molecules, which hinder gas-phase reaction. Pt/Ru or Pt/Re was identified as alloys that bind CO less strongly on the surface, thus mitigating the reaction inhibition in the presence of products. These catalysts will produce syngas by the reaction [34–36,44,47,51–57,65]:

$$C_3H_8O_3 \rightarrow 3CO + 4H_2 \tag{6.11}$$

$$C_3H_8O_3 \rightarrow 7/25C_8H_{18} + 19/25CO_2 + 37/25H_2O \tag{6.12}$$

The syngas produced at these low temperatures can be easily used for the subsequent conversion of syngas to liquid fuels by the FT synthesis. The increase in Re to carbon-supported Pt catalysts also promotes the water–gas shift reaction, which increases the H_2/CO ratio and decreases the CO/CO_2 ratio in syngas [34–36,44,47,51–57,65].

6.6.2 Monofunctional Groups

The literature results [4,32,48,58,59,65,68–86] also showed that for Pt/Re/C catalyst, an increase in pressure shifted the reaction away for reforming reaction to more in the direction of alkanes production. This shift also produced partially deoxygenated intermediates [70–78] such as alcohols and ketones. This suggests that it is possible to couple biomass reforming with hydrodeoxygenation to improve the energy density without an external source of hydrogen [4,32,48,58,59,65,68–86]. Thus, Pt–Re/C catalyst operating at low temperature, high pressure, and high oxygenate feed concentration will favor C–O bond breakage and partially deoxygenate polyols to produce monofunctional intermediates that are predominantly 2-ketones, secondary alcohols, heterocylces, and carboxylic acids [4,32,48,58,59,65,67–86]. These monofunctional groups provide a platform for a variety of upgrading strategies that allow the productions of fuel additives and fuels such as jet fuel, diesel, and gasoline.

Thus, C–C coupling (i.e., condensation reactions) can be employed along with oxygen removal to obtain larger hydrocarbons starting from biomass-derived C_5 and C_6 sugar compounds. Ketones are coupled via aldol condensation using basic catalysts such as $MgAlO_x$, MgAl, Pd–MgO/ZrO_2, $MgZrO_2$, La/ZrO_2, Y/ZrO_2, and Mg/TiO_2

[4,32,48,58,59,65,68–86]. Furthermore, the introduction of bifunctional metal basic catalysts allows for the coupling of secondary alcohols in the presence of hydrogen. More condensation reactions are also driven for ketones in the presence of hydrogen [32]. C–C coupling can also be enhanced by ketonization of carboxylic acid [4]. The complete hydrogenations of monofunctional groups can also produce alcohols. The alcohols can then be converted to gasoline using methanol-to-gasoline (MTG) technology of Mobil Oil Co. that uses H-ZSM-5 catalyst [4,32,48,58,59,65,68–86]. Alcohols can also be dehydrated to produce olefins.

Kunkes et al. [10,30,83] designed a process of converting monofunctional group to pentanol and hexanol and converting these alcohols to C_6^+ gasoline by H-ZSM-5 catalyst at 673 K. In a two-step process, alcohols can also be dehydrated by acidic niobia catalyst to form olefins that can be coupled over H-ZSM-5 to form branched olefins centered around C_{12} [1–6,16]. Less branched and more complex diesel fuel can also be created by using a mixed system of catalysts $CuMg_{10}Al_7O_x$, $Pd/CeZrO_x$, and $CeZrO_x$ to achieve ketonization and aldol condensation of biomass-derived monofunctional groups as shown in Figure 6.5 [4]. All of these strategies closely follow the details outlined in an excellent review by Alonso et al. [4] and they were the starting points for the development of a complete Virent's BioForming process described in Section 6.7 [60–63,66,67,87–89].

6.7 VIRENT'S BIOFORMING PROCESS

While the original work of Dumesic et al. [1–6,16] focused on the generation of hydrogen, syngas, alkanes, and monofunctional groups from the biomass-derived carbohydrates such as alcohols, glycerols, ketones, aldehydes, furans, and other polyols [1–6,16], more recently Virent Inc. (Madison, WI) developed a more complete BioForming process that integrates APR with proven catalytic upgrading technologies to generate hydrocarbons for direct use as a biofuel or as blending components for conventional liquid fuels such as gasoline, diesel, and jet fuels [60–63,66,67,87–89]. This process has been recently described by Blommel and Cortright [15]. Here, we briefly summarize their description of the process [15,60–63,66,67,87–89].

The overall objective of the Virent's bioforming process is to develop a continuous process of converting a wide variety of feedstock into various synthetic liquid fuels, fuel additives, and some useful chemicals. Blommel and Cortright [15] point out that for this process, based on the stoichiometry of the overall conversions of xylose and sucrose to C_6^+ hydrocarbons, carbon dioxide, and water by the following set of reactions,

$$3.7C_5H_{10}O_5 \rightarrow \text{iso } C_{12}H_{26} + 10.5CO_2 + 5.5H_2O \tag{6.13}$$

$$0.875C_{12}H_{22}O_{11} \rightarrow C_8H_{10} + 2.5CO_2 + 4.6H_2O \tag{6.14}$$

it is theoretically possible for the resulting hydrocarbons to capture 64% of the carbon from the carbohydrates and over 94% of the lower heating value of sugar. Since APR technology is the centerpiece of this process, the discussion and the studies reported

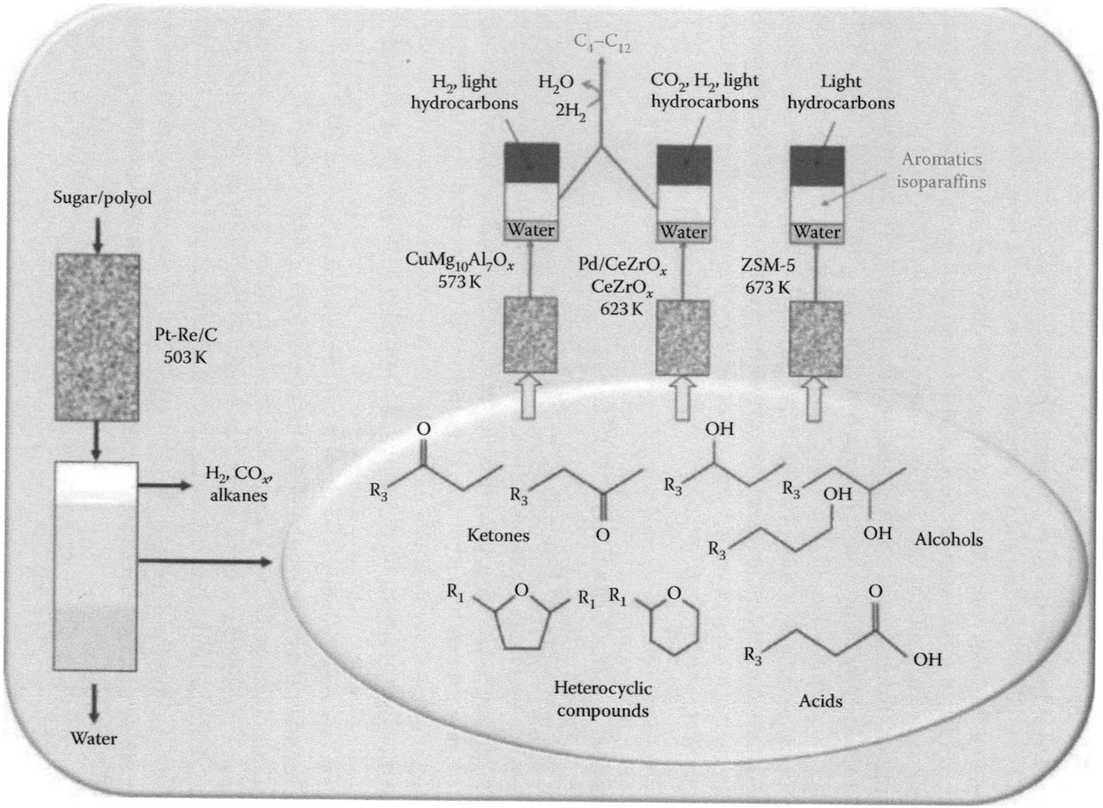

FIGURE 6.5 **(See color insert.)** Schematic pathways to convert sugar and polyols to biofuel through production of monofunctional intermediates. (Reprinted from *Green Chemistry*, 12, Alonso, D.M., Bond, J.Q., and Dumesic, J.A., Catalytic conversion of biomass to biofuels, 1493–1513, Copyright 2010, with permission from Elsevier.)

in Section 6.6 indicated that monofunctional groups generated by APR technology can be upgraded to mono-oxygenated hydrocarbons using conventional condensation and hydrotreating techniques. The discussion also indicated such upgrading of monofunctional groups to a variety of end products will require different catalytic strategies.

The development of bioforming process for the secondary feedstock outlined in Table 6.1 requires fractionation and pretreatment of these feedstock to separate hemicellulose, cellulose, and lignin. This fractionation can be carried out using various acidic and enzymatic hydrolyses, which result in the production of five- and six-carbon ring sugars such as xylose and glucose and other oxygenated compounds [15,60–63,66,67,87–89]. The separated lignin can be used for the process heating. The separated polysaccharides, C_5 and C_6 sugars, furans, phenolics, and acids are further upgraded by hydrogenation and hydrogenolysis to sugar alcohols such as sucrose and corn sugar, as well as water-soluble oxygenated compounds such as diols, glycerol, and sugar alcohols [15,60–63,66,67,87–89]. The required hydrogen for these processes can be generated *in situ*, recycled with excess hydrogen from the overall process, or provided by hydrogen from an external source [15].

The centerpiece of Virent's BioForming process is still APR technology, originally developed by Virent Inc. and Dumesic et al. [1–6,16], which utilizes heterogeneous catalysts at moderate temperatures (450–575 K) and pressures (10–90 bar) in a number of series and parallel reactions to reduce the oxygen content of the carbohydrate feedstock. As pointed out by Blommel and Cortright [15], a key feature of this method is the use of *in situ*-generated hydrogen for the defunctionalization of the highly reactive carbohydrates to a less reactive mono-oxygenated species.

While, as discussed earlier, the APR process can generate hydrogen, syngas, alkanes, and condensable monofunctional groups, for the purpose of BioForming process, the most important products are hydrogen and condensable monofunctional intermediates. Just like lignin, the lower alkanes can be used for the process heating purposes. Both hydrogen and condensable products can be formed using Pt–Re catalysts on ZrO_2. The literature shows the range of oxygenates generated from a sucrose solution through a consecutive deoxygenation and APR processing [1–6,15,16,60–63,66,67,87–89]. These results were generated by first hydrogenating aqueous solution of sucrose by Ru/C catalyst into sorbitol/mannitol mixture. This mixture was then subjected to an APR process using Pt/Re catalyst supported on ZrO_2[15]. The process generated 0.76 mol of hydrogen per mole of sugar monomer and 35% of feed carbon to CO_2 [1–6,15,16,60–63,66,67,87–89]. Besides C_1–C_6 alkanes, the process generated alcohols, ketones, acids, and cyclized components suitable for condensation to longer chain hydrocarbons [1–6,15,16,60–63,66,67,87–89].

The total amount of hydrogen generated by the APR process is governed by the reaction [15]:

$$C_6H_{14}O_6 + 6H_2O \rightarrow 13H_2 + 6CO_2 \tag{6.15}$$

The generated hydrogen is either recovered or used within the overall process. If the sorbitol is converted to xylitol, one will obtain a H_2/CO_2 ratio of 2/1 instead of

13/6 as indicated by the above reaction [15,60–63,66,67,87–89]. Very little CO will be produced due to very favorable conditions for forward water–gas shift reaction.

While the original APR process identified various strategies for upgrading monofunctional groups to different types of liquid fuels (Figure 6.5), bioforming process further developed these strategies to build a continuous process (Figure 6.6). The process tested various system operating conditions to produce different intermediate compounds appropriate for use in the downstream condensation reactions that generate different types of fuels or chemicals. The research showed that monofunctional groups can be converted to aromatics and isoalkanes via direct catalytic condensation over acid catalysts, such as solid acids and zeolites [32,48,58,59,64,65,67–86,90–98]. Zeolite ZSM-5 carries out a series of reactions that include the following [15,78,60–64,66,67,87–98]: (1) dehydration of oxygenates to alkenes, (2) oligomerization of the alkenes, (3) cracking, (4) cyclization and dehydrogenation of larger alkenes to form aromatics, (5) alkane isomerization, and (6) hydrogen transfer to form alkanes [95,96]. All of these reactions are important to produce liquid fuels of varying properties such as gasoline, diesel, and jet fuel. The heavier components are generally separated by distillation and blended into jet fuel [15].

Based on further research and development of BioForming process, Blommel and Cortright at Virent [15,60–63,66,67,87–89] proposed a unified continuous process for the conversion of sucrose and xylose into gasoline-range hydrocarbons using proprietary APR catalyst and ZSM-5 (Figure 6.6). The new integrated process, which uses four different types of catalyst beds with no intermediate separation, is schematically described in Figure 6.7 [15]. In this process, each catalyst bed carries out different set of reactions and hydrogen is added with sucrose/xylose mixture in the first reactor. The first reactor (with two stages) operates with aqueous mixtures. The APR process is carried out at 523 K that generates hydrogen, light alkanes, and monofunctional groups. The product from the first reactor is heated to 648 K and passed over two different catalyst beds, both containing different types of acid catalysts. The final carbon number distribution coming out of the second reactor includes C_6^+, which is necessary for the liquid fuel productions. About 60% of hydrogen used in the first reactor is recovered by the APR process [15].

Blommel and Cortright [15,60–63,66,67,87–89] pointed out that this transformation requires numerous types of condensation reactions such as (1) aldol condensation to form beta-hydroxy ketone and aldehydes; (2) dehydration of these products to form enone; (3) hydrogenation of conjugated enone to ketone, aldehyde, or alcohol; and finally (4) dehydration/hydrogenation or hydrogenolysis to form alkanes. This multifunctional process allows the formation of longer chain and branched hydrocarbons needed to produce gasoline, diesel, and jet fuels with subsequent distillation [15,60–64,66,67,78,87–89,98]. Many oxygenates such as alcohols, carbonyls, and acids can form C–C bonds through aldol and decarboxylative condensation reactions [15,64,78,98]. Further analysis and details on various types of condensation reactions and the role of different catalysts are given by Blommel and Cortright [15] along with some other published reports [60–63,66,67,87–89]. Currently, Virent Inc. is building a pilot plant to demonstrate the viability of the BioForming process at a larger scale with the aim of making a commercial process.

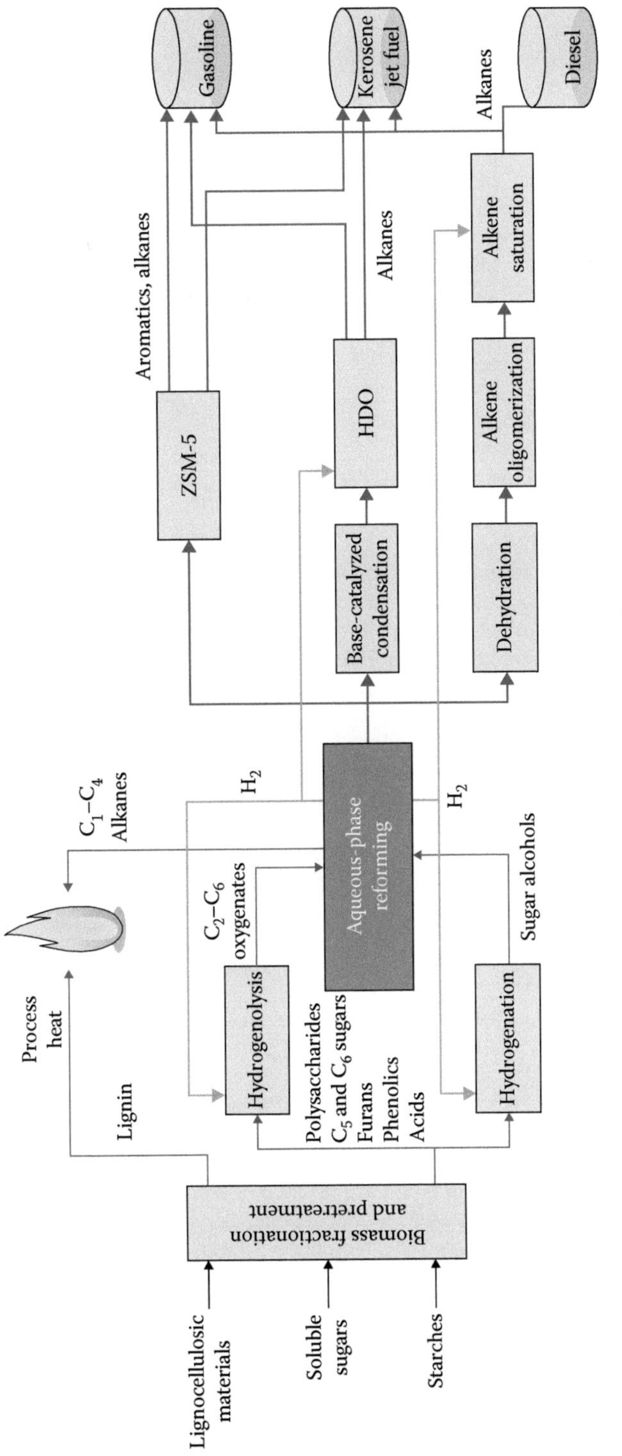

FIGURE 6.6 **(See color insert.)** Virent's BioForming® process to produce liquid transportation fuels from biomass feedstock. APR enables the process to partially defunctionalize carbohydrate feedstock for further catalytic upgrading. (Reprinted from White Paper for European Platform on Biofuels, Blommel, P.G. and Cortright, R.D., Production of conventional liquid fuels from sugars, Copyright 2012, with permission from Elsevier.)

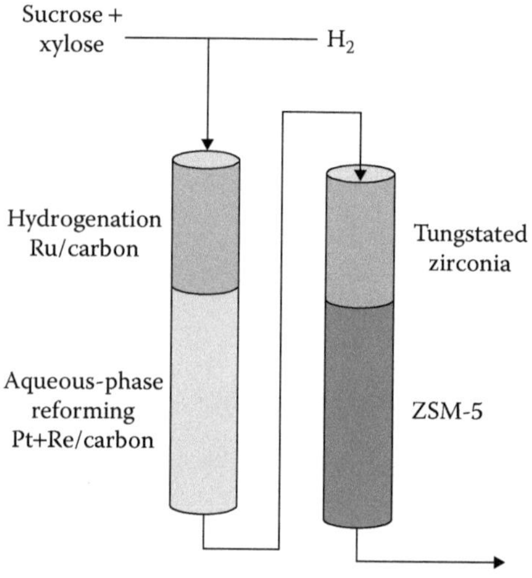

(a)

	Carbon (%)	LHV (%)
Feeds		
Sucrose + xylose	100	87
Hydrogen	–	12
Products		
CO_2	25	–
H_2	–	3
C_1–C_4	26	31
C_5–C_9	43	53
C_{10+}	5	6

(b)

FIGURE 6.7 **(See color insert.)** Detailed two-stage reactor setup for BioForming process: Panel (a) illustrates the catalytic steps used to convert glucose and xylose to gasoline-range hydrocarbons; panel (b) summarizes the molar carbon and heating value yields of the resulting products. (Reprinted from White Paper for European Platform on Biofuels, Blommel, P.G. and Cortright, R.D., Production of conventional liquid fuels from sugars, Copyright 2012, with permission from Elsevier.)

REFERENCES

1. Huber, G.W., Cortright, R.D., and Dumesic J.A., "Renewable alkanes by aqueous phase reforming of biomass derived oxygenates," *Angewandte Chemie International Edition*, 43, 1549–1551 (2004).
2. Davda, R. and Dumesic, J., "Catalytic reforming of oxygenated hydrocarbons for hydrogen with low levels of carbon monoxide," *Angewandte Chemie International Edition*, 42, 4068 (2003).
3. Tao, J., Shishi, C., and Fahai, C., "Advances in H_2 production from the aqueous-phase reforming of biomass-derived polyols: A review," *Chemical Industry and Engineering Progress*, 31 (5), 1010–1017 (2012).
4. Alonso, D.M., Bond, J.Q., and Dumesic, J.A., "Catalytic conversion of biomass to biofuels," *Green Chemistry*, 12, 1493–1513 (2010).
5. Cortright, R., Davda, R., and Dumesic, J., "Hydrogen from catalytic reforming of biomass derived hydrocarbons in liquid water," *Nature*, 418, 964–967 (2002).
6. Huber, G. and Dumesic, J., "An overview of aqueous phase catalytic processes for production of hydrogen and alkanes in a biorefinery," *Catalysis Today*, 111 (1–2), 119–132 (2006).
7. King, D., "Hydrogen production via aqueous phase reforming," Paper presented by Pacific Northwest National Laboratory at NIChE Catalysis Conference, Washington, DC, September 21 (2011).
8. Cortright, R.D., "Hydrogen generation from sugars via aqueous-phase reforming," *Proceedings of the 16th World Hydrogen Energy Conference*, June 13–16, Lyon, France (2006).

9. D'Angelo, M., Ordomsky, V., van der Schaaf, J., Schouten, J., and Nijhuis, T., "Aqueous phase reforming in a microchannel reactor: The effect of mass transfer on hydrogen selectivity," *Catalysis Science & Technology*, 3, 2834–2842 (2013).
10. Kunkes, E.L., Simonetti, D.A., West, R.M., Serrano-Ruiz, J.C., Gartner, C.A., and Dumesic, J.A., *Science*, 322, 417–421 (2008).
11. Dunn, S., "Hydrogen futures: Toward a sustainable energy system," *International Journal of Hydrogen Energy*, 27, 235–264 (2002).
12. Huber, G., Iborra, S., and Corma, A., "Synthesis of transportation fuels from biomass: Chemistry, catalysts, and engineering," *Chemical Reviews*, 106, 4044–4098 (2006).
13. Soares, R., Simonetti, D., and Dumesic, J., *Angewandte Chemie International Edition*, 45, 3982–3985 (2006).
14. King, D., "Biomass-derived liquids distributed (aqueous phase) reforming," 2012 DOE Hydrogen and Fuel Cells Program Review, Pacific Northwest National Laboratory, May 17, Richland, WA (2012).
15. Blommel, P.G. and Cortright, R.D., "Production of conventional liquid fuels from sugars," a White Paper for European Platform on Biofuels, UK (2012).
16. Davda, R., Shabaker, J., Huber, G., Cortright, R., and Dumesic, J., "A review of catalytic issues and process conditions for renewable hydrogen and alkanes by aqueous-phase reforming of oxygenated hydrocarbons over supported metal catalysts," *Applied Catalysis B: Environmental*, 56, 171–186 (2005).
17. Huber, G., Shabaker, J., and Dumesic, J., "Raney Ni-Sn catalyst for H_2 production from biomass-derived hydrocarbons," *Science*, 300, 2075–2077 (2003).
18. Vos, B., Poels, E., and Bliek, A., *Journal of Catalysis*, 198, 77 (2001).
19. Zhu, X., Huo, P., Zhang, Y., Cheng, D., and Liu, C., *Applied Catalysis B: Environmental*, 81, 132 (2008).
20. Zhang, J., Xu, H., Jin, X., Ge, Q., and Li, W., *Applied Catalysis A: General*, 290, 87 (2005).
21. Alberton, A., Souza, M., and Schmal, M., *Catalysis Today*, 123, 257 (2007).
22. Bartholomew, C.H. and Pannell, R.B., *Journal of Catalysis*, 65, 390 (1980).
23. Mustard, D. and Bartholomew, C., *Journal of Catalysis*, 67, 186 (1981).
24. Lehnert, K. and Claus, P., *Catalysis Communications*, 9, 2543 (2008).
25. Wang, Z., Pan, Y., Dong, T., Zhu, X., Kan, T., Yuan, L., Torimoto, Y., Sadakata, M., and Li, Q., *Applied Catalysis A: General*, 320, 24 (2007).
26. Minowa, T., Fang, Z., Ogi, T., and Varhegyi, G., *Journal of Chemical Engineering of Japan*, 31, 131 (1998).
27. Trimm, D., "Coke formation and minimisation during steam reforming reactions," *Catalysis Today*, 37, 233–238 (1997).
28. Shabaker, J., Huber, G., and Dumesic, J., "Aqueous-phase reforming of oxygenated hydrocarbons over Sn-modified Ni catalysts," *Journal of Catalysis*, 222, 180–191 (2004).
29. Park, H.J., Kim, H.-D., Kim, T.-W., Jeong, K.-E., Chae, H.-J., Jeong, S.-Y., Chung, Y.-M., Park, Y.-K., and Kim, C.-U. "Production of biohydrogen by aqueous phase reforming of polyols over platinum catalysts supported on three-dimensionally bimodal mesoporous carbon," *ChemSusChem*, 5, 629–633 (2012).
30. Kunkes, E., Simonetti, D., Dumesic, J., Pyrz, W., Murillo, L., Chen, J., and Buttrey, D., *Journal of Catalysis*, 260, 164–177 (2008).
31. D'Angelo, M.F.N., Ordomsky, V., Paunovic, V., van der Schaaf, J., Schouten, J.C., and Nijhuis, T.A., "Hydrogen production through aqueous-phase reforming of ethylene glycol in a washcoated microchannel," *ChemSusChem*, 6, 1708–1716 (2013).
32. Huber, G., Chheda, J., Barrett, C., and Dumesic, J., "Chemistry: Production of liquid alkanes by aqueous phase processing of biomass-derived carbohydrates," *Science*, 308 (5727), 1446–1450 (2005).

33. Serrano-Ruiz, J. and Dumesic, J., "Catalytic production of liquid hydrocarbon transportation fuels," in Guczi, L. and Erdohelyi, A. (eds.), *Catalysis for Alternative Energy Generation*. Springer, New York, pp. 29–56 (2012).
34. Wen, G., Xu, Y., Ma, H., Xu, Z., and Tian, Z., "Production of hydrogen by aqueous phase reforming of glycerol," *International Journal of Hydrogen Energy*, 33, 6657–6666 (2008).
35. Boonyanuwat, A., Jentys, A., and Lercher, J.A., "Hydrogen production by aqueous phase reforming of glycerine on noble supported catalysts," DGMK-Tagungsbericht, October 4–6, Dresden, Germany (2006).
36. Luo, N., Fu, X., Cao, F., Xiao, T., and Edwards, P., "Glycerol aqueous phase reforming for hydrogen generation over Pt catalyst—Effect of catalyst composition and reaction conditions," *Fuel*, 87, 3483–3489 (2008).
37. Valenzuela, M., Jones, C., and Agrawal, P., "Batch aqueous phase reforming of woody biomass," *Energy & Fuels*, 20, 1744–1752 (2006).
38. Chu, X., Liu, J., Sun, B., Dai, R., Pei, Y., Qiao, M., and Fan, K., "Aqueous phase reforming of ethylene glycol on Co/ZnO catalysts prepared by the coprecipitation method," *Journal of Molecular Catalysis A: Chemical*, 335, 129–135 (2011).
39. Wen, G., Xu, Y., Xu, Z., and Tian, Z., "Characterization and catalytic properties of the Ni/Al2O3 catalysts for aqueous phase reforming of glucose," *Catalysis Letters*, 129, 250–257 (2009).
40. Ni, M., Leung, D.Y.C., Leung, M.K.H., and Sumathy, K., *Fuel Processing Technology*, 87, 461 (2006).
41. Davda, R. and Dumesic, J., "Renewable hydrogen by aqueous phase reforming of glucose," *Chemical Communications*, 36 (2004).
42. Tanksale, A., Wong, Y., Beltramini, J.N., and Lu, G.Q., "Hydrogen generation from liquid phase catalytic reforming of sugar solutions using metal-supported catalysts," *International Journal of Hydrogen Energy*, 32, 717 (2007).
43. Xu, Y.P., Tian, Z.J., Wen, G.D., Xu, Z.S., Qu, W., and Lin, L.W., *Chemistry Letters*, 35, 216 (2006).
44. Manfro, R., Pires, T., Ribeiro, N., and Souza, M., "Aqueous phase reforming of glycerol using Ni-Cu catalysts prepared from hydrotalcite like precursors," *Catalysis Science & Technology*, 3, 1278–1287 (2013).
45. Tungal, R. and Shende, R., "Subcritical aqueous phase reforming of wastepaper for biocrude and H_2 generation," *Energy & Fuels*, 27 (6), 3194–3203 (2013).
46. Manfro, R., de Costa, A., Ribeiro, N., and Souza, M., "Hydrogen production by aqueous phase reforming of glycerol over nickel catalysts supported on CeO_2," *Fuel Processing Technology*, 92, 330–335 (2013).
47. Vaidya, P. and Rodrigues, A., "Glycerol reforming for hydrogen production: A review," *Chemical Engineering & Technology*, 32, 1463–1469 (2009).
48. Shabaker, J., Davda, R., Huber, G., Cortright, R., and Dumesic, J., "Aqueous phase reforming of methanol and ethylene glycol over alumina-supported platinum catalysts," *Journal of Catalysis*, 215, 344–352 (2003).
49. Shabaker, J., Huber, G., Cortright, R., and Dumesic, J., "Aqueous-phase reforming of ethylene glycol over supported platinum catalysts," *Catalysis Letters*, 88, 1–8 (2003).
50. Cruz, I., Ribeiro, N., Aranda, D., and Souza, M., "Hydrogen production by aqueous-phase reforming of ethanol over nickel catalysts prepared from hydrotalcite precursors," *Catalysis Communications*, 9, 2606–2611 (2008).
51. Behr, A., Eilting, J., Irawadi, K., Leschinski, J., and Lindner, F., "Improved utilization of renewable resources: New important derivatives of glycerol," *Green Chemistry*, 10, 13–30 (2008).
52. Iriondo, A., Barrio, V., Cambra, J., Arias, P., Güemez, M., Navarro, R., Sanchez-Sanchez, M., and Fierro, J., "Hydrogen production from glycerol over nickel catalysts supported on Al_2O_3 modified by Mg, Zr, Ce or La," *Topics in Catalysis*, 49, 46–58 (2008).

53. Buffoni, I., Pompeo, F., Santori, G., and Nichio, N., "Nickel catalysts applied in steam reforming of glycerol for hydrogen production," *Catalysis Communications*, 10, 1656–1660 (2009).
54. Zhang, B., Tang, X., Li, Y., Xu, Y., and Shen, W., "Hydrogen production from steam reforming of ethanol and glycerol over ceria-supported metal catalysts," *International Journal of Hydrogen Energy*, 32, 2367–2373 (2007).
55. Wawrzetz, A., Peng, B., Hrabar, A., Jentys, A., Lemonidou, A., and Lercher, J., "Towards understanding the bifunctional hydrodeoxygenation and aqueous phase reforming of glycerol," *Journal of Catalysis*, 269, 411–420 (2010).
56. Chheda, J. and Dumesic, J., "An overview of dehydration, aldol condensation, and hydrogenation processes for production of liquid alkanes from biomass-derived carbohydrates," *Catalysis Today*, 123 (1–4), 59–70 (2007).
57. Cho, S.H. and Moon, D.J., "Aqueous phase reforming of glycerol over Ni-based catalysts for hydrogen production," *Journal of Nanoscience and Nanotechnology*, 11 (8), 7311–7314 (2011).
58. Davda, R., Shabaker, J., Huber, G., Cortright, R., and Dumesic, J., *Applied Catalysis B: Environmental*, 56, 171–186 (2005).
59. Davda, R.R., Shabaker, J.W., Huber, G.W., Cortright, R.D., and Dumesic, J.A., *Applied Catalysis B: Environmental*, 43, 13–26 (2003).
60. Cortright, R.D. and Blommel, P.G., "Synthesis of liquid fuels and chemicals from oxygenated hydrocarbons," Publication No. US2008/0216391 (September 11, 2008).
61. Cortright, R.D. and Blommel, P.G., "Synthesis of liquid fuels and chemicals from oxygenated hydrocarbons," Publication No. WO2008/109877 (September 12, 2008).
62. Cortright, R.D. and Blommel, P.G., "Synthesis of liquid fuels and chemicals from oxygenated hydrocarbons," US Application No. 12/044908 (December 4, 2008).
63. Cortright, R.D. and Blommel, P.G., "Synthesis of liquid fuels and chemicals from oxygenated hydrocarbons," US Application No. 12/044876 (December 4, 2008).
64. Ipatieff, V., Corson, B., and Egloff, G., "Polymerization: A new source of gasoline," *Industrial & Engineering Chemistry Research*, 27 (9), 1077–1081 (1935).
65. Tuza, P., Manfro, R., Ribeiro, N., and Souza, M., "Production of renewable hydrogen by aqueous-phase reforming of glycerol over Ni–Cu catalysts derived from hydrotalcite precursors," *Renewable Energy*, 50, 408–414 (2013).
66. Cortright, R.D., "Bioforming process—Production of conventional liquid fuels from sugar," ACS/EPA Green Chemistry Conference, June 23, College Park, MD (2009).
67. Held, A., "Catalytic conversion of renewable plant sugars to fungible liquid hydrocarbon fuels using Bioforming process," TAPPI, IBCC Session 3, Conversion Pathways, October 15, Memphis, TN (2009).
68. Alonso, D., Gallo, J., Mellmer, M., Wettstein, S., and Dumesic, J., "Direct conversion of cellulose to levulinic acid and gamma-valerolactone using solid acid catalysts," *Catalysis Science Technology*, 3, 927–931 (2013).
69. Cortright, R., Davda, R., and Dumesic, J., *Nature*, 418, 964–967 (2002).
70. West, R., Kunkes, E., Simonetti, D., and Dumesic, J., *Catalyst Today*, 147, 115–125 (2009).
71. Sasaki, M.K., Goto, K., Tajima, K., Adschiri, T., and Arai, K., *Green Chemistry*, 4, 285–287 (2002).
72. Climent, M., Corma, A., Iborra, S., Epping, K., and Velty, A., *Journal of Catalysis*, 225, 316–326 (2004).
73. Barrett, C., Chheda, J., Huber, G., and Dumesic, J., *Applied Catalysis B: Environmental*, 66, 111–118 (2006).
74. Di Cosimo, J., Torres, G., and Apesteguia, C., *Journal of Catalysis*, 208, 114–123 (2002).
75. Nikolopoulos, A., Jang, B., and Spivey, J., *Applied Catalysis A: General*, 296, 128–136 (2005).

76. Renz, M., *European Journal of Organic Chemistry*, 6, 979–988 (2005).
77. Corma, A., Renz, M., and Schaverien, C., *ChemSusChem*, 1, 739–741 (2008).
78. Dooley, K., Bhat, A., Plaisance, C., and Roy, A., "Ketones from acid condensation using supported CeO_2 catalysts: Effect of additives," *Applied Catalysis A: General*, 320, 122–133 (2007).
79. Serrano-Ruiz, J. and Dumesic, J., *Green Chemistry*, 11, 1101–1104 (2009).
80. Gayubo, A., Aguayo, A., Atutxa, A., Aguado, R., and Bilbao, J., *Industrial & Engineering Chemistry Research*, 43, 2610–2618 (2004).
81. West, R., Braden, D., and Dumesic, J., *Journal of Catalysis*, 262, 134–143 (2009).
82. Blommel, P., Keenan, G., Rozmiarek, R., and Cortright, R., *International Sugar Journal*, 110, 672 (2008).
83. Kunkes, E., Gurbuz, E., and Dumesic, J., *Journal of Catalysis*, 266, 236–249 (2009).
84. Gurbuz, E., Kunkes, E., and Dumesic, J., *Green Chemistry*, 12, 223–227 (2010).
85. Gurbuz, E., Kunkes, E., and Dumesic, J., *Applied Catalysis B: Environmental*, 94, 134–141 (2010).
86. Huber, G., Cortright, R., and Dumesic, J., *Angewandte Chemie International Edition*, 43, 1549–1551 (2004).
87. Blommel, P., "Catalytic conversion of carbohydrates to hydrocarbons," DOE Biomass R&D TAC Meeting, May 19, Washington, DC (2011).
88. Held, A., "Production of renewable aromatic chemicals using Virent's catalytic bioforming process," Frontiers in Biorefining, October 19–22, St. Simons Island, GA (2010).
89. "Hydrogen generation from waste antifreeze via the BioForming process," VIRENT Co., 11th Electrochemical Power Sources, July 16, Baltimore, MD (2009).
90. Chang, C., "Mechanism of hydrocarbon formation from methanol," *Studies in Surface Science and Catalysis*, 36, 127–143 (1988).
91. Chang, C., "Methanol to gasoline and olefins," *Chemical Industries*, 57, 133–173 (1994).
92. Chang, C., Lang, W., and Silvestri, A., "Conversion of carbonyl compounds to aromatics," US Patent No. 3,907,915 (1975).
93. Chang, C. and Silvestri, A., "MTG: Origin, evolution, operation," *CHEMTECH*, 17 (10), 624–631 (1987).
94. De Klerk, A., Nel, R., and Schwarzer, R., "Oxygenate conversion over solid phosphoric acid," *Industrial & Engineering Chemistry Research*, 46 (8), 2377–2382 (1998).
95. Goguen, P., Xu, T., Barich, D.H., Skloss, T.W., Song, W., Wang, Z., Nicholas, J.B., and Haw, J.F., "Pulse-quench catalytic reactor studies reveal a carbon pool mechanism in methanol to gasoline chemistry on zeolite HZSM-5," *Journal of the American Chemical Society*, 120 (11), 2650–2651 (1998).
96. Chheda, J., Huber, G., and Dumesic, J., "Liquid phase catalytic processing of biomass derived oxygenated hydrocarbons to fuels and chemicals," *Angewandte Chemie International Edition*, 46 (38), 7164–7183 (2007).
97 Holmgren, J. and Arena, B., "Solid base as catalysis in aldol condensation," US Patent No. 5,254,743 (1993).
98. King, F., Kelly, G., and Stitt, E., "Improved base catalysts for industrial condensation reactions," *Studies in Surface Science and Catalysis*, 145, 443–446 (2003).

7 Biofine Hydrolysis Process and Derivative Product Upgrading Technologies

7.1 INTRODUCTION

Cellulose is the most abundant raw material on the Earth exceeding at any given time known fossil fuel reserves. Its annual production is estimated to be around 100 billion tons [1–8] (Fitzpatrick, 2011, pers. comm.). Unlike fossil fuels, cellulose is renewable: Using modern forestry techniques to grow short-rotation hybrid tree species such as willow or poplar, sustainable wood yields up to 10 dry tons per acre per year have been predicted [2–5]. Furthermore, in the United States, about 250 million tons of municipal solid waste (MSW) per year is discarded, 50% of which is cellulosic in nature. Globally, this number will shortly exceed one billion tons per year. Thus, the biomass could be a primary source of energy, chemicals, and materials for the United States and the rest of the world. It is estimated that using the current one-third of forest and marginally arable land for the production of short-rotation hybrid species or grassy energy crops such as switchgrass, it would be possible to supply all transportation needs and a large fraction of petrochemical needs from woody biomass sources [5–7] (Fitzpatrick, 2011, pers. comm.). Replacing fossil fuels by these sources will also have enormous environmental benefits because the use of biomass is carbon dioxide neutral and will be favorable to the issue of global warming. Also, the cellulosic fraction of MSW is most difficult to recycle, and making its use for energy and products will also help the environment (Fitzpatrick, 2011, pers. comm.).

In this chapter, we outline a novel aqueous-phase "Biofine process," which is an acid hydrolysis process to convert cellulose to levulinic acid (LA), a platform chemical with dozens of known potential use for both fuels and chemicals. Formic acid is a coproduct of the LA. Furfural is also produced if the feedstock contains hemicellulose pentosan polymer. The basic features of the Biofine process are as follows [8–11]:

1. Biofine is a simple thermochemical process allowing the conversion of cellulose and hemicellulose from a wide variety of sources.
2. Biofine is an acid hydrolysis process, which can be used for the feedstock containing up to 50% water without significantly affecting the overall

system economics. This also alleviates the need for water removal from the feedstock, which can be costly and cumbersome.

3. The process does not require lengthy pretreatment, enzymatic hydrolysis, or fermentation. The reaction occurs in minutes rather than days resulting in lower capital costs, smaller physical footprint, and larger production rate.
4. No specially designed bugs or microorganisms are required, thus lowering the costs and eliminating the risks from contamination and biological stability. The entire process is purely a chemical process.
5. One of the greatest strengths of the process is its ability to process a variety of feedstock. Any input furnished with sufficient cellulose or other carbohydrates is a suitable feedstock including low-value forest residues, whole tree chips, agricultural residues, food wastes, recycled paper, and sorted MSW.
6. Gasification processes that convert biomass into gas and then catalyze the gas into liquid fuels via Fischer–Tropsch synthesis can be hindered by the high natural variability in biomass. The Biofine process, however, can handle most cellulosic-based biomass without significant changes in the process.
7. One of the drawbacks of the fermentation technology is that a very effective enzyme and microbes for conversion of five-carbon sugars such as xylose and pentose has not yet been found. The Biofine process works well for both six-carbon (glucose) and five-carbon (xylose) sugars.
8. The cellulose fraction is broken down to form two coproducts: LA and formic acid.
9. The hemicellulose fraction is broken down into furfural, which can be delivered as a product with many other applications or can be chemically converted to LA.
10. Lignins, along with some degraded cellulose and hemicellulose and any inert components of the feed, come out of the process as a carbon-rich char mixture that can be burned to produce steam and power for the process or can be further converted to products such as carbon black, activated carbon, or carbon fiber.

In sum, the Biofine process is operated as a two-stage continuous process that allows the complete breakdown of cellulosic and starchy feedstock to LA, formic acid, furfural, and ligneous char in sufficiently high yield to be economically attractive. The typical operating conditions of the Biofine process are as follows: the temperature in the range of 190°C–220°C, the acid concentration in the range of 1–5 wt%, and the residence time in the order of 15 min overall [8–11] (Fitzpatrick, 2011, pers. comm.). The primary products are potent "platforms" for other valuable products including fuels and chemicals. The major features of the Biofine process are schematically illustrated in Figure 7.1 [12].

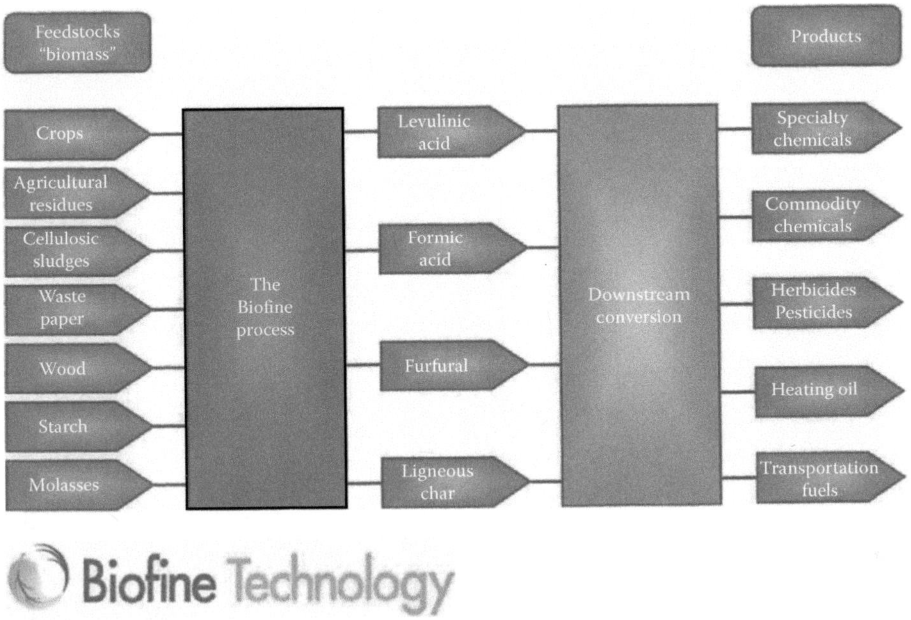

FIGURE 7.1 **(See color insert.)** A schematic of the overall Biofine process that includes product upgrading. (From Fitzpatrick, S. and Nace, P., "Biofine Technology, LLC: Renewable chemicals and biofuels," Paper presented for Sustainable Bioplastics Council of Maine, 2012. With permission.)

7.2 THE HYDROLYSIS PROCESS

The Biofine process uses one of the most advanced and commercially viable lignocellulosic-fractionating technologies that are currently available. The process involves the acid hydrolysis of polysaccharides to their monomeric constituents, and these are then used to produce valuable platform chemicals such as furfural, LA, and gamma-valerolactone (GVL). The major polysaccharides of importance in biomass are the glutans and hemicelluloses. Glucans (which are carbohydrate homopolysaccharides consisting of repeated D-glucopyranose units) largely contain starch and cellulose [8–11] (Fitzpatrick, 2011, pers. comm.).

The hydrolysis of starch using alpha-amylase and glucoamylase enzymes can be carried out with relative ease and high efficiency. This is because, as shown in Figure 7.2, alpha($1 \rightarrow 4$) linkages in the amylose component of starch and alpha($1 \rightarrow 6$) amylopectin branches in the starch are easy to break as there is no internal hydrogen bonding preventing the breakage [8]. This has allowed the production of ethanol from grains (94% from corn) to the level of 1.48 billion gallons in the United States in 2011. The hydrolysis and fermentation of cellulose, however, is about 100 times more difficult than that of starch due to the presence of hydrogen bonding as shown in Figure 7.2. Cellulose is much more abundant than starch and requires less energy to produce than starch crops [1–8] (Fitzpatrick, 2011, pers. comm.).

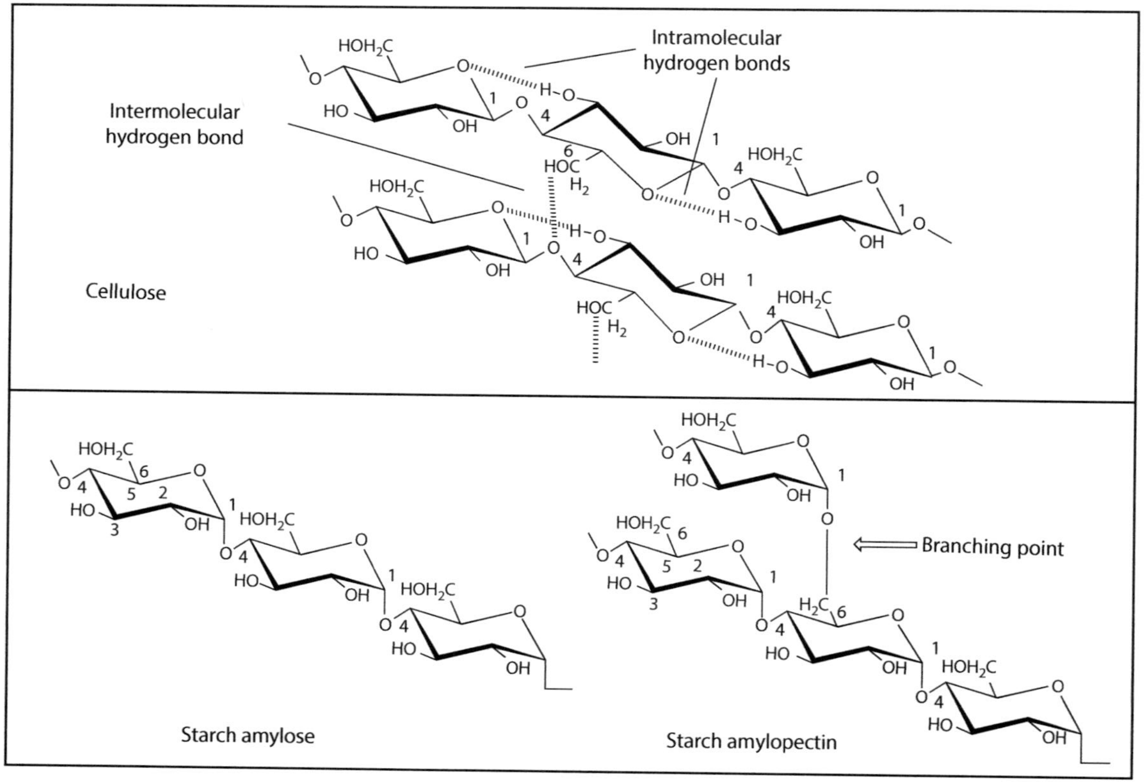

FIGURE 7.2 Physical structure of cellulose and starch amylose and amylopectin. (From Hayes, D., Ross, J., Hayes, M., and Fitzpatrick, S., "The Biofine process: Production of levulinic acid, furfural and formic acid from lignocellulosic feedstocks," in *Biorefinery (8b) Industrial Processes and Products: Status Quo and Future Directions*. Wiley, New York, 2008. With permission; Fitzpatrick, S. and Nace, P., "Biofine Technology, LLC: Renewable chemicals and biofuels," Paper Presented for Sustainable Bioplastics Council of Maine, 2012. With permission.)

The hydrolysis of cellulose can be carried out through attack by the electrophilic hydrogen atoms in the water on the glucosidic oxygen (Figure 7.2). This is, however, a very slow reaction. The reaction can be accelerated using high temperatures and pressures, and acid (dilute or concentrated) as a catalyst, or by highly selective enzymes such as cellulases. The reaction path for acid-catalyzed hydrolysis is identified by Sjostrom [13] and Hayes et al. [8], and it generally involves the protonation of the glycosidic oxygen. In this process, H^+ ions equilibrate between the O atoms in the system such that there is an equilibrium concentration of protonated glucoside. The equilibrium shifts more toward the protonated form of glucoside as the temperature increases. The protonated conjugate acid then slowly breaks down to the carbonium ion, and after a rapid addition of water, free sugar is liberated [8]. Because sugar competes with water, a small amount of disaccharides is also produced by the reverse reaction. In general, the reaction requires a longer time, but this can be reduced with the use of larger concentrations of acids. The temperature, pressure, time, and acid concentration can be economically optimized. The ash content of feedstock is important because it lowers the acidity of the mixture, thus requiring higher amount of acid making the process less economical [13,14].

The reaction paths of the two-stage hydrolysis process are schematically described in Figure 7.3 [8,12]. As shown in the figure, the hydrolysis process follows a fast set

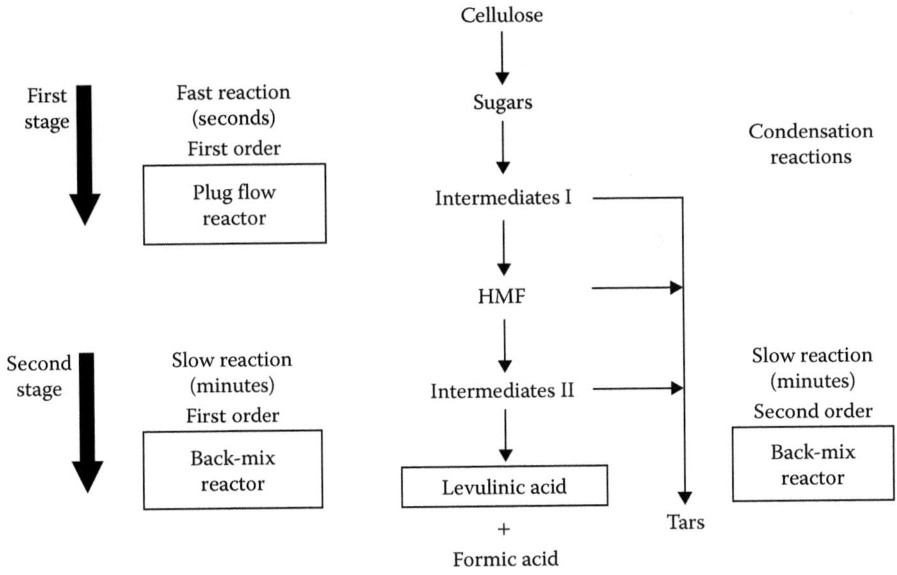

FIGURE 7.3 Chemical conversion of cellulose to LA (major product), formic acid (byproduct), and tars (minor condensation products) in the two-stage Biofine hydrolysis process. (From Fitzpatrick, S. and Nace, P., "Biofine Technology, LLC: Renewable chemicals and biofuels," Paper presented for Sustainable Bioplastics Council of Maine, 2012. With permission; Hayes, D., Ross, J., Hayes, M., and Fitzpatrick, S., "The Biofine process: Production of levulinic acid, furfural and formic acid from lignocellulosic feedstocks," in *Biorefinery (8b) Industrial Processes and Products: Status Quo and Future Directions*. Wiley, New York, 2008. With permission.)

of reactions followed by a slow set of reactions. Both steps generate a large number of intermediates. These intermediates produce tars via the second-order condensation reactions along with the main products (hydroxymethylfurfural [HMF] in the first stage and LA and formic acid in the second stage). Short reactor residence time is important for the first rapid kinetic step and good mixing is important for the kinetically controlled second reaction.

The detailed flow diagram for the entire process is illustrated in Figure 7.4 [12]. The process can use a variety of biomass such as wood and crops; cellulosic wastes such as waste paper, cellulosic sludges, and agriculture residues; and carbohydrates such as starch and molasses. The solid feedstock is shredded to reduce its particle size to 0.5–1.0 cm. This ensures an efficient hydrolysis and an optimum yield of the desired products. This shredded feedstock is mixed with fresh and recycled dilute sulfuric acid (1.5–3.0 wt%) solution. The acid concentration can be adjusted depending on the requirement based on the nature of the feedstock and rest of the process conditions. Sulfuric acid acts as a catalyst for the hydrolysis process. The hydrolysis process differs from other similar processes in that free monomeric sugars are not the products. Instead, six- and five-carbon monosaccharides undergo multiple acid-catalyzed reactions to give the platform chemicals furfural, LA, and formic acid along with ligneous char.

As shown in Figure 7.4, the hydrolysis process is carried out in two distinct acid-catalyzed stages that are operated to give optimum yields with a minimum degradation of products and tar formation. The first fast reaction produces HMF and is carried out in a plug flow reactor, whereas the second slow reaction of HMF hydration to form LA is carried out in a back-mixed reactor. The first stage is carried out at 210°C–220°C and pressure of 25 bar, and the reaction lasts only for 12 s. The reaction is the first-order acid hydrolysis of carbohydrate polysaccharides to their soluble intermediates. The second reactor is operated at 190°C–200°C and pressure of 14 bar, and the reaction takes about 20 min [8]. As mentioned earlier, along with the desired products, the intermediates in both reactors and HMF produce tars by the second-order condensation reactions. After the second stage, furfural and other volatile products are removed, and levulinic and formic acids are separated from water by the dehydration unit. The separated acid solution is recycled back to the feed unit. LA (4-oxopentanoic acid) is recovered by boiling under reduced pressure and further purified in a product refining stage. Ligneous char is bone-dried and recovered separately both from product separation and product refining stage. The maximum theoretical yield of LA from a hexose is 71.6 wt% and the remainder is formic acid [15].

HMF is an intermediate product formed in the first stage of the process. A series of consecutive reactions that occur to produce HMF have been established by numerous studies that aimed at the identification of intermediate products and analyses of pathways for their further transformation [9]. These reactions involve dehydrations of six-carbon compounds such as D-glucose, D-mannose, and D-fructose to form enediol, which undergoes a series of further dehydration reactions to form 3,4-dideoxyglucosulosene. This substance is readily converted to dienediol that eventually forms HMF [16,17].

The five-carbon sugar, namely, xylose and pentose, is produced by substituting CH_2OH group of the hexoses by hydrogen. The hydrolysis of xylose and pentose

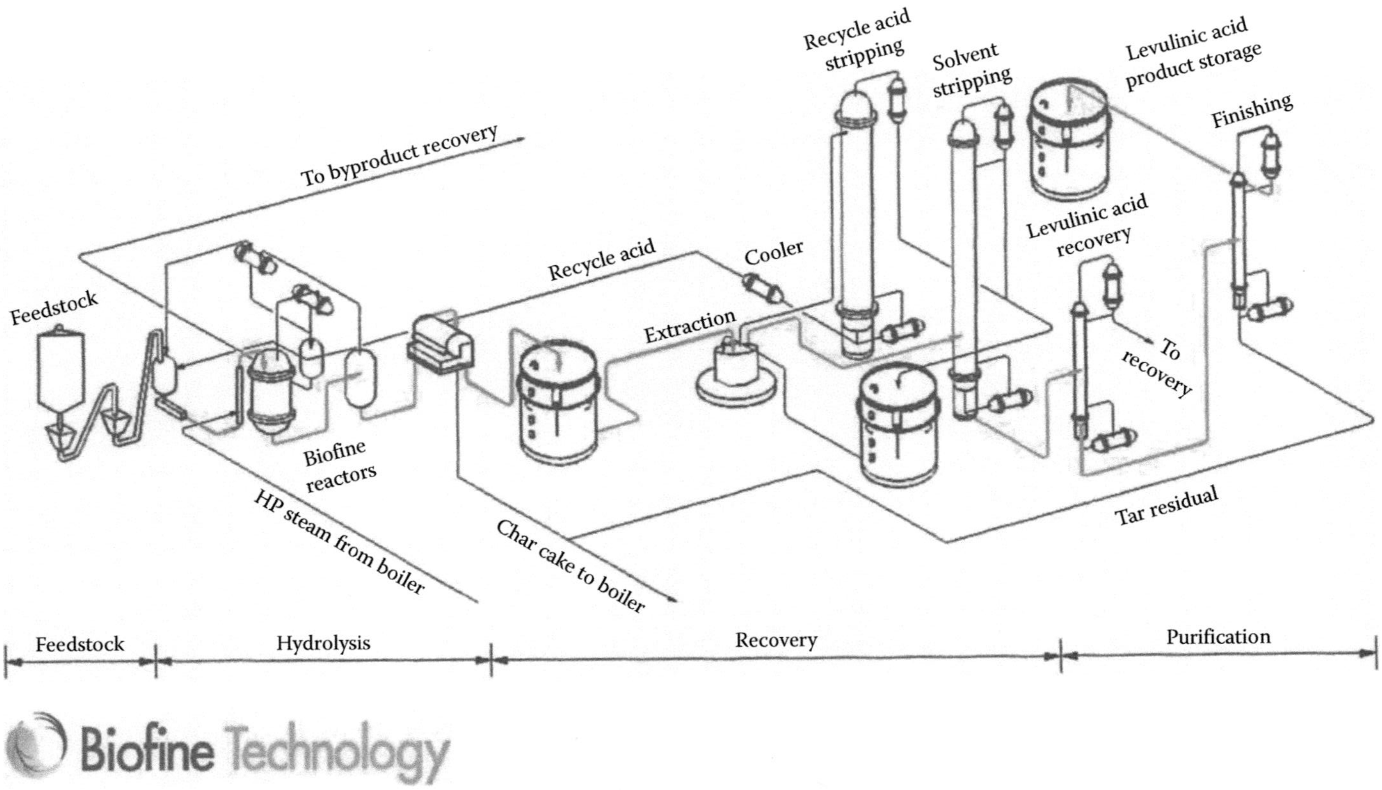

FIGURE 7.4 **(See color insert.)** A complete process flow diagram of the Biofine process. (From Fitzpatrick, S. and Nace, P., "Biofine Technology, LLC: Renewable chemicals and biofuels," Paper presented for Sustainable Bioplastics Council of Maine, 2012. With permission.)

will result in the final product HMF. Hydration of HMF by addition of a water molecule to the C_2–C_3 olefinic bond of the furan ring leads to an unstable tricarbonyl intermediate [18], which decomposes to the LA [13] and formic acid [8]. While the intermediates proposed and some of the reaction steps identified by Sjostrom [13], Klass [18], and Hayes et al. [8] are not completely proven, they were proposed by Horvat et al. [19,20] based on the ^{13}C nuclear magnetic resonance (NMR) spectra of the reaction mixture formed in the hydration of HMF. The reaction paths discussed earlier indicate that both five- and six-carbon sugars can be converted to LA by the appropriate hydrolysis process chemistry. This makes the process more attractive than sugar fermentation process in which the conversion of five-carbon sugar by enzymatic fermentation is problematic [8,21].

The actual hydrolysis process involves many degradation reactions producing many intermediates. Some authors have estimated more than 100 such intermediates [19,20]. These intermediates tend to cross-react and ultimately coalesce (partly by a series of condensation reactions) to form an acid-resistant tar, which incorporates many insoluble residues such as humins. The overall objective of the Biofine process is to minimize the degradation and subsequent condensation reactions that produce tar and increase the yield of LA. An improved reactor system and the use of polymerization inhibitors can provide LA yields of up to 70%–80% of the theoretical yield. This means that a typical product distribution will have about 50% LA, 20% formic acid, and about 30% tar for six-carbon sugars. The mass yield of furfural from five-carbon sugars is about 50% of the original mass, the remainder being incorporated in the Biofine char. These data are supported by the pilot-scale experiments from the Biofine process at Glens Falls, New York [8]. The pilot plant that is in operation since 1996 has used numerous feedstock including paper sludges from the paper mill.

Biofine char contains ash and acid-insoluble ligneous materials. The properties of char can be changed and optimized using high-temperature and high-pressure cracking. Feedstock that contains high amount of extractive such as barks (that may contain up to 25% fats, waxes, and terpenes) or a large amount of water-soluble carbohydrates will have those components largely end up in Biofine char [8,22]. While these components reduce the overall yield per unit biomass processed, they improve the heating value of char when the char is combusted [8,22].

The hydrothermal conversion of biomass to LA in the presence of homogeneous acid catalysts was also examined by Galletti et al. [23]. They examined different types of cheap raw materials such as poplar sawdust, paper mill sludge, tobacco chops, wheat straw, and olive tree pruning. The yield of LA was improved by optimization of the operating parameters such as the type and amount of acid catalysts, temperature, reactor residence time, biomass concentration, and electrolyte addition. The catalytic performances were also improved by the use of microwave radiation for heating the system. The microwave heating required less time for heating and was more energy efficient. The hydrothermal conversion of inulin and wheat straw was also examined in the presence of niobium phosphate catalyst.

The experimental data reported by Galletti et al. [23] showed that for both hydrochloric and sulfuric acid solutions in water, the favorable yields of LA were obtained. For wheat straw with hydrochloric acid at 200°C and residence time of 1 h, the yields for LA based on the cellulose content varied from 49% to 55% and the theoretical yields varied from 69% to 77%. For tobacco chops, lower yields were obtained for sulfuric acid solution (about 13%–14% based on the cellulose content and 17%–21% based on the theoretical yield) than for hydrochloric acid solution (about 21% to as high as 59% based on the cellulose content and 29%–82% based on the theoretical yield). The optimization of the main operating parameters mentioned earlier allowed an increase of yield up to 83% of the theoretical yield. The use of microwave improved the catalyst performance with significant energy and time saving. The hydrothermal conversion of soluble inulin and wheat straw/water slurry to LA in the presence of a heterogeneous niobium phosphate catalyst also gave favorable results.

7.3 UPGRADING OF INTERMEDIATE PRODUCTS FROM THE BIOFINE PROCESS

Three most important products resulting from the hydrolysis of cellulosic materials are furfural (or HMF), LA, and GVL. The process also produces two coproducts: formic acid and ligneous char. Both of them have significant market values. Furfural is produced from the hydrolysis of five-carbon sugars and LA is produced from the hydrolysis of six-carbon sugars as well as from HMF. The GVL is produced from LA and it is a very good feedstock for various kinds of fuels. LA produced from the Biofine process has two highly reactive functional groups (as shown in Figure 7.6) that allow a great number of synthetic transformation. LA is readily soluble in water, alcohols, esters, ketones, and ethers. It can react as both a carboxylic acid and a ketone [24–27]. Due to the special relationship of the carboxylic and ketone groups, many of the reactions proceed with cyclization to form heterocyclic molecules such as methyltetrahydrofuran (MTHF). A vast number of derivatives are possible from the LA as a platform chemical [24–27]. As shown by Hayes et al. [8] and others [23–28], the intermediate products from the Biofine process, namely, LA, formic acid, furfural, and char containing lignin, can be upgraded to numerous products that can be used in five separate markets [8,23–28]:

1. ***Energy and Fuel Industries***
   The products can be upgraded and used for heating and turbine fuels, gasifier fuels, and electric power. In this respect, Biofine char has a significant heating value. The products can also be upgraded to make MTHF, methyl and ethyl levulinate (fuel additives), jet fuel, fuel esters, and hydrocarbons useful for fuel industries.
2. ***Specialty Chemical, Pharmaceutical, and Polymer Industries***
   LA, angelica lactone, delta-aminolevulinic acid (DALA), ketals, and others are used for specialty chemical and pharmaceutical products. The products can also be upgraded to monomers and polymers such as epoxies, polycarbonates, diphenolic acid (DPA), GVL, tetrahydrofuran (THF), succinic acid, and furans.
3. ***Agricultural Industry***
   The coproducts such as formic acid and the upgraded products such as DALA and DPA have agricultural usages. Biochar can be used as a soil conditioner.

4. ***Transportation Industry***
   A number of products such as CMA (calcium magnesium acetate), carbon, sodium, succinic acid, DALA, and levulinate can also be used in the transportation industries.
5. ***Chemical Industry***
   Both coproducts and upgraded products have significant values as general chemical solvents. These include coproducts such as formic acid and furfural, as well as upgraded and byproducts such as *N*-methylpyrrolidinone (NMP), pyridine, gamma-butyrolactone (GBL), pentanediol, THF, succinic acid, and ethyl formate.

Various end products formed from the conversion of LA, formic acid, and ligneous char are graphically illustrated in Figure 7.5 [12]. In Sections 7.3.1 through 7.3.5, we briefly examine some of the upgrading strategies for the products (LA, formic acid, furfural, GVL, and char) of the Biofine hydrolysis process. These strategies closely follow the excellent article by Hayes et al. [8] and others [12,23–42].

7.3.1 Transformation of Levulinic Acid

As shown in Figure 7.6 [12], the Department of Energy has identified LA as one of the 12 important platform chemicals produced from biomass. LA can be transformed into a number of important industrial chemicals, fuel additives, or platform chemicals for fuels. For example, the reaction of LA with two molecules of phenol produces DPA [27], a material that can be substituted for bisphenol A (BPA) in polycarbonates, epoxy resins, polyarylates, and other polymers. The use of DPA can also reduce the cost of lubricants, adhesives, and paints [10]. The succinic acid obtained from the oxidation of LA is very useful for food additives, soldering fluxes, and pharmaceutical products. Succinic acid can also be used to produce THF, 1,4-butanediol, and GBL. Both THF and GBL are also very important intermediate chemicals [8]. THF, a cyclic ether, is useful for the production of polytetramethylene ether glycol, a component of polyurethane stretch fibers, and it is also used as a reaction solvent for the poly(vinyl chloride) (PVC) cements, pharmaceuticals, and coatings. GBL is used for the production of pyrrolidone solvents, pesticides, herbicides, and plant growth regulators [8].

Another important product from LA is DALA, which is an active ingredient in a broad spectrum of herbicides. It is also used as an insecticide and for cancer treatment [44]. The most efficient way of making DALA from LA is [8,10]

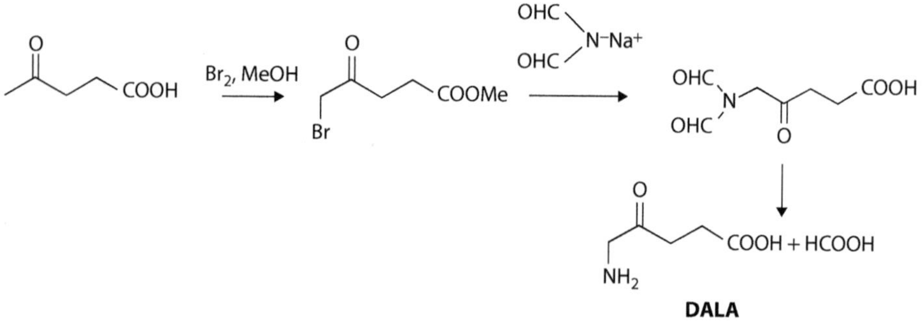

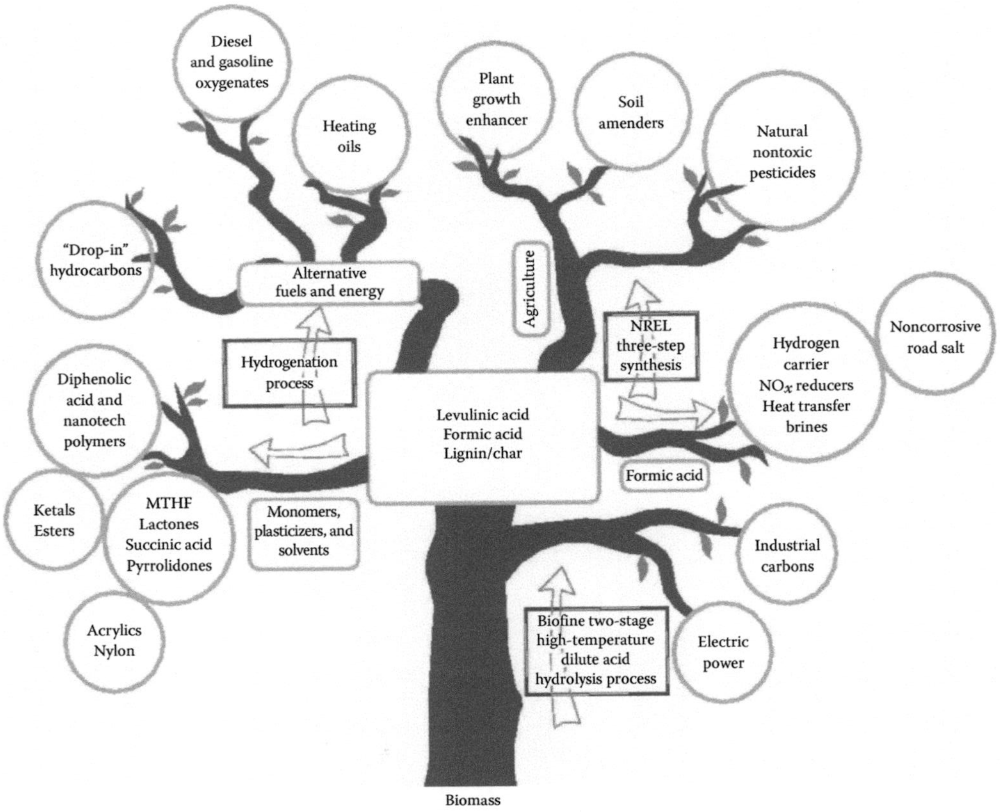

FIGURE 7.5 **(See color insert.)** Biorefinery products "family tree." (From Fitzpatrick, S. and Nace, P., "Biofine Technology, LLC: Renewable chemicals and biofuels," Paper presented for Sustainable Bioplastics Council of Maine, 2012. With permission.)

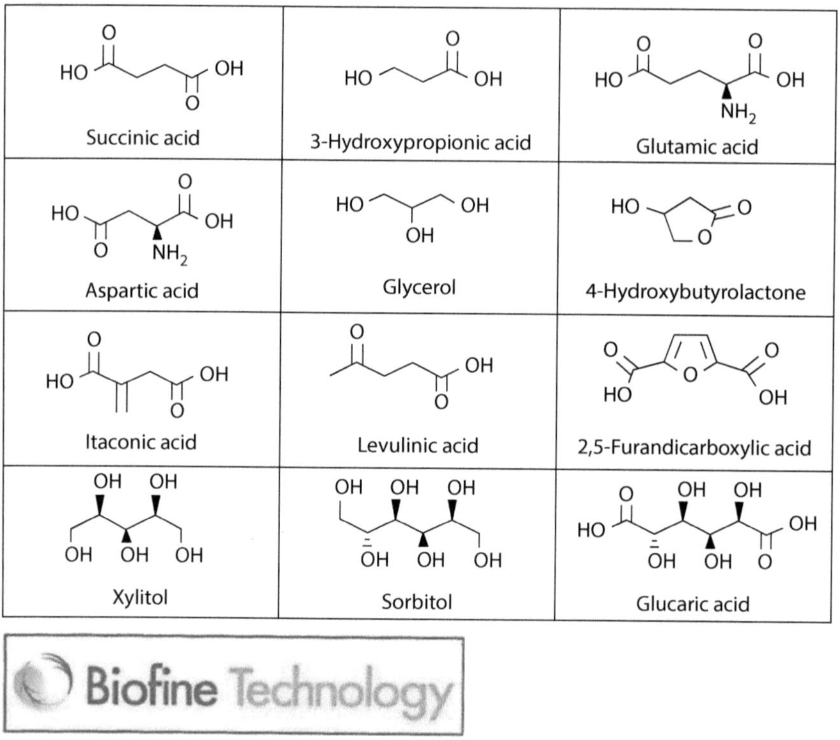

FIGURE 7.6 **(See color insert.)** "Select 12" platform chemicals from biomass as identified by the Department of Energy. (From Fitzpatrick, S. and Nace, P., "Biofine Technology, LLC: Renewable chemicals and biofuels," Paper presented for Sustainable Bioplastics Council of Maine, 2012. With permission.)

The National Renewable Energy Laboratory (NREL) mechanism produces two moles of formic acid per mole of DALA. Significant research for the production of cheap DALA is being pursued because it has a large potential in agricultural and horticultural industries.

While LA is an important platform chemical for many products of industrial values, its greatest potential is in the production of fuel additives. MTHF obtained from LA can be added up to 30% by volume with petroleum with no adverse effects on performance and requires no engine modifications. Esters of LA produced by methanol or ethanol have a significant potential as blend components in diesel formulation. LA esters (ethyl and methyl levulinate) are similar to fatty acid methyl esters (FAMEs), and addition of ethyl or methyl levulinate to FAME alleviates cold flow properties and gum formation of FAME [45]. Ethyl levulinate made from LA and fuel-grade ethanol is one of the most important oxygenate additives to diesel fuel, and its addition in diesel (by 20%) gives a significantly cleaner burning diesel fuel [46]. The ethyl levulinate and diesel blend also gives lower sulfur emission and higher lubricity compared to regular diesel. The levulinate esters can also replace kerosene as a home-heating oil and can be used as a fuel for the direct firing of gas turbines for electrical generation [47]. The production of LA esters from LA produced by the Biofine process has an added advantage that this method does not produce glycerol as coproduct that needs to be disposed.

LA can be converted to GVL by dehydration to angelica lactone and subsequent reduction or by reduction to 4-hydroxy-pentanoic acid and subsequent dehydration. These reductions are carried out at relatively low temperatures (373–543 K) and high pressures (50–150 bar), and both homogeneous and heterogeneous catalysts can be employed [48]. The highest yield of GVL (97%) was obtained at 423 K and 34.5 bar using Ru/C catalyst and dioxane as solvent [49]. The external hydrogen is often replaced by formic acid that acts as a hydrogen donor solvent. More recently, GVL has also been produced by integrating hydrolysis/dehydration/hydrogenation of carbohydrates in a single vessel [48,50]. LA can also be catalytically hydrogenated to GVL, which upon further hydrogenation yields 1,4-pentanediol and finally MTHF [10,32]. The reaction is carried out at an elevated temperature of 240°C and a pressure of 100 atm. This method uses trifluroacetic acid as a hydrolysis medium due to the poisoning of Ru/C catalyst by sulfuric acid. Fructose and sucrose gave better yields of GVL than glucose and cellulose when formic acid as a hydrogen donor solvent and external hydrogen are used [8].

MTHF insertion in a blend (gasoline and ethanol) has led to the creation of P-series fuels. These types of fuels can be either used alone or mixed with any proportions with gasoline [8]. These types of fuels reduce ozone-forming potential and reduce emission of non-methane hydrocarbons and total hydrocarbons. MTHF is also an excellent solvent (better than THF) and can also be produced from furfuryl alcohol [8]. Dimethyl THF can also be produced from HMF.

7.3.2 Gamma-Valerolectone

GVL is a versatile platform chemical, which can be used as a fuel additive, a solvent, or a reactant for diverse upgrading strategies for the production of fuels and chemicals [51]. GVL's low-energy density, blending limits, and high solubility in water limit its use as a direct fuel. GVL needs to be separated from water, or an aqueous solution of GVL should be processed to produce hydrophobic liquid alkanes with an appropriate molecular weight to be used as liquid fuels.

Dumesic et al. [29–42] have outlined some of the alternatives for converting GVL to liquid hydrocarbons. Serrano-Ruiz et al. [52] have shown that the aqueous solution of GVL (50 wt%) can be upgraded to C_9 hydrocarbons by ring opening to produce pentenoic acids and subsequent hydrogenation to produce pentanoic acids [53]; both of these reactions can be catalyzed by water-soluble Pb/Nb_2O_5 catalysts. The yield of pentanoic acid is controlled by the metal content in the catalyst and the partial pressure of hydrogen. The best yields of pentanoic acid (92%) were obtained with 0.1% Pd at 598 K and 35 bar (50% H_2 and 50% He) [32,54].

The pentanoic acid can be upgraded to 5-nonanone by ketonization over $CeZrO_x$ at 698 K and pressures from 1 to 20 bar [55]. The hydrogenation/dehydration of 5-nonanone over Pt/Nb_2O_5 at 528–568 K and 60 bar produces nonane [56]. In the overall process, lower ketones are converted to C_6–C_7 alkanes that can be removed in the gas phase, and nonane remains in the liquid phase to be used as a blender in diesel fuels [29–42].

Another alternative is to hydrogenate ketones to produce alcohols that can be dehydrated to produce nonene which can be coupled by acid-catalyzed oligomerization [32,57]. Smaller ketones can be converted to alkenes, which also undergo oligomerization to produce the final mixture of C_6–C_{27} alkenes that can be hydrogenated over Pt/Nb_2O_5 to produce liquid alkenes to be used as jet fuels or diesel blenders [29–42].

Bond et al. [58] reported that GVL can undergo ring opening to produce pentenoic acid and isomers which subsequently undergo decarboxylation to produce equimolar mixture of butenes and carbon dioxide. Both reactions occur on solid acid catalyst SiO_2/Al_2O_3. The butene monomers products can be coupled by oligomerization over an acid catalyst to form C_8^+ alkenes that can be converted to jet fuels upon hydrogenation. More details of this reaction chemistry are described by Alonso et al. [32].

7.3.3 Furfuryl and Hydroxymethyl Furfuryl

Furfuryl is produced from the hemicellulose pentose fractions of biomass. Xylose is the predominant pentose and hemicellulosic arabinose is found to a lesser extent in most of the feedstock. Furfuryl can be sold as a solvent or converted to furfuryl alcohol, which in turn can be converted to THF and LA as shown by Hayes et al. [8]. Furfuryl alcohol is a monomer of furan resins that are mainly used as foundry binders. It is produced by hydrogenation of furfuryl. THF is produced by decarbonylation of furfuryl to furan followed by catalytic hydrogenation [40]. Furfuryl alcohol, when boiled in ethyl methyl ketone in the presence of HCl, gives rise to 90%–93% yield of LA [17].

HMF and furfuryl are also precursors of liquid hydrocarbon fuels and are an option for the production of linear alkanes in the molecular weight range appropriate for diesel and jet fuels. Since furans can be produced from both cellulose and hemicellulose, they utilize the larger fraction of available lignocellulosic feedstock. Furfuryl and HMF can be produced with good selectivity (90%) from xylose and fructose in biphasic reactors; the yields for glucose are lower. The addition of dimethyl sulfoxide (DMSO) improves the selectivity of HMF from fructose. In the presence of water, HMF is readily hydrated to LA and formic acid. Furfuryl can be extracted from water using solvents such as THF, butanol, and methyl isobutyl ketone (MIBK), and by adding salts to the aqueous phase [8,32].

Dumesic et al. [29–42] have shown different strategies to upgrade HMF to liquid fuels. HMF can be converted to DMF over Cu–Ru/C catalyst by hydrogenolysis. DMF can be used as a blender in transportation fuels. Higher hydrocarbons are produced by aldol condensation with ketones. Single condensation of HMF produces C_9 intermediates that can react with HMF again to produce C_{15} intermediates [32]. The condensation products are hydrogenated and dehydrated over a bifunctional catalyst with metal and acid sites to produce linear C_9 or C_{15} alkanes that can be easily separated from water [32]. Aldol condensation can be coupled with hydrogenation steps using a bifunctional catalyst such as $Pd/MgO–ZrO_2$ leading to high yields of condensation products at 326–353 K [59]. The selective hydrogenation of HMF and furfuryl can also be converted to C_{12} and C_{10} alkanes through a series of reaction steps involving self-condensation and hydrogenation/dehydration, respectively [32].

7.3.4 Formic Acid

As shown earlier, the conversion of cellulose to LA produces formic acid as a byproduct. Formic acid produced in this manner can be either directly sold as a commodity or further purified by distillation. As pointed out by Hayes et al. [8], formic acid is a very versatile product and can be used in a number of different ways as indicated below:

1. It can be used as road salt in the form of calcium magnesium formate. It can also be used as a silage additive and a decalcifier as well as an acidulating agent in textile dyeing and finishing and leather tuning [60].
2. Formic acid is a hydrogen donor solvent. It can also be used in catalyst preparation and regeneration of catalyst metals that are poisoned by sulfur.
3. It can be used in the manufacture of organic esters, drugs, dyes, insecticides, and refrigerants. Esters of formic acid can be fuel components and platform chemicals.

Thus, formic acid is a very valuable byproduct of Biofine process.

7.3.5 Biofine Char

Since Biofine char largely contains lignin, the amount of residual char in the Biofine process depends on the acid-insoluble lignin content, along with the ash content, any insoluble protein present, and the amount of degradation and reversion products formed from cellulose and hemicellulose fractions during LA production [8]. As mentioned earlier, extractives and any water-soluble carbohydrates can also be a part of Biofine char. Since boiling of volatiles and LA cracks char, it is difficult to predict the final composition of char [8]. The Biofine char, however, has a significant heating value. This heating value can significantly exceed the heating value of the original biomass with its water content. If the feedstock lignin content is about 25% of biomass, and if the process is larger than 270 tons of feedstock per day, it has been estimated [8] that the energy provided by the residual char is greater than that needed to completely fuel the steam and electric power needs of the biorefinery.

Biofine char is a good soil conditioner. The chars from straw and paper show significant carbonyl/carboxyl and acidic functionalities [8]. The char can also be used to produce syngas via steam gasification/reforming process. This syngas can be converted to various fuels and chemicals via Fischer–Tropsch synthesis.

7.4 COMPARISON OF BIOFINE PROCESS WITH OTHER TECHNOLOGIES

7.4.1 DIBANET Project

The Development of Integrated Biomass Approaches Network (DIBANET) is a collaborative research project between the European Union and Latin America to produce sustainable diesel miscible biofuels (DMBs) from the residues and wastes

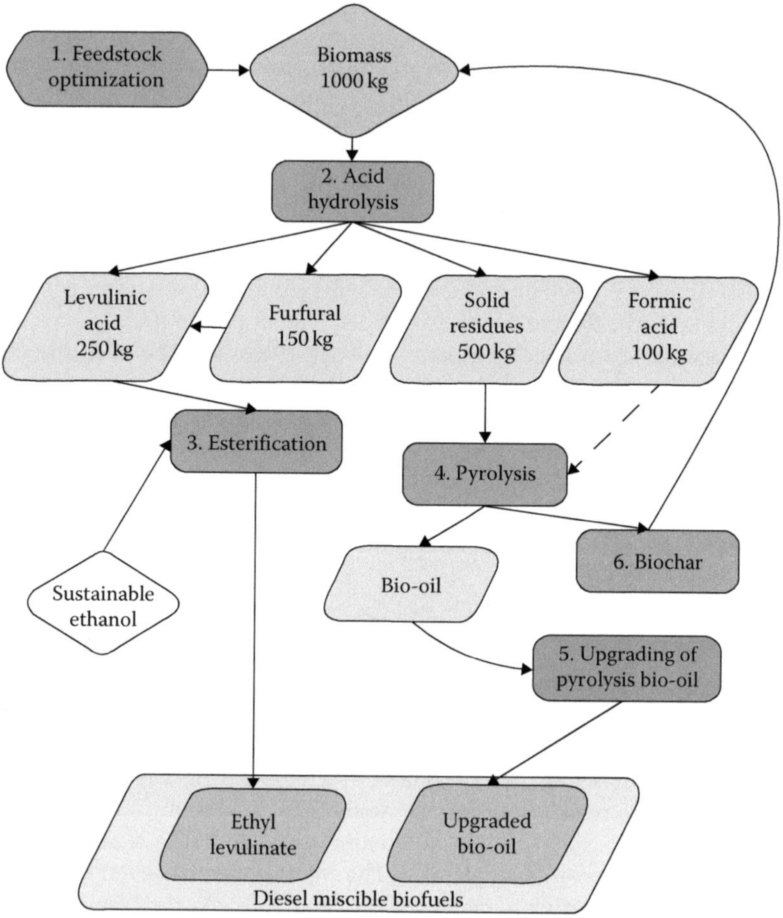

FIGURE 7.7 **(See color insert.)** Process chain to produce maximum yields of DMB from organic waste and residue. (From Hayes, D., DiBANET project, 2013. With permission.)

from these places [61–66]. This project is coordinated by Carbolea at the University of Limerick in Ireland. The basic process is graphically illustrated in Figure 7.7. While the process is similar to the Biofine process for the hydrolysis step to produce LA, the subsequent part of the process focuses on the production of ethyl levulinate and bio-oil (via pyrolysis) that will be upgraded to a DMB, which is in compliance with EN 590 requirements. Ethyl levulinate is produced from LA and ethanol via the esterification process. The overall process shown in Figure 7.7 has the following objectives:

1. Optimize the sourcing, selection, and preparation of the feedstock followed by hydrolysis and subsequent degradation of biomass. This step is very similar to that in the Biofine process, and it produces LA, furfural, formic acid, and residue.
2. Produce ethyl levulinate by esterification in the presence of ethanol.

3. Convert solid residue to bio-oil and biochar by pyrolysis. This process can be enhanced by the use of formic acid produced earlier as a coproduct.
4. Upgrade bio-oil with the use of a catalyst to produce upgraded bio-oil that is miscible with diesel.
5. Utilize the biochar as a soil conditioner or to provide fuel for the process.

This project is thus an application of the overall concept of the Biofine process with specifically tailored upgrading of the hydrolysis products. The process is well described by Hayes et al. [61–66] and at http://www.carbolea.ul.ie/project.php?=dibanet.

7.4.2 Biofine Process versus Fermentation Process

As indicated earlier, Biofine process is a chemical process and does not rely on microorganism or enzymes [68,69]. The biochemical process generally produces only alcohols and can use only a limited range of feedstock, whereas the Biofine process can use a variety of feedstock (containing both five- and six-carbon sugars and starch) and deliver a host of products by a suitable transformation of platform chemicals LA and HMF.

The fermentation process takes about seven days to generate ethanol from cellulose, whereas the Biofine process takes two days for hydrolysis and about 30 min for the production of LA. The fermentation process often gives poor yields due to the inhibition effect of products on enzymes and microbes, whereas no such inhibition occurs in the Biofine process. The fermentation process is also very difficult and economically unattractive for five-carbon sugars such as xylose, whereas these sugars can give up to 50% yield to an important intermediate HMF or furfural in the Biofine process. The contaminants in feedstock such as those in MSW and sewage can significantly inhibit fermentation, whereas the experiments in New York plant has shown that these feedstock can be easily processed by the Biofine process [8]. Finally, the lignin content in biomass can affect the effectiveness of enzymatic process due to stearic hindrance caused by lignin–polysaccharide linkages, whereas the same lignin content has no effect on the Biofine process. The lignin content in the enzymatic process limits the access of fibrolytic enzymes to specific carbohydrate moieties and requires steam explosion pretreatment, which adds cost to the overall fermentation process [67]. Fundamentally, all these differences are inherent partly due to the difference in the basic nature of a biochemical (i.e., fermentation) and a chemical (i.e., Biofine process) process.

7.4.3 Biofine Process versus Bioforming Process

Another competing technology is the most recently developed "Bioforming process" described in Chapter 6. The basic difference between these two technologies is the reaction path chosen to obtain fuels, fuel additives, and chemicals. As discussed earlier, the Bioforming process can generate both gaseous (hydrogen and syngas) and liquid fuels and chemicals, and in that sense, it offers more product upgrading possibilities. However, the intermediate platform chemicals produced by the Biofine process,

FIGURE 7.8 Commercial plant in Caserta, Italy. Recovery vessels (top left); outside of building (bottom left); mixing tank (top middle); second reactor (bottom middle); aerial view (right). (From "Fitzpatrick, S., Renewable chemicals and biofuels for sustainable bioplastics council of Maine," Paper presented by Biofine Technology, LLC, 2012. With permission; Hayes, D., Ross, J., Hayes, M., and Fitzpatrick, S., "The Biofine process: Production of levulinic acid, furfural and formic acid from lignocellulosic feedstocks," in *Biorefinery (8b) Industrial Processes and Products: Status Quo and Future Directions*. Wiley, New York, 2008. With permission.)

namely, LA and HMF, are extremely versatile in producing a host of end products mentioned earlier. When both the Bioforming and Biofine processes are fully commercialized, they together will offer a wide range of possibilities to generate synthetic fuels and chemicals. Water plays the most important role in both of these processes.

7.5 LARGE-SCALE BIOFINE PROCESS

The Biofine technology is commercially viable. A commercial plant processing 50 dry tons of feedstock per day has been operating in Caserta, Italy [8]. The primary feedstock of this plant is paper sludge, agricultural residue, and waste paper with the major products LA and ethyl levulinate (for use as fuel). The process char is gasified to produce a fuel gas for the process boilers. The images of the various parts of the plant are illustrated in Figure 7.8 and this plant has been successfully operating for several years. A number of larger-scale (250 and 1000 tons per day) plants are under considerations in Ireland, the United Kingdom, and the United States.

REFERENCES

1. Bozell, J.J., *Chemicals and Materials from Renewable Resources*. American Chemical Society, Washington, DC (2001).
2. *Renewable Energy Resources: Opportunities and Constraints 1990–2020*. World Energy Council, London (1993).

   Referred in Bull, S. and Billman, L., "Renewable energy: Ready to meet its promise," NREL report (December 7, 1998). www.nrel.gov/docs/legosti/old/25890.pdf

3. Sterzinger, G., "Making biomass energy a contender," *Technology Review*, 98, 34 (1995).
4. Spedding, C., *Biofuels and the Future*. British Association for Bio Fuels and Oils, Gloucester, MA (2002).
5. Borjesson, P., *Biomass & Bioenergy*, 16, 291, (1999).
6. Paul, S., *Plant/Crop-Based Renewable Resources 2020*, DOE/GO-10098-385. US Department of Energy, Washington, DC (1998).
7. Jarnefeld, J. et al., *Biomass*, 16, 33–49 (1998).
8. Hayes, D., Ross, J., Hayes, M., and Fitzpatrick, S., "The Biofine process: Production of levulinic acid, furfural and formic acid from lignocellulosic feedstocks," in Kamm, B., Gruber, P., and Kamm, M. (eds.), *Biorefinery (8b) Industrial Processes and Products: Status Quo and Future Directions*. Wiley, New York, pp. 139–164 (2008).
9. "Biofine's voyage from the lab to the plant," *Proceedings of the 7th National Bioenergy Conference*, SE Regional Biomass Energy Program, Vol. 2, 1083 (1996).
10. Bozella, J., Moensa, L., Elliottb, D.C., Wangb, Y., Neuenscwanderb, G., Fitzpatrickc, S.W., Bilskid, R.J., and Jarnefelde, J.L., "Production of levulinic acid and use as a platform chemical for derived products," *Resources, Conservation and Recycling*, 28, 227–239 (2000).
11. Fitzpatrick, S. and Framingham, M.A., "Levulinic acid in organic synthesis," *Russian Chemical Reviews*, 68 (1), 73–84 (1999).
12. Fitzpatrick, S. and Nace, P., "Biofine Technology, LLC: Renewable chemicals and biofuels," Paper presented for Sustainable Bioplastics Council of Maine, April 11, Bangor, ME (2012).
13. Sjostrom, E., *Wood Chemistry: Fundamentals and Applications*. Academic Press, London (1981).
14. Zerbe, J.I. and Baker, A.J., "Investigation of fundamentals of two-stage, dilute sulphuric acid hydrolysis of wood," in Klass, D.L. (ed.), *Energy from Biomass and Wastes X*. Institute of Gas Technology, Chicago, IL, 927–947 (1987).
15. Leonard, R.H., "Levulinic acid as a basic chemical raw materials," *Industrial & Engineering Chemistry*, 48 (8), 1331–1334 (1956).
16. Kooherkov, N.A., Bochkov, A.F., Dimitriev, B.A., Usov, A.I., Chizbov, O.S., and Shibaev, Y.N., *Khimiya Uglevodov (The Chemistry of Carbohydrates)*. Khimiya, Moscow (1967).
17. Timokhin, B.V., Baransky, V.A., and Eliseeva, G.D., "Levulinic acid in organic synthesis," *Russian Chemical Reviews*, 68 (1), 73–84 (1999).
18. Klass, D.L., *Biomass as a Nonfossil Fuel Source*. American Chemical Society, Washington, DC (1981).
19. Horvat, J., Klaic, B., Metelko, B., and Sunjic, V., "Mechanism of levulinic acid formation," *Tetrahedron*, 26, 2111–2114 (1985).
20. Horvat, J., Klaic, B., Metelko, B., and Sunjic, V., "Mechanism of levulinic acid formation in acid-catalyzed hydrolysis of 2-hydroxymethylfuran and 5-hydroxymethylfuran-2-carbaldehyde," *Croatica Chemica Acta*, 59, 429–438 (1986).
21. Lee, S. and Shah, Y.T., *Biofuels and Bioenergy: Processes and Technologies*. CRC Press, New York (2012).
22. USDA, *Bark and Its Possible Uses*. USDA, Madison, WI (1971).
23. Galletti, A., Antonetti, C., de Luise, V., Licursi, D., and Di Nasso, N., "Levulinic acid production from waste biomass," *BioResources*, 7 (2), 1824–1835 (2012).
24. Oono, T., Saito, S., Shinohara, S., and Takakuwa, K., "Fluxes for electric circuit board soldering and electric circuit boards," Japanese Patent No. 0824378 (1996).
25. Shimizu, A., Nishio, S., Wada, Y., and Metoki, I., "Photographic processing method for processing silver halide photographic light-sensitive material," European Patent No. 704756 (1996).
26. Adams, P.E., Lange, R.M., Yodice, R., Baker, M.R., and Dietz, J.G., "Intermediates useful for preparing dispersant-viscosity improvers for lubricating oils," European Patent No. 882745 (1998).

27. Isoda, Y. and Azuma, M., "Preparation of bis(hydroxyaryl)pentanoic acids," Japanese Patent No. 08053390 to Honshu Chemical Industries (1996).
28. Sheldo-Coulson, G., *Production of Levulinic Acid in Urban Biorefineries*. Department of Technology and Policy, MIT, Cambridge, MA (2011).
29. Huber, G.W., Cortright, R.D., and Dumesic, J.A., "Renewable alkanes by aqueous phase reforming of biomass derived oxygenates," *Angewandte Chemie International Edition*, 43, 1549–1551 (2004).
30. Davda, R. and Dumesic, J., "Catalytic reforming of oxygenated hydrocarbons for hydrogen with low levels of carbon monoxide," *Angewandte Chemie International Edition*, 42, 4068 (2003).
31. Cortright, R.D., Davda, R., and Dumesic, J., "Hydrogen from catalytic reforming of biomass derived hydrocarbons in liquid water," *Nature*, 418, 964–967 (2002).
32. Alonso, D.M., Bond, J.Q., and Dumesic, J.A., "Catalytic conversion of biomass to biofuels," *Green Chemistry*, 12, 1493–1513 (2010).
33. Huber, G. and Dumesic, J., "An overview of aqueous phase catalytic processes for production of hydrogen and alkanes in a biorefinery," *Catalysis Today*, 111 (1/2), 119–132 (2006).
34. King, D., "Hydrogen production via aqueous phase reforming," Paper presented by Pacific Northwest National Laboratory at NIChE Catalysis Conference, September 21, Washington, DC (2011).
35. Cortright, R.D., "Hydrogen generation from sugars via aqueous phase reforming," *Proceedings of the 16th World Hydrogen Energy Conference*, June 13–16, Lyon, France (2006).
36. Huber, G.W., Shabaker, J.W., and Dumesic, J.A., *Science*, 300, 2075 (2003).
37. Kunkes, E.L., Simonetti, D.A., West, R.M., Serrano-Ruiz, J.C., Gartner, C.A., and Dumesic, J.A., *Science*, 322, 417 (2008).
38. Dunn, S., "Hydrogen futures: Toward a sustainable energy system," *International Journal of Hydrogen Energy*, 27, 235–264 (2002).
39. Huber, G., Iborra, S., and Corma, A., "Synthesis of transportation fuels from biomass: Chemistry, catalysts, and engineering," *Chemical Reviews*, 106, 4044–4098 (2006).
40. Soares, R., Simonetti, D., and Dumesic, J., *Angewandte Chemie International Edition*, 45, 3982–3985 (2006).
41. King, D., "Biomass derived liquids distributed (aqueous phase) reforming," 2012 DOE Hydrogen and Fuel Cells Program Review, Pacific Northwest National Laboratory, Richland, WA (May 17, 2012).
42. Blommel, P.G. and Cortright, R.D., "Production of conventional liquid fuels from sugars," White Paper for European Platform on Biofuels (2012).
43. Rebeiz, C.A., Gut, L.J., Lee, K., Juvik, J.A., Rebeiz, C.C., and Bouton, C.E., "Photodynamics of porphyric insecticides," *Critical Reviews Plant Science*, 14, 329–366 (1995).
44. Bedwell, J., McRoberts, A.J., Phillips, D., and Brown, S.G., "Fluorescence distribution and photodynamic effect of ALA-induced PP IX in the DMF rat colonic tumor model," *British Journal of Cancer*, 65, 818–824 (1992).
45. Huang, C. and Wilson, D., "Improving the cold flow properties of biodiesel," Paper presented at 91st American Oil Chemists' Society Annual Meeting, April 26, San Diego, CA (2000).
46. Texaco/NYSERDA/Biofine, *Ethyl Levulinate D-975 Diesel Additive Test Program*. Biofine, Glenham, NY (2000).
47. Erner, W.E., "Synthetic liquid fuel and fuel mixtures for oil burning devices," US Patent No. 4364743 (1982).
48. Heeres, H., Handana, R., Chunai, D., Rasrendra, B., Girisuta, B., and Heeres, H., *Green Chemistry*, 11, 1247–1255 (2009).
49. Manzer, L., *Applied Catalysis A: General*, 272, 249–256 (2004).

50. Deng, L., Li, J., Lai, D., Fu, Y., and Guo, Q., *Angewandte Chemie International Edition*, 48, 6529–6532 (2009).
51. Horvath, I., Mehdi, H., Fabos, V., Boda, L., and Mika, L., *Green Chemistry*, 10, 238–242 (2008).
52. Serrano-Ruiz, J., Wang, D., and Dumesic, J., *Green Chemistry*, 12, 574–577 (2010).
53. Ayoub, P. and Lange, J. "Process for converting levulinic acid into pentanoic acid," World Patent No. WO/2008/142127 (2008).
54. Renz, M., "Ketonization of carboxylic acids by decarboxylation mechanism and scope," *European Journal of Organic Chemistry*, 6, 979–988 (2005).
55. Serrano-Ruiz, J. and Dumesic, J., *Green Chemistry*, 11, 1101–1104 (2009).
56. Werst, R.M., Liu, Z., Peter, M., and Dumesic, J., *ChemSusChem*, 1, 417–424 (2008).
57. Alonso, D., Bond, J., Serrano-Ruiz, J., and Dumesic, J., *Green Chemistry*, 12, 992–999 (2010).
58. Bond, J., Alonso, D., Wang, D., West, R., and Dumesic, J., *Science*, 327, 1110–1114 (2010).
59. Barrett, C., Chheda, J., Huber, G., and Dumesic, J., *Applied Catalysis B: Environmental*, 66, 111–118 (2006).
60. Bizzari, S. and Ishikawa, Y., *CEH Report: Formic Acid*. SRI, Menlo Park, CA (2001).
61. Girisuta, B., Kalogiannis, K.G., Dussan, K., Leahy, J.J., Hayes, M.H.B., and Stefanidis, S., "An integrated process for the production of platform chemicals and diesel miscible fuels by acid-catalyzed hydrolysis and downstream upgrading of the acid hydrolysis residues with thermal and catalytic pyrolysis," *Bioresource Technology*, 126, 92–100 (2012).
62. Melligan, F., Dussan, K., Auccaise, R., Novotny, E.H., Leahy, J.J., Hayes, M.H.B., and Kwapinski, W., "Characterisation of the products from pyrolysis of residues after acid hydrolysis of Miscanthus," *Bioresource Technology*, 108, 258–263 (2012).
63. Melligan, F., Auccaise, R., Novotny, E.H., Leahy, J.J., and Hayes, M.H.B., "Pressurised pyrolysis of *Miscanthus* using a fixed bed reactor," *Bioresource Technology*, 102 (3), 3466–3470 (2011).
64. Hayes, D.J.M., "DIBANET, an integrated approach for making the best use of biomass," *1st Iberoamerican Congress on Biorefineries*, October 24–26, Los Cabos, Mexico (2012).
65. Hayes, D.J.M., "Review of biomass feedstocks and guidelines of best practice," DIBANET WP2 Report, 150p (2012).
66. Hayes, D.J., "Protection of NIR calibration equations and their application for biomass analysis," DIBANET WP2 Report, 71p (2012).
67. Donaldson, L.A., Wong, K.K.Y., and Mackie, K.L., "Ultrastructure of steam-exploded wood," *Wood Science and Technology*, 22, 103–114 (1988).
68. Fitzpatrick, S.W., "Lignocellulose degradation to furfural and levulinic acid," US Patent No. 4897497 (1990).
69. Fitzpatrick, S.W., "Production of levulinic acid from carbohydrate-containing materials," US Patent No. 5608105 (1997).

8 Anaerobic Digestion of Aqueous Waste for Methane and Hydrogen

8.1 INTRODUCTION

The global energy usage is growing rapidly due to increasing demands from countries like China, India, Russia, Brazil, Mexico and other developing nations. The report from International Energy Agency (IEA [1]) predicts that global energy demand during this century will increase by two fold. Currently 86% of world's energy demand is supplied by fossil energy such as coal, oil, and natural gas. However, over next several decades the supply of oil may go down and the major suppliers of oil and gas are in the politically unstable regions of Middle East. Furthermore, fossil energy is also causing more environmental problems due to emissions of greenhouse gases (GHGs) such as carbon dioxide and lower volatile hydrocarbons. According to the Intergovernmental Panel on Climate Change (IPCC) [2], GHG emissions must be reduced to less than half of global GHG emissions level of 1990.

An alternate to fossil energy, renewable energy from biomass has a significant potential. Biogas from wastes, residues, energy crops, and many other organic materials is a versatile renewable energy source. Methane-rich biogas can be used for heat and power applications, as fuel for vehicles, and for the production of a variety of chemicals and materials. Fehrenbach et al. [3] showed that biogas generated by anaerobic digestions of numerous different types of biomass and effluent wastes is the most energy-efficient and environment-friendly source of bioenergy. Like natural gas, methane in biogas can be used in a variety of ways. Biogas will drastically reduce the emission of GHG compared to fossil fuels by utilizing locally available sources of wastes and other forms of biomass. The digestate from the biogas facilities is an improved fertilizer in terms of its availability to crops than conventional mineral fertilizers.

In Europe, biogas is the fastest growing bioenergy and it reached six million tons of oil equivalent (Mtoe) in 2007 with a yearly increase of >20% [4]. Germany is the biggest producer of biogas with about 4000 agricultural biogas production plants by the end of 2008. Within the agriculture sector of the European Union, 1500 million tons of biomass could be digested anaerobically each year, half of which will come from energy crops [5,6]. Besides the agriculture sector, biogas coming from landfills is also becoming more important source of power generation. Biogas is generated not only from various wastewater effluent streams with solids concentration <10 wt%, but also from various streams with high solids concentration (25–35 wt%) such as municipal solid waste (MSW) and animal wastes. The amount of MSW generated in the United States is close to 250 million tons per year, and globally, this number will soon reach one billion tons per year. Biogas is an effective way to convert this waste into useful and environmentally acceptable form of energy for growing waste industry. Since every country in the world has waste problem, biogas industry is universally applied [5].

The literature on biogas deals with both biomethanation and biohydrogenation. As will be discussed later, the hydrolysis of organic waste followed by anaerobic digestion can produce hydrogen or methane depending on the nature of operating conditions, the nature of microorganisms present, and the nature of feedstock. It should also be noted that methane can be converted to hydrogen by reforming reactions.

8.2 BASIC PRINCIPLES OF ANAEROBIC DIGESTION

Anaerobic digestion (in the absence of oxygen) with anaerobic bacteria or methane fermentation is used worldwide for disposal of domestic, municipal, agricultural, and industrial biomass wastes. This reaction generally produces methane and carbon dioxide, and it also occurs in the ecosystem as well as in the digestive tract. As shown by the following reactions, hydrogen along with acetic and butyric acids can be produced by dark fermentation processes using anaerobic and facultative anaerobic chemoheterotrophs [5–8]:

$$C_6O_6H_{12} + 2H_2O \rightarrow 2CH_3COOH + 4H_2 \quad (8.1)$$

$$C_6O_6H_{12} \rightarrow CH_3CH_2CH_2COOH + 2CO_2 + 2H_2 \quad (8.2)$$

Different types of waste materials can be used for hydrogen fermentation. Hydrogen production highly depends on the pH, retention time, and gas partial pressure along with the nature of microbes [5,9]. Generally, hydrogen production increases with the retention time. Hydrogen production is important for its use in fuel cell or microbial electrolytic cell. Wang [10] described the use of low-cost cathode catalysts for high-yield biohydrogen production in microbial electrolytic cell [10–32]. Cheng and Logan [27,32] and Logan et al. [28,29] evaluated both catalysts and membranes for high-yield biohydrogen production via electrohydrogenesis in microbial electrolytic cells.

Fan et al. [33] examined the possible pathways of fermentative hydrogen evolution and other byproducts during biohydrogen fermentation of wheat straw waste by cow dung compost. They found the hydrogen content in the biogas to be 52% with very little methane. Their experimental results showed that the pretreatment of the substrate plays a key role in the conversion of wheat straw waste into biohydrogen by the compost generating hydrogen.

Ding et al. [34] evaluated the effect of protein on biohydrogen production from carbohydrates, particularly starch. They used two model compounds: rice as starch-rich and soybean as protein-rich food waste. They found that the maximum hydrogen production potential was 0.99 mol of H_2/mol of initial starch as glucose and the maximum hydrogen production rate occurred at a starch/protein ratio of 1.7. The protein content in the food waste increased the hydrogen production in two ways. First, it provided the buffering capacity to neutralize the volatile fatty acids as concurrent products.

Second, it provided the readily available organic nitrogen such as soluble proteins and amino acids to microorganisms. Thus, the existence of protein in the substrate of biohydrogen production is important. To get the maximum hydrogen production from carbohydrates, the protein content in feedstock should be optimized. Organic nitrogen in proteins is transformed into inorganic ammonia nitrogen in anaerobic degradation. Ammonia and amino groups released from proteins neutralize the potential pH decrease imposed by volatile fatty acids. Thus, proteins can maintain a suitable pH by the production of bicarbonate, which is given by the following reaction:

$$R - NH_{2^-} \rightarrow NH_{4^-} + HCO_{3^-} + xCO_2 + yH_2O \tag{8.3}$$

The pH stabilization by these two counteracting effects requires the protein-to-starch ratio to be at least 2 to have a pH decrease within 0.5 limit. Lay [35] showed that the pH window for optimal hydrogen production from carbohydrates may be so narrow that a half-unit decrease in pH can cause a 50% decrease in hydrogen production from optimum.

Biogas produced from landfills generally contains methane (about 55%) and carbon dioxide with traces of hydrogen, ethane, and other impurities. The description of the sequence of biochemical reactions that occur to convert complex molecules to methane given here closely follows the excellent review by Weiland [5].

In general, methane fermentation can be divided into four phases: hydrolysis, acidogenesis, acetogenesis/dehydrogenation, and methanation. As shown by Weiland [5], the degradation of complex polymers such as polysaccharides, proteins, and lipids results in the formation of monomers and oligomers such as sugars, amino acids, and long-chain fatty acids (LCFAs). The individual degradation steps are carried out by different consortia of microorganisms, which place different requirements on the environment [36–40]. Initial conversion of polymers and monomers to acetate, hydrogen, and different amount of fatty acids is carried out by hydrolyzing and fermenting microorganisms [5]. *Hydrolytic microorganisms* such as *Bacteroides*, Clostridia, and Bifidobacteria (all of them are strict anaerobes) excrete hydrolytic enzymes such as cellulase, cellobiase, xylanase, amylase, lipase, and protease, which participate in the hydrolysis and fermentation of organic materials [5]. The higher volatile fatty acids are converted into acetate and hydrogen by obligate hydrogen-producing acetogenic bacteria. The maintenance of an extremely low partial pressure of hydrogen is very important for the acetogenic and hydrogen-producing bacteria. The current state of knowledge indicates that hydrogen may be a limiting substrate for methanogens [30], because an addition of hydrogen-producing bacteria to the natural biogas-producing consortium increases the biogas production [5].

The studies have shown that only two groups of methanogenic bacteria produce methane from acetate, hydrogen, and carbon dioxide [5]. These bacteria are strictly anaerobes and require a lower redox potential for growth than most other anaerobic bacteria. Only few species are able to degrade acetate into CH_4 and CO_2, for example, *Methanosarcina barkeri*, *Methanococcus mazei*, and *Methanothrix soehngen*, whereas all methanogenic bacteria are able to convert hydrogen to methane [5]. The first and second groups of microbes and the third and fourth groups of microbes are linked closely with each other [38], allowing the overall process to be divided into two stages.

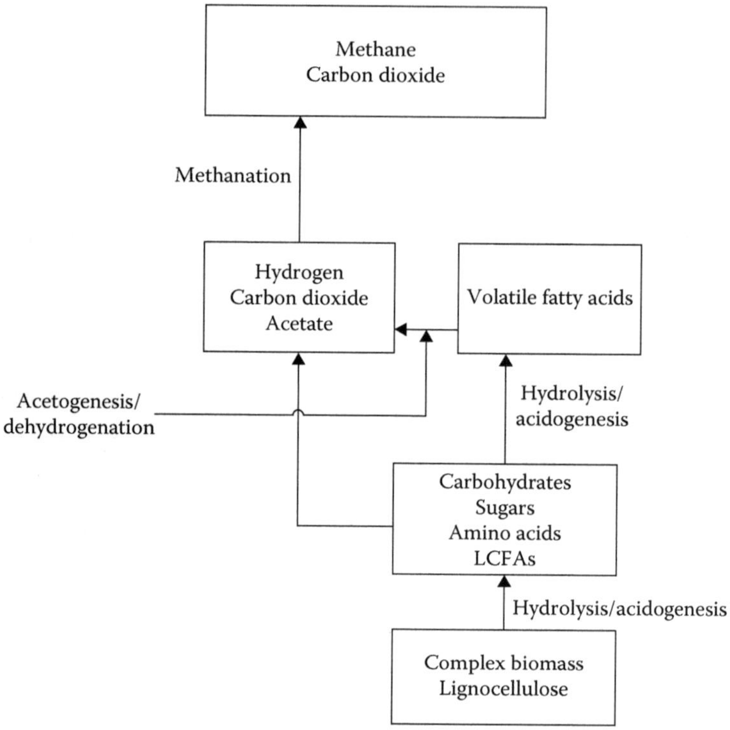

FIGURE 8.1 The stages of methane fermentation process. (Modified from Weiland, P., *Applied Microbiology Technology*, 85, 849–860, 2010.)

The above description of the four steps of waste-to-methane conversion (methane fermentation) is graphically illustrated in Figure 8.1. This process involves two stages: in the first stage, waste is converted to acetate, hydrogen, and carbon dioxide, and in the second stage, acetate and hydrogen are converted to methane. A balanced anaerobic digestion process demands that in both stages, the rates of degradation must be equal in size. If the first degradation step runs too fast, the acid concentration rises and the pH drops below 7.0 that inhibits methanogenic bacteria for conversion to methane. If the second phase runs too fast, methane production is limited by the rate of the hydrolytic stage to produce hydrogen and acetates.

Thus, the rate-limiting step depends on the compounds of the substrate that is used for the biogas production. Undissolved compounds such as cellulose, proteins, and fats take several days to crack, whereas soluble carbohydrates crack in few hours. The overall process design must take into account the substrate properties for achieving complete degradation without process failure. Each step of the overall process described in Figure 8.1 requires an independent assessment. For example, hydrolysis of complex insoluble substrate depends on the parameters such as particle size, productions of enzymes, pH, and temperature. The conversion of acetate and hydrogen to methane depends on the effectiveness of the methanogenic bacteria [5].

In the following discussion, we examine the effects of microbes, operating conditions (Section 8.3), nature of feedstock (Section 8.4), methods of harvesting, storage and pretreatment (Section 8.6), co-digestion (Section 8.5), and digester technology (Section 8.7) on the biogas production. Some details of biogas purification (Section 8.9) and the usage of produced biogas and digestate (Section 8.10) are also briefly examined.

8.3 MICROBES AND THE EFFECTS OF OPERATING CONDITIONS

While the success of anaerobic treatment depends on the effectiveness of various microbes, very little is known about how they work and the interactions between them, and this lack of knowledge sometimes results in malfunction and failure of biogas digestive process. Furthermore, only few percent of bacteria and archaea have been isolated. However, with new molecular techniques, more information about the community structure in the anaerobic processes can be obtained [41–44]. The quantification of methanogens can be carried out by fluorescence *in situ* hybridization technique. Klocke et al. [43] detected 68 taxonomic groups by 16SrDNA analysis of samples from agricultural biogas plants and showed that hydrogenotrophic methanogens dominate most of the agricultural biogas plants [5]. The effectiveness of microbes in anaerobic digestion process depends on the temperature, ammonia inhibition, pH, and presence of nutrients. The effects of these operating variables on the digestive process are briefly described in Sections 8.3.1 through 8.3.3.

8.3.1 Effects of Temperature and Ammonia Inhibition

The digestive process can be operated at lower temperature, that is, mesophilic conditions (temperature range of 35°C–42°C), or at high temperature, that is, thermophilic conditions (temperature range of 45°C–60°C). Generally, temperature fluctuation decreases the biogas productivity. In general, under thermophilic conditions, the growth rate of methanogenic bacteria is higher, and the process is more efficient and faster [42,45]. However, these bacteria are more temperature sensitive and have difficult time adjusting to temperature variations. The faster rate allows the operations to run at lower hydraulic retention time (HRT) than in mesophilic operations. Mesophilic bacteria, however, can tolerate temperature fluctuation of ±3° variations without a significant variation in methane production.

Since ammonia toxicity increases with temperature, thermophilic operations are more susceptible to ammonia inhibition (particularly for the ammonia concentration above 80 mg/l). An increase in ammonia concentration is, however, accompanied by an increase in volatile fatty acid concentration [5]. This can lower the pH and thus counterbalance the effect of ammonia. Many strategies to reduce the ammonia inhibition effects have been examined [46,47]; the most stable digestive process was observed when biomass was diluted with reactor effluent.

8.3.2 pH Effect

The pH of the reacting solution has also significant effects on the effectiveness of bacteria and methane production. The anaerobic digestion process best operates between pH of 6.5 and 8.5 with an optimum value between 7 and 8 [5]. The process is severely affected when pH drops below 6 or increases above 8.5 [5]. While ammonia accumulation increases the pH, and VFA (volatile fatty acid) accumulation decreases the pH, the latter is not always valid due to buffer capacity of some substrate. For example, animal manure has surplus alkalinity that counteracts the increase in VFA concentration [5]. While acetic acid is always present in larger amount than volatile fatty acids, only propionic and butyric acids are more inhibitory to methanogens [48,49]. The inhibition effect of VFA is higher in the reacting systems with lower pH values [5].

8.3.3 Nutrients Effect

Besides temperature and pH, the availability of several macro- and micronutrients is also very important for the growth and survival of specific groups of microorganisms. Very low amount of macronutrients such as carbon, nitrogen, phosphorus, and sulfur is needed (C:N:P:S = 600:15:5:1) because only a small amount of biomass is developed [5]. Micronutrients such as iron, nickel, cobalt, selenium, molybdenum, and tungsten are important for the growth rate of microorganisms, and they must be added, particularly if the energy crops are the only substrate for biogas production [5]. Nickel is important for all methanogenic bacteria because it is necessary for the cell component cofactor F430, which is involved in the methane formation [5]. For optimum growth, the cell requires cobalt to build up the Co-containing corrinoid factor III [5]. The growth of only few methanogens depends on the trace elements such as selenium, molybdenum and tungsten. The required concentration is only 0.05–0.06 mg/l [5]. The iron is, however, necessary in higher concentration of 1–10 mg/l [50]. These micronutrients are very important for the stable process and high loading for energy crops [51]. While the addition of manure reduces the need for micronutrients addition, even with 50% manure in the reaction medium, the addition of micronutrients can increase the biogas production rate [5].

8.4 FEEDSTOCK EFFECTS

All substrates containing carbohydrates, proteins, fats, cellulose, and hemicellulose as major components can produce biogas by anaerobic digestion. The composition of biogas and methane yield depends on carbohydrates, proteins, and fat content of biomass feedstock (Table 8.1) along with other operating parameters. As pointed out by

TABLE 8.1
Maximal Gas Yields and Theoretical Methane Contents

Substrate	Biogas (Nm^3/t TS)	CH_4/CO_2
Carbohydrates (not including inulins and single hexoses)	790–800	1/1
Raw proteins	700	~70/30
Raw fat	1200–1250	~67/33
Lignin	0	Both 0

Source: Weiland, P., *Applied Microbiology Technology*, 85, 849–860, 2010; Baserga, U., Landwirtschaftliche Co-vergarungs-Biogasanlagen, FAT-Berichte No. 512, Tanikon, 1998. With permission.

TS, total solids.

Weiland [5], while the data shown in Table 8.1 need to be corrected for solubilization of CO_2 in digestate, they clearly indicate that the biogas production follows the order: raw fat > carbohydrates > protein. Lignin cannot be digested by the anaerobic process.

Protein generates more methane in biogas. Thus, the properties of the feedstock play a very important role in the rate and composition of the biogas production. For example, wood undergoes very slow anaerobic decomposition and therefore not suitable for anaerobic decomposition. However, as shown in Table 8.2, several plants, plant materials, and energy crops produce significant biogas by the anaerobic digestion process [52–54].

The data shown in Table 8.2 are the arithmetic averages of the ranges for each plant identified by Braun et al. [52], Braun [53], and Braun and Wellinger [54],

TABLE 8.2
Average Methane Yields from Various Energy Crops, Plants, and Plant Materials

Materials	Average Methane Yield (m^3/kg Volatile Solid)
Barley	0.56
Triticale	0.49
Leaves	0.48
Alfalfa	0.46
Wheat (grain)	0.45
Peas	0.43
Grass	0.42
Hemp	0.42
Clover	0.38
Potatoes	0.37
Sorghum	0.37
Rapeseed cake	0.36
Maize (whole crop)	0.36
Sugar beet	0.34
Kale	0.31
Straw	0.31
Sunflower	0.30
Oats (grain)	0.30
Sudan grass	0.28
Flax	0.23
Miscanthus	0.22

Source: Braun, R., Weiland, P., and Wellinger, A., "Biogas from energy crop digestion," IEA Bioenergy Task 37-Energy from Biogas and Landfill gas, 2011. With permission; Braun, R., "Potential of co-digestion," 2002, http://www.novaenergie.ch/iea-bioenergy-task37/Dokumente/final.PDF; Braun, R. and Wellinger, A., "Potential for co-digestion," IEA Bioenergy Report-Task 37, Energy from Biogas and Landfill gas, 2002. With permission.

Note: These data are calculated from the arithmetic averages of the ranges.

and they show nearly twofold variation in biogas production even within different types of plants and plant materials. Energy crops are extensively used as pure or co-substrate for anaerobic digestion. In general, easily degradable biomass results in biogas production. The biogas generated from landfills generally contains about 50%–55% methane and the remaining composition consists of largely CO_2 and traces of water, hydrogen, and other impurities. Thus, the nature of feedstock makes a significant difference in the level and composition of biogas production.

A vast amount of literature on the effect of feedstock on biogas production is available. A brief summary of this literature is outlined in Table 8.3. Biogas production from some of these feedstock is further discussed below in Sections 8.4.1 through 8.4.12.

8.4.1 Coir Pith

Coir pith is a lignocellulosic agro residue that is produced as a byproduct in coir industry in large quantities. Kunchikannan et al. [56] examined the production of methane from this waste material by anaerobic digestion. The study indicated that the yield of methane is 38.1% per kilogram of dry pith weight in 44 days; the yield can be improved by about 1.5 times by the reduction of particle size. The increase in acidity during the digestion process decreases the methane yield, whereas an increase in alkalinity does not significantly change the methane yield.

8.4.2 Whey

The wastes from various food industries are capable of generating methane due to their high organic content. Whey is normally used as a component of dairy products or as an additive for food product. Beszedes et al. [60] examined biogas generation from membrane-separated fractions, that is, permeate and concentrate of whey. The study examined the effects of pH, thermal, and microwave pretreatment, and their combinations on the biogas yield. The pretreatment had a significant effect on the biogas yield. The hydrolysis of large molecules enhanced the biodegradability of whey, thereby increasing the productions of biogas and methane. The long-time classical heat treatment and the microwave radiation in an acidic medium significantly increased the methane production. The concentrate of whey was more adaptable to anaerobic digestion than the permeate or the whole whey.

8.4.3 Distillery Spent Wash

Distillery spent wash is a major pollutant in water. In recent years, methane is generated from anaerobic digestion of distillery spent wash by fixed-film systems and two-phase anaerobic systems. Pathe et al. [58] showed that a treatment option that involved two-stage aerobic oxidation processes (activated sludge and extended aeration) followed by a physical chemical treatment using lime, polyaluminum chloride, polyelectrolyte, and carbon adsorption as the tertiary treatment can be the most efficient method for methane generation from the distillery spent wash. The treated effluent can be used for green belt development and in the agriculture industry.

TABLE 8.3
Some Typical Literature Studies on Anaerobic Digestion of Waste Materials

Types of Waste	Authors
Swine waste	Chen et al. [55]
Coir path	Kunchikannan et al. [56]
Wastewater and organic kitchen waste	Weichgrebe et al. [57]
Distillary spent wash	Pathe et al. [58]
Biodiesel byproducts	Kolesarova et al. [59]
Whey (a component of dairy product or an additive for food product)	Beszedes et al. [60]
Palm oil effluent	Yusoff et al. [61]
Tofu wastewater	Zheng et al. [62]
Starch of food waste	Ding et al. [34]
MSW	Abderrezaq [63]
Solid organic waste and energy crops	Angelidaki et al. [64]
Food residuals	Shin et al. [65]; Haug et al. [66]
Dairy effluent	Desai et al. [67]
Organic solid waste	Zhang [68]; Mata-Alvarez et al. [69]
Household organic waste	Narra et al. [70]
Distillery spent waste	Nandy et al. [71]
LCFAs	Alves et al. [72]
Horse and cow dung	Yusuf et al. [73]
Agricultural and industrial wastes	Kujawski and Steinmetz [74]
MSW/FOG (fats, oils, and greases) wastes	Martin-Gonzalez et al. [75]
Nonedible oil cake and cow dung	Singh and Mandal [76]
Food wastes	Zhu et al. [77]; Chen et al. [46]
Maize grains and maize silage	Hutnan et al. [78]
Co-digestion of olive mill wastewater and swine manure	Azaizeh and Jadoun [79]
Cow dung and water hyacinth	Yusuf and Ify [80]
Cattle manure and slaughterhouse waste	Bagge et al. [81]
Co-digesting swine manure with three crop residues	Wu et al. [82]
Biomass (IFBB) and whole crop digestion (WCD)	Buhle et al. [83]
Organic solid poultry slaughterhouse waste	Salminen and Rintala [84]
MSW/agricultural waste/dairy cow manure	Macias-Corral et al. [85]
Animal manure	Holm-Nielsen et al. [86]
Maize hybrids	Oslaj et al. [87]
Byproducts of sugar production/cow manure	Fang et al. [88]
Biomass	Gunaseelan [89]
Fruit waste	Kaparaju and Rintala [90]; Lopez et al. [91]

IFBB, integrated generation of solid fuel and biogas from biomass.

Nandy et al. [71] treated high-strength distillery spent wash in the fixed-film, fixed-bed, two-stage anaerobic reactors using cheaper and abundantly available pebbles as media. The experiments were carried out in the laboratory as well as in the pilot-scale operation. The results showed that the overall chemical oxygen demand (COD) removal was about 80% with a specific biogas yield of 0.3 m^3 CH_4/kg COD. For a two-stage system, while the packed bed reactor can be easily fed with the spent wash, the detention period for each reactor has to be increased to obtain the COD removal close to 80%. In both reactors, the biogas yield decreased for HRT beyond 2.43 days. Feeding to the reactors can only be stopped for a maximum of three days, and the reactors become sour and need to be reenergized for a pH level <6.0. Most efficient operation was obtained at a temperature of 35°C–40°C and a pH of 7.0. Greater depth of reactor gave poorer performance and no clogging of the reactor was observed for 18 months.

8.4.4 Swine Waste

Chen et al. [55] examined various engineering options of conversion of swine waste to biomethanol. They applied target costing method in the development of marketable and environment-friendly product such as biomethanol from swine waste. Biomethanol is produced from methane, which is generated by anaerobic digestion of swine waste.

8.4.5 Byproducts of Biodiesel Production

The process of biodiesel production is predominantly carried out by catalyzed transesterification. Besides the desired methyl esters, the process produces several byproducts such as crude glycerol, oil-pressed cakes, and washing water. Crude glycerol or g-phase is a heavier, separate liquid phase, composed mainly of glycerol. Numerous types of oil cakes such as canola, rapeseed, coconut, cottonseed, groundnut oil, mustard oil, olive oil, palm kernel, sesame oil, soybean, and sunflower are also created in this process. Although their composition widely varies depending on the parameters and substrates used for biodiesel production, all these byproducts provide valuable feedstock for biogas production. The study by Kolesarova et al. [59] leads to the following conclusions:

1. Crude glycerol from biodiesel production is a valuable substrate for anaerobic degradation and the production of biogas using g-phase as a single substrate.
2. G-phase also has a great potential as a co-substrate by anaerobic treatment of different types of organic wastes such as organic fraction of MSW, mixture of olive mill wastewater, slaughterhouse wastewater, corn maize, maize silage, and swine manure.
3. Olive cakes and olive meals along with rapeseed and sunflower oil cakes can be used for anaerobic digestion to produce biogas and methane. High stability of the anaerobic digestion of sunflower oil cake under mesophilic conditions was obtained. With the increased amount of oil gained from

rapeseed meal by the extraction process, the possible biogas production from rapeseed cake decreased. Pretreatment (thermal and chemical) of sunflower and rapeseed residues did not enhance the methane yield.
4. Washing water from biodiesel production is a good candidate for anaerobic degradation due to its high content of biodegradable organic substances.
5. Specific inhibition effects resulting from the substrate composition should be considered during anaerobic treatment of biodiesel byproducts. In the case of anaerobic digestion of crude glycerol, high salinity of the substrates may negatively affect the methanogenic microorganisms. The concentration of ammonium should also be monitored. Since nitrogen is an essential nutrient for microorganisms, its low concentration in the crude glycerol and washing water has to be compensated by the ammonium supplement. However, nitrogen-rich substances have high concentration in rapeseed cake, which may cause ammonium accumulation in the reactor, thereby inhibiting the digestion process.

The use of byproducts of the biodiesel process as a potential source of energy producer makes the process of biodiesel more economically attractive.

8.4.6 Palm Oil Mill Effluent

Yusoff et al. [61] examined the effects of HRT and volatile fatty acids produced during fermentation on biohydrogen production from palm oil mill effluent. Both HRT and volatile fatty acid concentration played a vital role in the biohydrogen concentration, rate, and yield. The results were obtained for HRT of two, three, and five days, and two days gave the optimum operation with a maximum biohydrogen yield, rate, and concentration of 30%. The VFA as soluble metabolites reduced the amount of biohydrogen production by 8%–10%. The study concluded that HRT and VFA affect the biohydrogen production and should be considered in biohydrogen fermentation.

8.4.7 LCFAs in Wastewater

As shown in Table 8.1 [5,92], the potential for biogas and methane production from lipids is much higher than that from proteins and carbohydrates. LCFAs commonly found in wastewaters include lauric acid, myristic acid, palmitic acid, palmitoleic acid, stearic acid, oleic acid, and linoleic acid among others [72]. An extensive number of studies for the treatment of wastewater containing lipids and LCFAs in different types of anaerobic reactors have been reported. These studies are evaluated by Alves et al. [72].

The high-rate anaerobic technology (HR-AnWT) for the wastewater treatment requires the expansion of suitable substrates, in particular better treatment of the wastewater with high-lipid content. Waste lipids are good candidates for substrates needed to improve biogas and methane production, compared to proteins and carbohydrates. Alves et al. [72] presented a review of how LCFA degradation is accomplished by syntrophic communities of anaerobic bacteria and methanogenic archaea. For optimal performance, these syntrophic communities need to be clustered in compact aggregate, which is often difficult to achieve with wastewater that contains fats and lipids.

Alves et al. [72] proposed a new reactor concept that provides the primary biomass retention through floatation and the secondary biomass retention through settling.

The types of bacteria involved in methanogenic conversion of LCFA are known and the biochemical mechanism of LCFA degradation by beta-oxidation is well understood. The initial steps in the anaerobic conversion of unsaturated LCFA are, however, unclear. Besides the obligate hydrogen-producing acetogens (OHPAs) that degrade the unsaturated LCFA, bacteria exist which have the ability to hydrogenate unsaturated LCFA to saturated LCFA. This conversion can be coupled to growth and these bacteria may compete with hydrogenotrophic methanogens for hydrogen.

LCFAs require the syntrophic cooperation of OHPA and methanogens. These synthropic communities perform optimally when they are organized in microcolonies; the interspecies hydrogen transfer is enhanced with a short intermicrobial distance. It is yet not clear how microcolonies are developed in a fatty matrix and what is the effect of hydrogen transfer. Since hydrogen is poorly soluble in water, hydrogen transfer is increased when the matrix is LCFA. More work in this area is needed.

8.4.8 Food and Kitchen Organic Waste

Significant efforts have been made to generate biogas (biomethane) from different types of organic wastes [64,65,68,70,71]. Anaerobic digestion is a preferred method for energy resource recovery from organic residuals because this method (1) generates biomethane, (2) reduces the volume of the waste, and (3) stabilizes the waste. Shin et al. [65] showed how this method has been successfully applied to food waste from restaurants, markets, institutions, and households. They described a multi-step sequential batch two-phase anaerobic composting (MUSTAC) process that was stable, reliable, and effective in treating food residuals. The process can remove 82.4% of volatile solids and convert 84.4% of biomethane potential into methane in 10 days. The output from the posttreatment can be used as a soil amendment. The MUSTAC process was simple to operate and had high performance. Haug et al. [66] described the use of Los Angeles Wastewater Hyperion Treatment plant to anaerobically digest the food residual from the Los Angeles airport and the surroundings serving airline industry and passengers. The plant was cost effective and handled waste in an environmentally acceptable way.

Weichgrebe et al. [57] examined the energy and CO_2 reduction potentials of anaerobic treatment of wastewater and organic kitchen wastes. They considered three different scenarios: (1) the classical waste treatment and the composting of the organic waste fraction, (2) the anaerobic treatment of wastewater combined with deammonification and the digestion of the organic waste fraction, and (3) a mutual anaerobic treatment of wastewater and waste as co-digestion with deammonification. Scenario 2 was found to be the best. With the today's state of the art concerning the wastewater and waste treatment, both energy surplus and simultaneous CO_2 emission reduction was accomplished for scenario 2 at 20°C without the use of the dissolved methane into the reactor's effluent. If in the future an economical process for the usage of dissolved methane is developed, GHG emission can be further lowered. A further positive effect of scenario 2 is that the dissolved nutrients can be reused. Since a small part of these nutrients is needed for the anaerobic metabolism (<20%), a majority of mineral fertilizer can be substituted by using the effluent for irrigation. Furthermore, energy and GHGs would be additionally saved and the wastewater treatment costs will be reduced.

8.4.9 Wastewater Treatment

Cowan [93] points out that one way to reduce the emission of CO_2 and increase the energy production from bioprocess technologies for wastewater treatment is to use an integrated algae pond system (IAPS) to address the range of wastewater treatment problems. The growth of algae requires the use of CO_2 that minimizes the emission to the environment. Furthermore, the IAPS system produces a quality effluent suitable for irrigation, negates the food versus fuel debate, and reduces the demand for fossil fuel-derived energy and fertilizers.

Ryan et al. [94] addressed the issue of wastewater treatment from ethanol-producing biorefineries. They suggested that inorder to treat the effluent from these refineries efficiently and economically which meets the local requirements and minimizes the net water consumption, a process integration that (1) improves the existing secondary (i.e., biological) treatment to maximize COD reduction, (2) incorporates a tertiary "polishing" stage to remove color, and (3) uses the reverse osmosis membrane technology to recover process water would be desirable. They also showed that the energy required for the secondary and tertiary treatment stages can be obtained from biogas-derived power from the anaerobic digester. Thus, this type of an integrated approach of postbiological treatment of ethanol stillage can address the issues of efficient refinery operation with minimum net water consumption.

8.4.10 Dairy Effluent

Energy generation potential from dairy effluent was recently evaluated by Desai et al. [67]. India is the largest milk producer in the world (100 MMT). In an organized sector, which produces only 30% of the total milk generated in the country, the 140 dairy processing plants generate a very significant amount of effluent that is rich in organic waste. Desai et al. [67] described an anaerobic digestion system for one dairy processing 100,000 l/day milk to generate biogas (biomethane) that can provide energy for the aerators of the existing aerobic treatment system (mostly activated sludge system). The study presented the details of the anaerobic filter system. The 40 million liters of milk handled by the organized sector of milk industry in India has potential of generating 11 MW power from methane produced by the anaerobic digestion filter system.

8.4.11 Tofu Wastewater

Zheng et al. [62] examined the hydrogen production from organic wastewater from tofu production by photo bacteria. While this is a very useful process, NH_4^+, which is normally the integrant in organic wastewater, is the inhibitor for hydrogen production with photo bacteria. They showed that the concentration of NH_4^+ at $\geq$2 mmol/l significantly affected the hydrogen production of wild-type sphaeroides because NH_4^+ concentration inhibited the nitrogenase activity. They generated the mutant named AR-3 that can produce hydrogen in the medium containing even 4 mmol/l NH_4^+ due to the release of the inhibition of NH_4^+ to the nitrogenase activity. Under suitable conditions, they also showed that the hydrogen generation rate of AR-3 from tofu wastewater could reach 14.2 ml/l/h. It was increased by >100% compared to that of wild-type R sphaeroides.

8.4.12 Fruit Waste

Methane can be produced from waste orange peel. The thermophilic anaerobic digestion of industrial orange waste pulp and peel with a subsequent aerobic post-treatment of the digestate has been successfully demonstrated by Kaparaju and Rintala [90]. In this study, in anaerobic batch cultures, the methane production rate of 0.49 m^3/kg volatile solids, and in a semicontinuous process, the methane production rate as high as 0.6 m^3/kg VS was generated. This did require the pH adjustment from 3.2 to 8.0 by $CaCO_3$ addition. An aerobic follow-up treatment with activated sludge produced CO_2 and water and converted ammonia into nitrate. The removal of nitrogen required an additional denitrification step. The process can be adapted to other fruit and vegetable wastes such as mango, pineapple, tomato, jackfruit, banana, and whole orange. The methane production rate from these fruits can be improved by using selected strains of *Sporotrichum*, *Aspergillus*, *Fusarium*, and *Penicillium* [5,90,91]. These fungal pretreatments enhanced the availability of nutrient in the medium, decreased the concentrations of antimicrobial components, and enabled the higher loading rate utilization.

Besides other fruits and vegetable wastes, the wastewater from pressing of these byproducts is also good substrate for methane production [91]. This wastewater is generated by pressing the rind of orange peel and it contains a large amount of organic matter and alkalinity because in the pressing process, $Ca(OH)_2$ is used as binder. Before the anaerobic treatment, the waste is pretreated by aluminum phosphate flocculent to remove solids that can hinder the anaerobic treatment and to reduce the pH from 11.21 to 5.5. In the batch process, this treatment removed 84% of soluble COD and generated 295 ml of methane per gram of COD removal. The presence of antimicrobial components reduced the methane production when COD loadings were high [5,91].

8.5 CO-DIGESTION

Co-digestion is the simultaneous digestion of a homogeneous mixture of multiple substrates. The most common situation is when a major amount of basic substances is mixed and digested together with minor amounts of a single or a variety of additional substrates. As pointed out by Braun et al. [52], Braun [53], and Braun and Wellinger [54], co-digestion can improve the overall nutrient balance and digestion, and create an additional biogas and fertilizer. It can also equalize the particulates, floating and settling materials, and acidity in settlers by a suitable dilution by manure and sewage sludge to the agricultural waste. The co-digestion can, however, create an increased COD in the digester effluent, may require more pretreatment of the waste, and increase the mixing requirement in the digester. It may also cause an increased hygienic requirement and restriction on the available land use. Finally, its economics is very much dependent on the crop cost and yield.

Recent research results by Wu et al. [82] and numerous others [53,54,73,78,79,80, 83–85,88,95–108] demonstrate that using co-substrates in the anaerobic digestive systems improves the biogas yields through positive synergisms established in the digestion medium and the supply of missing nutrients by the co-substances. This subject is under an extensive investigation in the anaerobic digestion industry.

Historically, anaerobic digestion was carried out for animal manure and sewage sludge from aerobic wastewater treatment. In the recent years, agricultural biogas plants use pig, cow, and chicken manure with co-substrates, which increase the organic content of the total substrate. The co-wastes can be organic wastes from the agriculture-related industries, food waste, collected municipal biowaste from households, energy crops, tops and leaves of sugar beets, and so on. Fats provide the largest biogas yield but require high retention time. Carbohydrates and proteins have faster conversion rates but lower yields. If pathogens or other organisms are present, pasteurization at 70°C and sterilization at 130°C of feed materials are needed prior to fermentation. The carbon/nitrogen (C/N) ratio should be between 15 and 30 to avoid the process failure by ammonia accumulation. The fermentation residue should be used as an fertilizer.

8.6 EFFECTS OF HARVESTING, STORAGE, AND PRETREATMENT

8.6.1 Effect of Harvesting

The specific methane yield obtained from this material depends on its age [5,109,110,111]. Harvesting time and its frequency are important for biogas yield. Crops can be grown as preceding crop, main crop, or succeeding crop, each leading to a different biogas yield [5]. Weiland [5] pointed out that maize crops harvested after 97 days of milk ripeness produced 37% more methane yield than those at full ripeness.

8.6.2 Storage

Easy storage is an important factor in the selection of energy crops. The storage of energy crops by ensiling converts soluble carbohydrates into lactic acid, acetate, propionate, and butyrate, which inhibit the growth of detrimental microorganisms by a strong drop in pH between 3 and 4 [112]. The starter cultures, enzymes, and easily degradable carbohydrates can control and accelerate the acid formation. The optimum ensiling conditions are obtained by cutting particle length between 10–20 mm, and maintaining the total solid contents between 25% and 35%. Often, the storage by ensiling can be considered as a pretreatment process [5,113].

The structural polysaccharides of plant material are partly degraded during storage. They lose about 8%–20% of energy due to aerobic degradation, which is largely caused by oxygen, pH, and growth of yeasts that are responsible for heat upon exposure to oxygen. During storage, a plastic wrap should cover the plant material to minimize degradation.

8.6.3 Pretreatment

Muller et al. [5,114] showed that thermal, chemical, mechanical, or enzymatic processes as pretreatment can alter the degradation rate. Particle size reduction accelerates the biogas production rate but not the methane yield [5,115]. This is often achieved by the use of a crushing device or an ultrasonic treatment of the feed stream [5]. Also, thermal pressure hydrolysis at 230°C and 20–30 atm splits polymers into short-chain compounds, which can give better biogas yields with reduced retention time in the digesters [116,117].

The addition of an enzyme to the feed can have a mixed effect. While it can increase the biogas yield up to about 20% [5] by the acceleration of decomposition of polysaccharides, if added in excess, the protease of anaerobic microorganisms can degrade the enzyme, thus limiting its effectiveness [5,118]. In general, the enzyme reduces the viscosity of the substrate mixture and increases the degradation rates by avoiding the formation of floating layers [5]. For wheat grass, the addition of an enzyme improved the biogas production, but at the end of the digestion period, no significant improvement of methane yield or degradation rate was observed [119].

8.7 TYPES OF FERMENTATION AND ASSOCIATED DIGESTER CONFIGURATIONS

The nature of digester configuration depend on the method of fermentation. Four types of fermentation and associated digester configurations have been most widely used: wet fermentation, dry fermentation, batch fermentation, and two-stage fermentation. Solid concentration plays an important role in the choice of the fermentation. Wet fermentation is used for low solids concentration (<10%), dry fermentation is used for an intermediate concentration (between 15 and 35 wt%), and batch fermentation is used for solids concentration as high as 70 wt%. More details on the different types of digester are briefly described in Sections 8.7.1 through 8.7.4.

8.7.1 Wet Fermentation

In the wet fermentation process, the solids concentration is <10% and generally carried out in a vertical, continuous stirred slurry fermenter [5]. Low slurry concentration allows the stirring at lower power cost. The digested material is spread on the fields for the fertilization. For energy crops, the feed must be mixed with recycled process water or liquid manure to make the slurry pumpable. In the agriculture sector, wet fermentation is the preferred mode of operation.

Often the fermenter is covered with gas-tight, single- or double-membrane roof to store the gas before utilization. To achieve the uniform temperature and good contact between microbes and feedstock, good mixing of the slurry is very important. While mixing can be provided by different types of stirrer (mechanical, hydraulic, or pneumatic), mechanical stirrers are often used. In order to obtain maximum mixing in the reactor, the number, size, direction, and depth of the stirrer paddles depend on the nature of feedstocks, solids concentration, and height of the slurry. The stirring can be slow if it is continuous or it can be high if it is intermittent.

Most wet fermenters are operated at temperature between 38°C and 42°C (mesophilic condition). Since higher temperature gives faster degradation rate requiring lower HRT and reactor volume, some fermenters do operate under thermophilic conditions. Ultimate methane yield is, however, not influenced by higher temperature. Under mesophilic conditions, both ammonia toxicity and growth rate of microorganisms are reduced which can lead to the washout problem of microbial population [5,120]. In general, mesophilic processes are more energy efficient than thermophilic processes. The energy crops require very high retention time (weeks to months) and are generally fed at lower solids concentration (2–4 wt%) in wet fermenter [5,121].

8.7.2 Dry Fermentation

Dry solids fermentation is carried out with solids concentration between 15% and 35% [5]. Dry fermentation is operated either batchwise or in a continuous mode. For dry fermentation of slurry containing >25% solids, a horizontal mechanically mixed fermenter or a vertical plug flow reactor can be used [5]. These have been used for the anaerobic treatment of municipal organic solids [5,122,123]. In the vertical fermenter, the substrate flows from top to bottom by gravity only. The substrate fed at the top is mixed with the digestate coming from the bottom. This recycling and mixing of digestate with fresh feedstock prevents the accumulation of VFA and allows the high organic loading rate in the fermenter. The typical vertical fermenter volume in dry fermentation varies from 1000 to 4000 m^3 [5,122,123].

8.7.3 Batch Fermentation

For energy crops, dry fermentation in batch processes is preferred. For these processes, sometimes no mixing is required and solid inoculum up to 70% is necessary [5,124]. The batch process is operated with gas-tight lids, and operated for several weeks of digestion period.

8.7.4 Two-Stage Fermentation

Horizontal digesters are generally a part of a two-stage system in which high solids concentration flows in a horizontal plug flow mode with a low rotating horizontal paddle mixer. The reactor volume of such a reactor is limited to 700 m^3. For energy crops and processing of high solids concentration slurry, a two-stage digester system includes a high-loaded main fermenter followed by a low-loaded secondary fermenter.

The two-stage process generally gives higher biogas production and a lower methane potential of the final digestate [5,121,125]. In the two-stage process, hydrolysis and methanation take place in both reactors, although it is possible to use the first bed only for hydrolysis and treat the leachate coming out of the first bed in a second fixed-bed methanation reactor [5,126,127]. For achieving better metabolization of solid organic compounds, a two-stage reactor system with a separate hydrolysis stage can be advantageous because the ideal pH required for hydrolysis (5.5–6.5) is different from that required for methanation (6.8–7.2) [5,126,128]. This technology is mainly applied to MSW, industrial solid wastes and solid manure and seldom to energy crops. The control of operation and process parameters for the two-stage fermentation system is generally difficult. Furthermore, if the hydrolysis stage does not work properly, methane and hydrogen can escape in the environment [5,129].

8.7.5 Novel Digester Technology

The University of California at Davis developed a new anaerobic digester technology called anaerobic-phased solids (APS) digester for biogasification of organic waste solids that are normally difficult to process using conventional anaerobic digesters. A variety of feedstock including crop residues, animal manures, feed processing residuals, paper sludge, and MSW can be processed by APS digesters. The digester has been used to generate power for the University of California. The first commercial APS digester was built in Boynton Beach, Florida, to process 80 tons/day horse stable wastes. The possible benefits of this plant are renewable energy generation, odor control, pathogen and insect control, truck traffic reduction, and production of high-quality soil amendment.

The APS digester combines the favorable features of both batch and continuous operations in one system. Solids to be digested are handled in batches while biogas production is continuous. This allows the solids to be loaded and unloaded without disrupting an anaerobic environment for bacteria. The typical APS digester system consists of four hydrolysis reactors and one biogasification reactor. Liquid is recirculated intermittently between each hydrolysis reactor and the biogasification reactor. The solids are housed in the hydrolysis reactor, whereas the bacteria (methanogens) are housed in the biogasification reactor. The solids are broken down and liquefied in soluble compounds, which are mainly organic acids, and transferred to the biogasification reactor to generate biogas. The four hydrolysis reactors are operated in different time schedules so that biogasification reactor is constantly fed with the dissolved organic acid. High bacteria concentration in the biogasification reactor is maintained to get the optimum performance. More details on the APS digester are given by Zhang [68].

8.8 SIMULATION, MODELING, SCALE-UP, AND CONTROL OF FERMENTATION PROCESS

Angelidaki et al. [39,40] and Gavala et al. [130] gave a systematic assessment of complex kinetic models for organic waste digestion. They described the degradation by a simple first-order reaction that can be applied knowing the yield of substrate and the specific reaction rate [5]. Their kinetic models also depended on the nature of feedstock and the temperature range of the digestion process. The kinetic of biogas production from energy crops and manure was reported extensively by Mahnert [131]. Several kinetic models were developed for low-temperature (35°C–42°C) mesophilic conditions as well as high-temperature (45°C–60°C) thermophilic conditions by Andara and Esteban [132], Linke [133], and Biswas et al. [134].

As the interest in using the anaerobic digestion technique to generate biogas and biomethane increases due to economical and environmental reasons, it is important to determine the ultimate methane potential for a given solid substance. In fact, this parameter determines to some extent both design and economic analysis of a biogas plant. The ultimate methane potential thus identifies the "thermodynamic limit" for a given substance. Furthermore, to compare the potentials of various substrates, the definition of common units to be used in anaerobic assays is becoming increasingly important. Angelidaki et al. [39,40,64] presented some guidelines for biomethane assays of the anaerobic digestion prepared by the specialists group of the International Water Association. The guidelines include the considerations of biodegradability, bioactivity, inhibition, and matrices for biostability.

Narra et al. [70] evaluated a model for anaerobic digestion of household organic waste in high solids concentration (25 wt%) in an urban city in India and showed with pilot-scale experiments that the biogas production of 209 l/kg of the total solids is possible in a 30-day incubation period. High solids concentration reduces the water requirement and slurry handling problems. Composting takes 35 days and yields a quality product that can be used either as manure or a part of chemical fertilizer. A batch pilot plant was developed. Abderrezaq [63] evaluated the use of anaerobic digester for the MSW generated in Jordan. They showed that the digester technology can generate the energy from waste without generating GHG.

It is difficult to find a suitable and simple control parameter to control the complex fermentation process. Furthermore, only few parameters can be measured on-line. In agricultural biogas plants, the methane production is the only continuously measured parameter. However, complex and variable process dynamics make the interpretation of data difficult [5,135]. Only VFA can serve as an efficient indicator of process imbalances. Weiland [5,135] proposed that an indicator for process failure is the propionic acid/acetic acid ratio of >1.

Ahring et al. [136] suggested that if the propionic acid concentration is >1000 mg/l, the concentration of both butyrate and isobutyrate could be a reliable tool for indication of process failure. Nielsen et al. [137] suggested that propionate is the key parameter for process control and optimization. VFA analysis by manual sampling and the subsequent analysis by gas chromatography or high-pressure liquid chromatography is a slow process. On-line measurement is a difficult process [138]. A fast control of the process stability is possible by determining the ratio of total VFA to total inorganic carbonate. If this ratio is >0.3, the process is stable.

8.9 PURIFICATION OF BIOGAS

Biogas mainly contains methane and carbon dioxide with some impurities of hydrogen sulfide (with sulfur concentration from 100 to 3000 ppm) and ammonia, and it is generally saturated with water vapor. Before it can be used for heat and electricity generation, sulfur concentration should be reduced to the level below 250 ppm [5]. This will prevent the excessive corrosion and expensive deterioration of lubrication oil.

H_2S removal is carried out by biological desulfurization either within digester or outside digester. For this type of desulfurization, *Sulfobacter oxydans* bacteria and

2%–5% air must be present in the digester, which converts H_2S to elemental sulfur and sulfurous acid [5]. These bacteria are often present in the digester or added in the headspace of the digester. An efficient desulfurization requires a high contact area for microorganisms' fixation, which can be accomplished by an installation of specific wood or fabric support at the top of the fermenter.

For biological desulfurization outside the fermenter, trickling filter installations filled with plastic support materials on which the microorganisms can grow are used [139]. Raw biogas and air are injected at the bottom of the column, and the aqueous solution of nutrients is circulated to wash out the acidic products and supply the nutrients to microorganisms. The process is carried out at 35°C (mesophilic condition), and the support material is washed with air/water mixture at regular intervals to prevent sulfur deposits on the filters.

Desulfurization can also be done by adding commercial ferrous solution to the digester. In this expensive method, the production of hydrogen sulfide is prevented because ferrous binds sulfur to produce compounds which are insoluble in the liquid phase.

8.10 UTILIZATION OF BIOGAS AND DIGESTATE

The purified biogas can be used to generate electricity with about 43% efficiency [5]. It can be used in microgas turbine or fuel cell. While it is used in microgas turbine with a lower (25%–31%) efficiency, it gives good loading efficiency and long maintenance intervals for the turbines [5]. Furthermore, the exhaust heat from microgas turbine can be used to generate the process heat. The use of clean biogas in various fuel cells, which are operated at temperatures between 80°C and 800°C, gives higher efficiency. The investment costs for such applications are, however, higher. In the recent years, significant efforts are being made to upgrade the biogas and inject it into the grid or utilize it as a vehicle fuel [5].

The injection of biogas into natural gas grid requires further removal of all contaminants and carbon dioxide such that the final product must contain at least 95% methane. Both bacteria and molds must also be removed to make the use of biogas environmentally acceptable. The carbon dioxide is absorbed with the use of polyethylene glycol or mono- or diethanolamines. Carbon dioxide can also be removed using cryogenic separation, pressure swing adsorption, or membrane separation technology [5].

The process of anaerobic digestion reduces 80% of odor of the feedstock. The digestate generated from anaerobic digestion process possess valuable properties as fertilizers. Both nitrogen (in the form of ammonia) and carbon are useful as fertilizers. The nitrogen content in the digestate depends on the feedstock; it can be increased by a factor of 3 when only energy crops as substrate are used [5,52]. The faster permeation of digestate with improved flow properties can reduce loss of ammonia in air, thereby making "digestate fertilizer" more effective. While the "digestate fertilizer" inactivates weed seeds, bacteria, viruses, fungi, and parasites, their decay rates depend on the temperature, pH, treatment time, and VFA concentration. The best and faster results are obtained at higher temperature (>50°C) [5]. For certain wastes, while a separate pasteurization after digestion (at 70°C) is effective, digestate is prone to recontamination [5].

REFERENCES

1. IEA, *World Energy Outlook*. International Energy Agency, Paris, France (2006).
2. Davidson, O., Mertz, B., and cochairs, IPCC Special report on emission scenarios, Intergovernmental Panel on Climate Change (IPCC) working group III (2000).
3. Fehrenbach, H., Giegrich, J., Reinhardt, G., Sayer, U., Gretz, M., Lanze, K., and Schmitz, J., "Kriterien einer nachhaltigen Bioenergienutzung im globalen Mabstab," *UBA-Forschungsbericht*, 206, 41–112 (2008).
4. EurObserv'r Report, The state of renewable energies in Europe, 47–51 (2008) http://www.eurobserver.org.
5. Weiland, P., "Biogas production: Current state and perspectives," *Applied Microbiology Technology*, 85, 849–860 (2010).
6. Amon, T., Hackl, E., Jeramic, D., Amon, B., and Boxberger, J., "Biogas production from animal wastes, energy plantsand organic wastes," in van Velsen, A. and Verstraete, W. (eds.), *Proceedings of the 9th World Congress on Anaerobic Digestion*, Antwerp, Belgium, September 2–6, 381–386 (2001).
7. Gujer, W. and Zehnder, A.J.B., "Conversion processes in anaerobic digestion," *Water Science and Technology*, 15, 127–167 (1983).
8. Fachverband Biogas, "Biogas dozentral erzeugen, regional profitieren, international gewinnen," *Proceedings of the 18th Jahrestagung des Fachverbandes Biogas*, February 3–5, Hannover, Germany (2009).
9. Rader, G. and Logan, B., "Multi-electrode continuous flow microbial electrolysis cell for biogas production from acetate," *International Journal of Hydrogen Energy*, 35, 8848–8854 (2010).
10. Wang, A., Liu, W., Cheng, S., Xing, D., Zhou, J., and Logan, B.E., "Source of methane and methods to control its formation in single chamber microbial electrolysis cells," *International Journal of Hydrogen Energy*, 34, 3653–3658 (2009).
11. Liu, H., Grot, S., and Logan, B.E., "Electrochemically assisted microbial production of hydrogen from acetate," *Environmental Science & Technology*, 39, 4317–4320 (2005).
12. Selembo, P.A., Merrill, M.D., and Logan, B.E., "The use of stainless steel and nickel alloys as low-cost cathodes in microbial electrolysis cells," *Journal of Power Sources*, 190, 271–278 (2009).
13. Call, D.F., Merrill, M.D., and Logan, B.E., "High surface area stainless steel brushes as cathodes in microbial electrolysis cells (MECS)," *Environmental Science & Technology*, 43, 2179–2183 (2009).
14. Clauwaert, P. and Verstraete, W., "Methanogenesis in membraneless microbial electrolysis cells," *Applied Microbiology and Biotechnology*, 82, 829–836 (2009).
15. Parameswaran, P., Zhang, H., Torres, C.I., Rittmann, B.E., and Krajmalnik-Brown, R., "Microbial community structure in a biofilm anode fed with a fermentable substrate: The significance of hydrogen scavengers," *Biotechnology and Bioengineering*, 105 (1), 69–78 (2010).
16. Lee, H.-S. and Rittman, B.E., "Significance of biological hydrogen oxidation in a continuous single-chamber microbial electrolysis cell," *Environmental Science & Technology*, 44, 948–954 (2010).
17. Ye, Y., Wang, L., Chen, Y., Zhu, S., and Shen, S., "High yield hydrogen production in a single-chamber membrane-less microbial electrolysis cell," *Water Science and Technology*, 61 (3), 721–727 (2010).
18. Call, D.F. and Logan, B.E., "Hydrogen production in a single chamber microbial electrolysis cell lacking a membrane," *Environmental Science & Technology*, 42, 3401–3406 (2008).
19. Chae, K.-J., Choi, M.-J., Kim, K.-Y., Ajayi, F.F., Chang, I.-S., and Kim, I.S., "Selective inhibition of methanogens for the improvement of biohydrogen production in microbial electrolysis cells," *International Journal of Hydrogen Energy*, 35 (24), 13379–13386 (2010).

20. Lalaurette, E., Thammannagowda, S., Mohagheghi, A., Maness, P.-C., and Logan, B.E., "Hydrogen production from cellulose in a two-stage process combining fermentation and electrohydrogenesis," *International Journal of Hydrogen Energy*, 34 (15), 6201–6210 (2009).
21. Selembo, P.A., Perez, J.M., Lloyd, W.A., and Logan, B.E., "High hydrogen production from glycerol or glucose by electrohydrogenesis using microbial electrolysis cells," *International Journal of Hydrogen Energy*, 34, 5373–5381 (2009).
22. Hu, H., Fan, Y., and Liu, H., "Hydrogen production using single-chamber membrane-free microbial electrolysis cells," *Water Research*, 42, 4172–4178 (2008).
23. Zhang, Y., "The use and optimization of stainless steel mesh cathodes in microbial electrolysis cells," Masters thesis, Pennsylvania State University, University Park, PA (2010).
24. Liu, H., Cheng, S., Huang, L., and Logan, B.E., "Scale-up of membrane-free single-chamber microbial fuel cells," *Journal of Power Sources*, 179, 274–279 (2008).
25. Shimoyama, T., Komukai, S., Yamazawa, A., Ueno, Y., Logan, B.E., and Watanabe, K., "Electricity generation from model organic wastewater in a cassette-electrode microbial fuel cell," *Applied Microbiology and Biotechnology*, 80, 325–330 (2008).
26. Tartakovsky, B., Manuel, M.-F., Wang, H., and Guiot, S.R., "High rate membrane-less microbial electrolysis cell for continuous hydrogen production," *International Journal of Hydrogen Energy*, 34 (2), 672–677 (2009).
27. Cheng, S. and Logan, B.E., "Ammonia treatment of carbon cloth anodes to enhance power generation of microbial fuel cells," *Electrochemistry Communications*, 9 (3), 492–296 (2007).
28. Logan, B., Cheng, S., Watson, V., and Estadt, G., "Graphite fiber brush anodes for increased power production in air-cathode microbial fuel cells," *Environmental Science & Technology*, 41, 3341–3346 (2007).
29. Logan, B.E., Call, D., Cheng, S., Hamelers, H.V.M., Sleutels, T.H.J.A., Jeremiasse, A.W., and Rozendal, R.A., "Microbial electrolysis cells for high yield hydrogen gas production from organic matter," *Environmental Science & Technology*, 42 (23), 8630–8640 (2008).
30. Bagi, Z., Acs, N., Balint, B., Horvath, L., Dobo, K., Perei, K., Rakhely, G., and Kovacs, K., "Biotechnological intensification of biogas production," *Applied Microbiology and Biotechnology*, 76, 473–482 (2007).
31. Xu, K., Liu, H., and Chen, J., "Effect of classic methanogenic inhibitors on the quantity and diversity of archaeal community and the reductive homoacetogenic activity during the process of anaerobic sludge digestion," *Bioresource Technology*, 101 (8), 2600–2607 (2010).
32. Cheng, S. and Logan, B.E., "Evaluation of catalysts and membranes for high yield biohydrogen production via electrohydrogenesis in microbial electrolysis cells (MECS)," *Water Science and Technology*, 58, 4, 853–857 (2008).
33. Fan, Y., Zhang, Y., Zhang, S., Hou, H., and Ren, B., "Efficient conversion of wheat straw wastes into biohydrogen gas by cow dung compost," *Bioresource Technology*, 97 (3), 500–505 (2006).
34. Ding, H.B., Liu, X.Y., Stabnikova, O., and Wang, J.-Y., "Effect of protein on biohydrogen production from starch of food waste," *Water Science and Technology*, 57, 7, 1031–1036 (2008).
35. Lay, J., "Modelling and optimization of anaerobic digested sludge converting starch to hydrogen," *Biotechnology Bioengineering*, 68 (3), 269–278 (2000).
36. Ange Abdoun, E. and Weiland, P., *Bermimer Agrartecnisclre Berichte*, 68, 69–78 (2009).
37. Ahrens, T. and Weiland, P., "Biomethane for future mobility," *Landbauforschung Volkenrode*, 57, 71–79 (2007).

38. Schink, B., "Energetics of syntrophic cooperation in methanogenic degradation," *Microbiology Molecular Biology Review*, 61, 262–280 (1997).
39. Angelidaki, I., Ellegard, L., and Ahring, B., "A comprehensive model of anaerobic bioconversion of complex substrates to biogas," *Biotechnology Bioengineering*, 63, 363–372 (1999).
40. Angelidaki, I., Ellegaard, L., and Ahring, B.K., "A mathematical model for dynamic simulation of anaerobic digestion of complex substrates: Focusing on ammonia inhibition," *Biotechnology Bioengineering*, 42, 159–166 (1993).
41. Elferink, S., van Lis, R., Heilig, H., Akkermans, A., and Stams, A., "Detection and quantification of microorganisms in anaerobic bioreactors," *Biodegradation*, 9, 169–177 (1998).
42. Karakashev, D., Bastone, D., and Angelidaki, I., "Influence of environmental conditions on methanogenic compositions in anaerobic biogas reactors," *Applied and Environmental Microbiology*, 71, 331–338 (2005).
43. Klocke, M., Nettman, E., Bergmann, I., Mundt, K., Souidiu K., Mumme, I., and Linke, B., "Characterization of the methanogenic archaea within two-phase biogas reactor systems operated with plant biomass," *Systematic and Applied Microbiology*, 31, 190–205 (2008).
44. Yu, Y., Lee, C., Kim, J., and Hwangs, S., "group specific primer and probe sets to detect methanogenic communities using quantitative real time polymerase chain reaction," *Biotechnology Bioengineering*, 89, 670–679 (2005).
45. Leven, L., Eriksson, A., and Schnurer, A., "Effect of process temperature on bacterial and archaeal communities in two methanogenic bioreactors treating organic household waste," *FEMS Microbiology Ecology*, 59, 683–693 (2007).
46. Chen, Y., Cheng, J., and Creamer, K., "Inhibition of anaerobic digestion process: A review," *Bioresource Technology*, 99 (10), 4044–4064 (2008).
47. Nielsen, H. and Angelidaki, I., "Strategies for optimizing recovery of biogas process following ammonia inhibition," *Bioresource Technology*, 99, 7995–8001 (2008).
48. Wang, Q., Kuninobu, M., Ogawa, H., and Kato, Y., "Degradation of volatile fatty in highly efficient anaerobic digestion," *Biomass & Bioenergy*, 16, 407–416 (1999).
49. Mosche, M. and Jordening, H., "Comparison of different models of substrate and product inhibition in ammonia digestion," *Water Research*, 33, 2545–2554 (1999).
50. Bischoff, M., "Erkenntnisse beim Einsatz von Zusatz-und Hilfsstoffen sowie von Spurene lementen in Biogasanalangen," *VDI-Ber*, 2057, 111–123 (2009).
51. Friedmann, H. and Kobe, J., "Optimierung der Biogasproduktionaus nachwachsenden Rohstoffen durch den Einsatz von Mikronahrstoffen-ein Erfahrungsbericht," in Tagungsband 17, Jahrestagung des Fachverbandes Biogas, Nuremberg, Germany, 125–130 (2008).
52. Braun, R., Weiland, P., and Wellinger, A., "Biogas from energy crop digestion," IEA Bioenergy Task 37 Energy from Biogas and Landfill Gas IEA, Paris, France (2011).
53. Braun, R., "Potential of co-digestion," (2002), http://www.novaenergie.ch/iea-bioenergy-task37/Dokumente/final.PDF.
54. Braun, R. and Wellinger, A., "Potential for co-digestion," IEA Bioenergy Task 37 Energy from Biogas and Landfill Gas (2002).
55. Chen, Y.S., Zuckerman, G.J., and Zering, K., "Applying target costing in the development of marketable and environmentally friendly products from swine waste," *The Engineering Economics*, 53, 156–170 (2008).
56. Kunchikannan, L.K.N.V., Mande, S.P., Kishore, V.V.N., and Jain, K.L., "Coir pith: A potential agro residue for anaerobic digestion," *Energy Sources, Part A*, 29, 293–301 (2007).
57. Weichgrebe, D., Urban, I., and Friedrich, K., "Energy and CO_2 reduction potentials by anaerobic treatment of wastewater and organic kitchen wastes in consideration of different climate conditions," *Water Science and Technology*, 58, 2, 379–384 (2008).

58. Pathe, P.P., Rao, N.N., Kharwade, M.R., Lakhe, S.B., and Kaul, S.N., "Performance evaluation of a full scale effluent treatment plan for distillery spent wash," *Journal of Environmental Studies*, 59 (4), 415–437 (2002).
59. Kolesarova, N., Hutnan, M., Bodik, I., and Spalkova, V., "Utilization of biodiesel by-products for biogas production," *Journal of Biomedicine and Biotechnology*, 126798, 1–15 (2011).
60. Beszedes, S., Laszlo, Z., Szabo, G., and Hodur, C., "The possibilities of bioenergy production from whey," *Journal of Agricultural Sciences Technology*, 4 (1), 62–68 (2010).
61. Yusoff, M., Rahman, N., Abd-Aziz, S., Ling, C.M., Hassan, M., and Shirai, Y., "The effect of hydraulic retention time and volatile fatty acids on biohydrogen production from palm oil mill effluent under non-sterile condition," *Australian Journal of Basic and Applied Sciences*, 4 (4), 577–587 (2010).
62. Zheng, G.H., Kang, Z.H., Qian, Y.F., Wang, L., Zhou, Q., and Zhu, H.G., "Biohydrogen production from tofu wastewater with glutamine auxotrophic mutant of Rhodobacter sphaeroides," in Tohji, K., Tsuchiya, N., and Jeyadevan, B. (eds.), *Proceedings of the 5th International Workshop on Water Dynamics*, Sendal, Japan, American Institute of Physics, 143–148 (2008).
63. Abderrezaq, I.A., "Employment of anaerobic digestion process of municipal solid waste for energy," *Energy Sources, Part A*, 29, 657–668 (2007).
64. Angelidaki, I., Alves, M., Bolzonella, D., Borzacconi, L., Campos, L., Guwy, A., Kalyuzhnyi, S., Jenicek, P., and van Lier, J.B., "Defining the biomethane potential (BMP) of solid organic wastes and energy crops: A proposed protocol for batch assays," *Water Science and Technology*, 59, 5, 927–934 (2009).
65. Shin, H.K., Han, S., Song, Y., and Hwan, E., "Biogasification of food residuals," *Biocycle*, 82–86 (2000).
66. Haug, R.T., Hernandez, G., Sarullo, T., and Gerringer, F., "Using wastewater digesters to recycle food residuals into energy," *Biocycle*, 74–77 (2000).
67. Desai, H., Nagori, G., and Vahora, S., "Energy generation potential of dairy effluent—A case study of Vidya dairy," *Proceedings of the International Conference on Energy and Environment*, March 19–21, Chandigarh, India (2009).
68. Zhang, R., "Biogasification of organic solid wastes," *Biocycle*, 43, 56–59 (2002).
69. Mata-Alvarez, J., Mace, S., and Llabres, P., "Anaerobic digestion of organic solid wastes. An overview of research achievements and perspectives," *Bioresource Technology*, 74, 3–16 (2000).
70. Narra, M., Nagori, G.P., and Pushalkar, S., "Biomethanation of household organic waste at high solid content a package for waste disposal and energy generation," *Proceedings of the International Conference on Energy and Environment*, March 19–21, Chandigarh, India (2009).
71. Nandy, T., Kaul, S.N., Pathe, P.P., Deshpande, C.V., and Daryapurkar, R.A., "Pilot plant studies on fixed bed reactor system for biomethanation of distillery spent wash," *International Journal of Environmental Studies*, 41, 87–107 (1992).
72. Alves, M., Pereira, M., Sousa, D., Cavaleiro, A.J., Picavet, M., Smidt, H., and Stams, A., "Waste lipids to energy: How to optimize methane production from long-chain fatty acids (LCFA)," *Microbial Biotechnology*, 2 (5), 538–550 (2009).
73. Yusuf, M., Debora, A., and Ogheneruona, D., "Ambient temperature kinetic assessment of biogas production from co-digestion of horse and cow-dung," *Research in Agricultural Engineering*, 57 (3), 97–104 (2011).
74. Kujawski, O. and Steinmetz, H., "Development of instrumentation systems as a base for control of digestion process stability in full-scale agricultural and industrial biogas plants," *Water Science and Technology*, 60 (8), 2055–2063 (2009).

75. Martin-Gonzalez, L., Castro, R., Pereira, M., Alves, M., Font, X., and Vicent, T., "Thermophilic co-digestion of organic fraction of MSW with FOG wastes from a sewage treatment plant: Reactor performance and microbial community monitoring," *Bioresource Technology*, 102 (7), 4734–4741 (2011).
76. Singh, R. and Mandal, S., "The utilization of non-edible cake along with cow dung for methane enriched biogas production using mixed inoculums," *Energy Resources, Part A, Recovery, Utilization and Environmental Effects*, 33 (5), 449–458 (2011).
77. Zhu, H., Parker, W., Basnar, R., Proracki, A., Falletta, P., Beland, M., and Seto, P., "Buffer requirements for enhanced hydrogen production in acidogenic digestion of food wastes," *Bioresource Technology*, 100 (21), 5097–5102 (2009).
78. Hutnan, M., Spalkova, V., Bodik, I., Kolesarova, N., and Lazor, M., "Biogas production from maize grains and maize silage," *Polish Journal of Environmental Studies*, 19 (2), 323–329 (2010).
79. Azaizeh, H. and Jadoun, J., "Co-digestion of olive mill wastewater and swine manure using up-flow anaerobic sludge blanket reactor for biogas production," *Journal Water Resource and Protection*, 2, 314–321 (2010).
80. Yusuf, M. and Ify, N., "The effect of waste paper on the kinetics of biogas yield from the co-digestion of cow dung and water hyacinth," *Biomass & Bioenergy*, 35 (3), 1345–1351 (2011).
81. Bagge, E., Peterson, M., and Johansson, K.E., "Diversity of spore forming bacteria in cattle manure, slaughterhouse waste and samples from biogas plants," *Journal of Applied Microbiology*, 109 (5), 1549–1565 (2010).
82. Wu, X., Yao, W., Zhu, J., and Miller, C., "Biogas and CH_4 productivity by co-digesting swine manure with three crop residues as an external carbon source," *Bioresource Technology*, 101 (11), 4042–4047 (2010).
83. Buhle, L., Stulpagel, R., and Wachendorf, M., "Comparative life cycle assessment of the integrated generation of solid fuel and biogas from biomass (IFBB) and whole crop digestion (WCD) in Germany," *Biomass & Bioenergy*, 35 (1), 363–373 (2011).
84. Salminen, E. and Rintala, J., "Anaerobic digestion of organic solid poultry slaughterhouse waste: A review," *Bioresource Technology*, 83 (1), 13–26 (2002).
85. Macias-Corral, M., Samani, Z., Hanson, A., Smith, G., Funk, P., Yu, H., and Longworth, J., "Anaerobic digestion of municipal solid waste and agricultural waste and the effect of co-digestion with dairy cow manure," *Bioresource Technology*, 99 (17), 8288–8293 (2008).
86. Holm-Nielsen, J., Seadi, T., and Oleskowicz-Popiel, P., "The future of anaerobic digestion and biogas utilization," *Bioresource Technology*, 100 (922), 5478–5484 (2009).
87. Oslaj, M., Mursec, B., and Vindis, P., "Biogas production from maize hybrids," *Biomass & Bioenergy*, 34 (11), 1538–1545 (2010).
88. Fang, C., Boe, K., and Angelidaki, I., "Anaerobic co-digestion of by products from sugar production with cow manure," *Water Research*, 45 (11), 3473–3480 (2011).
89. Gunaseelan, V.N., "Anaerobic digestion of biomass for methane production: A review," *Biomass & Bioenergy*, 13, 83–114 (1997).
90. Kaparaju, P. and Rintala, J., "Anaerobic co-digestion of potato tuber and its industrial by-products with pig manure," *Resources, Conservation and Recycling*, 43, 175–188 (2005).
91. Lopez, J., Li, Q., and Thompson, I., "Biorefinery of waste orange peel," *Critical Reviews in Biotechnology*, 30 (1), 63–69 (2010).
92. Baserga, U., Landwirtschaftliche Co-vergarungs-Biogasanlagen, FAT-Berichte No. 512, Tanikon, Switzerland (1998).
93. Cowan, A.K., "Bio-refineries: Bioprocess technologies for waste water treatment, energy and product valorization," in Tarasenko, O. (ed.), *Proceedings of the 4th Bionanotox Conference*, Little Rock, AK, American Institute of Physics, College Park, MD, November 4–5, 80–86 (2010).

94. Ryan, D., Gadd, A., Kavanagh, J., and Barton, G., "Integrated biorefinery wastewater design," *Chemical Engineering Research and Design*, 87, 1261–1268 (2009).
95. Agdag, O.N. and Sponza, D.T., "Co-digestion of mixed industrial sludge with municipal solid wastes in anaerobic simulated landfilling bioreactors," *Journal of Hazardous Materials Volume*, 140, 75–85 (2007).
96. Alvarez, R. and Liden, G., "Semi-continuous co-digestion of solid slaughterhouse waste, manure, and fruit and vegetable waste," *Renewable Energy*, 33 (4), 726–734 (2007).
97. Davidsson, A., Lovstedt, C., la Cour Jansen, J., Gruvberger, C., and Aspegren, H., "Co-digestion of grease trap sludge and sewage sludge," *Waste Management*, 28 (6), 986–992 (2008).
98. Fernandez, A., Sanchez, A., and Font, X., "Anaerobic co-digestion of a simulated organic fraction of municipal solid wastes and fats of animal and vegetable origin," *Biochemical Engineering*, 26, 22–28 (2005).
99. Fezzani, B. and Cheikh, R., "Thermophilic anaerobic co-digestion of olive mill wastewater with olive mill solid wastes in a tubular digester," *Chemical Engineering Journal*, 132 (1–3), 195–203 (2007).
100. Gelegenis, J., Georgakakis, D., Angelidaki, I., and Mavris, V., "Optimization of biogas production by co-digesting whey with diluted poultry manure," *Renewable Energy*, 32, 2147–2160 (2007).
101. Gomez, G., Cuetos, M., Cara, J., Moran, A., and Garcia, A., "Anaerobic co-digestion of primary sludge and the fruit and vegetable fraction of the municipal solid wastes: Conditions for mixing and evaluation of the organic loading rate," *Renewable Energy*, 31, 2017–2024 (2006).
102. Lehtomäki, A., Huttunen, S., and Rintala, J., "Laboratory investigations on co-digestion of energy crops and crop residues with cow manure for methane production: Effect of crop to manure ratio," *Resources, Conservation and Recycling*, 51, 591–609 (2007).
103. Murto, M., Björnsson, L., and Mattiasson, B., "Impact of food industrial waste on anaerobic co-digestion of sewage sludge and pig manure," *Journal of Environmental Management*, 70, 101–117 (2004).
104. Neves, L., Oliveira, R., and Alves, M., "Anaerobic co-digestion of coffee waste and sewage sludge," *Waste Management*, 26, 176–181 (2006).
105. Romano, R.T. and Zhang, R., "Co-digestion of onion juice and wastewater sludge using an anaerobic mixed biofilm reactor," *Bioresource Technology*, 99 (3), 631–637 (2007).
106. Zupancic, G.D., Uranjek-Zevart, N., and Ros, M., "Full-scale anaerobic co-digestion of organic waste and municipal sludge," *Biomass & Bioenergy*, 32 (2), 162–167 (2008).
107. Alvarez, J., Otero, L., and Lema, J., "A methodology for optimizing feed composition for anaerobic co-digestion of agro-industrial wastes," *Bioresource Technology*, 101, 1153–1158 (2010).
108. McDonald, N., "Experiences with anaerobic digestion of energy crops and high solids manures," OWS Report, AgSTAR, Mankato, MN (May 2011).
109. Dohler, H., Eckel, H., and Frisch, J., Energiepflanzen, KTBL, Darmstadt (2006).
110. KTBL/FNR Faustazahlen Biogas, Kuratorium fur Technik und Bauwesen in der Landwirtschaft, Darmstadt, Germany, 49–51 (2007).
111. Karpenstein-Machan, *Energiepflanzenbau fur Biogasanlagenbetreiber*. DLG-Verlag, Frankfurt, Germany (2005).
112. Weinberg, Z., Muck, R., and Weimer, P., "The survival of silage inoculant lactic acid bacteria in rumen fluid," *Journal of Applied Microbiology*, 93, 1066–1071 (2003).
113. Banemann, D. and Nelles, M., Von der Emte bis in den Fermenter, VDI-Ber 2057, 29–46 (2009).
114. Muller, J., Tiehm, A., Eder, B., Günthert, F., Hruschka, H., Kopp, J., Kunz, P., Otte-Witte, R., Schmelz, K., and Seiler, K., "Thermische, Chemische, und biochemische Desintegrationsverfahren," *Korresp Abwasser*, 50, 796–804 (2003).

115. Mahandete, A., Bjornsson, L., Kivaisi, A., Rubindamayugi, M.S.T., and Matthiasson, B., "Effect of particle size on biogas yield from sisal fiber waste," *Renewable Energy*, 31, 2385–2392 (2006).
116. Prechtel, S., Anzer, T., Schneider, R., and Faulstich, M., "Biogas production from substrate with high amount of organic nitrogen," *Proceedings of the 10th World Congress—Anaerobic Digestion 2004*, Montreal, QC, August 29–September 2 (2004).
117. Mladenovska, Z., Hartmann, H., Kvist, T., Sales-Cruz, M., Gani, R., and Ahring, B., "Thermal pretreatment of the solid fraction of manure: Impact of the biogas reactor performance and microbial community," *Water Science and Technology*, 53, 59–67 (2006).
118. Romano, R., Zhang, R., Teter, S., and McGarry, J., "The effect of enzyme addition on anaerobic digestion of Jose Tall Wheat Grass," *Bioresource Technology*, 100, 4564–4571 (2009).
119. Morgavi, D., Beauchemin, K., and Nsereko, L., "Resistance of feed enzymes to proteolytic inactivation by rumen microorganisms and gastrointestinal proteases," *Journal of Animal Science*, 79, 1621–1630 (2001).
120. Angelidaki, I., Ellegaard, L., and Ahring, B., "Application of the anaerobic digestion process," in Ahring, B. (ed.), *Biomethanation II, Advances in Biochemical Engineering/Biotechnology*, Springer, Berlin, Germany, 82, 2–33 (2003).
121. Gemmeke, B., Rieger, C., and Weiland, P., Biogas-Messprogramm II, 61 Biogaanlangen im Vergleich, FNR, Gulzow (2009).
122. Schon, M., *Verfahren zur Vergarung organischer Ruckstande in der Abfallwirtschaft.* Erich Schmidt Verlag, Berlin, Germany (1994).
123. de Baere L. and Mattheeuws, B., "State of the art 2008-anaerobic digestion of solid waste," *Waste Management World*, 9, 1–8 (2008).
124. Kusch, S., Oechsner, H., and Jungbluth, T., "Vergarung landwirtschaft-licher substrtae in diskontinuierlichen Feststoffermentern," *Agratechnische Forschung* 11, 81–91 (2007); Biogashandbuch Bayern-Materialband. Bayerisches Landesamt fur Umwelt, Augsburg, Germany (2005).
125. Lethomaki, A., "Biogas production from energy crops and crop residues," *Jyvask Studies in Biological and Environmental Science*, 163, 1–91 (2006).
126. Parawira, W., Read, J., Mattiasson, B., and Bjornsson, L., "Energy production from agricultural residues: High methane yields in a pilot scale two-stage anaerobic digestion," *Biomass & Bioenergy*, 32, 44–50 (2008).
127. Busch, G., Grossmann, J., Sieber, M., and Burkhardt, M., "A new and sound technology for biogas from solid waste and biomass," *Water, Air, & Soil Pollution: Focus*, 9, 89–97 (2009).
128. Vieitez, E. and Gosh, S., "Biogasification of solid wastes by two phase anaerobic fermentation," *Biomass & Bioenergy*, 16, 299–309 (1999).
129. Oechsner, H. and Lemmer, A., "Was Kann die Hydrolyse bei der Biogasvergarung leisten?" VDI-Ber 2057, 37–46 (2009).
130. Gavala, H., Angelidaki, I., and Ahring, B., "Kinetics and modeling of anaerobic digestion process," in Scheper, T. and Ahring, B.K. (eds.), *Biomethanation I.* Springer, Berlin, Germany, pp. 57–93 (2003).
131. Mahnert, P., "Kinetic der Biogas produktion aus nachwachsenden Rohstoffen und Gulle," Dissertation, Humboldt-Universitat, Berlin, Germany, 202 (2007).
132. Andara, A.R. and Esteban J.M.B., "Kinetic study of the anaerobic digestion of the solid fraction of piggery slurries," *Biomass & Bioenergy*, 17, 435–443 (1999).
133. Linke, B., "Kinetic study of thermophilic anaerobic digestion of solid wastes from potato processing," *Biomass & Bioenergy*, 30, 892–896 (2006).
134. Biswas, L., Chowdhury, R., and Battacharya, P., "Mathematical modeling for the prediction of biogas generation characteristics of an anaerobic digester based on food/vegetable residues," *Biomass & Bioenergy*, 31, 80–86 (2007).

135. Weiland, P., "Wichtige Messdaten fur den Prozessablauf und Stand der Technik in der Praxis," *Gulzower Fachgesprache*, 27, 17–31 (2008).
136. Ahring, B., Sandberg, M., and Angelidaki, I., "Volatile fatty acids as indicators of process imbalance in anaerobic digesters," *Applied Microbiology and Biotechnology*, 34, 559–565 (1995).
137. Nielsen, H., Uellendahl, H., and Ahring, B., "Regulation and optimization of biogas process: Propionate as a key factor," *Biomass & Bioenergy*, 31, 820–830 (2007).
138. Boe, K., Bastone, D., and Angelidaki, I., "Online headspace chromatographic method for measuring VFA in biogas reactor," *Water Science and Technology*, 52, 473–478 (2005).
139. Schneider, R., Quicker, P., Anzer, T., Prechtl, S., and Faulstich, M., "Grundlegende Untersuchungen zur effektiven, kostengunstingen Entfernung von Schwefelwasserstoff aus Biogas," in *Biogasanlagen Anforderungen zur Luftreinhaltung*, Bayerisches Landesamt fur Umweltschutz, Augsburg, Germany (2002).
140. Klocke, M., Netterman, E., and Bergmann, I., "Monitoring der methanbildenden Mikroflora in Praxis-Biogasanlagen im landlichen Raum: Analyse des Ist-zustandes und Entwicklung eines quantitativen Nachweissystems," *Bornimer Agrartechnische Berichte*, 67, (2009).

9 Hydrolysis and Fermentation Technologies for Alcohols

9.1 INTRODUCTION

Water plays an essential role in the hydrolysis and fermentation of lignocellulosic materials to produce alcohols. While most of the efforts have been made to generate ethanol, in the recent years, some research has been pursued to generate higher alcohols such as butanol via fermentation technologies [1]. In this chapter, we mainly focus on ethanol with a brief update on the recent advances in the generation of butanol via hydrolysis and fermentation technologies.

Ethanol is the second largest solvent (next to water) in the world [2]. While ethanol has been generated from sugars via fermentation technology for more than 2000 years in all parts of the world, its expanded use as a fuel or fuel additive has substantially increased its production by the use of innovative technologies. Ethanol is a very versatile material with the following usages [2]:

1. It is a raw material for the manufacture of plastics, lacquers, polishes, plasticizers, perfumes, and cosmetics.
2. Ethyl acetate and ethyl acrylate (i.e., ethyl esters produced by the reaction of ethanol with carboxylic acid) are raw materials for acrylate polymers.
3. Vinegar can be produced by *Acetobacter* bacteria in ethanol solutions.
4. It is a raw material for polylactic acid and polylactide (PLA). Polylactic acid is a biodegradable polymer and can also be used with other polymers as a composite.
5. It is a source for hydrogen via processes of reforming or supercritical water gasification.
6. It can be a substitute for methanol for transesterification of triglycerides to make biodiesel.
7. It is a very good substitute for reformulated gasoline and additive as well as a source of renewable fuel.

Ethanol can be used as E10 (10% ethanol in gasoline), E22 (gasohol used in Brazil), E85 (85% ethanol used in flexible fuel automobiles in Brazil as well in the United

States), or E100 (100% ethanol) [2]. It has a high research octane number, makes the engine more efficient, has reasonable vapor pressure, and helps reduce emissions of NO_x, volatile organic compound (VOC), CO, and CO_2. It fits well the new Clean Air Act Amendments (CAAA) of 1990, which requires (1) 2% of oxygen by weight in gasoline, (2) maximum benzene content of 2%, and (3) maximum of 25% by volume aromatic hydrocarbons. While ethanol can be used up to 20% in gasoline for conventional cars, it is also very useful for flexible cars and two-cycle engines. Up to 25% ethanol can also be added in acetylene-based dual-fuel systems. Cellulose ethanol will be an important contributor to 32 billion of renewable fuel mandate by the US government by 2022 [2].

During the past several decades, the hydrolysis and fermentation technologies to convert sugar or starch materials have been commercialized very extensively in the United States and Brazil among other countries. While Brazil has chosen sugarcanes as starting raw materials, in the United States corn (starch such as wheat, barley, and rice) has been the major feedstock for the ethanol production. Corn refining in the United States has a relatively long history going back in time of the Civil War with the development of cornstarch hydrolysis process. The first cornstarch plant was built in Jersey City, NJ. By 1857, the cornstarch industry accounted for a significant portion of the US starch industry (along with starch from wheat and potatoes). The industrial production of dextrose (sugar) from cornstarch started in 1866. This led to the production of corn sugar, corn sweeteners, corn syrup, and so on [2–7]. After the World War II, ethanol was produced by the fermentation of corn sugar, but major quantities of ethanol from corn were produced only after 1970. While starch and glucose are important parts of corn refineries, about 13.2 billion gallons of ethanol was produced from corn in the United States in 2010. The United States is the world's leading producer of corn, totaling about 331 million tons of corn (worth $66.7 billion) in 2010. As shown in Figure 9.1, about 34.9% of this production was used to make ethanol [2–7].

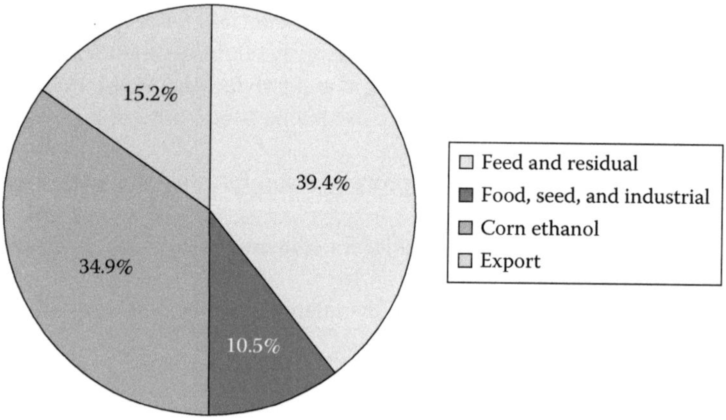

FIGURE 9.1 Breakdown by categories of 2010 end uses of corn in the United States. (Adapted from Lee, S. and Shah, Y., *Biofuels and Bioenergy—Processes and Technologies*, CRC Press, Boca Raton, FL, 2012; US Grains Council, 2011.)

Today, there are two types of ethanol depending on the source of feedstock: grain ethanol [8–10] and cellulosic ethanol [11,12]. Grain ethanol is made from starch or sugar feedstock such as corn, wheat, barley, rice, and sugarcane, while cellulosic ethanol is produced from lignocellulosic materials. The process for cellulosic ethanol is much more complex than that for grain ethanol due to the complexity of the feedstock in the former case. Here, we describe the processes for both types of ethanol in detail.

9.2 GRAIN (CORN) ETHANOL

Grain ethanol suffers from the fact that the feedstock has food value, and the use of corn, wheat, and so on for the purpose of fuel may become challenging in times of drought or when the corn is in short supply. The need for food is always higher than that for fuel, and these competitive usages for corn may make the price of corn too high for an economical ethanol process. The advantages of grain ethanol are that the starting feedstock is starch (or sugar) and it only requires hydrolysis and fermentation steps and not the pretreatment step as it is required in cellulosic ethanol.

Ethanol production facilities for corn (grain) ethanol are classified into two broad categories: wet milling [2,13–15] and dry milling operations [2,16,17]. As the terms indicate, in dry milling operation, the entire corn kernel is pulverized into flour called "corn meal," and it is then mixed with water and processed for hydrolysis and fermentation. Dry mills are usually smaller in size and are built primarily to manufacture ethanol only. The remaining stillage from ethanol purification undergoes a different process treatment to produce highly nutritious animal feedstock (often called dried distillers grains [DDGs]). In 2008, a total of 86% of corn ethanol was produced by this method in about 150 dry milling plants [17].

Wet milling processes are called "corn refineries," which along with ethanol also produce high-value coproducts such as high-fructose corn syrup, dextrose, cornstarch, DDG, and Splenda. They are larger and have more capital and operating costs. While the wet milling process is more versatile and produces many co- and byproducts, it is less efficient than the dry milling process. Thermal energy and electricity are the main types of energy used in both dry and wet milling processes. Dry milling uses natural gas in several parts of the process such as generating steam for mash cooking, distillation, and evaporation. In many new ethanol plants, the use of combined heat and power (CHP) has been very popular due to its increased production efficiencies and expanded fuel capabilities. A CHP system improves the efficiency by 10%–30% more than 50% efficiency obtained in conventional operations [2].

The process of hydrolysis and fermentation of starch involves the following steps:

$$\text{Starch} \xrightarrow{\text{Hydrolysis}} \text{D-glucose} \xrightarrow{\text{Fermentation}} 2C_2H_5OH + 2CO_2$$

The theoretical yield of ethanol from sugar (D-glucose) is 51% by weight basis.

9.2.1 Starch Hydrolysis

Starch is regarded as a long-chain polymer of glucose (i.e., many glucose molecular units are bonded in a polymeric chain similar to a condensation polymerization product) [18]. This starch is first broken down to simple sugar units by the hydrolysis process. In this process, starch feedstock is ground and mixed with water containing about 15%–20% starch. The mash is then cooked at or above its boiling point and subsequently treated with two enzyme preparations. The first enzyme hydrolyzes the starch into short-chain molecules and the second enzyme hydrolyzes the short-chain molecules into glucose. The first enzyme is amylase, which liberates "maltodextrin" by the liquefaction process. These maltodextrins are very sweet and contain a group of low-molecular-weight carbohydrates called dextrins and oligosaccharides (a polymer of small number of simple sugars, monosaccharides). The dextrins and oligosaccharides are further hydrolyzed in the second step by an enzyme called pullulanase and glucoamylase in a process known as saccharification. Complete saccharification converts all the dextrans to glucose, maltose, and isomaltose. The mash is then cooled and subjected to yeast fermentation.

9.2.2 Yeast Fermentation

Yeasts convert sugar into ethanol via a biochemical process called fermentation. The yeasts of primary interest to industrial fermentation of ethanol include *Saccharomyces cerevisiae*, *Saccharomyces uvarum*, *Schizosaccharomyces pombe*, and *Kluyueromyces* sp. Under anaerobic conditions, the yeasts metabolize glucose to ethanol primarily via the Embden–Meyerhof pathway. This pathway for glucose metabolism is the series of enzymatic reactions in the anaerobic conversion of glucose to lactic acid or ethanol, resulting in the energy in the form of adenosine triphosphate (ATP) [19]. Generally, the yield is about 90%–95% of the stoichiometric relationship mentioned earlier. About 1716 kg of fermentable sugar is required for the production of 1000 l of ethanol. When the fermentation is completed, the remaining solution is called distilled mash or stillage that contains a large amount of nonfermentable portions of fibers or proteins.

9.2.3 Ethanol Purification and Product Separation

Ethanol is separated from the mash by distillation. Unfortunately, conventional distillation process works only up to 95.63% ethanol because water and ethanol form an azeotrope that will not allow any further concentration of ethanol. The minimum boiling point temperature of 78.2°C is attainable at the azeotropic concentration and not at the pure ethanol concentration. The additional concentration of ethanol is carried out by dehydration by one of the two methods. In the first method, a third component (such as benzene) is used to change the boiling characteristics of the solution. The third component breaks azeotrope and allows conventional distillation to be carried out in a tertiary system to achieve the desired separation. The second method uses the molecular sieves that absorb water selectively and therefore concentrate ethanol further. There are different forms of molecular sieves that are based on the dimensions of effective pore opening, which include 3A, 4A, 5A, and 13X. Commercial molecular sieves are typically available in powder, bead, granular, and extrudate forms.

9.2.4 Byproducts and Coproducts

The nonfermentable solids in distilled mash (stillage) contain variable amounts of proteins and fibers depending on the feedstock. The recovery of protein and other nutrients in stillage for use as an animal feedstock is essential for making the overall ethanol production process profitable. Corn and barley yield solid byproducts called DDGs. The protein content of DDG typically ranges from 25% to 30% by mass and makes an excellent feedstock for the animals. Byproducts and coproducts are very important for corn refineries (wet milling process).

9.2.5 Environmental Implications

The liquid effluent generated from ethanol process may contain some harmful chemicals and other pollutants that must be discarded properly. About 9 l of liquid effluent is generated for every liter of ethanol produced. The biological oxygen demand (BOD) of effluent can be high and the effluent can be acidic. Both of these factors require additional treatments before discarding the effluents to fields or water streams.

9.3 CORN TO ETHANOL PROCESS TECHNOLOGIES

As mentioned earlier, the conversion of corn to ethanol can be carried out as either (1) wet milling corn ethanol (or corn refinery) technology or (2) dry milling corn ethanol technology. We briefly describe both of these technologies in Sections 9.3.1 and 9.3.2.

9.3.1 Wet Milling Technology for Conversion of Corn to Ethanol

The corn wet milling process to produce ethanol separates corn into its four basic components: starch, germ, fiber, and protein. There are eight basic steps involved to accomplish this corn refining and alcohol fermentation process [10].

Step 1: This step inspects the incoming corn visually and removes cob, dust, chaff, and any other foreign unwanted materials before the next processing step of steeping. The inspected and screened corn is then conveyed to storage silos holding up to 350,000 bushels.

Step 2: This step carries out the steeping process in which about 2,000–13,000 bushels of corn is soaked in water at 50°C–52°C for 20–48 h in a stainless steel tank. A series of tanks are used. During this process, the kernel of corn (Figure 9.2) absorbs water from 15% to 45% by weight and swells by more than double its original size. The addition of 0.1% sulfur dioxide to water suppresses the excessive bacterial growth in the warm water environment.

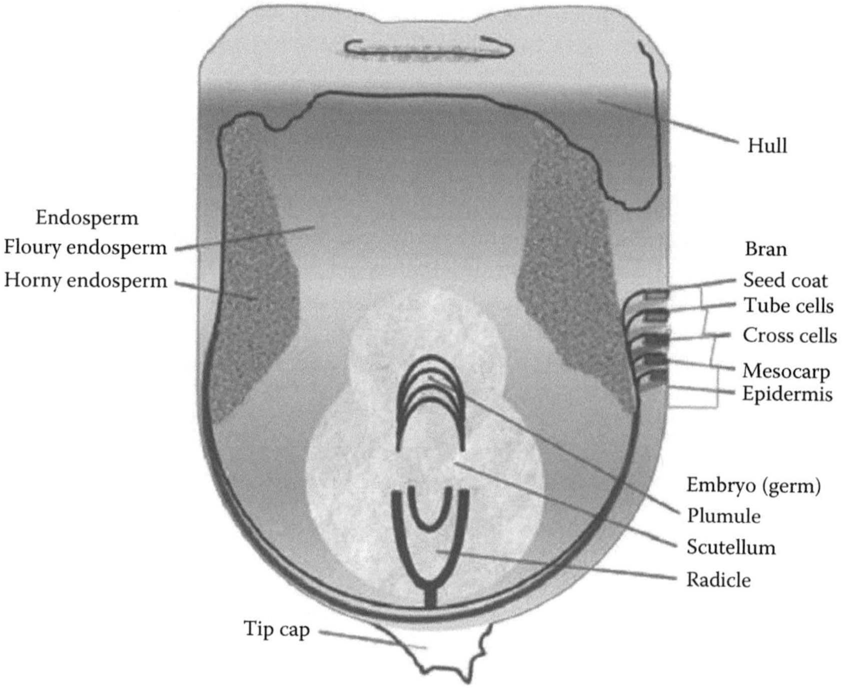

FIGURE 9.2 Corn kernel. (Adapted from Lee, S. and Shah, Y., *Biofuels and Bioenergy—Processes and Technologies*, CRC Press, Boca Raton, FL, 2012.)

As the corn swells, the mild acidity of the steeping water loosens the gluten bond within the corn eventually releasing the starch [10]. Thus, this step initiates polymeric bond cleavage of starch and protein into simpler molecules.

Step 3: The third step is the germ separation. The coarse grinding of corn in the slurry separates germ (Figure 9.2) from corn. This germ separation is accomplished by cyclone separator that removes the low-density corn germ from the slurry. The germs are repeatedly washed to remove any leftover starch, and then with the use of mechanical and solvent processes, oil from the germ is extracted. The oil is then refined and filtered into finished corn oil. The germ residue is saved as another important component of animal feed. Both corn oil and germ residues are important byproducts of the process.

Step 4: In this step, the remaining slurry containing fiber, starch, and protein is finely ground and screened to separate the fiber from starch and protein. A thorough grinding in impact or attrition-impact mill releases the starch and gluten from the fiber in the kernel. Fiber is separated from starch and gluten using concave screens. Fiber is collected and slurried again to reclaim any residual starch and protein, and then sent to the feed house as a major ingredient for animal feed. The starch–gluten suspension (called mill starch) is sent to starch separators [20].

Step 5: In this step, starch is separated from gluten by hydrocyclones. Separated gluten that contains proteins and is called corn gluten meal (CGM) is used for animal feed. CGM can also be used as an organic herbicide. The last 1%–2% protein remained in the starch is further removed by repeated washings and the high-quality starch is now called unmodified cornstarch. While most of cornstarch is converted to corn syrups and dextrose, the cornstarch is also used for a variety of industrial and domestic uses [20].

Step 6: This is a starch-to-sugar conversion step. The starch–water suspension is liquefied in the presence of acid or enzymes. Enzymes help convert the starch to dextrose that is soluble in water as an aqueous solution. If needed, the solution is also treated with another enzyme. The process of acid and enzymatic reactions is controlled according to the desired mixtures of sugars such as dextrose and maltose (a disaccharide) for syrup. The reaction time is used to control the concentration of dextrose and maltose in the final product. Once the conversion is completed, the syrup is passed through filters, centrifuges, or ion exchange columns, and the excess water is evaporated. Syrup can be sold directly as is, crystallized into pure dextrose, or processed further to produce high-fructose corn syrup. Across the corn wet milling industry, about 80% of starch slurry goes to corn syrup, sugar, and fermentation.

Step 7: In this step, corn syrup is converted to several products through a fermentation process. Dextrose (called corn sugar or grape sugar) also known as D-glucose is easily fermentable. The process of fermentation can be carried out either in a continuous way in a series of fermenters to give higher throughput or in a batch fermenter for about 48 h to get a better quality product.

Step 8: The resulting broth from step 7 is distilled to recover ethanol or concentrated through membrane separation to produce other byproducts. Carbon dioxide generated from fermenter is recaptured to produce dry ice for sale, and nutrients still remaining in the broth after fermentation are used as components of animal feed ingredients. These byproducts contribute significantly to the overall economics of the corn refineries. A schematic of corn refinery or wet milling corn-to-ethanol process is described in Figure 9.3.

9.3.2 Dry Milling Corn-to-Ethanol Process

This process also contains eight steps, which are as follows:

Step 1: In this step, corn is received and stored in silos designed to hold grain supply for 7–12 days of plant operation.

Step 2: The grain is inspected and screened to remove corn cobs, stalks, finer materials, stones, and other foreign objects by a blower and screen. The cleaned material is coarse grinded using hammer mill. The grinded material is combined with hot water to form slurry.

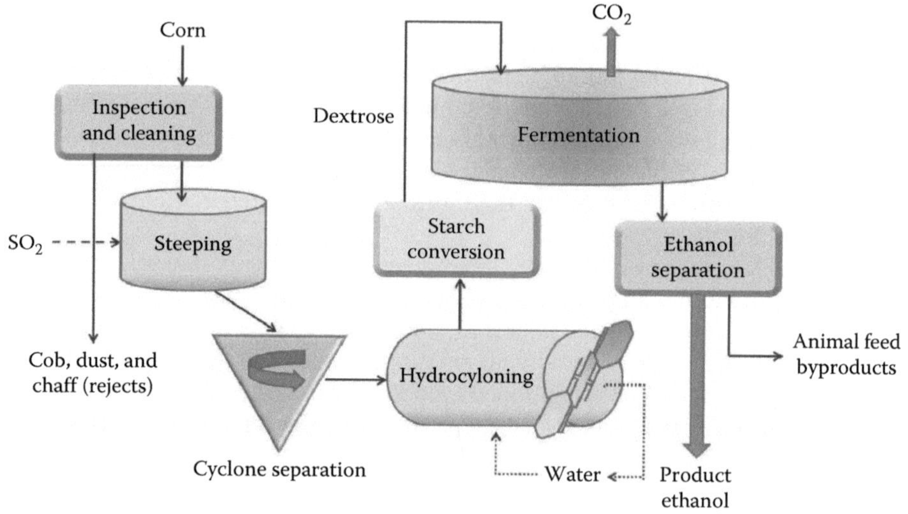

FIGURE 9.3 A schematic of a typical wet milling corn-to-ethanol process. (Adapted from Lee, S. and Shah, Y., *Biofuels and Bioenergy—Processes and Technologies*, CRC Press, Boca Raton, FL, 2012.)

Step 3: This step has three parts and involves the cooking process, which is also called hot slurry primary and secondary liquefaction. In this process, the starch in the flour is physically prepared and chemically modified for fermentation. In the first part, coarsely ground grain is soaked in hot process water, the pH is adjusted to about 5.8, and alpha-amylase enzyme is added. The agitated slurry is heated to 82°C–88°C for 30–60 min. In the second part of primary liquefaction, the slurry is pumped through a pressurized jet cooker at 105°C and held there for about 5 min. The mixture is then cooled by an atmospheric or vacuum flash condenser. Within jet cooker, the steam rapidly heats the slurry and evenly hydrolyzes. The fluid dynamic relationship between the jet cooker's steam injector and the condensing tube produces a pressure drop to help maximize shear action to improve starch conversion [14]. In the third part of secondary liquefaction, the mixture is held for 1–2 h at 82°C–88°C to give alpha-amylase enzyme sufficient time to break down starch into short-chain low-molecular-weight dextrins. This chemical conversion is called gelatinization. As the conversion of starch proceeds, the viscosity of slurry decreases. Dextrins are a mixture of polymers of D-glucose units. After pH and temperature adjustment, a second enzyme glucoamylase is added as the mixture is pumped into the fermentation tanks. Glucoamylase is an amylase enzyme that cleaves the last alpha-1,4-glycosidic linkages at the nonreducing end of amylase and amylopectin to yield glucose. The cleavages of the bonds near the ends of long-chain starches release maltose as well as glucose. Maltose, or malt sugar, is a disaccharide that is formed from the two units of glucose joined with alpha($1 \rightarrow 4$) bond.

Step 4: The fourth step is called simultaneous saccharification fermentation. Once the mixture of milled kernel and water, now known as mash, is inserted in the fermentation tank, the glucoamylase enzyme breaks down the dextrins and oligosaccharides to form simple sugars that are monosaccharides. Yeast is added to convert sugar into ethanol and carbon dioxide. The mash is allowed to ferment for 50–60 h, resulting in a mixture that contains about 15% ethanol as well as solids from the grain and added yeast [16,20].

Step 5: The fermented mash is pumped into the distillation system to separate ethanol from water at a concentration of up to 95% ethanol by volume (a level of azeotropic mixture). The residue from this process called stillage contains nonfermentable solids and water, and is pumped out of the bottom of the distillation columns into the centrifuges.

Step 6: The near-azeotropic binary mixture of 95% ethanol and 5% water is dehydrated by a molecular sieve that physically separates the remaining water from the ethanol based on the size difference between the two molecules [16]. The process produces nearly 100% ethanol.

Step 7: The produced ethanol is stored up to 7–12 days. The ethanol is appropriately used as a fuel blend with gasoline.

Step 8: Ethanol production process creates two coproducts: carbon dioxide and distillers grains. These coproducts are captured and sold as dry ice and animal feed, respectively, to improve the overall economics of the process.

9.4 CELLULOSIC ETHANOL

While starch and sugar produce grain ethanol, the feedstock obtained from this method is also used as food [2,21–54]. In the recent years, more efforts are made to convert all lignocellulosic materials such as hardwood, softwood, agricultural waste, and energy crops into ethanol. Unlike corn, this material is not useful for food purposes. The ethanol produced from lignocellulosic material is called cellulosic ethanol.

Lignocellulosic materials are composed of four ingredients: cellulose, hemicellulose, lignin, and extractives. As shown in Figure 9.4, a generalized plant cell wall structure is like a composite material in which rigid (and crystalline) cellulose fibers are embedded in a cross-linked matrix of lignin and hemicellulose that binds the cellulose fibers. Generally, the dry weight of a typical cell wall consists of approximately 30%–50% cellulose, 20%–35% hemicellulose, and 10%–25% lignin [12]. The exact percentages vary with the nature of the feedstock. For example, for woody biomass, cellulose accounts for 40%–50%, and lignin and hemicellulose each account for about 20%–30%. Lignin is aromatic in nature and provides higher heating value than cellulose or hemicellulose. The chemicals in the biomass matrix include extractives such as resins, phenols, and other chemicals and minerals such as calcium, magnesium, and potassium. These extractives are left behind in ash when biomass is combusted. The trace minerals and major elements in lignocellulosic materials display a high degree of variability for most of the elements between different species and between different organs within a given plant, depending on the growing conditions including the soil characteristics [21].

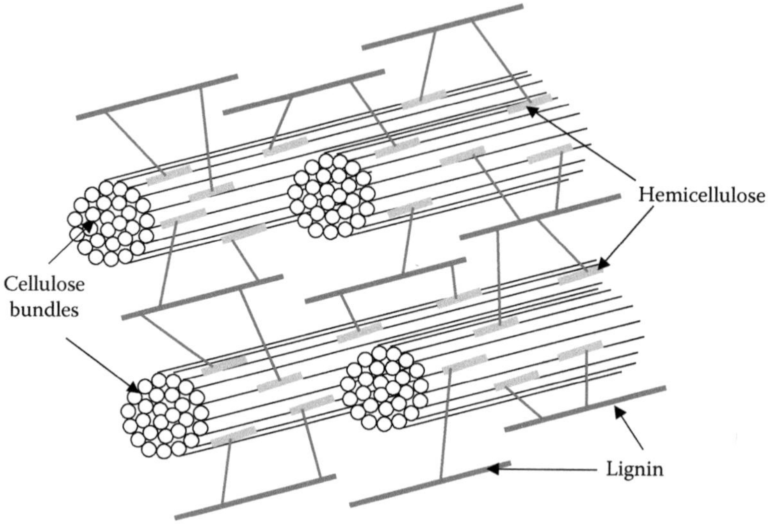

FIGURE 9.4 A universal description of plant cell wall. (Adapted from Lee, S. and Shah, Y., *Biofuels and Bioenergy—Processes and Technologies*, CRC Press, Boca Raton, FL, 2012.)

Cellulose is a large polymeric molecule composed of many hundreds or thousands of monomeric sugar (glucose) molecules, and in this regard, it can be considered as polysaccharide. The molecular linkages in cellulose form linear chains that are rigid, highly stable, and resistant to chemical attack. It is also crystalline and may be somewhat soluble in a suitable solvent [18]. However, cellulose molecules (which are the predominant source of glucose for ethanol) in their crystalline form are packed so tightly that even small molecules of water cannot easily permeate the structure. It is even more difficult for large enzyme molecules to permeate and diffuse into the cellulose structure. To break the crystalline structure of cellulose and make them more exposed to enzymatic hydrolysis, all processes of cellulosic ethanol require pretreatments. As discussed earlier, this step was not required in the production of grain ethanol.

Starch and sugar can also come from hemicellulose that consists of short and highly branched chains of sugar molecules. It contains both five-carbon sugars (such as D-xylose and L-arabinose) and six-carbon sugars (such as D-galactose, D-glucose, and D-mannose as well as uronic acid. For example, galactan found in hemicellulose is a polymer of sugar galactose. Since hemicellulose is amorphous due to highly branched structures, it is relatively easy to hydrolyze to its constituents—five- and six-carbon sugars [18]. While both five- and six-carbon sugars are in principle fermentable to ethanol, the fermentation chemistry, yeast requirement, and process chemistry for six- and five-carbon sugars (pentose and xylose) are considerably different. In general, five-carbon sugars are more difficult to ferment than six-carbon sugars.

Lignin molecule is a complex and highly cross-linked aromatic polymer that is covalently linked to hemicellulose (Figure 9.4). Lignin contributes to the stabilization of mature cell walls. Due to its high calorific value, it provides more energy than cellulose or hemicellulose, but it cannot be fermented to ethanol. Lignin is a

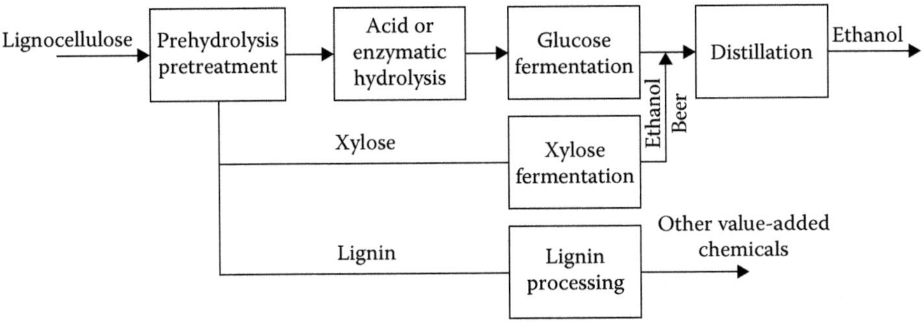

FIGURE 9.5 Conversion of lignocellulose to ethanol. (Adapted from Lee, S. and Shah, Y., *Biofuels and Bioenergy—Processes and Technologies*, CRC Press, Boca Raton, FL, 2012.)

macromolecule whose typical molecular weight exceeds 10,000. Because of its cross-linked structure, it is difficult to process, extract, and hydrolyze. The major purpose of the pretreatment step of the cellulosic ethanol process is to degrade the cross-linked structure so that both cellulose and hemicellulose are more exposed for subsequent hydrolysis and fermentation steps. An efficient conversion of lignin results in a substantial increase in the overall fuel yield of the cellulosic ethanol process.

A general scheme for the conversion of lignocellulose to ethanol is shown in Figure 9.5. The lignocellulose is pretreated to separate the xylose and sometimes the lignin from the crystalline cellulose. This step is very important because the efficiency of the pretreatment affects the efficiency of the subsequent steps. The xylose can then be fermented to ethanol, whereas the lignin can be further processed to produce other liquid fuels and valuable chemicals. Crystalline cellulose, the largest (about 50%) and most useful fraction, remains behind as solids after pretreatment and is sent to an acid or enzymatic hydrolysis process to break down the cellulose to glucose. Enzymatic hydrolysis (which is more popular now) is very specific and does not break down further sugars. Enzymatic processes are capable of achieving 100% yield. The glucose is then fermented to ethanol and combined with the ethanol from xylose fermentation. This dilute ethanol–water solution is further concentrated by distillation and other dehydration processes.

For an overall efficiency of the conversion process, it is important to convert hemicellulose (which can be up to 25% of lignocellulose) to xylose and xylose to ethanol. Hemicellulose is primarily composed of xylan that can be easily converted to xylose. Xylose constitutes about 17% of woody angiosperms and accounts for a substantially higher percentage of herbaceous angiosperms. Though the fermentation of xylose to ethanol is more difficult than that of glucose, it is very essential for the overall efficiency of the process. Significant new yeast developments for this purpose are currently pursued. Methods have been identified using new strains of or metabolically engineered yeasts [22], bacteria, and processes containing enzymes and yeasts.

Lignin (around 25% of lignocellulose) is a large random phenolic polymer. In lignin processing, the polymer is broken down into fragments containing one or two phenolic rings. Extra oxygen and side chains are stripped from the molecules by the catalytic methods and the resulting phenol groups are reacted with methanol to produce methyl aryl ethers. These substances are high-value octane enhancers and can be blended with gasoline. We now examine each of the steps outlined in Figure 9.5 in detail.

9.4.1 Pretreatment

Unlike in the production of grain ethanol, in the production of cellulosic ethanol, pretreatment is essential to achieve the reasonable rates of yields in the enzymatic hydrolysis of biomass [2]. Pretreatment has generally been practiced to reduce the crystallinity of cellulose, to lessen the average degree of polymerization of the cellulose and the lignin–hemicellulose sheath that surrounds the cellulose, and to alleviate the lack of the available surface area for enzymes to attack. The importance of pretreatment can be better understood by examining the hydrolysis process in which the interaction between the enzymes and the substrates must occur. The hydrolysis of cellulose into sugars and other oligomers is a solid-phase reaction in which enzymes must bind to surface to catalyze the reaction. Cellulase enzymes (which are commonly used) are large proteins with molecular weight ranging from 30,000 to 60,000 and are thought to be ellipsoid with major and minor dimensions of 30°A–200°A. The internal surface area of wood is very large; however, only about 20% of the prevolume is accessible to cellulose-sized molecules. By breaking down the tight hemicellulose–lignin matrix, hemicellulose or lignin can be separated and the accessible volume of cellulose can be greatly increased. This removal of materials greatly enhances the enzymatic digestibility.

A typical pretreatment consists of size reduction, pressure sealing, heating, reaction, pressure release, surface area increase, and hydrolyzate/solids separation [23]. Mechanical pretreatments such as intensive ball milling and roll milling to expose more surface area have been found to be very expensive. The hemicellulose–lignin sheath can be disrupted by either acidic or basic catalysts. While basic catalysts simultaneously remove lignin and hemicellulose, its consumption is very large due to its use in neutralization by ash and acidic groups in the hemicellulose. In the recent years, more acidic catalysts such as mineral acids and organic acids generated *in situ* by autohydrolysis of hemicellulose have been tested.

The five important pretreatment processes that are currently being examined and implemented are as follows [2]:

1. Rapid steam hydrolysis (RASH) or autohydrolysis steam explosion
2. Dilute acid prehydrolysis
3. Organosolv pretreatment
4. Combined RASH and organosolv pretreatment
5. Ionic liquid pretreatment

Most pretreatment approaches are not intended to actually hydrolyze cellulose to soluble sugars, but rather to generate the pretreated cellulosic residue that is more hydrolyzable by cellular enzymes than native biomass. Here we examine each pretreatment process in detail.

9.4.1.1 Rapid Steam Hydrolysis

This process was recently described by Lee and Shah [2]. A typical autohydrolysis process uses compressed liquid hot water at a temperature of about 200°C and pressure above the saturation pressure [48]. Thus, the liquid water can hydrolyze hemicellulose in minutes. While hemicellulose recovery is high in this noncatalytic process, wet pyrolysis results in the production of inhibitory compounds. A well-controlled process at high temperature with small particles, however, gives high xylose yields and is desirable. Dekker and Wallis [24] showed that for bagasse, this process gave 90% solubilization of hemicellulose and the product that was highly susceptible to hydrolysis by cellulases from *Trichoderma reesei*. In general, however, the xylose yield in the RASH process is low (30%–50%).

Steam consumption in autohydrolysis strongly depends on the moisture content of the starting material. An important advantage of autohydrolysis is that it breaks down lignin into smaller fragments that can be easily solubilized in either base or organic solvents. This process was first developed in 1925 for hardwood application and more recently for aspen wood (in the 1980s). At a high pressure of 20–50 atm and temperature of 210°C–290°C, the water molecules diffuse into the microporous structure of lignocellulose and the steam condenses at high pressure, thereby wetting the materials [23]. The wetted material is then driven out of the reactor by a small nozzle using a pressure difference. The term "explosion" is used because of the process characteristics of the ejection driven by a sudden large pressure drop of steam.

9.4.1.2 Dilute Acid Prehydrolysis

The pretreatment process can be operated at a lower temperature with reduced sugar degradation by adding a small amount of mineral acid in the pretreatment process. The acid increases the reaction rates at a given temperature, and the ratio of hydrolysis rate to the degradation rate is also increased. The reaction rate can be optimized between the temperature and the reaction time. Higher temperature (200°C) can take 10 s, whereas lower temperature (100°C) may take several hours. Generally, the acid concentration (sulfuric acid) between 0.5 and 4 wt% is used. While sulfuric acid gives the xylose yields of 70%–95%, it produces more condensed lignin. Sulfur dioxide is often used as a catalyst. Numerous reports have indicated good results using this method [2,25]. While acid hydrolysis has been used for more than 100 years, the replacement of dilute acid hydrolysis by more concentrated acid prehydrolysis was found to be more expensive. While sulfuric acid is the most widely used catalyst in this pretreatment, other mineral acids such as hydrochloric, nitric, and trifluoroacetic (CF_3COOH) acids have also been used.

9.4.1.3 Organosolv Pretreatment

This process is a pulping technique that uses an organic solvent to solubilize lignin and hemicellulose. A process developed by Kraft pulping produces high-quality lignin for added values and easy recovery and recycling of solvents used in the process. The organic solvents such as ethanol, butanol, and methanol are added to the pretreatment reaction to dissolve and remove lignin fraction. The internal lignin and hemicellulose bonds are broken, and both fractions are solubilized while cellulose remains intact as solid. Careful steps are taken such that the process cleanly separates the feedstock into a solid cellulose residue, a solid lignin that has undergone a few condensation reactions and a liquid stream containing xylon. The process is carried out at high temperature (140°C–230°C) and under pressure to achieve the desired bond cleavages. Ethanol is the most desired solvent due to its price, availability, and easy recovery. In general, organosolv processes have higher xylose yields than other processes due to the influence of the organic solvent on hydrolysis kinetics [26].

9.4.1.4 Combined RASH and Organosolv Pretreatment

Attempts have been made to improve the pretreatment process by combining RASH and organosolv treatments [27]. A schematic diagram of this process is shown in Figure 9.6. As shown, for the organosolv pretreatment, in this process the steam generator is disconnected and the condensate valve is closed. The rest of the reactor setup is similar to the typical RASH process. The combination of these two processes, which requires high temperature, leads to an increased solubilization of lignin and hemicellulose. RASH temperature is the major factor in maximizing the percentage of cellulose in the final product. The maximum yield of solubilized lignin was obtained at a temperature of 240°C for RASH and 160°C for the organosolv process.

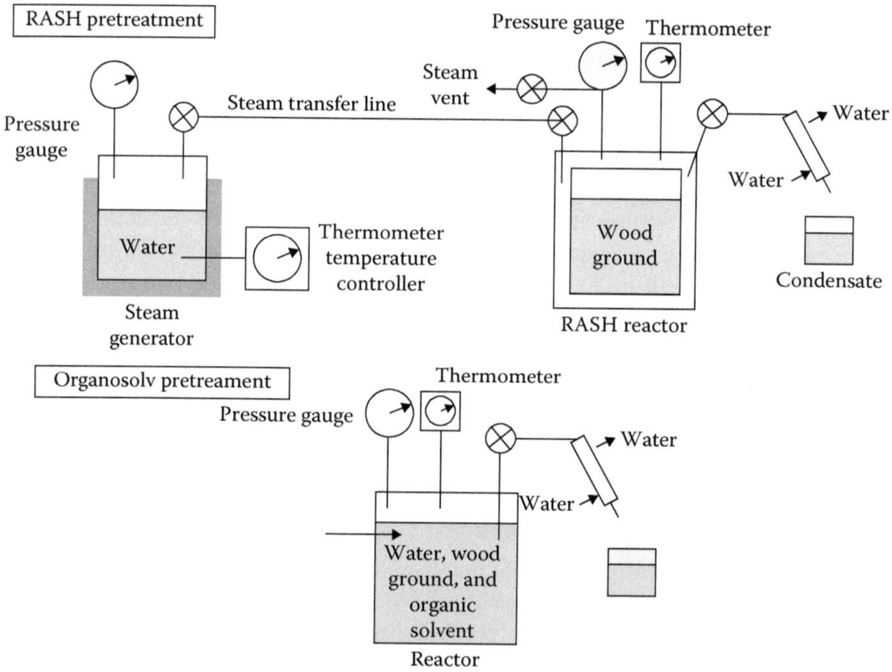

FIGURE 9.6 A combined RASH and organosolv pretreatment scheme. (Adapted from Lee, S. and Shah, Y., *Biofuels and Bioenergy—Processes and Technologies*, CRC Press, Boca Raton, FL, 2012; Rughani, L. and McGinnis, G.D., *Biotechnology and Bioengineering*, 33, 681–686, 1989.)

9.4.1.5 Ionic Liquid Pretreatment

An ionic liquid is a salt composed of anions and cations that are poorly coordinated and that has a melting point below 100°C. Ionic liquids have been demonstrated as very efficient solvents for hydrogenation, esterification, nanomaterial synthesis, biocatalysis, and selective extraction of aromatics [28,29]. The first demonstration of an ionic liquid as a cellulose solvent under relatively mild operating conditions was reported in 2002 by Swatloski [29]. The treatment used a range of anions and 1-butylmethylimidazolium cations; some ionic liquids were able to completely dissolve microcrystalline cellulose and cellulose was recovered through the addition of an antisolvent such as water or ethanol. The most effective cellulose solvents were the ionic liquids that contain chloride anions. An important finding associated with this novel pretreatment method is that enzymes can more efficiently hydrolyze into glucose, an amorphous cellulose produced by ionic liquids, than the microcrystalline cellulose found in lignocellulose naturally [28,30]. More research on this treatment is needed.

9.4.2 Hydrolysis

There are two types of hydrolysis processes for lignocellulose. The old process that has been practiced for a long time is acid or chemical hydrolysis and the new and novel process is enzymatic hydrolysis. Here we briefly examine both of these processes.

9.4.2.1 Acid or Chemical Hydrolysis

Important parameters in acid or chemical hydrolysis are the surface-to-volume ratio of particles, acid concentration, temperature, and time. The surface-to-volume ratio is especially important because it determines the magnitude of yield of glucose. Smaller particles result in better hydrolysis [31]. An increase in the liquid-to-solid ratio also gives a faster reaction. However, higher ratio requires larger equipment and more capital cost. For chemical hydrolysis, a liquid-to-solid ratio of 10/1 seems to be most suitable [31].

The chemical hydrolysis is carried out by first pulverizing lignocellulose or waste into a fine particle size. The powdered waste is mixed with aqueous solution of weak acid (0.2%–10%) at about 180°C–230°C and moderate pressure. The acid solution converts waste into glucose, but the extent of yield depends on the nature of the waste (i.e., Kraft paper will give 84%–86% yield, whereas ground refuse will give 38%–53% yield). The yield will increase with temperature. Generally, 0.5% H_2SO_4 concentration is used.

The Tennessee Valley Authority (TVA) developed a two-stage, low-temperature, atmospheric pressure process that utilizes the separate unit operations to convert hemicellulose and cellulose to sugars [32]. An experimental pilot plant was designed and built in 1984. The process showed a very low level of inhibitor concentration. The results of this study are briefly summarized as follows:

1. The size of ground corn stover of 2.5 cm was adequate for the hydrolysis of hemicellulose.
2. The time required for optimum hydrolysis in 10% acid at 100°C was 2 h.

3. For 1 and 3 h reaction times, the overall xylose yields were 86% and 93%, respectively.
4. Recycle leachate, dilute acid, and prehydrolysis acid solutions were stable during the storage for several days.
5. Vacuum drying was adequate in the acid concentration step.
6. Cellulose hydrolysis by cooking stover containing 66%–78% acid for 6 h at 100°C resulted in 75%–99% cellulose conversion to glucose.
7. Fiberglass-reinforced plastics of vinyl ester resin were used for the construction of process vessels and piping.

More detailed description of the process is described by Lee and Shah [2].

9.4.2.2 Enzymatic Hydrolysis

Cellulose differs from other carbohydrates that are generally used as a substrate for fermentation in that cellulose is insoluble and polymerized as beta-1,4 glycosidic linkages. Each cellulose molecule is an unbranched polymer of 15–10,000 D-glucose units. Hydrolysis of crystalline cellulose is the rate-limiting step in the overall conversion of biomass to ethanol because aqueous enzyme solutions have difficulty acting on insoluble, impermeable highly structured cellulose. Cellulose needs to be efficiently solubilized such that an entry can be made into cellular metabolic pathways. Solubilization is brought about by enzymatic hydrolysis catalyzed by the cellulose system of certain bacteria and fungi. Cellulase is a class of enzyme produced primarily by fungi, bacteria, and protozoans that catalyze the hydrolysis of cellulose commonly known as cellulolysis.

The discussion on enzymatic hydrolysis is broken into three parts: enzyme system, enzyme production and inhibition, and mechanism of cellulose hydrolysis that considers cellulase enzyme adsorption on the substrate.

9.4.2.2.1 Enzyme System

There are mechanistically and structurally different types of cellulases. Each cellulolytic microbial group has an enzyme system unique to it. The capabilities of enzyme can vary from hydrolysis of soluble derivatives of cellulose to disrupting the cellulose complex. Cellulase is actually composed of a number of distinctive enzymes based on the specific types of reactions catalyzed. In fact, cellulase can be characterized into five general groups:

1. Endocellulase cleaves the internal bonds to disrupt the crystalline structure of cellulose and expose individual polysaccharide chains.
2. Exocellulase detaches two or four saccharide units from the ends of exposed chains produced by endocellulase, resulting in the production of disaccharides or tetrasaccharides, such as cellobiose. There are two principal types of exocellulases or cellobiohydrolases (CBHs): (1) CBH-I that works processively from the reducing end and (2) CBH-II that works processively from the nonreducing end of cellulose. Here the processivity implies the ability of enzyme to continue repetitively its catalytic

function without dissociating from its substrate. The chance for reaction is significantly increased by an active enzyme adsorbed onto the surface of the substrate.
3. Beta-glucosidase or cellobiase hydrolyzes the disaccharides and tetrasaccharides into individual monosaccharides.
4. Oxidative cellulase depolymerizes cellulose by the free radical reactions as in the case of a cellobiose dehydrogenase (acceptor), an enzyme that catalyzes the dehydrogenation of cellobiose.
5. Cellulose phosphorylase depolymerizes cellulose using phosphates instead of water.

In most cases, the enzyme complex breaks down cellulose to beta-glucose. This type of cellulose enzyme is produced mainly by symbiotic bacteria. Enzymes that break down hemicellulose are called hemicellulase, which are still classified under cellulases. The principal challenge in building an enzyme system is how to make these different enzymes work together for hydrolytic degradation of biomass.

The enzymes described above can also be classified as progressive (or processive) and nonprogressive (or nonprocessive). Progressive cellulase will continue to interact with a single polysaccharide strand, whereas nonprogressive cellulase will interact once, disengage, and then engage another polysaccharide strand. Based on the enzymatic capability, cellulase enzyme is characterized into two groups: C1 enzyme (or factor) and Cx enzyme (or factor). The C1 factor is regarded as an "affinity" or prehydrolysis factor that transforms crystalline cellulose (i.e., cotton fiber or Avicel) into linear and hydroglucose chains. The C1 factor has very little effect on the soluble derivatives. The Cx (hydrolytic) factor breaks down the linear chains into soluble carbohydrates, usually cellobiose and glucose. Microbes rich in the C1 factor are more useful in the production of glucose from the cellulose. This is generally a rate-controlling step. *Trichoderma reesei* microbes contain the best amount of C1 factor. This is an industrially important fungus that is capable of screening large amounts of cellulases and hemicellulases [33]. The site of action of cellulolytic enzymes is important in the design of Cx factor. If the enzyme is within cell mass, the material to be reacted must diffuse into the cell mass. Therefore, the enzymatic hydrolysis of cellulose usually takes place extracellularly, where enzyme is diffused from the cell mass into the external medium.

Another important factor in the enzymatic reaction is whether the enzyme is adaptive or constitutive. A constitutive enzyme is present in a cell at all times, whereas an adaptive enzyme is only found in the presence of a given substance and the synthesis of enzyme is triggered by an inducing agent. Most fungal cellulases are adaptive [31,34]. Cellobiose is an inducing agent for microbes *T. reesei*. For high concentration (>0.5%–1%), it can also be an inhibitor. In most practical situations, it acts as an inducing agent.

9.4.2.2.2 Enzyme Production and Inhibition

As mentioned earlier, the most useful enzyme system for hydrolysis of cellulose is cellulase. Cellulase is a multicomponent enzyme system consisting of *endo*-beta-1,4-glycanases, *exo*-beta-1,4-glucan glucohydrolases, and *exo*-beta-1,4-glucan cellobiohydrolases. Cellobiose is the dominant product of this system, but it is highly inhibitory to the enzymes and is not usable by most organisms. Cellobiase hydrolyzes cellobiose to glucose, which is much less inhibitory and highly fermentable. Many fungi produce this cellobiase and most of the work that is presently conducted is on *T. reesei* (*viride*). The cellulase produced by *T. reesei* is much less inhibited than other cellulases that have the major advantages for industrial purposes [35].

Cellulases can inhibit competitively [36–41], noncompetitively [39,42–44], or uncompetitively [37]. Uncompetitive inhibition takes place when an enzyme inhibitor binds only to the complex formed between the enzyme and the substrate, whereas noncompetitive inhibition takes place when an enzyme inhibitor and the substrate may both be bound to the enzyme at any given time. For purified cellulose, wheat straw and bagasse, *T. reesei* produced enzyme is competitively inhibited by glucose and cellobiose. However, some enzyme is noncompetitively inhibited by cellobiose using other substrates such as rice straw and Avicel (microcrystalline cellulose). *Trichoderma viride* is uncompetitively inhibited by glucose in a cotton waste substrate [37].

Besides *T. reesei*, other mutants such as Rut C-30 [45] and *Clostridium thermocellum* have also been extensively examined. Cellulases isolated from *C. thermocellum* have high specific activities [46], especially against crystalline forms of cellulose that have proven to be resistant to other cellulase preparations. Low-cost but efficient enzymes for the lignocellulosic ethanol technology is continued to be developed to reduce the operational cost and improve the productivity of the process.

9.4.2.3 Mechanism of Cellulose Hydrolysis

The overall cellulose hydrolysis is based on the synergistic action of three distinct cellulase enzymes and depends on the concentration ratio and the adsorption ratio of the component enzymes: *endo*-beta-gluconases, *exo*-beta-gluconases, and beta-glucosidases. The *endo*-beta-gluconases attack the interior of the cellulose polymer in a random fashion [47], exposing new chain ends. This enzyme is strongly but reversibly adsorbed to the microcrystalline cellulose commonly known as Avicel and catalyzes the solid-phase reaction. The strength of the adsorption is greater at the lower temperatures. This enzyme is necessary for the hydrolysis of crystalline substrates of cellulose, resulting in a considerable accumulation of reducing sugars, mainly cellobiose, because the extracellular cellulase complex does not possess the cellobiose activity. Sugars that contain aldehyde groups that are oxidized to carboxylic acids are classified as reducing sugars.

The *exo*-beta-gluconases remove disaccharide cellobiose units from the nonreducing ends of cellulose chains. The *exo*-beta-gluconases adsorb strongly on both crystalline and amorphous substrates, and carry out the solid-phase reaction. The mechanism of the reaction is complex because there are two distinct forms of both endo- and exoenzymes, each with a different type of synergism with other members of the complex. The concentration of cellobiose in the solution increases as these enzymes continue to split off the cellobiose units. The action of *exo*-beta-gluconases may be severely hampered (or stopped) by the accumulation of cellobiose in the solution.

Beta-glucosidase converts cellobiose to glucose by hydrolysis. In general, glucosidase is any enzyme that catalyzes the hydrolysis of glucoside. Beta-glucosidase catalyzes the hydrolysis of terminal, nonreducing beta-D-glucose residues with the release of beta-D-glucose. Kadam and Demain [48] determined the substrate specificity of the beta-glucosidase and demonstrated that its addition to the cellulase complex enhances the hydrolysis of Avicel, specifically by removing the accumulated cellobiose. They used *C. thermocellum* that is expressed in *Escherichia coli* to determine the surface specificity of the enzyme. The hydrolysis of cellobiose to glucose is a liquid-phase reaction. The action of beta-glucosidase on this reaction can be slowed or halted by the inhibitive action of glucose accumulated in the solution. The accumulation may also induce the entire hydrolysis to a halt as inhibition of the beta-glucosidase results in the buildup of cellobiose, which in turn inhibits the action of exogluconases. Thus, the hydrolysis of the cellulosic materials depends on the presence of all three enzymes in proper amounts. If any of these enzymes is present in the amount less than the required amount, the other enzymes will be inhibited or lack the necessary substrates upon which to act.

While higher temperature increases the rate of hydrolysis, the high temperature can inactivate or destroy the enzyme. To strike a balance between the increased activity and the simultaneous deactivation rate, enzymatic hydrolysis is generally operated at ~40°C–50°C. While enzymatic reactions are carried out at low temperatures, as mentioned earlier, dilute acid hydrolysis is generally carried out at high temperatures (195°C–215°C).

One of the issues that need to be addressed in enzymatic hydrolysis is the loss of enzyme that is left on the lignocellulose residues, on the cellulose substrate, or in the solution. The enzyme adsorption capacity of the lignocellulose residue decreases as the pretreatment temperature is increased, whereas the capacity of cellulose increases with higher temperature. The reduction of enzyme on the residue is essential for the overall economics of the process.

An enzymatic hydrolysis process involving solid lignocellulosic materials can be designed in many ways. Generally, substrate and enzymes are fed into the process, and sugar solution along with the solid residue leaves the process at various points. The enzyme adsorbed on the residue is lost and this hurts the economics of the process. The recycling and reuse of the enzyme adsorbed on the residue is essential. In essence, the enzymatic process should be designed in such a way that the loss of enzymes is minimal.

9.4.3 Fermentation

The hydrolysis and fermentation of cellulose can be carried out in sequence often called as separate hydrolysis and fermentation (SHF) process or simultaneously called as simultaneous saccharification and fermentation (SSF) process. Here we briefly examine both of these processes.

9.4.3.1 Separate Hydrolysis and Fermentation

In the SHF process, hydrolysis and fermentation are carried out in two separate vessels. The most expensive items in the overall process costs are the cost of feedstock, enzyme production, hydrolysis, and utilities. The feedstock and utility costs are high because only about 73% of the cellulose is converted to ethanol in 48 h and the remainder of the cellulose, hemicellulose, and lignin is burned or gasified. Enzyme production is expensive due to a large amount of enzymes that are used in the attempt to overcome the end-product inhibition and the slow reaction rate. The hydrolysis step is also expensive due to the large capital and operating costs. The most important parameters are the hydrolysis section yield, the product quality, and the required enzyme loading, all of which are interrelated. Generally, the process should be operated at the minimum required enzyme loading. Um and Hanley [49] examined the effect of cellulose loading on the performance of the SHF process.

Generally, hydrolysis is carried out at 50°C and fermentation requires a lower temperature (around 30°C). The SHF process accommodates both of these requirements. The fermentation step takes about 48 h.

9.4.3.2 Simultaneous Saccharification and Fermentation

The operating cost of the SSF process is generally lower than that of SHF process as long as the process integration is synergistically done. Since in the SSF process both hydrolysis and fermentation are carried out in the same vessel, yeast ferments glucose to ethanol as soon as the glucose is produced, thus preventing the sugars from accumulating and causing end-product inhibition. Even in the SSF process, cellobiose inhibition occurs to an appreciable extent. The enzyme loading for SSF is only 7 IU/g of cellulose compared to 33 IU/g in SHF. The cost of energy and feedstock is somewhat reduced because of the improved yield and the increased ethanol concentration, which also considerably reduce the cost of distillation and utilities. The cost of the SSF process is slightly less than the combined cost of hydrolysis and fermentation in the SHF process. The longer reaction time for SSF (about seven days) versus two days for hydrolysis and two days for fermentation for SHF is offset by the reactor volume and high ethanol concentration. In earlier studies, fermentation was the rate-limiting step, but with recent advances in recombinant yeast strains that are capable of effectively fermenting both glucose and xylose, the rate-liming step may have changed to hydrolysis.

The hydrolysis is carried out at 37°C and an increase in temperature (up to 50°C) increases the reaction rate. However, in the SSF process, the ceiling temperature is usually limited by the yeast cell viability. The concentration of ethanol is also a limiting factor (a periodic removal of ethanol improves the productivity up to 44%). Recycling the residual solids may also increase the process yield. However, enzyme recycling may be limited by the increase in lignin concentration causing handling difficulties.

Two types of enzyme recycling schemes have been examined: in one scheme, enzymes are recovered in the liquid phase and in the other, enzymes are recovered by recycling unreacted solids [47]. The first scheme works well with the SHF process in which hydrolysis is carried out at higher temperatures (50°C). The increase in temperature allows more enzymes to remain in the liquid phase. At lower temperature, more enzymes are adsorbed on the surface, and therefore, for SSF solids recycling becomes a more effective option.

9.4.3.3 Comparison between SSF and SHF Processes

Due to low end-product inhibition of the cellulase enzyme complex, the SSF systems offer many advantages over the SHF processes. The SSF process shows higher yields (88% vs. 73%) compared to the SHF process [2] and greatly improves the product glucose concentration (10% vs. 4.4%). The most significant advantage of the SSF process is the enzyme loading that is reduced from 33 to 7 IU/g cellulose, which considerably cuts down the cost of ethanol production. A comparative study of the approximate cost of the two processes reported in the literature [47] showed that based on the ethanol selling price from a production capacity of 25,000,000 gallons per year, the SSF process is found to be more cost effective than the SHF process by a factor of 1.49. These estimates may change with new developments on enzymes and yeasts.

A hybrid hydrolysis fermentation (HHF) process may also gain some acceptance. This process will begin with a separate prehydrolysis step and ends with a simultaneous saccharification and fermentation step. In the first step of hydrolysis, higher temperature enzymatic cellular saccharification takes place, whereas in the second stage of the SSF process, mesophilic (moderate temperature) enzymatic hydrolysis and sugar fermentation take place simultaneously. The optimized process scheme may have to change if a specific enzyme that is proven to be highly efficient and cost effective but also found to be intolerant against certain inhibitors that are associated with any of these processing steps.

9.4.3.4 Xylose Fermentation

For certain feedstock such as hardwood and herbaceous biomass, xylose amounts to 30%–60% of fermentable sugars. The efficient fermentation of xylose is therefore very important for the overall economics of ethanol from these feedstock. Co-fermentation of both glucose and xylose is most desirable. Xylose fermentation using pentose yeasts is difficult due to (1) the requirement of O_2 during ethanol production, (2) the acetate toxicity, and (3) the production of xylitol as byproduct. Xylitol is a naturally occurring low-calorie sugar substitute with anticarcinogenic properties. Other approaches to xylose fermentation include conversion of xylose to xylulose using xylose isomerase prior to fermentation by *S. cerevisiae* and the development of genetically engineered strains [50].

A method of integrating xylose fermentation into the overall process is illustrated in Figure 9.7. In this method, dilute acid hydrolysis is adapted as a pretreatment step. The liquid stream is neutralized to remove any mineral or organic acid liberated in the pretreatment process, and is then sent to the xylose fermentation. Water is added before the fermentation, if necessary, so that organisms can make full use of the substrate without having the yield limited by end-product inhibition. The dilute ethanol stream from xylose fermentation is then used to provide the dilution water for the cellulose–lignin mixture entering the SSF process. Thus, the water that enters during the pretreatment process is used in both the xylose fermentation and the SSF process. The conversion of xylose to ethanol using *E. coli* in pH-controlled batch fermentation was investigated [51]. The results showed high concentrations of ethanol (56 g/l) produced from xylose with good efficiencies. Recombinant *E. coli* also gave good conversions of glucose, mannose, arabinose, and galactose to ethanol.

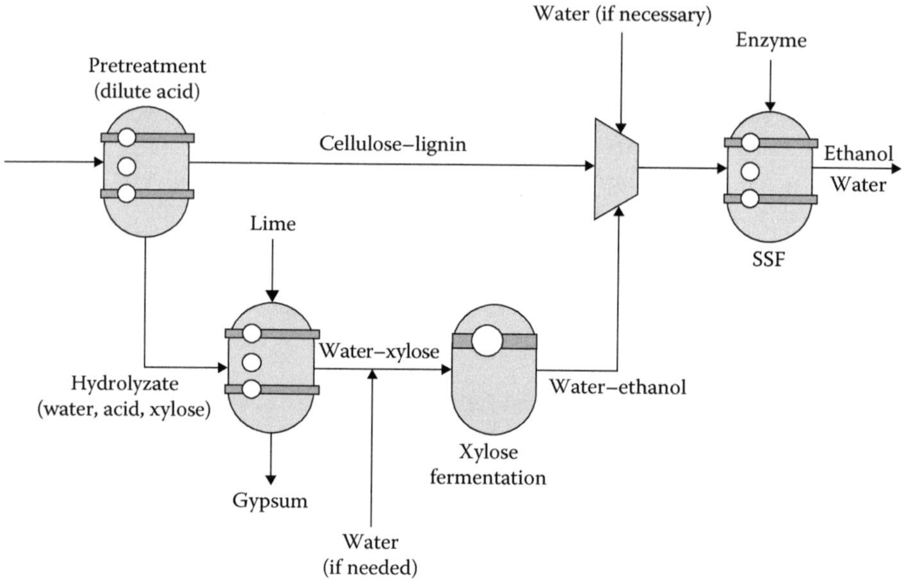

FIGURE 9.7 Integration of xylose fermentation and SSF. (Adapted from Lee, S. and Shah, Y., *Biofuels and Bioenergy—Processes and Technologies*, CRC Press, Boca Raton, FL, 2012; Wright, J.D., *Chemical Engineering Progress*, 84, 62–74, 1988.)

Slower fermentation was observed for pH <6 and addition of metal ions such as calcium, magnesium, and ferrous ions stimulated ethanol production [51].

Xylose fermentation does not require precise temperature control as long as the temperature is between 25°C and 40°C. Higher concentration of xylose slows down the fermentation. Ingram et al. [51–54] showed that *E. coli* of special type can efficiently convert both hexose and pentose sugars to ethanol. Ethanologenic *E. coli* strains require simpler fermentation conditions, produce higher concentration of ethanol, and are more efficient than pentose-fermenting yeasts for ethanol production from xylose and arabinose [55].

Sedlak et al. [56] successfully developed a genetically engineered *Saccharomyces* yeast that can ferment both glucose and xylose simultaneously to ethanol, although xylose was metabolized more slowly than glucose. Ideally, xylose should be consumed simultaneously with glucose at similar efficiency and speed [57]. This new co-fermentation process has a very bright future. They also found that ethanol was the most abundant product from glucose and xylose metabolism, but small amounts of the metabolic byproducts glycerol and xylitol were also obtained [56].

9.4.4 Ethanol Extraction during Fermentation

Significant research for concentration of dilute ethanol product to pure ethanol has been carried out to reduce high energy consumption for purification of dilute end products. Conventional distillation suffers from high energy cost and azeotropic phenomenon that allows only about 95% pure ethanol. In the recent years, along with improved unit operations, liquid–liquid extraction with biocompatible organic solvents, distillation under vacuum, and selective adsorption on the solids have demonstrated the technical feasibility of the extractive fermentation concept. Finally, membrane separation processes that decrease the biocompatibility constraints have been proposed, which include dialysis [58] and reverse osmosis [43].

More recently, the concept of supported liquid membranes has been reported. This method minimizes the amount of organic solvents involved and permits simultaneous realization of the extraction and recovery phases. Enhanced volumetric productivity and high substrate conversion yields have been reported [59] via the use of a porous "Teflon" sheet soaked with isotridecanol as support for the extraction of ethanol during semicontinuous fermentation of *Saccharomyces bayanus*. This selective process results in ethanol purification and combines fermentation, extraction, and re-extraction (stripping). Such a novel process idea can further accomplish maximized alcohol production and energy savings, and reduce the cost in production.

9.4.5 Lignin Conversion

In the United States, about 250 billion pounds per year of lignin is produced largely as a byproduct of paper and pulp industry. Lignins are complex amorphous phenolic polymers that are not sugar based and fermentable. The phenol in lignins may be either a guaiacyl or a syringyl unit. These units are bonded by alpha- or beta-ether linkages. A variety of C–C linkages may also be present and these are less common [2]. The distribution of linkages in lignin is random and highly resistant to chemical, enzymatic, and microbial hydrolysis due to extensive cross-linking. Lignin protects cellulose and needs to be removed to carry out hydrolysis and fermentation of cellulose. Lignin monomer units are similar to gasoline that has high octane number. The removal of oxygen and the breaking down of lignin molecules make it a suitable transportation fuel. Hydrotreating of lignin will produce a mixture of phenolic and hydrocarbon compounds, which can then be converted to methyl aryl ether by reaction with methanol. The conversion of lignin can be carried out by dual function catalysts. Metals such as molybdenum and molybdenum/nickel catalyze deoxygenation and acidic alumina support promote carbon–carbon bond cleavage. Lignin chemicals have applications in drilling muds, binders for animal feeds, base for artificial vanilla, and surfactants for oil recovery [60]. For the last usage, lignosulfonates are blended with tallow amines and conventional sulfonates. Lignin can react with hydrogen or carbon monoxide to form new class of chemicals called lignin phenols. These phenols are soluble in organic solvents but not in water, and they are good candidates for further conversion to produce chemicals that may be useful in enhanced oil recovery.

9.4.6 Coproducts of Cellulosic Ethanol Technology

Potential coproducts for cellulosic ethanol technology include hemicellulose hydrolyzate (xylose), cellulose hydrolyzate (glucose of mixed sugars), cell mass, enzymes, soluble and insoluble lignins, lignin-derived chemicals and fuels, solid residues, and so on. Other valuable coproducts include xylitol (which is sugar alcohol sweetener) and is produced by hydrogenation of xylose (an aldehyde) into a primary alcohol.

9.4.7 Future Directions for Cellulosic Ethanol

While the future for cellulosic ethanol is very bright, the future efforts need to address following issues:

1. While each step, pretreatment, hydrolysis, and fermentation are separate and the needs to further develop their interconnections separately are also very important. For optimization of the cost, an integration of these steps (such as the SSF process described earlier) needs to be further evaluated.
2. More work on the development of enzymes and yeasts that are more tolerant to the product inhibition needs to be carried out. This can be helped by the use of genomics, proteomics, and metabolic engineering techniques for plant systems that are applied to other living systems.
3. From the cost point of view, full use of all parts of plants, namely, cellulose, hemicellulose, and lignin for coproduct development needs to be further considered. The development of the enzyme and yeasts that can simultaneously convert both glucose and xylose needs to be further evaluated. More efficient lignin separation and refining should be further explored.
4. The energy consumption for various unit operations such as distillation and extraction should be further optimized. The transport and storage of biomass feedstock is also an issue that needs to be addressed. Larger-scale operations need to be considered.
5. The work on cellulosic ethanol should be extended to other alcohols, especially butanol, which is discussed in Section 9.5.

9.5 FERMENTATION OF SUGAR TO ISOBUTANOL

Recently, Quereshi et al. [1] presented a review of recent advances in fermentation of isobutanol from various carbohydrates and starch materials. They examined the effectiveness of a number of microbes for the fermentation of various feedstock such as wheat and barley straws, corn stover, switchgrass and dried distillation grains and solubles. Isobutanol is produced in two phases and always found in the mixture of acetone–butanol–ethanol (ABE). Some of their conclusions are outlined as follows:

1. The experiments performed so far gave low productivity due to the toxicity of butanol to the culture.
2. *Clostridium beijerinckii* was found to be the best overall culture followed by *Clostridium actobutylicum* for butanol production.
3. *Escherichia coli* strains and *S. cerevisiae* microbes have also been examined, but they gave low butanol production.
4. Simultaneous removal of butanol while fermentation significantly improved the production rate of butanol (from 1.2 g/l to 461 g/l in batch operation).
5. More butanol-tolerant strains using genetic engineering techniques need to be pursued.

REFERENCES

1. Quereshi, N., Liu, S., and Ezeji, T.C., "Cellulosic butanol production from agricultural biomass residues: Recent advances in technology," in Lee, J. (ed.), *Advanced Biofuels and Bioproducts*. Springer, New York, 247–265 (2012).
2. Lee, S. and Shah, Y., *Biofuels and Bioenergy—Processes and Technologies*. CRC Press, Boca Raton, FL (2012).
3. Lichts, F.O., "Industry statistics: 2010 World fuel ethanol production," Renewable Fuels Association (2011), http://www.ethanolrfa.org/pages/statistics# E.
4. Speight, J.G. and Lee, S., *Handbook of Environmental Technologies*. Taylor & Francis, New York (2000).
5. American Coalition for Ethanol, "All about ethanol" (October 2010), http://www .ethanol.org/.
6. Lee, S., Speight, J.G., and Loyalka, S.K., *Handbook of Alternative Fuel Technologies*. CRC Press, Boca Raton, FL (2007).
7. BBI International, *The Ethanol Plant Development Handbook*, 4th ed. BBI International, Salida, CO (2003).
8. Shapouri, H., Duffield, J.A., McAloon, A., and Wang, M. "The 2001 net energy balance of corn ethanol," Technical Report No. AER-814, US Department of Agriculture, Washington, DC (July 2002).
9. Mueller, S. and Copenhaver, K., "News from corn ethanol: Energy use, coproducts, and land use," *Near-Term Opportunities for Biorefineries Symposium*, October 11–12, Champaign, IL (2010).
10. Corn Refiners Association, "The corn refining process," Vol. 2010 (2010).
11. Li, A., Antizar-Ladislao, B., and Khraisheh, M., "Bioconversion of municipal solid waste to glucose for bio-ethanol production," *Bioprocess and Biosystem Engineering*, 30, 189–196 (2007).
12. Saha, B.C., "Lignocellulose biodegradation and applications in biotechnology," in Saha, B.C. and Hayashi, K. (eds.), *Lignocellulose Biodegradation*, ACS Publication, New York, pp. 2–34 (2004).
13. Dickey, L.C., Kurantz, M.J., and Parris, N., "Oil separation from wet-milled corn germ dispersions by aqueous oil extraction and aqueous enzymatic oil extraction," *Industrial Crops and Products*, 27, 303–307 (2008).
14. Prosonix, "AP-40 starch processing for wet milling," (2011), http://www.prosonix.corn/files/AP-40_Starch_Wet_Milling_20101210.pdf.
15. Singh, V.J. and Eckhoff, S., "Effect of soak time, soak temperature, and lactic acids on germ recovery parameters," *Cereal Chemistry*, 73, 716–720 (1996).
16. ICM, "ICM's dry milling ethanol production."
17. Mueller, S., *Detailed Report: 2008 National Dry Mill Corn Ethanol Survey*. University of Illinois, Chicago, IL (2008), http://ethanolrfa.3cdn.net/2e04acb7ed88d08d21_99m6idfc1.pdf.
18. Odian, G., *Principles of Polymerization*. Wiley, Hoboken, NJ (2004).
19. Dorland, W.A.N., *Dorland's Illustrated Medical Dictionary*, 30th ed. W.B. Saunders, Philadelphia, PA (2003).
20. Knauf, M. and Krau, K., "Specific yeasts developed for modem ethanol production," *Sugar Industry*, 131, 753–775 (2006).
21. Cohen, D., "Form and distribution of trace elements in biomass for power generation," Technical Report No. 48, QCAT Technology Transfer Center, Pullenvale, QLD (July 2004).
22. Eliasson, A., Christensson, C., Wahlbom, C.F., and Hahn-Hagerdal, B., "Anaerobic xylose fermentation by recombinant *Saccharomyces cerevisiae* carrying XYL1, XYL2, and XKS1 in mineral medium chemostat cultures," *Applied and Environmental Microbiology*, 66, 3381–3386 (2000).

23. Cort, J.B., Pschorn P., and Stromberg, B., "Minimize scale-up risk," *Chemical Engineering Progress*, 106, 3–49 (2010).
24. Dekker, R.F.H. and Wallis, A.F.A., "Enzymic saccharification of sugarcane bagasse pretreated by autohydrolysis-steam explosion," *Biotechnology and Bioengineering*, 25, 3027–3048 (1983).
25. Ojumu, T.V. and Ogunkunle, O.A., "Production of glucose from lignocellulosics under extremely low acid and high temperature in batch process—Autohydrolysis approach," *Journal of Applied Sciences*, 5, 15–17 (2005).
26. Pan, X.J., Xie, D., Yu, R.W., Lam, D., and Saddler, J.N., "Pretreatment of lodgepole pine killed by mountain pine beetle using the ethanol organosolv process: Fractionation and process optimization," *Industrial & Engineering Chemistry Research*, 46, 2609–2617 (2007).
27. Rughani, L. and McGinnis, G.D., "Combined rapid steam hydrolysis and organosolv pretreatment of mixed southern hardwoods," *Biotechnology and Bioengineering*, 33, 681–686 (1989).
28. Simmons, B.A., Singh, S., Holmes B.M., and Blanch, H.W., "Ionic liquid pre-treatment," *Chemical Engineering Progress*, 106, 50–55 (2010).
29. Swatloski, R.P., "Dissolution of cellulose with ionic liquids," *Journal of the American Chemical Society*, 124, 4974–4975 (2002).
30. Dadi, A.P., "Mitigation of cellulose recalcitrance to enzymatic hydrolysis by ionic liquid pretreatment," *Applied Biochemistry and Biotechnology*, 137, 407–421 (2007).
31. Diaz, L.F., Savage, G.M., and Golueke, C.G., "Critical review of energy recovery from solid wastes," *Critical Reviews in Environmental Control*, 14, 285–288 (1984).
32. Farina, G.E., Barrier, J.W., and Forsythe, M.L., "Fuel alcohol production from agricultural lignocellulosic feedstocks," *Energy Sources*, 10, 231–237 (1988).
33. Kumar, R., Singh, S., and Singh, O.V., "Bioconversion of lignocellulosic bio-mass: Biochemical and molecular perspectives," *Journal of Industrial Microbiology and Biotechnology*, 35, 377–391 (2008).
34. Bailey, J.E. and Ollis, O.F., *Biochemical Engineering Fundamentals*, 2nd ed. McGraw-Hill, New York (1986).
35. Holtzapple, M.T., Cognata, M., Shu, Y., and Hendrickson, C., "Inhibition of *Trichoderma reesei* cellulase by sugars and solvents," *Biotechnology and Bioengineering*, 38, 296–303 (1991).
36. Blotkamp, P.J., Takagi, M., Pemberton, M.S., and Emert, G.H., "Biochemical engineering: Renewable sources of energy and chemical feedstocks," in Nystrom, J.M. and Barnett, S.M. (eds.), *Renewable Sources of Energy and Chemical Feedstocks*, AIChE Symposium Series No. 181, New York (1978).
37. Beltrame, P.L., Carniti, P., Focher, B., Marzetti, A., and Sarto, V., "Enzymatic hydrolysis of cellulosic materials: A kinetic study," *Biotechnology and Bioengineering*, 26, 1233–1238 (1984).
38. Ohmine, K., Ooshima, H., and Harano, Y., "Kinetic study on enzymatic hydrolysis of cellulose by cellulase from *Trichoderma viride*," *Biotechnology and Bioengineering*, 25, 2041–2053 (1983).
39. Okazaki, M. and Young, M., "Kinetics of enzymatic hydrolysis of cellulose: Analytic description of mechanistic model," *Biotechnology and Bioengineering*, 20, 637–663 (1978).
40. Ryu, D.Y. and Lee, S.B., "Enzymatic hydrolysis of cellulose: Determination of kinetic parameters," *Chemical Engineering Communications*, 45, 119–134 (1986).
41. Gonzales, G., Caminal, G., de Mas, C., and Santin, J.L., "Kinetic models for pretreated wheat straw saccharification by cellulose," *Journal of Chemical Technology* and *Biotechnology*, 44 (4), 275–288 (1989).
42. Vallander, L. and Erikkson, K., "Enzymatic hydrolysis of lignocellulosic materials: I. Models for the hydrolysis process—A theoretical study," *Biotechnology and Bioengineering*, 38, 135–138 (1991).

43. Garcia, A., Lannotti, E.L., and Fischer, J.L., "Butanol fermentation liquor production and separation by reverse osmosis," *Biotechnology and Bioengineering*, 28, 785–791 (1986).
44. Vallander, L. and Erikkson, K., "Enzymatic hydrolysis of lignocellulosic materials: II. Experimental investigation of theoretical hydrolysis process models for an increased enzyme recovery," *Biotechnology and Bioengineering*, 38, 139–144 (1991).
45. Szengyel, Z., Zacchi, G., Varga, A., and Reczey, K., "Cellulase production of *Trichoderma reesei* Rut C 30 using steam-pretreated spruce. Hydrolytic potential of cellulases on different substrates," *Applied Biochemistry and Biotechnology*, 84–86, 679–691 (2000).
46. Moses, V., Springham, D.G., and Cape, R.E., *Biotechnology—The Science and the Business.* Harwood Academic, London (1991).
47. Wright, J.D., "Ethanol from biomass by enzymatic hydrolysis," *Chemical Engineering Progress*, 84, 62–74 (1988).
48. Kadam, S. and Demain, A., "Addition of cloned beta-glucosidase enhances the degradation of crystalline cellulose by the *Clostridium thermocellum* cellulase complex," *Biochemical and Biophysical Research Communication*, 161, 706–711 (1989).
49. Um, B.H. and Hanley, T.R. "High-solid enzymatic hydrolysis and fermentation of solka floc into ethanol," *Journal of Microbiology and Biotechnology*, 18, 1257–1265 (2008).
50. Sarthy, A., McConaughy, L., Lobo, Z., Sundstorm, A., Furlong E., and Hall, B., "Expression of the *Escherichia coli* xylose isomerase gene in *Saccharomyces cerevisiae*," *Applied and Environmental Microbiology*, 53, 1996–2000 (1987).
51. Beall, D.S., Ohta, K., and Ingram, L.O., "Parametric studies of ethanol production from xylose and other sugars by recombinant *Escherichia coli*," *Biotechnology and Bioengineering*, 38, 296–303 (1991).
52. Ohta, K., Beall, D.S., Mejia, J.P., Shanmugam, K.T., and Ingram, L.O., "Genetic improvement of *Escherichia coli* for ethanol production: Chromosomal integration of *Zymomonas mobilis* genes encoding pyruvate decarboxylase and alcohol dehydrogenase II," *Applied and Environmental Microbiology*, 57, 893–900 (1991).
53. Ingram, L.O. and Conway, T., "Expression of different levels of ethanologenic enzymes from *Zymomonas mobilis* in recombinant strains of *Escherichia coli*," *Applied and Environmental Microbiology*, 54, 404 (1988).
54. Alterthum, F. and Ingram, L.O., "Efficient ethanol production from glucose, lactose, and xylose by recombinant *Escherichia coli*," *Applied and Environmental Microbiology*, 55, 1943–1948 (1989).
55. Skoog, K. and Hahn-Hagerdal, B., "Xylose fermentation," *Enzyme and Microbial Technology*, 10, 66–80 (1988).
56. Sedlak, M., Edenberg, H.J., and Ho, N.W.Y., "DNA microarray analysis of the expression of the genes encoding the major enzymes in ethanol production during glucose and xylose co-fermentation by metabolically engineered *Saccharomyces* yeast," *Enzyme and Microbial Technology*, 33, 19–28 (2003).
57. US DOE Office of Science and Office of Energy Efficiency and Renewable Energy, "Breaking the biological barriers to cellulosic ethanol: A joint research agenda," A Research Roadmap Resulting from the Biomass to Biofuels Workshop, Technical Report DOE/SC-0095, December 7–9, Rockville, MD (2006).
58. Kyung, K.H. and Gerhardt, P., "Continuous production of ethanol by yeast immobilized in membrane-contained fermenter," *Biotechnology and Bioengineering*, 26, 252 (1984).
59. Christen, P., Minier, M., and Renon, H., "Enhanced extraction by supported liquid membrane during fermentation," *Biotechnology and Bioengineering*, 36, 116–123 (1990).
60. Naee, D.G., *200th National Meeting of the American Chemical Society*, August, 17 (1990).
61. U.S. Grains council, World Corn Production and Trade (October, 2011). http://www.grains.org/corn.

10 Fuel Production by Supercritical Water

10.1 INTRODUCTION

In recent years, the interest in the use of supercritical water (SCW) for the production of fuels and chemicals as well as for waste treatment has been rapidly expanding. The main reason for this is the unique properties of SCW that allow a variety of organic reactions to occur in SCW, where water not only plays a benign role of solvent but also plays a role as an active reactant or a catalyst. Water under these conditions possesses properties such that important organic reactions can be carried out in a homogeneous medium [1–13] (Aljishi et al., 2010, pers. comm.). SCW can provide five different functions: (1) a medium in which numerous types of organic chemical synthesis occur, (2) a medium for partial or complete oxidation of numerous hazardous or nonhazardous materials, (3) a medium in which complex materials decompose and produce liquids and gases, (4) a medium for thermal or catalytic gasification of simple and complex materials to produce fuels like methane and hydrogen, and (5) a medium to generate hydrogen by catalytic reforming of various carbonaceous materials. This chapter examines the role of SCW in each of these functions, with a special emphasis on the functions that generate synthetic fuels.

SCW technologies offer many advantages [1–13] (Aljishi et al., 2010, pers. comm.):

1. The energy efficiency for SCW gasification of biomass is generally high, particularly for the feedstock containing large water content, because no drying is required.
2. Most organic materials of biomass and other carbonaceous feedstock can be dissolved in SCW due to their high solubility in SCW and high dielectric constant of SCW. These features make the gasification in SCW a homogeneous reaction, with no mass transfer resistance between the two phases.
3. While the SCW requires high pressure of 22.1 MPa and high temperature of 374°C, these conditions are still milder than what is required for conventional gasification and pyrolysis to obtain the same level of conversion efficiency. For example, conventional steam gasification generally requires 1000°C, whereas the complete gasification of glucose can be achieved at 650°C and 35.4 MPa pressure in SCW.
4. SCW gasification produces very little impurities; no NO_x and SO_x and low CO concentration are generated. The use of catalyst to enhance water–gas shift reaction further reduces the gas-phase impurities.

5. High pressure used in the supercritical gasification helps downstream operations such as storage and transportation of the product gases, carbon capture, and purification of the product gases by reforming or pressure swing adsorption.
6. As shown by Savage and others [1–13] (Aljishi et al., 2010, pers. comm.), SCW provides a homogeneous medium to carry out numerous organic chemical reactions such as decomposition, partial and complete oxidation, hydration/dehydration, hydrogenation/dehydrogenation, hydrolysis, elimination and rearrangement, and C–C bond formation with ease in which water acts as a benign medium, a reactant, or a catalyst.
7. With the use of a suitable catalyst, SCW gasification can be easily accompanied by a reforming reaction.

The major disadvantages deal with the operational issues such as the use of high-pressure water, which may carry some toxic and corrosive substances. The processing of supercritical operations may require the use of special materials that may be expensive and demand substantial maintenance and replacements costs. The capital and operating costs associated with high-pressure operations may be considerably larger than those for low-pressure gasification and pyrolysis operations. In recent years, however, the prices of high-pressure equipment have come down.

10.2 PROPERTIES OF SCW

Hydrothermal treatment of carbonaceous materials in supercritical conditions has taken a significant momentum ever since the pioneering work of Modell and coworkers from the Massachusetts Institute of Technology (MIT) in the late 1970s [1–14] (Aljishi et al., 2010, pers. comm.). Figure 10.1 illustrates the thermodynamic region (in terms of pressure–temperature diagram) of SCW treatment of the carbonaceous materials. The three regions shown in the figure take advantage of substantial changes in the properties of water that occur in the vicinity of its critical point at 374°C (T_c) and 22 MPa (P_c). The behavior of the important properties of water such as density, ion dissociation constant, dielectric constant, and solubility limits of various salts as a function of temperature was described in Chapter 5 and will not be repeated here [1–5]. In that chapter, we examined the role of water as a chemical reactant under subcritical conditions. In this chapter, we focus on the role of SCW for carrying out various organic chemical reactions. In SCW, more chemically and energetically favorable pathways to gaseous and liquid fuels can be achieved by better control of the rate of hydrolysis and phase partitioning and solubility of components.

Water at ambient conditions (25°C and 0.1 MPa) is a good solvent for electrolytes due to its high dielectric constant [1–10] (Aljishi et al., 2010, pers. comm.), whereas most organic matter are sparingly soluble [1–10] (Aljishi et al., 2010, pers. comm.). As water is heated, the H-bonding starts weakening, allowing dissociation of water into acidic hydronium ions (H_3O^+) and basic hydroxide ions (OH^-). The structure of water changes significantly near the critical point because of the breakage of infinite networks of hydrogen bonds and water exists as separate clusters with a chain structure. In fact, the dielectric constant of water decreases considerably near the critical point,

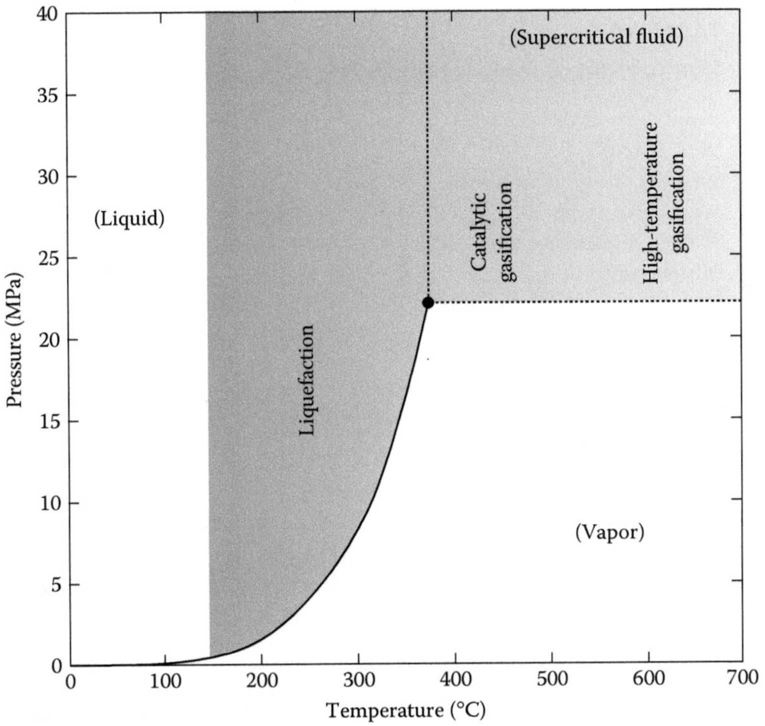

FIGURE 10.1 Hydrothermal processing regions referenced to the pressure–temperature phase diagram of water. (After Peterson, A., Vogel, F., Lachance, R., Frolling, M., Antal, M., and Tester, J., *Energy & Environmental Science*, 1, 32–65, 2008.)

which causes a change in the dynamic viscosity and an increase in the self-diffusion coefficient of water [1–10] (Aljishi et al., 2010, pers. comm.).

SCW has liquid-like density and gas-like transport properties and behaves very differently than water at room temperature. For example, it is highly nonpolar, permitting complete solubilization of most organic compounds. The resulting single-phase mixture does not have many of the conventional transport limitations that are encountered in multiphase reactors. However, the polar species such as inorganic salts, are no longer soluble and they start precipitating. The physical properties of water, such as viscosity, density, and heat capacity, also change dramatically in the supercritical region. A small change in the temperature or pressure, results in a substantial increase in the rates of chemical reactions.

It is important to mention that the dielectric behavior of 200°C water is similar to that of ambient methanol, 300°C water is similar to ambient acetone, 370°C water is similar to methylene chloride, and 500°C water is similar to ambient hexane [1–10] (Aljishi et al., 2010, pers. comm.). In addition to the unusual dielectric behavior, as shown in Table 10.1 the transport properties of water are significantly different than those of ambient water.

Supercritical water also offers some interesting possibilities for catalytic processes. Supercritical water can dissolve unwarranted hydrocarbons from the

TABLE 10.1
Comparison of Ambient and SCW

	Ambient Water	SCW
Dielectric constant	78	<5
Solubility of organic compounds	Very low	Fully miscible
Solubility of oxygen	6 ppm	Fully miscible
Solubility of inorganic compounds	Very high	~0
Diffusivity (cm^2/s)	10^{-5}	10^{-3}
Viscosity (g cm/s)	10^{-2}	10^{-4}
Density (g/cm)	1	0.2–0.9

Source: Lee, S. and Shah, Y., *Biofuels and Bioenergy—Processes and Technologies.* CRC Press, New York, 2012. With permission.

catalytic surface. Supercritical water has better capacity to handle heat due to high heat capacity. The adsorption/desorption phenomena can be better handled in supercritical water due to higher solubility of absorbing/describing species. The oligomeric coke precursors or sulfur species can be easily dissolved in the supercritical water.

As mentioned earlier, the number and persistence of hydrogen bonds under supercritical conditions are both diminished. The dissociation constant for water at supercritical conditions is about 3 orders of magnitude higher than it is for ambient liquid water. This constant, however, decreases as temperature increases further in supercritical conditions. SCW is an excellent solvent for all organic compounds. It can also have higher H^+ OH^- ion concentrations than liquid water under certain conditions. Thus, it becomes an effective medium for acid- and base-catalyzed reactions of organic compounds. In fact, the dissociation constant at supercritical conditions generates such high H^+ concentrations that organic compounds can undergo acid-catalyzed reactions without addition of acid. Gases are also miscible in SCW, thus creating a homogeneous medium for any multiphase reaction. Since there are no interphase mass and heat-transfer resistances, higher concentration of reactants is obtained in a supercritical medium, leading to higher reaction rates.

Recently, Savage [11], Watanabe et al. [12], Matsumura et al. [13], and Ding et al. [15], among others, have shown that SCW provides an excellent medium for chemical synthesis, decomposition, and/or partial or total oxidation of organic materials and compounds. They have shown that a broad range of chemical transformations can be affected in the SCW medium. These transformations include C–C bond formation, dehydration, decarboxylation, hydrodehalogenation, partial oxidation, and hydrolysis. The rates and selectivities of these different reactions can be manipulated by judicious selection of temperature, pH, catalyst, and water density, which controls the functional group transformations in SCW. Catalyst role in SCW can be subtle and may involve participation of water molecules in transition states for elementary reactions.

10.3 ROLE OF SCW IN CHEMICAL SYNTHESIS

Due to the unique properties possessed by SCW in which numerous types of organic reactions can be carried out with ease, this medium has been widely exploited for a variety of chemical synthesis [11–13,15–28]. Parsons [16], Katritzky et al. [17], An et al. [18], Leif and Simoneit [19], and Savage [11] provide good reviews of the types of synthetic organic chemistry that can be carried out in SCW. Savage [11] provides a brief account of the types of chemical synthesis that are possible in SCW. These include the following:

1. Hydrogenation/dehydrogenation reactions using transition metal complexes
2. C–C bond formation reaction such as Friedel–Crafts alkylation reactions
3. Rearrangement reactions such as formation of ketones by rearrangement of pinacol and two different bicyclic diols
4. Hydration and dehydration reactions such as conversion of alcohols to olefins (e.g., conversion of *tert*-butyl alcohol to isobutylene)
5. Elimination reaction such as facile decarboxylation of carboxylic acid
6. Hydrolysis such as conversion of esters to carboxylic acids and alcohols
7. Partial oxidation such as conversion of methane to oxygenates or higher hydrocarbons
8. H–D exchange such as substitute of H by D in alpha positions of ketone carboxyl groups

Savage [11] gives numerous examples of these different types of chemical synthesis. He also points out that future research should be more focused on the use of SCW to carry out these and other novel chemical synthesis. While not all chemical synthesis are targeted toward synthetic fuels, many such as hydrogenation/dehydrogenation, C–C bond formation, hydration/dehydration, hydrolysis, and partial oxidation play important role in the generations of synthetic fuels or various important additives to the synthetic fuels. This subject will be under intense future research investigation.

Some details of the specific examples quoted by Savage [11] as they relate to fuels are worth noting. As an example of C–C bond formation, both phenol and *p*-cresol can be successfully alkylated with *tert*-butyl alcohol and 2-propanol at 275°C in the absence of any added acid catalyst to produce sterically hindered phenols [18]. Water in these alkylation reactions serves as both catalyst and reactant. Xu and Antal [21,22] were successful in converting *tert*-butyl alcohol to isobutylene by dehydration reaction. In the absence of an added acid, hydronium ions formed by the dissociation of water are the primary catalytic agents. The dehydration of other alcohols such as cyclohexanol, 2-methylcyclohexanol, and 2-phenylethanol to form alkenes is also reported [23–25]. Esters can undergo an autocatalytic hydrolysis to form carboxylic acids and alcohols [17,18]. Partial oxidation of methane in SCW at 400°C to form methanol has been explored with both homogeneous free radical reactions [26,27] and heterogeneous catalytic reactions [28]. High selectivities for oxygenates, but very low methane conversions, have been obtained. More research to synthesize fuel components or fuel additives in SCW continues to be pursued.

Savage [11] also presented an excellent review of some other organic reactions in SCW. These reactions include decomposition of complex materials, individual hydrocarbons, nitrogen-containing compounds, sulfur-containing compounds, oxygen-containing compounds, compounds with two heteroatoms, and chlorine- and fluorine-containing compounds. He noted that the rates and selectivity of these and other reactions can be manipulated by judicious selection of temperature, pH, catalyst, and water density; one can thus control the functional group transformation in SCW.

10.4 OXIDATION IN SCW

Catalytic oxidation that has been used for many wastewater treatment, wet air oxidation, and photolysis is now being used for oxidation of organic compounds in SCW medium [12,14,15,29–72]. As mentioned earlier, water in supercritical conditions behaves like many organic solvents, and it is miscible with these solvents. Thus, SCW provides a homogeneous, benign, and nontoxic environment for many organic reactions in the presence or absence of a catalyst.

Oxidation in SCW (SCWO) is a rapidly developing technology for the destruction of organic wastes [34–39]. Hazardous organic pollutants can be destroyed by SCWO at temperatures around 500°C in less than 1 min [34–47]. The world's first commercial SCWO facility for treating industrial wastewater became operational in 1995 [48,49]. In order to increase process capacity and handle more stubborn refractory compounds and condensation products with an ease, catalytic oxidation in supercritical conditions has become more important. The SCW allows maximum concentration driving forces for the reaction because there are no interfacial mass or heat-transfer resistances to hinder the reaction rate.

In 2000, General Atomics was selected by DOE's hydrogen program to carry out SCW partial oxidation (SWPO) of biomass, municipal solid waste (MSW), and high sulfur coal to generate hydrogen. SWPO carries out oxidative reactions in the SCW environment akin to high-pressure steam in the presence of substoichiometric oxygen or air. SWPO forms more hydrogen and less char and tar than the similar operation in the subcritical conditions. Furthermore, SWPO eliminates the formations of particulates NO_x, SO_x, and hazardous air pollutants. High-density aqueous environment is also ideal for reacting and gasifying organics. The high density also allows utilization of compact equipment that minimizes capital cost and the plant footprint requirements.

SCW has density one-tenth of the liquid water and solubility behavior that of high-pressure steam, hydrogen bonding in SCW is totally disrupted, and polarity and many thermal properties are such that they facilitate mass and heat-transfer operations along with many different types of chemical reactions. The effectiveness of SCWO has been demonstrated at the laboratory and pilot scale on hundreds of feedstock, which include sewage sludge; coal slurry; pig manure; various biomass slurries including pulp mill sludge, pulverized wood with ground plastic, rubber, and charcoal; fermentation waste; ground cereal; highly refractory hazardous wastes such as hexachloro-benzene; and many more [29]. Maximum gaseous hydrogen yield that can be obtained can vary to as high as 26.1 g H_2/100 g dry feed for ethanol and 42.9 g H_2/100 g dry feed for polyethylene to as low as 13.7 g H_2/100 g dry feed for cornstarch. Some of the practical results obtained in SCW conditions are described by Johanson et al. [30] and Hong and Spritzer [29]. One of the earliest patents on processing methods for the oxidation of organics in SCW was published by Modell [14].

In recent years, more efforts have been made to find (1) suitable catalysts to carry out SCWO most efficiently, (2) novel reactor designs to obtain clean syngas through oxidation, and (3) novel approaches to convert methane to methanol in economically viable way under supercritical conditions. Numerous compounds such as alcohols, acetic acid, ammonia, benzene, benzoic acid, phenol, pyridine, quinolone, MEK (methylethyl ketone), and dichlorobenzene have been catalytically oxidized in SCW [15]. Special applications have been targeted to various aromatic and aliphatic organic compounds, inorganic compounds, and various wastewaters and sludges. The most notable catalysts used for these purposes are oxides of copper, zinc, vanadium, manganese, as well as noble metal such as platinum. Additional data are reported by Savage et al. [26,53,54], Savage [11], Subramaniam and McHugh [9], Thomason et al. [59], and Tester et al. [48]. Various mechanisms for oxidation reactions are outlined by Ding et al. [15] and Savage [11].

A two-stage approach to SCWO has also been investigated. In the first stage, contaminated waste is exposed to hydrothermal carbonization or liquefaction to extract harmful substances (such as chlorinated and other toxic components) from waste. Biocoal, biocrude, or biochar produced from this first stage then undergo oxidation and reforming in SCW to decompose organic compounds and generate syngas containing hydrogen, carbon dioxide, carbon monoxide, and may be some lower hydrocarbons depending on the temperature of the gasification and the nature of the catalyst. Some practical examples of multistage operations are outlined by Brunner [10].

The most extensive and critical review of oxidation of methanol in SCW was carried out by Vogel et al. [31]. This study is very important for treating aqueous effluents containing methanol by SCWO (an exothermic reaction) and for performing hydrothermal reforming under autothermal (i.e., in the presence of partial oxidation) conditions. They critically evaluated all existing literature data and concluded that there are important differences in the reported kinetics of methanol oxidation. The factors responsible for these differences are (1) the methanol feed concentration, (2) insufficient reaction heat removal from tubular or coiled flow reactors, and (3) inherent difference in apparent kinetics of autocatalytic reactions in continuous stirred-tank reactor (CSTR) and in plug flow reactor (PFR) due to recirculation of radicals (i.e., mixing effect) in a CSTR. The study indicated that the best kinetic data for methanol SCWO cannot be recommended because of lack of information on (1) induction time and (2) influence of wall catalysis on the apparent reaction rate.

Watanabe et al. [41,42] showed that NaOH and ZrO_2 have catalytic effects for partial oxidation of *n*-hexadecane and lignin in SCW. The experiments were carried out at 400°C. For both compounds, ZrO_2 catalyst gave hydrogen yield twice higher than those obtained without the catalyst. With NaOH, the yield increase was four times. Both catalysts enhanced the decomposition of aldehyde and ketone intermediates into CO. For lignin, both catalysts enhanced the decomposition of carbonyl compounds, which in turn inhibited the char formation and promoted the formations of CO and H_2.

10.4.1 Catalysts for SCWO

Ding et al. [15] showed that it is possible to develop effective catalysts for SCWO applications. These catalysts can be used either to enhance oxidation rates of organic compounds or to increase destruction of refractory products. The catalyst can be designed to increase the selectivity of certain products. Because of a wide variation in the nature of aqueous wastes, understanding the unique characteristics of SCW and its effect on the catalyst surface, reaction activity, and mechanisms, and the knowledge of preferred crystalline phases of metal oxides is essential for the development and design of an effective SCWO catalytic system. Oxides of Ce, Co, Fe, Mn, Ti, and Zn may be used as catalysts, and their supports can be selected from the oxides of Al, Hf, Zr, and Ti. These supports have been found to be stable in SCWO environments. The additives that can increase the physical strength or stabilize the activity of a catalyst may be an oxide of Bi, Cd, Ga, Ir, K, Mo, Ta, or W. An effective SCWO catalyst must have large surface area and be able to withstand larger surface area changes.

Catalyst activity and stability is affected by the preparation methods. Traditionally, catalysts are produced by coprecipitation, impregnation (coating), fused alloy, fused metal oxide, and crystal growth processes [62–72]. Coprecipitation and impregnation are two of the most popular methods for the preparation of metal and metal oxide catalysts [17,20,62–72]. Many commercial oxidation catalysts are prepared by coating noble metals on metal oxide supports to modify catalyst surface structure and active sites that can result in the increase in catalyst activity and stability. While the physical conditions of these catalysts are adequate for the gas-phase oxidation, they may not be completely suitable for the SCW conditions. Since transition metal oxide catalysts are major components of ceramics, the common methods of ceramic preparation such as sol-gel, coprecipitation, polymeric sponge, and high-temperature aerosol decomposition methods have been adapted for the preparation of metal oxide catalysts. The structure and properties of ceramic catalysts depend on the process parameters such as solvents, pH, temperature, and aging time. Numerous reported studies have evaluated these effects [62–72]. In the final analysis, preparation method must be chosen that gives the desired activity, selectivity, stability, and prepares catalyst that can handle refractory materials and possible poisons in the waste feed.

10.5 DECOMPOSITION AND EXTRACTION OF MATERIALS BY SCW

SCW is a good extracting and decomposition agent for many complex organic materials [73–118] (Kim and Mitchell, 2012, pers. comm.; Swanson et al., 2012, pers. comm.). This application generally produces useful liquids that can be either a fuel or raw materials for various downstream chemicals. Feedstock normally used for the liquid productions are coal, polymeric materials, rubber tires, cellulose among others, or mixtures of them.

The extraction of coals with SCW is a promising route for the production of liquid fuels and chemical feedstock from coal. Deshpande et al. [73] obtained high conversion for extraction of a German Brown coal and a Bruceton bituminous coal by SCW at 375°C and 23 MPa. They reported conversions of 70%–79% for the brown coal and about 58% for the bituminous coal. Pauliatis et al. [74] reported 35% conversion and only 10% liquid yield for North Dakota lignite at 400°C and 28 MPa pressure. Deshpande et al. [73] also obtained low liquid yield with lignite coal with high sodium content. Other studies report low conversion for bituminous coal, particularly when solvent density is low [75–80,82–84] (Swanson et al., 2012, pers. comm.). Kershaw and Bagnell [78] showed that at 380°C and 22 MPa, the conversions of Australian brown coals were considerably higher for supercritical extraction of water than with toluene. The reverse was, however, true for black coals. In general, they found SCW extraction was more effective for low-rank coals than high-rank bituminous coals. The extraction by water was also more dependent on pressure presumably due to solvent density effect. The hydroxyl concentration of liquid yield by SCW extraction was higher than that obtained in the liquid produced by toluene extraction.

Swanson et al. (2012, pers. comm.) showed that for low-rank coals, the conversion and extract yields increased with increasing temperature and pressure. The conversion also decreased with increasing coal rank and correlated well with the percent volatile matter in the coals. The study also indicated that SCW extracts the volatile hydrogen-rich fraction of the coal. The extract was found to be highly polar in nature, with significant quantities of phenols and long-chain aliphatic fatty acids.

Numerous other studies have also addressed the behavior of coal, shale oil, biomass and mixtures of coal and biomass, polymers, rubber, algal oil, lignin, residual oil, and so on under SCW conditions [77–115] (Kim and Mitchell, 2012, pers. comm.; Swanson et al., 2012, pers. comm.). Three typical studies illustrating the coal decomposition in SCW are reported by Nonaka [89], Nonaka et al. [97], Li and Eglebor [106], Vostrikov et al. [84], and Cheng et al. [83]. These studies showed that as the temperature of SCW increases, more gas and less liquid are produced.

SCW has also been explored as a medium for the degradation of waste synthetic polymers [107–116]. Rubber tires were converted to a 44% oil yield by reaction in SCW at 400°C. When polystyrene-based ion exchange resins were subjected to SCW at 380°C for 1 h [107–116], less than 5% polymer decomposed and the products included styrene and several oxygenated arenes such as acetophenone and benzaldehyde. SCW is also used to extract oil and oil precursors from oil shale [98,100–105]. The process involved C–C bond cleavages, and in the presence of CO, higher hydrocarbon yields were obtained than those obtained in conventional pyrolytic treatment. Holliday et al. [115] showed that water near its critical point is a good medium for the hydrolysis of triglyceride-based vegetable oils into their fatty acid constituents.

A number of studies examined the decomposition of mixed feedstock under SCW conditions [89,95–99]. Veski et al. [98] examined the decomposition of a mixture of kukersite oil shale and pinewood and showed improved liquid and gas yields at 380°C temperature. The mixture indicated a synergistic effect and showed the product to be 1.5–2.0 times better than what would be predicted based on simple additive yields.

The liquid product was richer in heterocompounds including polar ones compared to that predicted from simple additive effects. Kim and Mitchell (2012, pers. com.) examined the decomposition of coal–biomass mixture. The results show that at 647.3 K and 220.9 atm pressure, small polar and nonpolar organic compounds released from the mixture were completely miscible with SCW. The hydrolysis of large organic molecules in SCW resulted in high concentrations of H_2, CO, CO_2, and low-molecular-weight hydrocarbons with very little tar, soot, and PAH (polyaromatic hydrocarbons) formation. Sulfur, nitrogen, and many trace elements in coals were oxidized to form insoluble salts in SCW. There were no gaseous emissions, and all products were dissolved in the SCW. The salts can be precipitated from fluid mixture and removed along with ash. Matsumara et al. [90] examined co-liquefaction of coal and cellulose in SCW at 673 K and 25 MPa. The coal used was Ishikari coal. Unlike the results of synergy reported by Veski et al. [98], in this study no synergy between coal and cellulose conversion was found. Simple additive method for each compound product distribution worked well for this system. More discussion on synergistic effects in mixture decomposition has been recently discussed by Lee and Shah [117].

Sunphorka et al. [96] examined co-liquefaction of coal and plastic mixtures containing high-density polyethylene, low-density polyethylene, polystyrene, and polypropylene. The experiments were performed in the temperature range of 450°C–480°C, 40–70 wt% plastic mixtures, and a water/feedstock ratio of 2 to 10. During co-liquefaction, all experimental variables had effects on liquid yield, but temperature did not have a significant effect on the conversion. Long residue in the oil product decreased with increasing temperature while it increased with increasing water/feedstock ratio. For the plastic mixture alone, only temperature had a significant effect on the oil yield. Maximum conversion and liquid yield of 99% and 66%, respectively, were obtained. Onsari et al. [95] examined co-liquefaction of lignite coal and rubber tires in the temperature range of 380°C–440°C and water/feedstock ratio of 4/1 to 10/1 by weight. Variable tire concentration was examined. The maximum conversion and oil yield were obtained at 400°C, 1 min residence time, water/feedstock ratio of 10% and 80% tire concentration. The co-liquefaction of coal and tire yielded a synergistically increased level of oil production. Moreover, the total conversion level with co-liquefaction was almost equal to that obtained in the presence of either Fe_2O_3 or Ni/Mo catalysts under the same conditions. The study concluded that SCW is a good medium for the dissolution of the volatile matter from a coal and used tire matrix.

Mitsubishi Materials Corp. [87] with the project support of Petroleum Energy Center, Japan, developed a thermal process that used SCW to crack vacuum distillation residue (VR) oil into clean lighter oil products. The final volume of solid waste generated was below 5%. The process was carried out in two stages in the same reactor. At the bottom of the reactor, heavy VR components (pitch) are decomposed into lighter components using 5% SCW at temperatures 400°C–450°C and pressures 200–250 atm. In the upper part of the reactor, lighter components are cracked at a slightly higher temperature with SCW and hydrogen to form lighter products. Untreated pitch was withdrawn at the bottom and sent to a reformer where it is partially oxidized at SCW at 1000°C to form hydrogen gas and soot. This hydrogen stream is passed onto the upper section of the cracking reactor. Overall, the process converted 70% of VR into lighter products, which included 15% gas, 7% liquefied petroleum gas (LPG), 11% naphtha,

13% light oil, 24% vacuum gas oil, 21% carbon dioxide and 1% soot, and 8% heavy oil. The process was proven in a test plant of size 1 bbl/day.

The conversion of glycerol in SCW was examined by May et al. [118]. They studied the conversion of glycerol in the temperature range of 510°C–550°C, 350 atm pressure in a bed of inert nonporous ZrO_2 particles as well as in a bed of 1% Ru/ZrO_2 catalyst for the residence time of 2–10 s. The feed solution contained 5 wt% glycerol. The experiments in the absence of a catalyst resulted in the formation of liquid products such as acetaldehyde, acetic acid, hydroxyacetone, allyl alcohol, propionaldehyde, acrolein, and acrylic acid, and gases such as H_2, CO and CO_2, and methane. The catalyst enhanced the formation of acetic acid and inhibited the formation of acrolein. In the catalytic experiments, the main products formed were hydrogen and carbon dioxide with little methane and ethylene. Complete glycerol conversion occurred at 510°C in 8.5 s and at 550°C in 5 s in the presence of the catalyst. This, however, did not result in complete gasification; some acetic acid and acetaldehyde were still present. At high residence times, methanol and acetaldehyde were formed. The hydrogen yield was only 50% of what is achievable by stoichiometry due to lack of high activity of the catalyst. A simplified reaction pathway for glycerol conversion in supercritical conditions is illustrated in Figure 10.2 [118].

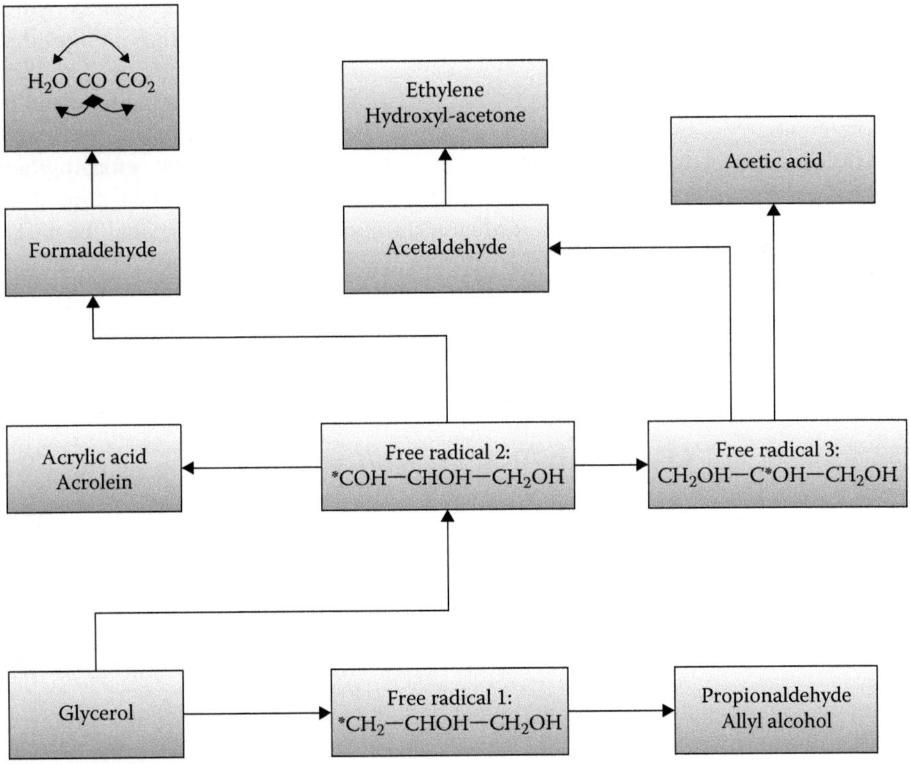

FIGURE 10.2 Simplified reaction pathways for hydrothermal transformation of glycerol in SCW. (Modified from May, A., Salvado, J., Torras, C., and Montane, D., *Chemical Engineering Journal*, 160, 751–759, 2010.)

All the studies described earlier indicate that near-critical conditions, complex carbonaceous materials tend to decompose into a mixture of liquids and gases. The amount of each phase depends on the nature of feedstock, pressure, and reaction time. The use of a suitable catalyst increases both liquid and gas yields. An increase in temperature generally produces more gas. In a case of a mixture, the synergistic effects between the decompositions of two components depend on the nature of the components.

10.6 GASIFICATION IN SCW

While steam gasification occurs at low pressure and high temperature, in recent years gasification of biomass in a pressurized water environment (called hydrothermal gasification) has gained significant support [83,84,119–153] (Antal and Xu, 2012, pers. comm.; Boukis, 2012, pers. comm.; Kruse, 2012, pers. comm.; Veriansyah et al., 2012, pers. comm.). The hydrothermal gasification can be divided into three regions depending on the range of temperature. Osada et al. [119] identified region 1 as the one with a temperature range of 500°C–700°C—a region in which biomass is decomposed in SCW in the presence of either activated carbon to avoid the formation of char or an alkali catalyst to facilitate water–gas shift reaction. In this region, very little solids remained and the main product of the gasification is hydrogen. In region 2 of SCW, where the temperature range is 374°C–500°C, biomass hydrolyzes and metal catalyst facilitates gasification. Here once again, the main product is hydrogen with some carbon dioxide, carbon monoxide, and methane. The third region was described in Chapter 5. Near the critical conditions, methane would be a preferential gas in the absence of a catalyst. However, at high temperature and/or in the presence of a suitable catalyst, hydrogen can be formed by reforming and water–gas shift reactions. The nature of the product will depend on the nature of the feedstock, temperature, pressure, feed concentration, residence time, and the nature of the catalyst (if any). The reported studies for SCW gasification of complex and simple materials are briefly described here.

The main steam gasification reactions under the SCW environment can be listed as follows:

$$C + H_2O \rightleftarrows CO + H_2 \quad \Delta H = 132\,\text{kJ/mol} \tag{10.1}$$

$$CO + H_2O \rightleftarrows CO_2 + H_2 \quad \Delta H = -41\,\text{kJ/mol} \tag{10.2}$$

$$CO + 3H_2 \rightleftarrows CH_4 + H_2O \quad \Delta H = -206\,\text{kJ/mol} \tag{10.3}$$

$$C + 2H_2O \rightleftarrows CO_2 + 2H_2 \quad \Delta H = 91\,\text{kJ/mol} \tag{10.4}$$

$$C + 2H_2 \rightleftarrows CH_4 \quad \Delta H = -87.4\,\text{kJ/mol} \tag{10.5}$$

$$C + CO_2 \rightleftarrows 2CO \quad \Delta H = 159.7\,\text{kJ/mol} \tag{10.6}$$

$$C + O_2 \rightleftarrows CO_2 \quad \Delta H = -405.9\,\text{kJ/mol} \tag{10.7}$$

Reactions 10.1 and 10.6 are important for gasification and are endothermic. The overall process is also endothermic. Reaction 10.7 is needed to provide the heat for autothermal conditions.

Li et al. [116] investigated coal gasification in the temperature range of 650°C–800°C and pressure 23–27 MPa, K_2CO_3 and Raney Ni as catalysts, and H_2O_2 as oxidant. Most experiments were performed with inlet slurry containing 16.5 wt% coal and 1.5 wt% CMC (carboxy methyl cellulose). The results showed that high temperature favors the gasification of coal in SCW, whereas pressure has a little effect on the gasification results. An optimum flow rate needs to be found to get the best results. Both gasification and carbon gasification efficiencies were improved by the catalysts; K_2CO_3 performed better than Raney Ni. Less char and tar were formed in the presence of catalysts. An increase in feed concentration decreased the hydrogen and gasification efficiencies. SCW desulfurizes the coal and the solid particles remained had less carbon and hydrogen than original coal. The data of Li et al. [116] indicate that for the entire range they studied, 90% of the gas-phase concentration was for hydrogen (60%) and carbon dioxide (30%).

Vostrikov et al. [84] examined coal gasification in the temperature range of 500°C–750°C, pressure of 30 MPa, and reaction time of 60–720 s with and without CO_2. Once again, the main gaseous products were CH_4, CO, CO_2, and H_2. Within the range of operating conditions examined, best carbon conversion was obtained at 750°C. The results show a significant temperature dependence on product compositions for temperatures below 650°C. BTX (benzene, toluene, and xylene), methane, and carbon dioxide were the main products below 650°C–700°C. Similar results were obtained by Cheng et al. [83] who studied gasification of lignite coals in the temperature range of 350°C–550°C and reaction time of 0–60 min in N_2 atmosphere. These data along with the data described earlier clearly indicate that product distribution during coal gasification in SCW will depend on the nature of coal along with all the operating parameters.

Battelle Pacific Northwest National Laboratory demonstrated that various alkali carbonate and Ni catalysts can convert wet biomass to methane-rich gas at temperatures between 400°C and 450°C and pressure as high as 34.5 MPa. Yu et al. [120] found that glucose at low concentration (0.1 M) can be completely gasified in 20 s at 600°C and 34.5 MPa, with major products being hydrogen and carbon dioxide. Higher concentration of glucose, however, reduces the product concentration of hydrogen and carbon dioxide and increases the concentration of methane. Xu et al. [129] showed that a wide range of carbons effectively catalyze the gasification of glucose in SCW at 600°C and 34.5 MPa pressure, with nearly 100% carbon gasification efficiency. The available surface area of carbon did not affect the effectiveness of the catalyst. For concentrated organic feeds in water, in the presence of a catalyst, the temperature above 600°C is needed to achieve high gasification efficiencies. Mass transfer resistances at high concentration (if any) can affect the equilibrium of water–gas shift reaction. In the presence of coconut shell, activated carbon, cellobiose, and various whole biomass feeds, as well as depithed bagasse liquid extract and sewage sludge were completely gasified. There was some deactivation of carbon catalyst after 4–6 h of operation.

Demirbas [2,6,131] examined the decomposition of olive husk, cotton cocoon shell, and tea waste by water under both sub- and supercritical conditions. He also observed an increase in hydrogen production with temperature, particularly for temperatures higher than the supercritical temperature. Demirbas [131] observed that as temperature increased from 600°C to 800°C, hydrogen production increased from 53 to 73 vol% in reaction time of 2–6 s. She indicated that hydrogen productions can be obtained from biomass such as bio-nutshell, olive husk, tea waste, crop straw, black liquor, MSW, crop grain residue, pulp and paper waste, petroleum-based plastic waste, and manure slurry.

An extensive amount of work on SCW gasification of organic wastes has been reported in the literature [129,130,139]. The studies have shown that the gasification generally produces hydrogen and carbon dioxide mixture with simultaneous decontamination of wastes, particularly at higher temperatures. The homogeneous solution of waste and water makes it easy to pump to the high-pressure reactor without pretreatment. Guo et al. [130] presented an excellent review of SCW gasification of biomass and organic wastes. They as well as Lu et al. [133] showed the equilibrium effects of temperature, pressure, and feed concentration of wood sawdust on hydrogen, carbon dioxide, carbon monoxide, and methane concentrations in SCW. The typical effects of temperature on product gas composition are illustrated in Table 10.2. The data showed that equilibrium favors the productions of hydrogen and carbon dioxide at high temperatures. The study also showed that an increase in pressure significantly decreased the product concentration of carbon monoxide and slightly decreased the product concentration of the hydrogen. The pressure change had very little effect on the product concentrations of carbon dioxide and methane. The complex effect of pressure on the product distribution was believed to be due to the complex interplay between hydrolysis and water–gas shift reactions. Besides temperature and pressure, other parameters that affected the gas yield were feedstock concentration, oxidant, reaction time, feedstock composition, inorganic impurities in the feedstock, and biomass particle size. Guo et al. [131] also concluded that alkali such as NaOH, KOH, Na_2CO_3, K_2CO_3, and Ca $(OH)_2$; activated carbon; metal oxides and metals such as noble metal catalysts (Ru/alpha-alumina > Ru/carbon > Rh/carbon > Pt/alpha-alumina, Pd/carbon, Pd/alpha-alumina); as well as Ni catalysts and metal oxides such as CeO_2 particles, nano-CeO_2, and nano-$(CeZr)_xO_2$ enhanced

TABLE 10.2
Equilibrium Gas Yield for 5 wt% Sawdust in SCW at 25 MPa Pressure

Temperature (°C)	Gas Yield (mol/kg)				
	Hydrogen	Carbon Dioxide	Methane	Carbon Monoxide	Methane/Hydrogen
400	13	24	20	10^{-3}	1.54
500	40	31	10	2.5×10^{-3}	0.25
600	80	40	~1	3.1×10^{-3}	0.0125
700	89	43	0	1.2×10^{-3}	0.0
800	89	43	0	0.5×10^{-3}	0.0

Source: Guo, L., Cao, C., and Lu, Y., "Supercritical water gasification of biomass and organic wastes," in Momba, M. and Bux, F. (eds.), *Biomass*, 165–182, 2010. With permission.

Note: These data are the best estimates from the graphical data presented.

the reactivity of biomass gasification in SCW. The last two are important for the reforming under supercritical conditions. These and other studies found that the yields of H_2O and CO increased with increasing water density. Yields of H_2 were 4 times better with NaOH and 1.5 times better with ZrO_2 compared to the reaction without a catalyst. Supercritical fluids gave increased pore accessibility, enhanced catalyst ability to coking, and increased desired product selectivity. While high-temperature SCW gasification produces hydrogen and carbon dioxide, Sinag et al. [152] showed that a combination of two technologies—SCW and hydropyrolysis on glucose in the presence of K_2CO_3—produces phenols, furfurals, organic acids, aldehydes, and gases. Xu and Antal [20,21], Antal and Xu (2012, pers. comm.), and Antal et al. [144] studied gasification of 7.69 wt% digested sewage sludge in SCW and obtained gas that largely contained H_2, CO_2, a smaller amount of CH_4, and a trace of CO. Other waste materials show a similar behavior.

Kong et al. [151] briefly summarized the reported work for the catalytic hydrothermal gasification of various types of biomass in SCW. They showed that in the literature, catalytic hydrothermal gasification in SCW has been examined for glucose, organic wastewater, cellulose, soft and hard wood, grass, lignin, sawdust, rice straw, alkylphenols, corn, potato starch gels, potato waste, glycerol, cellobiose, bagasse, sewage sludge, catchetol, vaniline, glycine, and many others. In all cases, the major products were hydrogen and methane depending on the operating conditions. The catalysts examined included Ni, Ru, Rh, Pd, Pt on alumina, NaOH, KOH, Na_2CO_3, K_2CO_3, ZrO_2, activated carbon, and Ni on carbon. The preference was given to the disposable or cheap catalysts or to the reforming catalysts if the objective was to carry out reforming along with gasification. The results show that except at low temperatures, the main product in all cases was hydrogen. Catalytic operations decrease the productions of char and tar and increase the production of hydrogen. Carbon and base catalysts play important roles in the increased gas yields and hydrogen production. Tanksale et al. [7] provided an extensive review of various catalytic and other processes to produce hydrogen from biomass. Supercritical gasification in water was one of these processes. Azadi and Farnood [5] reviewed heterogeneous catalysts for subcritical water and SCW gasification of biomass and wastes. The review provided an extensive information of carbon conversion and hydrogen and methane productions in sub- and supercritical conditions for a variety of biomass by various commercially available and laboratory-made catalysts that included supported and skeletal metal catalysts, activated carbon, metal wires, and other innovative catalysts.

The generation of hydrogen from waste has long-term and strategic implications since hydrogen is the purest form of energy and is very useful for product upgrading, fuel cell, and many other applications. Hydrogen can be produced from waste via numerous high-temperature technologies such as conventional or fast pyrolysis (e.g., olive husk, tea waste, crop straw, etc.), high-temperature or steam gasification (e.g., bio-nutshell, black liquor, wood waste, etc.), supercritical fluid extraction (e.g., swine manure, orange peel waste, crop grain residue, petroleum-based plastic waste, etc.), SCW gasification (e.g., all types of organic waste, agricultural and forestry waste, etc.) as well as low-temperature technologies such as anaerobic digestion and fermentation (e.g., manure slurry, agricultural residue, MSW, tofu wastewater, starch of food waste, etc.). For high-temperature technologies, SCW gasification generates more hydrogen at a lower temperature than pyrolysis or gasification [2,3,6,83,91] (Kim and Mitchell, 2012, pers. comm.). SCW gasification also does not require drying, sizing, and other methods of feed preparations, thereby costing less for the overall process. The temperature of the pyrolysis and gasification process can be reduced if the gases coming out of these processes are further steam reformed. This, however, adds to the overall cost. The rates for the low-temperature processes such as anaerobic digestion and fermentation can be enhanced with the use of suitable microbes and enzymes. The development of future hydrogen economy will require further research in the improvement of these technologies.

Biomass generally contains three important components: cellulose, hemicellulose, and lignin. Both cellulose and hemicellulose (collectively called homocellulose) are easy to hydrolyze, decompose, dehydrogenate, decarboxylate, and reformed as shown by numerous studies mentioned earlier. Lignin component is generally toughest to convert. Yamaguchi et al. [123,138] studied lignin gasification in SCW. They indicated that lignin gasification involves three steps: (1) lignin decomposition to alkylphenols and formaldehyde in SCW, (2) gasification of alkylphenols and formaldehyde over a catalyst, and (3) formation of char from formaldehyde. They showed that SCW gasification is a promising technique to reduce the lignin gasification temperature. They also studied lignin gasification with three different catalysts at 400°C—$RuCl_3/C$, $Ru(NO)(NO_3)_3/C$, and $RuCl_3/C$—and found that the order of gasification activity was $Ru/C = Ru(NO)(NO_3)_3/C > RuCl_3/C$. Extended x-ray absorption fine structure (EXAFS) analysis showed that during lignin gasification in SCW, ruthenium particle sizes in $Ru(NO)(NO_3)_3/C$ and Ru/C catalysts were smaller than that in the $RuCl_3/C$ catalyst. The study concluded that the ruthenium catalysts with smaller particle size of metal particles were more active for the lignin gasification.

Lignin is one of the major fractions of woody biomass that is a polymer of aromatic compounds such as coniferyl alcohol, sinapyl alcohol and, coumaryl alcohol, and it constitutes about 30 wt% and 40% of energy of woody biomass. Yamaguchi et al. [123,138] examined the effects of various noble and transition metal catalysts and titania and activated carbon supports on lignin conversion and hydrogen production rates. The results showed that for the lignin gasification, the activity order followed ruthenium > rhodium > platinum > palladium > nickel, whereas the hydrogen production rate followed the order palladium > ruthenium > platinum > rhodium > nickel. Both titania and activated carbon provided stable support. Hydrogen production rate from lignin increased with temperature and shorter residence time.

Byrd et al. [124] examined a two-stage process to obtain clean fuels from switchgrass. In the first stage, subcritical hydrothermal liquefaction of switchgrass was carried out to obtain biocrude that did not contain some of the inorganic and other undesirable elements. In the second stage, catalytic gasification of biocrude in SCW was carried out to obtain clean syngas dominant in hydrogen. Biocrude contained many oxygenated hydrocarbons of varying molecular structure and weights, including lignin-derived products and sugars and their decomposition products. The supercritical gasification of biocrude was carried out at 600°C and 250 atm pressure. Nickel, cobalt, and ruthenium catalysts were prepared on titania, zirconia, and magnesium aluminum spinel supports. Magnesium aluminum spinel structure did not work. Over time, zirconia-supported catalyst plugged the reactor, although Ni/ZrO_2

catalyst gave the best hydrogen production. Titania-supported catalysts gave lower hydrogen conversions but did not plug the reactor over time. All support materials suffered surface area loss due to sintering.

Glycerol ($HOCH_2$–CHOH–CH_2OH) is obtained as a by-product from biodiesel manufacturing by transesterification of vegetable oils. Nine grams of biodiesel generates approximately 1 g of glycerol. With increasing production of biodiesel, glycerol production will rise, and it can be used for food, oral and personal care, tobacco, polymers, pharmaceuticals, and replacements of petroleum feedstock. Kersten et al. [153] have reported gasification results for glycerol and other model compounds in a variety of catalytic and noncatalytic reactors in SCW and found that without addition of a catalyst, only very dilute concentrations of model biomass feeds could be completely gasified. The density of SCW is higher than that of steam, resulting in a higher space time yield. Higher thermal conductivity and specific heat were helpful in carrying out the endothermic reforming reactions. The formation of char and tar was also minimized because of the solubility of hydrocarbons in SCW. Importantly, hydrogen produced from SCW reforming was produced at high pressure, which can be stored directly, thus avoiding large expenses associated with compression.

The above-described studies and many others lead to some general conclusions [83,84,119–153] (Antal and Xu, 2012, pers. comm.; Boukis, 2012, pers. comm.; Kruse, 2012, pers. comm.; Veriansyah et al., 2012, pers. comm.). As the temperature increases above the critical temperature, more gases are generally produced from most carbonaceous materials. At lower temperatures, for higher feedstock concentration, and in the absence of a catalyst, the gas production rate tends to be lower and contain more methane. At high temperature, for lower feedstock concentration, and in the presence of an effective catalyst, hydrogen production rate rapidly increases. Higher temperature and the presence of a catalyst promote reforming of gas and favor reverse water–gas shift reaction, thus producing more hydrogen and carbon dioxide. Pressure also affects the equilibrium of water–gas shift reaction. Higher pressure favors methane formation as opposed to hydrogen production.

10.7 REFORMING IN SCW

SCW is an ideal medium to carry out reforming reactions for both single components and complex materials [154–176] (Barendregt 2012, pers. comm.; Cremers et al., 2012, pers. comm.; Veriansyah et al., 2012, pers. comm.). Besides all the positive features of the supercritical medium outlined earlier, SCW provides possibilities of lower temperature, lesser coking issues, and more active and stable catalytic reforming process. In Sections 10.7.1 through 10.7.7, we briefly assess important reported literature on the subject.

10.7.1 Liquid Fuels

Lee et al. [154] showed that reforming of JP-8 fuel and diesel fuel can be carried out in SCW in the absence of a catalyst. High enthalpy level of SCW and high solubilities of fuel in SCW allowed the reforming reactions to occur in the temperature range of 650°C–825°C and 220–330 atm pressure. The study examined the productions of hydrogen and methane as functions of reactor operating conditions and the possibility of autothermal operation by simultaneously carrying out partial oxidation reaction with reforming reaction. The process handled fuel with sulfur. The results were obtained at temperatures lower than conventional reforming temperature. The autothermal operation was achieved by adding oxygen into the reacting mixture. In a noncatalytic operation, hydrogen production of 14% of theoretical maximum was obtained. Cremers et al. (2012, pers. comm.) studied SCW reforming of logistic diesel fuel at 550°C in the absence of a catalyst and obtained significant hydrogen production.

Veriansyah et al. (2012, pers. comm.) examined reforming of gasoline in SCW. They used methanol and isooctane (2,2,4-trimethylpentane) as model compounds for gasoline for experimental and simulation studies. The study presented the following conclusions:

1. SCW reforming of hydrocarbons offers a possible way to convert hydrocarbons to hydrogen at a lower temperature. It does not require a steam reforming catalyst, although nickel in reactor wall can act as a catalyst. It avoids the poisoning and deactivation problems associated with the catalyst.
2. The reactor is much compact compared to conventional steam reforming reactors. It is scalable and the reaction time is in seconds. SCW provides dual functions—excellent reactant and homogeneous medium.
3. As reaction temperature, initial feed concentration, and residence time increase, hydrogen, carbon dioxide, and methane productions increase while carbon monoxide and ethane yields remain stable. At high temperature, methane yield is higher than hydrogen yield because at high temperature methanation reaction is favored. In order to increase the hydrogen yield, methanation reaction needs to be suppressed. High inlet feed temperature decreases yields of hydrogen, carbon monoxide, and carbon dioxide and increases the yields of methane and ethane. High inlet temperature also forms coke in the feed line, which may plug the inlet pipes.

Numerous other studies have also examined catalytic reforming of various hydrocarbons in SCW [122,156,162,163,171,172] (Barendregt 2012, pers. comm.). Shekhawat et al. [171] studied catalytic reforming of liquid hydrocarbon fuels for fuel cell applications. They concluded that supercritical reforming of hydrocarbons occurs at lower temperatures than those required in conventional industrial reforming process. They also showed that hydrogen yield increases by using commercial catalysts even if they are not optimized for these conditions. Acetone and diesel fuel produced black liquor and plugged the reactors. Pinkwart et al. [122] showed that under SCW, *n*-decane can be converted to hydrogen-rich gas. They also showed that reforming of diesel oil by four different commercial reforming catalysts can be carried out at a lower temperature than the conventional steam reforming process. The lower temperature also caused lower production of coke during reforming reaction. Ramasamy and T-Raissi [163] studied hydrogen production during reforming of lube oil in supercritical water. They also examined the role of Ni, carbon, and alkali catalysts on the hydrogen production. Very little catalyst deactivation was observed under supercritical conditions.

10.7.2 Biomass

A number of investigators have looked at glucose as a model for biomass reforming under SCW. The pertinent reaction in this case is

$$C_6H_{12}O_6 + 6H_2O \rightarrow 6CO_2 + 12H_2 \tag{10.8}$$

Generally, hydrogen yield is smaller than predicted from the above equation because varying amounts of methane are produced depending on the reaction conditions. Kruse ([132]; 2012, pers. comm.), Kruse and Gawlik [141], and Kruse and Henningsen [142] gave a simplified reaction mechanism for cellulose reforming. Since glucose (and fructose) is the main product of hydrolysis of cellulose, their reaction mechanism also applies to glucose. The reforming of glucose was accelerated by alkali catalysts such as K_2CO_3 and $KHCO_3$. Both of these catalysts increased the hydrogen production and decreased coke formation. For biomass with low salt content and high protein content, these catalysts can increase the hydrogen yield.

Antal and Xu (2012, pers. comm.) and Antal et al. [144] showed the effectiveness of SCW reforming for the production of hydrogen for numerous different types of biomass such as wood sawdust, cornstarch gel, digested sewage sludge, glycerol, glycerol/methanol mixture, poplar wood sawdust, potato starch gels, and potato waste. Once again, higher temperature and catalysts gave better hydrogen productions. The final product distribution did depend on the nature of the feedstock. Similar results were obtained by Boukis et al. (2012, pers. comm.) for biomass slurries and sludges. They also showed an improved heat exchange scheme in "VERENA" German pilot plant for these processes. The VERENA pilot facility successfully demonstrated high carbon and energy efficiency for the SCW reforming of ethanol and corn silage in the temperature range of 540°C–600°C for at least 10 h. On average, the hydrogen concentration in the product for these biomass was about 77 vol%. Zhang et al. [134] examined the SCW reforming of glucose solution (50–200 g/l), a simulated aqueous organic waste (composed of glucose, acetic acid, and guaiacol), and a real aqueous organic waste stream generated from a sludge hydrothermal liquefaction process. The experiments were performed using two different types of catalysts—0.1 RuNi/gamma-Al_2O_3 and 0.1 RuNi/activated carbon catalysts (10 wt% Ni with a Ru-to-Ni molar ratio of 0.1). While the first catalyst was very effective with glucose solutions and simulated aqueous organic waste giving hydrogen yield of 53.9 mol/kg dried feedstock at 750°C, 24 MPa, and weight hourly space velocity (WHSV) of 6 h^{-1}, it was not effective in resisting the alkali and nitrogen compounds in the real waste. The second catalyst supported on active carbon exhibited higher stability.

10.7.3 Glycerol

Reforming of glycerol for hydrogen production can be summarized by the following reactions [118,129,140,148,155,166].

First, the steam reforming of glycerol can be expressed as

$$C_3H_8O_3 \rightarrow 3CO + 4H_2 \tag{10.9}$$

followed by the water–gas shift reaction

$$CO + H_2O \rightarrow CO_2 + H_2 \tag{10.10}$$

The desired overall reaction is then summarized as

$$C_3H_8O_3 + 3H_2O \rightarrow 7H_2 + 3CO_2 \tag{10.11}$$

Some hydrogen is also lost via the methanation of CO and CO_2:

$$CO + 3H_2 \rightarrow CH_4 + H_2O \tag{10.12}$$

$$CO_2 + 4H_2 \rightarrow CH_4 + 2H_2O \tag{10.13}$$

As a result, the product stream is a mixture of the above gases. Furthermore, the yield of hydrogen depends on several process variables such as system pressure, temperature, and water-to-glycerol feed ratio.

Most recently, Knoef [140] studied the reforming of glycerol over Ru/Al_2O_3 catalyst in SCW conditions at a temperature range of 700°C–800°C, feed concentration up to 40 wt%, and reaction time less than 5 s. Under these conditions, glycerol was completely gasified to hydrogen, carbon dioxide, and methane along with a small amount of carbon monoxide. Xu and Antal [21,22], Antal and Xu (2012, pers. comm.), Xu et al. [129], and Antal et al. [144] showed that even in the absence of a catalyst glycerol decomposes in SCW to a hydrogen-rich gas, with almost no CO after 44 s at 600°C and 34.5 MPa. Higher temperature, more active reforming catalyst, and longer residence time result in higher gas and hydrogen productions.

10.7.4 Ethylene Glycol

de Vlieger et al. [164] studied catalytic reforming of ethylene glycol (5 and 15 wt%) in SCW at 450°C and 250 atm pressure. The results were obtained for Pt, Ir, and Ni containing mono- and bimetallic catalysts. The best catalyst was found to be Pt–Ni/Al_2O_3 having a metal loading of 1.5 wt% (Pt:Ni molar ratio of 1:1). With this catalyst, high hydrogen and carbon dioxide yields (selectivity of around 80%) were obtained by suppressing methanation reaction. The addition of Ni prevented sintering of Pt particles, thereby providing a stable performance by bimetallic catalysts. Ethylene glycol also produced more CH_4 and CO than what was produced in methanol reforming.

10.7.5 Methanol

Numerous studies have reported methanol reforming in SCW to produce hydrogen [158–160,170]. Compared to water that has a critical pressure of 22.1 MPa, a critical temperature of 374°C, and a critical density of 320 kg/m^3, methanol has a lower critical temperature of 239°C, a critical pressure of 8.1 MPa, and a critical density of 270 kg/m^3. Thus, reaction of methanol in SCW also implied that methanol is also under supercritical conditions. Methanol reforming can be described by five chemical reactions:

$$CH_3OH \rightleftarrows CO + 2H_2 \quad \Delta H^0_{298} = 91.7 \text{ kJ/mol} \tag{10.14}$$

$$CO + H_2O \rightarrow CO_2 + H_2 \quad \Delta H^0_{298} = -41 \text{ kJ/mol} \tag{10.15}$$

$$CH_3OH + H_2O \rightleftarrows CO_2 + 3H_2 \quad \Delta H^0_{298} = 50.7 \text{ kJ/mol} \tag{10.16}$$

$$CO + 3H_2 \rightleftarrows CH_4 + H_2O \quad \Delta H^0_{298} = -211 \text{ kJ/mol} \tag{10.17}$$

$$CO_2 + 4H_2 \rightleftarrows CH_4 + 2H_2O \quad \Delta H^0_{298} = -223 \text{ kJ/mol} \tag{10.18}$$

While both methanation reactions and water–gas shift reaction are exothermic, main methanol reforming reaction is endothermic and is favored at higher temperatures. Boukis et al. [158] showed that for reaction time as low as 4 s, at temperature of 600°C, and pressure of 25–45 MPa, high conversion rate of methanol can be obtained. The reaction can occur at temperature as low as 400°C. The heavy metal of the inner surface of Inconel 625 can influence the conversion and the product composition of the reforming reaction. Boukis et al. [158] examined the feed concentration from 5 to 64 wt% methanol. Methanol conversion up to 99.9% can be obtained in the absence of a catalyst. The major product is hydrogen (up to 70%–80%) with small amounts (<20%–30%) of carbon dioxide, carbon monoxide, and methane. An increase in temperature increases methanol conversion, decreases CO concentration, and increases CO_2 concentration in the product. Complete methanol conversion at 600°C is achieved [158].

Taylor et al. [159] also examined reforming of methanol under SCW conditions in the temperature range of 550°C–700°C and at 27.6 MPa in an Inconel 625 reactor. They also reported a product rich in hydrogen and low in CH_4 and near the equilibrium ratio of CO and CO_2. A comparison of the product gas composition with equilibrium predictions indicated that the reaction occurs in two steps. First methanol decomposes to CO and H_2 and subsequently CO is converted to CO_2 by water–gas shift reaction. Higher steam-to-carbon ratios gave lower CO in the product gas. Both methanol decomposition and water–gas shift reactions are kinetically limited at temperatures under 700°C. Also methanation reaction was kinetically limited. As shown by Gadhe and Gupta [160], high pressure favored the formation of methane.

10.7.6 Ethanol

Wenzel [157] studied SCW reforming of ethanol under noncatalytic conditions for the temperature range of 618°C–710°C and pressure of 24.2 MPa [169]. The ethanol feed rate was varied from 0.17 to 2.2 g/min and water flow rate was varied from 6.4 to 19.7 g/min in a 1 1 625 grade 1 alloy tubular reactor. A complete conversion of ethanol was obtained producing hydrogen, carbon dioxide, methane, ethane, and carbon monoxide in the descending order of their concentrations. Hydrogen was produced by two competing reactions: the direct reformation of ethanol into hydrogen and carbon oxides and the pyrolytic dehydrogenation of ethanol:

$$C_2H_5OH \rightarrow C_2H_4O + H_2 \tag{10.19}$$

where acetaldehyde goes through further decarbonylation as

$$C_2H_4O \rightarrow CH_4 + CO \tag{10.20}$$

This decomposition is fast with Rh–cerium oxide catalyst at temperatures above 650°C. The net result of the above two reactions is to generate hydrogen, methane, and carbon oxides. In this system, forward water–gas shift reaction is active even without the presence of a water–gas shift catalyst. An undesirable competing reaction of dehydration of ethanol to form ethylene occurs, which is subsequently hydrogenated to form ethane. This reaction not only consumes hydrogen but also produces the coking precursor ethylene. Both pyrolytic and direct reforming reactions were first-order reactions.

Byrd et al. [177] studied supercritical reforming of ethanol over Ru/Al_2O_3 catalyst. Experiments were conducted at various temperature, pressure, residence time, and water-to-carbon ratio to evaluate their effects on the hydrogen yield. The results showed that hydrogen formation was favored at high temperature and high water-to-ethanol ratios. Under the same conditions and for an optimum residence time, methane production was suppressed. Excellent conversions were obtained for the residence time as low as 4 s. Pressure had negligible effect on hydrogen yield above the critical pressure and there was negligible coke formation for ethanol concentration in the feed less than 10 wt%. The overall reforming reaction for ethanol can be expressed as

$$C_2H_5OH + 3H_2O \rightleftarrows 6H_2 + 2CO_2 \quad \Delta H^0_{298} = 174 \text{ kJ/mol} \tag{10.21}$$

In the presence of Ru/Al_2O_3 catalyst, high reforming performance may be due to the fact that intermediates formed during ethanol decomposition such as dimethyl ether and acetaldehyde were also gasified in the presence of SCW. In the subcritical steam gasification, formation of significant amount of carbon limits hydrogen production. Reaction products also contain acetaldehyde, diethyl ether, ethane, and ethylene. The gasification under supercritical conditions is accompanied by several complex reactions such as ethanol decomposition, steam reforming, water–gas shift reaction, and methanation reaction. The product distribution depended on the relative rates of these reactions. It was assumed that during reforming, ethanol dehydrogenates on the metal surface to give adsorbed intermediates before the cleavage of C–C and C–O bonds. The water–gas shift reaction reduces CO concentration, and the final products predominantly contain hydrogen and carbon dioxide.

Gadhe and Gupta [160] examined the strategies for the reduction of methane formation and thereby increased the production of hydrogen. Three strategies that were examined were (1) operation at a low residence time by having a smaller reactor length or a high feed flow rate, (2) addition of a small amount of K_2CO_3 or KOH in the feed, and (3) utilization of the surface catalytic activity of the reactor made of Ni–Cu alloy. All the three strategies worked, resulting in lower methane production and correspondingly higher hydrogen production. The methanation reactions were favored by high pressure, high residence time, and low steam-to-carbon ratio.

10.8 TRI-REFORMING IN SCW

Fundamentally, there are three types of high-temperature reforming processes: stream reforming, dry reforming, and partial oxidation [170–176]. The term "tri-reforming" is applied to the process in which all of these reforming processes are combined in a single use. The three reforming processes are expressed by the following set of chemical reactions:

$$\text{Steam reforming: } CH_4 + H_2O \rightarrow CO + 3H_2 \quad \Delta H^0_{298k} = 206\,\text{kJ/mol} \quad (10.22)$$

$$\text{Dry reforming: } CH_4 + CO_2 \rightarrow CO + 2H_2 \quad \Delta H^0_{298k} = 247\,\text{kJ/mol} \quad (10.23)$$

$$\text{Partial oxidation: } CH_4 + O_2 \rightarrow CO + 2H_2 \quad \Delta H^0_{298k} = -38\,\text{kJ/mol} \quad (10.24)$$

As mentioned above, the three reactions combined are called tri-reforming reactions. It has been established that nickel, cobalt, iron, and the platinum group metals can catalyze steam reforming reaction to the thermodynamic equilibrium. However, the nickel catalyst has emerged as the most practical catalyst because of its fast turnover rates, long-term stability, and cost. The major technical problem for the nickel catalysts is carbon deposition on the catalysts via the following reactions that can lead to rapid deactivation and breakup of the catalyst:

$$\text{Methane decomposition: } CH_4 \rightarrow C + 2H_2 \quad \Delta H^0_{298k} = 74.8\,\text{kJ/mol} \quad (10.25)$$

$$\text{CO decomposition: } 2CO \rightarrow C + CO_2 \quad \Delta H^0_{298k} = -172.5\,\text{kJ/mol} \quad (10.26)$$

Carbon deposition can be substantially reduced by the use of an excess of water and a temperature of about 800°C. Other drawbacks of stream reforming are as follows:

1. Expensive generation of superheated steam (in excess) at high temperature
2. The production of a significant amount of CO_2 in the product gas via the reverse water–gas shift reaction, that is,
   Reverse water–gas shift reaction:

$$CO_2 + H_2 \rightarrow CO + H_2O \quad \Delta H^0_{298k} = 41\,\text{kJ/mol} \quad (10.27)$$

3. The H_2-to-CO ratio is higher than the optimum required for the downstream synthesis gas conversion to methanol, acetic acid, or hydrocarbons

Partial oxidation offers some advantages over steam reforming. First, the reaction produces extremely high yields of syngas by an exothermic reaction, and, therefore, the reactor would be more economical to heat. Oxygen is often used in steam reforming to provide heat and high methane conversion. Second, partial oxidation also gives a better ratio of hydrogen to carbon monoxide for subsequent conversion processes. Third, the product gases from the reaction are low in carbon dioxide that must often be removed before the syngas can be used.

Steam reforming and partial oxidation produce syngas. The dry reforming has an added advantage that it simultaneously consumes two greenhouse gases:

hydrocarbons, and CO_2. The best reducing agent for CO_2 is hydrogen. As shown by Rozovskii et al. [170], the synthesis of methanol from CO and H_2 proceeds not by their direct interaction, but by the transformation of CO into CO_2:

$$\text{Water–gas shift reaction: } CO + H_2O \rightarrow CO_2 + H_2 \quad \Delta H^0_{298k} = -41\,\text{kJ/mol} \quad (10.28)$$

$$\text{Methanol synthesis: } CO_2 + 3H_2 \rightarrow CH_3OH + H_2O \quad \Delta H^0_{298k} = -49.3\,\text{kJ/mol} \quad (10.29)$$

Direct utilization of the last reaction meets opposition because it converts expensive hydrogen into inexpensive water. Thus, a CO_2 reduction by hydrocarbons is preferred with lower hydrocarbons and alcohols. While dry reforming of hydrocarbons converts carbon dioxide and hydrocarbons into useful syngas, the tri-reforming allows the process to produce the syngas with a variety of H_2/CO ratios.

The H_2/CO ratio in syngas is very important for its further use for a variety of chemical products. Syngas can be converted to acetone, acetic acid, and ethylene by an exothermic reaction, while pure CO can be used for the production of acetic acid, formic acid, polyurethane, polycarbonates, methyl acrylates, and so on. A H_2/CO ratio of about 1 is required for the productions of polycarbonates, oxo alcohol, formaldehyde, iron ore reduction reaction, and so on; a H_2/CO ratio of about 2 is required for methanol and Fischer–Tropsch (FT) synthesis, and a H_2/CO ratio of 3 or higher is required for ammonia synthesis and hydrogen production.

Tri-reforming also offers some other advantages. Since dry and steam reforming reactions are highly endothermic, a careful integration of these reactions into any process scheme that internally generates heat (like partial oxidation) is very important in order to make the overall process energy balance more efficient, thus avoiding the need for expensive external heating. Both dry and steam reforming reactions require very high temperatures (>600°C) to reduce the cooking. While steam reduces carbon deposition, an addition of oxygen provides the necessary heat that can jump-start dry and steam reforming reactions and maintain the catalyst in a clean and carbon-free state through oxidation of coke on the catalyst surface. The extent to which oxygenates are added to the reforming reactions is determined strictly by the process conditions and the catalyst employed. While the combination of dry reforming and partial oxidation has been studied by a number of investigators [170–176], these studies have been largely restricted to one or two hydrocarbons. Since dry reforming produces water, the steam reforming always accompanies dry reforming, making these studies relevant for tri-reforming.

The most extensive study of tri-reforming was carried out by Puolakka et al. [172] and Puolakka [173]. The study focused on the tri-reforming of five model compounds—methane, heptanes, *n*-dodecane, toluene, and ethanol—over a number of different catalysts. It was reported that 0.25% Rh on ZrO_2 catalyst gave the best results, and its performance was comparable to the results for commercial Ni catalyst. The five model compounds were chosen to represent different types of fossil/biofuels. Methane was chosen to represent natural gas, *n*-heptane to represent aliphatic component of gasoline, *n*-dodecane to represent aliphatic component of biodiesel, toluene to represent aromatic part of gasoline, and diesel oil and ethanol to represent oxygenated compounds in biofuel.

The use of SCW as reaction medium for conducting the reforming can be an attractive and novel method. The literature on gasification/reforming under SCW indicates that in general, SCW reduces coking, lowers the required temperature for the same level of conversion, and modifies the product distribution, particularly in favor of more production of hydrogen. These results imply the need for a study of tri-reforming under SCW (critical point 374°C and 22.1 MPa) conditions. It is expected that the supercritical conditions will bring about significant improvement on product distributions, reaction temperature severity, and catalyst activity, stability, and life. Under SCW gasification, syngas is produced directly at high pressure, which means that a smaller reactor volume and lower energy are needed to pressurize the gas in a storage tank.

While tri-reforming of methane in SCW has been investigated by a number of researchers [171–176], these studies have been carried out with conventional Ni or bimetallic catalysts. The studies have shown that the supercritical conditions lower the required temperature for gasification, and at high temperatures (>600°C), hydrogen and carbon dioxide are the dominating products. The studies [171–176] have also shown that the product composition from tri-reforming under supercritical conditions depends on a number of variables such as temperature, pressure, feedstock and oxygen concentrations, reaction time, biomass properties, presence of inorganic elements, biomass particle size, and the nature of the catalyst. In the SCW environment, the syngas composition will heavily depend on the effectiveness of the dry reforming reaction. More catalytic studies to improve dry reforming reaction are presently being pursued. The use of nanocatalysts is also very heavily examined. In future, more work on tri-reforming in SCW environment with practical feedstock needs to be carried out.

REFERENCES

1. Gullon, P., Conde, E., Moure, A., Dominguez, H., and Parajo, J., "Selected process alternatives for biomass refining: A review," *The Open Agriculture Journal*, 4, 135–144 (2010).
2. Demirbas, A., "Biorefineries: Current activities and future developments," *Energy Conversion and Management*, 50, 2782–2801 (2009).
3. Demirbas, A., "Sub and supercritical water depolymerization of biomass," *Energy Sources, Part A*, 32, 1100–1110 (2010).
4. Peterson, A., Vogel, F., Lachance, R., Frolling, M., Antal, M., and Tester, J., "Thermochemical biofuel production in hydrothermal media: A review of sub- and supercritical technologies," *Energy & Environmental Science*, 1, 32–65 (2008).
5. Azadi, P. and Farnood, R., "Review of heterogeneous catalysts for sub- and supercritical water gasification of biomass and wastes," *International Journal of Hydrogen Energy*, 36, 9529–9541 (2011).
6. Demirbas, A., "Progress and recent trends in biofuels," *Progress in Energy and Combustion Science*, 33, 1–18 (2007).
7. Tanksale, A., Beltramini, J., and Lu, G., "A review of catalytic hydrogen production processes from biomass," *Renewable & Sustainable Energy Reviews*, 14, 166–182 (2010).
8. Huber, G., Iborra, S., and Corma, A., "Synthesis of transportation fuels from biomass: Chemistry, catalysts and engineering," *Chemical Reviews*, 106, 1–51 (2006).
9. Subramaniam, B. and McHugh, M., "Reactions in supercritical fluids: A review," *Industrial & Engineering Chemistry Process Design and Development*, 25, 1–12 (1986).
10. Brunner, G., "Applications of supercritical fluids," *Annual Review of Chemical and Biomolecular Engineering*, 1, 1321–1342 (2010).
11. Savage, P., "Organic chemical reactions in supercritical water," *Chemical Reviews*, 99, 603–621 (1999).

12. Watanabe, M., Sato, T., Inomata, H., Smith, R.L., Jr., Arai, K., Kruse, A., and Dinjus, E., "Chemical reactions of C1 compounds in near-critical and supercritical water," *Chemical Reviews*, 104, 5803–5820 (2004).
13. Matsumura, Y., Sasaki, M., Okuda, K., Takami, S., Ohara, S., Umetsu, M., and Adschiri, T., "Supercritical water treatment of biomass for energy and material recovery," *Combustion Science and Technology*, 178, 509–536 (2006).
14. Modell, M., "Processing methods for the oxidation of organics in supercritical water," Patent No. 4338199 (July 6, 1982).
15. Ding, Z., Frisch, M., Li, L., and Gloyna, E., "Catalytic oxidation in supercritical water," *Industrial & Engineering Chemistry Research*, 35, 3257–3279 (1996).
16. Parsons, E., "Organic reactions in very hot water," *Chemtech*, 26 (6), 30–34 (1996).
17. Katritzky, A., Allin, S., and Siskin, M., *American Chemical Society*, 29 (8), 399–406 (1996).
18. An, J., Bagnell, L., Cablewski, T., Strauss, C., and Trainer, R., *Journal of Organic Chemistry*, 62 (8), 2505–2511 (1997).
19. Leif, R.N. and Simoneit, B., *Origins of Life and Evolution of the Biosphere*, 25, 119–140 (1995).
20. Chandler, K., Deng, F., Dillow, A., Liotta, C., and Eckert, C., *Industrial & Engineering Chemistry Research*, 36 (12), 5175–5179 (1997).
21. Xu, X. and Antal, M., *AIChE Journal*, 40 (9), 1524–1531 (1994).
22. Xu, X. and Antal, M., *Industrial & Engineering Chemistry Research*, 36 (1), 23–41 (1997).
23. Crittendon, R. and Parsons, E., *Organometallics*, 13 (7), 2587–2591 (1994).
24. Diminnie, J., Metts, S., and Parsons, E., *Organometallics*, 14 (8), 4023–4025 (1995).
25. Kuhlmann, B., Arnett, E., and Siskin, M., *Journal of Organic Chemistry*, 59 (11), 3098–3101 (1994).
26. Savage, P., Li, R., and Santini J., *The Journal of Supercritical Fluids*, 7 (2), 135–144 (1994).
27. Lee, J. and Foster, N., *The Journal of Supercritical Fluids*, 9 (2), 99–105 (1996).
28. Aki, S. and Abraham, M., *The Journal of Supercritical Fluids*, 7 (4), 259–263 (1994).
29. Hong, G. and Spritzer, M., "Supercritical water partial oxidation," *Proceedings of the 2002 US DOE Hydrogen Energy Program Review*, General Atomics, San Diego, CA, NREL/CP-610-32405, Golden, CO (2002).
30. Johanson, N., Spritzer, M., Hong, G., and Rickman, W., "Supercritical water partial oxidation," *Proceedings of the 2001 DOE Hydrogen Program Review*, General Atomics, San Diego, CA, NREL/CP-570-30535, Golden, CO (2001).
31. Vogel, F., DiNaro-Blanchard, J., Marrone, P., Rice, S., Webley, P., Peters, W., Smith, K., and Tester, J., "Critical review of kinetic data for the oxidation of methanol in supercritical water," *The Journal of Supercritical Fluids*, 34, 249–286 (2005).
32. Spritzer, M., "Supercritical water partial oxidation," Hydrogen Fuel Cells and Infrastructure Technologies, DOE progress report, General Atomics, San Diego, CA (2003).
33. Sato, T., Adschiri, T., Arai, K., Rempel, G., and Ng, F., "Upgrading of asphalt with and without partial oxidation in supercritical water," *Fuel*, 82 (10), 1231–1240 (2003).
34. Youssef, E., Elbeshbishy, E., Hafez, H., Nakhia, G., and Charpentier, P., "Sequential supercritical water gasification and partial oxidation of hog manure," *International Journal of Hydrogen Energy*, 35 (21), 11756–11767 (2010).
35. Gloyna, E. and Li, L., "Supercritical water oxidation: An engineering update," *Waste Management*, 13, 379 (1993).
36. Gloyna, E. and Li, L., "Supercritical water oxidation research and development update," *Environmental Progress*, 14, 182 (1995).
37. Jin, F.M., Ausushi, K., and Heiji, E., "Oxidation of garbage in supercritical water," *High Pressure Research*, 20, 525–531 (2001).
38. Goto, M., Nada, T., Ogata, A., Kodama, A., and Hirose, T., "Supercritical water oxidation for the destruction of municipal excess sludge and alcohol distillery wastewater of molasses," *The Journal of Supercritical Fluids*, 13, 277–282 (1998).

39. Noam, E. and Ronald, M.L., "Review of materials issues in supercritical water oxidation systems and the need for corrosion control," *Transactions of the Indian Institute of Metals*, 56, 1–10 (2003).
40. Takuya, Y. and Yoshito, O., "Partial oxidative and catalytic biomass gasification in supercritical water: A promising flow reactor system," *Industrial & Engineering Chemistry Research*, 43, 4097–4104 (2004).
41. Watanabe, M., Inomata, H., Osada, M., Sato, T., Adschiri, T., and Arai, K., "Catalytic effects of NaOH and ZrO_2 for partial oxidative gasification of *n*-hexadecane and lignin in supercritical water," *Fuel*, 82 (5), 545–552 (2003).
42. Watanabe, M., Inomata, H., Osada, M., Sato, T., Adschiri, T., and Arai, K., "Partial oxidation of *n*-hexadecane and polyethylene in supercritical water," *The Journal of Supercritical Fluids*, 20 (3), 257–266 (2001).
43. Oka, H., Yamago, S., Yoshida, J., and Kajimoto, O., "Evidence for a hydroxide ion catalyzed pathway in ester hydrolysis in supercritical water." *Angewandte Chemie International Edition*, 41, 623–625 (2002).
44. Jin, F.M., Takehiko, M., and Heiji, E., "Oxidation reaction of high molecular weight carboxylic acids in supercritical water," *Environmental Science & Technology*, 37, 3220–3231 (2003).
45. Shamsi, A., "Partial oxidation and dry reforming of methane over Ca/Ni/K (Na) catalysts," *Catalysis Letters*, 109 (3/4), 189–193 (2006).
46. Krietemeyer, S. and Wagner, T., *Supercritical Water Oxidation.* Risk Reduction Engineering Laboratory, Cincinnati, OH (1992).
47. Modell, M., "Supercritical water oxidation," in Freeman, H. (ed.), *Standard Handbook of Hazardous Waste Treatment and Disposal.* McGraw-Hill, New York (1989).
48. Tester, J., Holgate, H., Amellini, F., Webley, P., Killilea, W., Hong, G., and Barner, H., "Supercritical water oxidation technology: Process development and fundamental research," in Tedder, D. and Pohland, F. (eds.), *Emerging Technologies in Hazardous Waste Management III.* American Chemical Society, Washington, DC (1993).
49. McBrayer, R., "Design and operation of first commercial supercritical water oxidation facility," *First International Workshop on Supercritical Water Oxidation*, Jacksonville, FL, February 6–9 (1995).
50. Lourdes, C. and David, V., "Formation of organic acids during the hydrolysis and oxidation of several wastes in sub- and supercritical water," *Industrial & Engineering Chemistry Research*, 41, 6503–6509 (2002).
51. Hodes, M., Marrone, P.A., Hong, G.T., Smith, K.A., and Tester, J.W., "Salt precipitation and scale control in supercritical water oxidation—Part A: Fundamentals and research." *The Journal of Supercritical Fluids*, 29, 265–288 (2004).
52. Svensson, P., "Look no stack, supercritical water destroys organic wastes," *Chemical Technology Europe*, 2, 16 (1995).
53. Savage, P., Gopalan, S., Mizan, T., Martino, C., and Brock, E., "Reactions in supercritical conditions: Application and fundamentals," *AIChE Journal*, 41, 1723 (1995a).
54. Savage, P., Gopalan, S., Mizan, T., Martino, C., and Brock, E., "Oxidation in supercritical water: Pathways, kinetics and mechanisms," *First International Workshop on Supercritical Water Oxidation*, Jacksonville, FL February 6–9 (1995).
55. Watanabe, M., Adschiri,T., and Arai, K., "Polyethylene decomposition via pyrolysis and partial oxidation in supercritical water," *Kobunshi Ronbunshu*, 58 (12), 631 (2001) (Japanese).
56. Lilac, W. and Lee, S., "Kinetics and mechanisms of styrene monomer recovery from waste polystyrene by supercritical water partial oxidation," *Advances in Environmental Research*, 6 (1), 9–16 (2001).

57. Peter, K. and Eckhard, D., "An assessment of supercritical water oxidation (SCWO): Existing problems, possible solutions and new reactor concepts," *Chemical Engineering Journal*, 83, 207–214 (2001).
58. Li, R., Savage, P., and Szmukler, D., "2-Chlorophenol oxidation in supercritical water: Global kinetics and reaction products," *AIChE Journal*, 39 (1), 178–187 (1993).
59. Thomason, T., Hong, G., Swallow, K., and Killilea, W., "The MODAR supercritical water oxidation process," in Freeman, H. (ed.), *Innovative Hazardous Waste Treatments Technology Series*. Technomic Publishing Co., Lancaster, PA (1990).
60. Wantanabe, M., Sato, T., Inomata, H., Smith, R., Arai, K., Kruse, A., and Dinjus, E., "Chemical reactions of C1 compounds in near critical and supercritical water," *Chemical Reviews*, 104, 5803–5821 (2004).
61. Watanabe, M., Hirakoso, H., Sawamoto, S., Adschiri, T., and Arai, K., *The Journal of Supercritical Fluids*, 13, 247–252 (1998).
62. Stiles, A., *Catalyst Manufacture*. Marcel Dekker, New York (1983).
63. Attia, Y., *Sol-Gel Processing and Applications*. Plenum Press, New York (1994).
64. Messing, G., Zhang, S., and Yayanthi, G., "Ceramic powder synthesis by spray pyrolysis," *Journal of the American Ceramic Society*, 76, 2707 (1993).
65. Moser, W., "Hydrocarbon partial oxidation catalysts prepared by the high temperature aerosol decomposition process: Crystal and catalytic chemistry," in Oyama, T. and Hightower, J. (eds.), *Catalytic Selective Oxidation*. American Chemical Society, Washington, DC, 244 (1993).
66. Nanjangud, S., Brezny, R., and Green, D., "Strength and Young's modulus behavior of a partially sintered porous alumina," *Journal of the American Ceramic Society*, 78, 266 (1995).
67. Pugh, R. and Bergstrom, L. (eds.), *Surface and Colloid Chemistry in Advanced Ceramic Processing*. Marcel Dekker, New York (1994).
68. Saggio-Woyansky, J., Scott, C., and Minnear, W., "Processing of porous ceramics," *American Ceramic Society Bulletin*, 72, 1674 (1992).
69. Segal, D., *Chemical Synthesis of Advanced Ceramic Materials*. Cambridge University Press, New York (1983).
70. Vincenzini, P., *Ceramic Powders: Preparation, Consolidation and Sintering*. Elsevier Scientific Publishing Co., New York (1993).
71. Ward, D. and Ko, E., "Preparing catalyst materials by sol-gel method," *Industrial & Engineering Chemistry Research*, 34, 421 (1995).
72. Wold, A. and Dwight, K., *Solid State Chemistry, Synthesis, Structure, and Properties of Selected Oxides and Sulfides*. Chapman & Hill, New York (1993).
73. Deshpande, G., Holder, G., Bishop, A., Gopal, J., and Wender, I., *Fuel*, 63, 956 (1984).
74. Pauliatis, M., Penninger, R., Gray, D., and Davidson, P. (eds.), *Chemical Engineering at Supercritical Fluid Conditions*. Ann Arbor Science Publishers, Ann Arbor, MI, 385–407 (1983).
75. Modell, M., Reid, R., and Amin, S., "Gasification process," US Patent No. 4113446 (September 12, 1978).
76. Jezko, J., Gray, D., and Kershaw, J., *Fuel Processing Technology*, 5, 229 (1982).
77. Vasilakos, N., Dobbs, J., and Parasi, A., *American Chemical Society Division of Fuel Chemistry*, preprints, 28 (4), 212 (1983).
78. Kershaw, J. and Bagnell, L., "Extraction of Australian coals with supercritical water," *ACS Reprints*, 101–111 (1988).
79. Minowa, T., Zhen, F., and Ogi, T., "Cellulose decomposition in hot-compressed water with alkali or nickel catalyst," *The Journal of Supercritical Fluids*, 13, 253–259 (1998).
80. Sasaki, M., Kabyemela, B., Malaluan, R., Hirose, S., Takeda, N., Adschiri, T., and Arai, K., "Cellulose hydrolysis in sub-critical and super-critical water," *The Journal of Supercritical Fluids*, 13, 261–268 (1998).

81. Sasaki, M., Fang, Z., Fukushima, Y., Adschiri, T., and Arai, K.,"Dissolution and hydrolysis of cellulose in sub-critical and supercritical water," *Industrial & Engineering Chemistry Research*, 39, 2883–2890 (2000).
82. Wahyudiono, Shiraishi, T., Sasaki, M., and Goto, M., "Non catalytic liquefaction of bitumen with hydrothermal/solvothermal process," *The Journal of Supercritical Fluids*, 60, 127–136 (2011).
83. Cheng, L., Zhang, R., and Bi, J., "Pyrolysis of a low rank coal in sub and supercritical water," *Fuel Processing Technology*, 85 (8–10), 921–932 (2004).
84. Vostrikov, A., Psarov, S., Dubov, D., Fedyaeva, O., and Sokol, M., "Kinetics of coal conversion in supercritical water," *Energy & Fuels*, 21, 2840–2845 (2007).
85. Kumar, S. and Gupta, R., "Biocrude production from switchgrass using supercritical water," *Energy & Fuels*, 23, 5151–5159 (2009).
86. Kim, I.C., Park, S.D., and Kim, S., "Effects of sulfates on the decomposition of cellobiose in supercritical water," *Chemical Engineering Process*, 43, 997–1005 (2004).
87. Mitsubishi Material Corp., "Supercritical water cracks residue oil," *Chemical Engineering*, 14 (September 2007).
88. Ayhan, D., "Hydrogen-rich gas from fruit shells via supercritical water extraction," *International Journal of Hydrogen Energy*, 29, 1237–1243 (2004).
89. Nonaka, H., "Development of liquefaction process of coal and biomass in supercritical water," *Fuel and Energy Abstracts*, 39 (1), 18 (1998).
90. Matsumara, Y., Nonaka, H., Yokura, H., Tsutsumi, A., and Yoshida, K., "Co-liquefaction of coal and cellulose in supercritical water," *Fuel*, 78 (9), 1049–1056 (1999).
91. Demirbas, A., "Hydrogen production from biomass via supercritical water extraction," *Energy Sources*, 27, 1409–1417 (2005).
92. Phillip, E., "Organic chemical reactions in supercritical water," *Chemical Reviews*, 99, 603–621 (1999).
93. Duan, P. and Savage, P., "Upgrading of crude algal bio oil in supercritical water," *Bioresource Technology*, 102 (2), 1899–1906 (2011).
94. Duan, P. and Savage, P., "Catalytic hydrotreatment of crude algal bio-oil in supercritical water," *Applied Catalysis B: Environmental*, 104 (1/2), 136–143 (2011).
95. Onsari, K., Prasassarakich, P., and Ngamprasertsith, S., "Co-liquefaction of coal and used tire in supercritical water," *Energy and Power Engineering*, 2, 95–102 (2010).
96. Sunphorka, S., Prasassarakich, P., and Ngamprasertsith, S., "Co-liquefaction of coal and plastic mixture in supercritical water," The 5th Mathematics and Physical Sciences Graduate Congress, Chulalongkorn University, Thailand, December 7–9, 2009 (2011).
97. Nonaka, H., Matsumura, Y., Tsutsumi, A., Yoshida, K., Matsuno, Y., and Inaba, A., Development of liquefaction process of coal and biomass in supercritical water," *Sekitan Kagaku Kaigi Happyo Ronbunshu*, 33, 73–76 (1996).
98. Veski, R., Palu, V., and Kruusement, K., "Co-liquefaction of kukersite oil shale and pine wood in supercritical water," *Oil Shale*, 23 (3), 236–248 (2006).
99. Yokura, H., Nonaka, H., Matsumura, Y., Tsutsumi, A., and Yoshida, K., "Effect of catalyst addition on co-liquefaction process of coal and biomass in supercritical water," *Sekitan Kagaku Kaigi Happyo Ronbunshu*, 34, 69–72 (1997).
100. Missal, P. and Hedden, K., "Extraction of a Colorado oil shale by water in the sub- and supercritical phases," *Erdoel & Kohle, Erdgas, Petrochemie*, 42 (9), 346–352 (1989).
101. Funazukuri, T., Yokoi, S., and Wasao, N., "Supercritical fluid extraction of Chinese Maoming oil shale with water and toluene," *Fuel*, 67 (1), 10–14 (1988).
102. Hu, H., Zhang, J., Guo, S., and Chen, G., "Extraction of Huadian oil shale with water in sub- and supercritical states," *Fuel*, 78 (6), 645–651 (1999).
103. Canel, M. and Missal, P., "Extraction of solid fuels with sub- and supercritical water," *Fuel*, 73 (11), 1776–1780 (1994).

104. Johnson, D.K., Chum, H.L., Anzick, R., and Baldwin, R.M., "Lignin liquefaction in supercritical water," in Bridgwater, A.V. and Kuester, J.L. (eds.), *Research in Thermochemical Biomass Conversion.* Elsevier, London, 485–496 (1988).
105. Palu, V., Kruusement, K., and Veski, R. "Supercritical water extraction of biomass and oil shale," 29th Estonian Chemistry Days, Tallinn, Estonia, 77 (2005).
106. Li, L. and Eglebor, N., "Oxygen removal from coal during supercritical water and toluene extraction," *Energy & Fuels*, 6 (1), 35–40 (1992).
107. Chen, D., Perman, C., Riechert, M., and Hoven, J., "Depolymerization of tire and natural rubber using supercritical fluids," *Journal of Hazardous Materials*, 44, 53–60 (1995).
108. Park, S. and Gloyna, E., "Statistical study of the liquefaction of used rubber tyre in supercritical water," *Fuel*, 76 (11), 999–1003 (1997).
109. Su, X., Zhao, Y., Zhang, R., and Bi, J., "Investigation on degradation of polyethylene to oils in supercritical water," *Fuel Processing Technology*, 85, 1249–1258 (2004).
110. Moriya, T. and Enomoto, H., "Investigation of the basic hydrothermal cracking conditions of polyethylene in supercritical water," *Shigen to Sozai*, 115, 245 (1999) (Japanese).
111. Matsubara, W. et al., "Development of liquefaction process of plastic waste in supercritical water," *Mitsubishi Juko Giho*, 34, 438 (1997) (Japanese).
112. Broll, D., Kaul, C., Kramer, A., Krammer, P., Richter, T., Jung, M., Vogel, H., and Zehner, P., "Chemistry in supercritical water," *Angewandte Chemie International Edition*, 38, 2998 (1999).
113. Moriya, T. and Enomoto, H., "Characteristics of polyethylene cracking in supercritical water compared to thermal cracking," *Polymer Degradation and Stability*, 65, 373 (1999).
114. Moriya, T. and Enomoto, H., "Role of water in conversion of polyethylene to oils through supercritical water cracking," *Kagaku Kogaku Ronbunshu*, 25 (6), 940 (1999) (Japanese).
115. Holliday, R., King, J., and List, G., "Hydrolysis of vegetable oils in sub- and supercritical water," *Industrial & Engineering Chemistry Research*, 36 (3), 932–935 (1997).
116. Li, Y., Guo, L., and Zhang, X., "Hydrogen production from coal gasification in supercritical water with a continuous flowing system," *International Journal of Hydrogen Energy*, 35, 3036–3045 (2010).
117. Lee, S. and Shah, Y., *Biofuels and Bioenergy—Processes and Technologies.* CRC Press, New York (2012).
118. May, A., Salvado, J., Torras, C., and Montane, D., "Catalytic gasification of glycerol in supercritical water," *Chemical Engineering Journal*, 160 (2), 751–759 (2010).
119. Osada, M., Sato, T., Watanabe, M., Shirai, M., and Arai, K., "Catalytic gasification of wood biomass in subcritical and supercritical water," *Combustion Science and Technology*, 178 (1–3), 537–552 (2006).
120. Yu, D., Aihara, M., and Antal, M., "Hydrogen production by steam reforming glucose in supercritical water," *Energy & Fuels*, 7 (5), 574–577 (1993).
121. D'Jesus, P., Boukis, N., Kraushaar-Czarnetzki, B., and Dinjus, E., "Gasification of corn and clover grass in supercritical water," *Fuel*, 85, 1032–1038 (2006).
122. Pinkwart, K., Bayha, T., Lutter, W., and Krausa, M., "Gasification of diesel oil in supercritical water for fuel cells," *Journal of Power Sources*, 136, 211–214 (2004).
123. Yamaguchi, A., Hiyoshi, N., Sato, O., Osada, M., and Shirai, M., "EXAFS study on structural change of charcoal-supported ruthenium catalysts during lignin gasification in supercritical water," *Catalysis Letters*, 122, 188–195 (2008).
124. Byrd, A., Kumar, S., Kong, L., Ramsurn, H., and Gupta, R., "Hydrogen production from catalytic gasification of switchgrass biocrude in supercritical water," *International Journal of Hydrogen Energy*, 36, 3426–3433 (2011).
125. Sato, T., Osada, M., Watanabe, M., Shirai, M., and Arai, K., "Gasification of alkylphenols with supported noble metal catalysts in supercritical water," *Industrial & Engineering Chemistry Research*, 42, 4277–4282 (2003).

126. Takuya, Y. and Yukihiko, M., "Gasification of cellulose, xylan, and lignin mixtures in supercritical water," *Industrial & Engineering Chemistry Research*, 40, 5469–5474 (2001).
127. Takuya, Y., Yoshito, O., and Yukihiko M., "Gasification of biomass model compounds and real biomass in supercritical water," *Biomass & Bioenergy*, 26, 71–78 (2004).
128. Tang, H.Q. and Kuniyuki, K, "Supercritical water gasification of biomass: Thermodynamic analysis with direct Gibbs free energy minimization," *Chemical Engineering Journal*, 106, 261–267 (2005).
129. Xu, X., Matsumura, Y., Stenberg, J., and Antal, M.J., Jr., "Carbon-catalyzed gasification of organic feedstocks in supercritical water," *Industrial & Engineering Chemistry Research*, 35, 2522–2530 (1996).
130. Guo, L., Cao, C., and Lu, Y., "Supercritical water gasification of biomass and organic wastes," in Momba, M. and Bux, F. (eds.), *Biomass*. 165–182 (2010).
131. Demirbas, A., "Hydrogen production from biomass via supercritical water gasification," *Energy Sources, Part A*, 32, 1342–1354 (2010).
132. Kruse, A., "Supercritical water gasification," *Biofuels, Bioproducts and Biorefining*, 2, 415–437 (2008).
133. Lu, Y., Guo, L., Zhang, X., and Yan, Q., "Thermodynamic modeling and analysis of biomass gasification for hydrogen production in supercritical water," *Chemical Engineering Journal*, 131, 233–244 (2007).
134. Zhang, L., Champagne, P., and Xu, C., "Supercritical water gasification of an aqueous by-product from biomass hydrothermal liquefaction with novel Ru modified Ni catalysts," *Bioresource Technology*, 102 (17), 8279–8287 (2011).
135. Michael, J.A., Jr., Allen, S.G., Schulman, D., and Xu, X., "Biomass gasification in supercritical water," *Industrial & Engineering Chemistry Research*, 39, 4040–4053 (2000).
136. Paul, T.W. and Jude, O., "Composition of products from the supercritical water gasification of glucose: A model biomass compound," *Industrial & Engineering Chemistry Research*, 44, 8739–8749 (2005).
137. Peter, K, "Corrosion in high-temperature and supercritical water and aqueous solutions: A review," *The Journal of Supercritical Fluids*, 29, 1–29 (2004).
138. Yamaguchi, A., Hiyoshi, N., Sato, O., Bando, K., Osada, M., and Shirai, M., "Hydrogen production from woody biomass over supported metal catalysts in supercritical water," *Catalysis Today*, 146, 192–195 (2009).
139. Guo, Y., Wang, S., Xu, D., Gong, Y., Ma, H., and Tang, X., "Review of catalytic supercritical water gasification for hydrogen production from biomass," *Renewable & Sustainable Energy Reviews*, 14, 334–343 (2010).
140. Knoef, H., *Handbook Biomass Gasification*. Biomass Technology Group Press, Enschede, the Netherlands, 22–23 (2005).
141. Kruse, A. and Gawlik, A., "Biomass conversion in water at 330°C–410°C and 30–50 MPa. identification of key compounds for indicating different chemical reaction pathways," *Industrial & Engineering Chemistry Research*, 42, 267–279 (2003).
142. Kruse, A. and Henningsen, T., "Biomass gasification in supercritical water: Influence of the dry matter content and the formation of phenols," *Industrial & Engineering Chemistry Research*, 42, 3711–3717 (2003).
143. Lee, I.G., Kim, M.S., and Ihm, S.K., "Gasification of glucose in supercritical water," *Industrial & Engineering Chemistry Research*, 41, 1182–1188 (2002).
144. Antal, M., Allen, S., Schulman, D., and Xu, X., "Biomass gasification in supercritical water," *Industrial & Engineering Chemistry Research*, 39, 4040–4053 (2000).
145. Kruse, A., Meier, D., Rimbrecht, P., and Schacht, M., , "Gasification of pyrocatechol in supercritical water in the presence of potassium hydroxide," *Industrial & Engineering Chemistry Research*, 39, 4842–4848 (2000).

146. Hao, X.H., Guo, L.J., Mao, X., Zhang, X.M., and Chen, X.J., "Hydrogen production from glucose used as a model compound of biomass gasified in supercritical water," *International Journal of Hydrogen Energy*, 28, 55–64 (2003).
147. Tang, H. and Kitagawa, K., "Supercritical water gasification of biomass: Thermodynamic analysis with direct gibbs free energy minimization," *Chemical Engineering Journal*, 106 (3), 261–267 (2005).
148. Chakinala, A., Brilman, D., van Swaaij, W., and Kersten, S., "Catalytic and non catalytic supercritical water gasification of microalgae and glycerol," *Industrial & Engineering Chemistry Research*, 49 (3), 1113–1122 (2010).
149. Susanti, R., Nugroho, A., Lee, J., Kim, Y., and Kim, J., "Noncatalytic gasification of isooctane in supercritical water: A strategy for high-yield hydrogen production," *International Journal of Hydrogen Energy*, 36 (6), 3895–3906 (2011).
150. Goodwin, A. and Rorrer, G., "Conversion of glucose to hydrogen rich gas by supercritical water in a microchannel reactor," *Industrial & Engineering Chemistry Research*, 47 (12), 4106–4114 (2008).
151. Kong, L., Li, G., Zhang, B., He, W., and Wang, H., "Hydrogen production from biomass wastes by hydrothermal gasification," *Energy Sources, Part A*, 30, 1166–1178 (2008).
152. Sinag, A., Kruse, A., and Rathert, J., "Influence of the heating rate and the type of the catalyst on the formation of key intermediates and on the generation of gases during hydropyrolysis of glucose in supercritical water in a batch reactor," *Industrial & Engineering Chemistry Research*, 43, 502–508 (2004).
153. Kersten, S.R.A., Potic, B., Prins, W., and van Swaaij, W., "Gasification of model compounds and wood in hot compressed water," *Industrial & Engineering Chemistry Research*, 45, 4169–4177 (2006).
154. Lee, S., Lanterman, H., Wenzel, J., and Picou, J., "Noncatalytic reformation of JP-8 fuel in supercritical water for production of hydrogen," *Energy Sources, Part A: Recovery, Utilization, and Environmental*, 31 (19), 1750–1758 (2009).
155. Van Bennekom, J., Venderbosch, R., and Heeres, H., "Supermethanol: Reforming of crude glycerine in supercritical water to produce methanol for re-use in biodiesel plants," Report, University of Groningen, Groningen, the Netherlands (2011).
156. Vos, J., "Reforming of crude glycerine in supercritical water to produce methanol for re-use in biodiesel plants," Report, B.T.G. Biotechnology Group BV, Enschede, the Netherlands (2007).
157. Wenzel, J., "The kinetics of non-catalyzed supercritical water reforming of ethanol," PhD thesis, University of Missouri, Columbia, MO (May 2008).
158. Boukis, N., Diem, V., Habicht, W., and Dinjus, E., "Methanol reforming in supercritical water," *Industrial & Engineering Chemistry Research*, 42, 728–735 (2003).
159. Taylor, J., Herdman, C., Wu, B., Walley, K., and Rice, S., "Hydrogen production in a compact supercritical water reformer," *International Journal of Hydrogen Energy*, 28, 1171–1178 (2003).
160. Gadhe, J. and Gupta, R., "Hydrogen production by methanol reforming in supercritical water: Suppression of methane formation," *Industrial & Engineering Chemistry Research*, 44, 4577–4585 (2005).
161. Guerrero-Perez, M., Rosas, J., Bedia, J., Rodriguez-Mirosol, J., and Cordero, T., "Recent inventions in glycerol transformations and processing," *Recent Patents in Chemical Engineering*, 2, 11–21 (2009).
162. Huidong, W. and Renan, W., "Conversion and reforming of fossil fuel by supercritical water," *Chemical Industry and Engineering Progress*, 3, 1–7 (1999).
163. Ramaswamy, K. and T-Raissi, A., "Hydrogen production from used lube oil via supercritical water reformation," Report, Florida Solar Energy Center, Cocoa, FL (2010).

164. de Vlieger, D., Chakinala, A., Lefferts, L., Kersten, S., Seshan, K., and Brilman, D., "Hydrogen from ethylene glycol by supercritical water reforming using noble and base metal catalysts," *Applied Catalysis B: Environmental*, 111–112, 536–544 (2012).
165. Yamaguchi, A., Hiyoshi, N., Sato, O., Bando, K., Osada, M., and Shirai, M., "Hydrogen production from woody biomass over supported metal catalysts in supercritical water," *Catalysis Today*, 146 (1/2), 192–195 (2009).
166. Byrd, A., Pant, K., and Gupta, R., "Hydrogen production from glycerol by reforming in supercritical water over Ru/Al_2O_3 catalyst," *Fuel*, 87, 2956–2960 (2008).
167. Penninger, J.M.L., Maassa, G.J.J., and Rep, M., "Compressed hydrogen rich fuel gas (CHFG) from wet biomass by reforming in supercritical water," *International Journal of Hydrogen Energy*, 32 (10/11), 1472–1476 (2007).
168. Penninger, J. and Rep., M., "Reforming of aqueous wood pyrolysis condensate in supercritical water," *International Journal of Hydrogen Energy*, 31 (11), 1597–1606 (2006).
169. Therdthianwong, S., Srisinwat, N., Therdthianwong, A., and Croiset, E., "Reforming of bioethanol over Ni/Al_2O_3 and Ni/CeZrO catalysts in supercritical water for hydrogen production," *International Journal of Hydrogen Energy*, 36 (4), 2877–2886 (2011).
170. Rozovskii, A.Ya. and Lin, G.I., *Theoretical Basis of Methanol Synthesis*, Khimiya, Moscow, 1990.
171. Shekhawat, D., Berry, D., Gardner, T., and Spivey, J., "Catalytic reforming of liquid hydrocarbon fuels for fuel cell applications," *Catalysis*, 19, 184 (2006).
172. Puolakka, K.J., Juutilainen, S., and Krause, A.O.I., "Combined CO_2 reforming and partial oxidation of n-heptane on noble metal zirconia catalysts," *Catalysis Today*, 115 (1–4), 217–221 (2006).
173. Puolakka, J., "CO_2 reforming," Thesis for the Degree of Licentiate of Science and Technology, University of Technology, Helsinki, Finland (2007).
174. Kang, J.S., Kim, D.H., Lee, S.D., Hong, S.I., and Moon, D.J., "Nickel-based tri-reforming catalyst for the production of synthesis gas," *Applied Catalysis A*, 332 (1), 153–158 (2007).
175. Antal, M.J., Jr., Manarungson, S., and Mok, W.S.-L., "Hydrogen production by steam reforming glucose in supercritical water," in Bridyewater, A.V. (ed.), *Advances in Thermochemical Biomass Conversion*. Blackie Academic & Professional, London, 1367–1377 (1994).
176. Shah, Y. and Gardner, T., "Dry reforming of hydrocarbon feedstocks," *Catalysis Review – Science and Engineering* (in press).
177. Byrd, A., Pant, K., and Gupta, R., "Hydrogen production from ethanol by reforming in supercritical water using Ru/Al_2O_3 catalyst," *Energy & Fuels*, 21 (6), 3541–3547 (2007).

11 Water Dissociation Technologies for Hydrogen

11.1 INTRODUCTION

The dissociation of water to produce hydrogen reversibly requires a supply of energy as follows:

$$H_2O \rightleftarrows H_2 + O_2 \tag{11.1}$$

$$\Delta H^0_{289\,K} = 241.93 \text{ kJ/mol}, \; \Delta G^0_{298\,K} = 228.71 \text{ kJ/mol}, \; T\Delta S^0_{298\,K} = 13.22 \text{ kJ/mol}$$

This means that work = ΔG^0 and heat = $T\Delta S$ are required to split the water at 25°C and 1 atm. Here T is the temperature and ΔS is the change in entropy. ΔH and ΔG are changes in heat of formation and free energy of formation. The superscript ^{0} denotes standard conditions of 1 atm and 25°C. If the reaction does not proceed reversibly, more work is required. The energy needed for this work can be provided in a number of different ways, and these are evaluated in this chapter.

There are three major ways water can be dissociated to produce hydrogen. The first method is electrolysis in which the water is dissociated electrochemically using electrochemical cell. The cell can be operated in a number of different ways (such as high temperature and high pressure), but all of them require significant amount of energy to dissociate water. The second method is the use of photosynthesis and photocatalysis to dissociate water. This method also requires photonic energy with or without a catalyst. The energy can, however, be provided using a solar cell. The third method is thermal or thermochemical dissociation of water in which water is dissociated either thermally or thermochemically. The latter method uses a chemical substance (or substances) to carry out dissociation using a series of chemical reactions. This method not only separates hydrogen and oxygen upon dissociation, but also reduces the temperature required for the thermal dissociation. In the recent years, this method has been heavily explored. Besides these three major methods, some miscellaneous methods such as chemical oxidation, magmalysis, and radiolysis are also explored for water dissociation. All of these are briefly discussed in this chapter.

As discussed in earlier chapter 4, the use of solar energy in steam gasification, reforming, and solar cracking of fuels such as coal, biomass, and natural gas has been gaining more acceptance. Similarly, three major technologies—electrochemical, photochemical/photobiological, and thermochemical—for water dissociation can also be carried with the use of solar energy. In electrochemical processes, solar electricity made from photovoltaic or concentrating solar thermal systems can be used for electrolytic process. In photochemical/photobiological processes, direct use of solar photon energy carries out photochemical and photobiological processes. Finally, in thermochemical processes, solar heat at high temperature supports endothermic thermochemical water dissociation reactions. While thermochemical route offers some intriguing thermodynamic advantages over other options, in general, irrespective of the type of fuel produced, higher temperature gives higher conversion efficiency but also leads to greater losses by reradiation from the solar cavity receiver. A summary of all the thermochemical processes described earlier to produce solar fuels such as hydrogen is given in Figure 11.1.

The recent report of the International Energy Agency (IEA) shows that a measure of how well solar energy is converted to chemical energy stored in solar fuels is called exergy efficiency (Figure 11.2) [1]. The thermochemical route offers the potential of exergy efficiency to exceed 50%, a number higher than that obtained by all other methods. In solar fuel productions, half of the total investment cost is solar concentrating system. Higher exergy efficiency means lower power required to generate the same level of chemical energy in solar fuels. Thus, high exergy efficiency makes the process economically more attractive.

Numerous excellent reviews on various methods for hydrogen production from water are reported in the literature [2–10]. They examined different methods of hydrogen productions [2,5,8], energy efficiencies of various methods [3,4], economics of various alternatives [2,8], and use of solar energy for hydrogen productions [6,7,9,10].

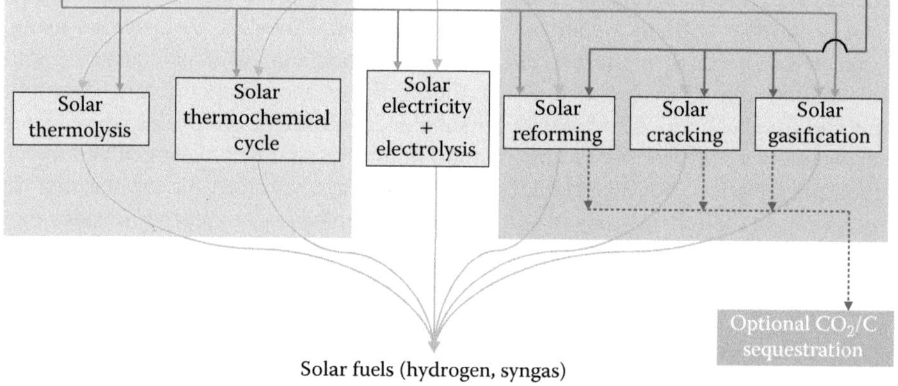

FIGURE 11.1 **(See color insert.)** Thermochemical routes for solar hydrogen production. (From Meier, A. and Sattler, C., "Solar fuels from concentrated sunlight," SolarPACES, Solar Power and Chemical Energy Systems, IEA report, 2009. With permission.)

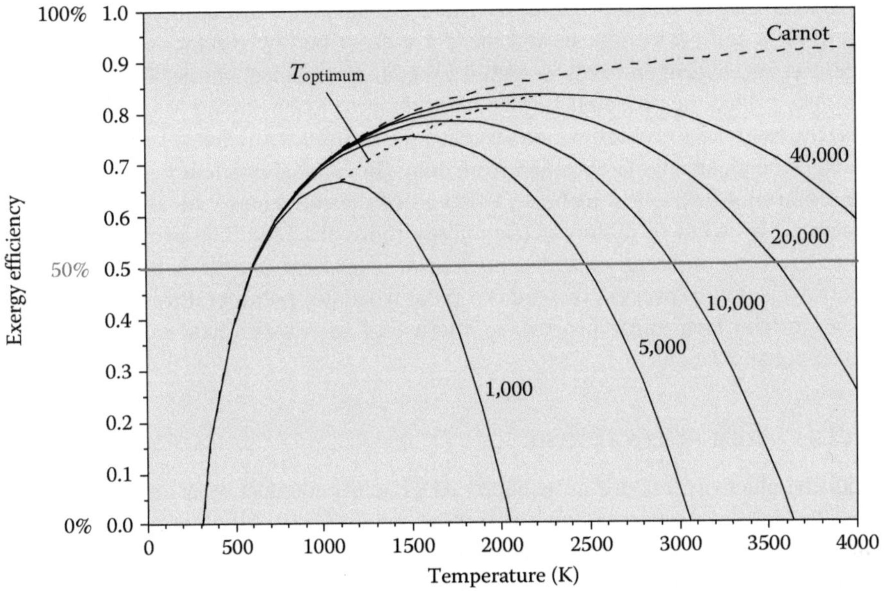

FIGURE 11.2 **(See color insert.)** Exergy efficiency—Variation of the exergy efficiency as a function of the process operating temperature for a blackbody cavity receiver converting concentrated solar energy into chemical energy. (From Meier, A. and Sattler, C., "Solar fuels from concentrated sunlight," SolarPACES, Solar Power and Chemical Energy Systems, IEA report, 2009. With permission; Fletcher, E.A. and Moen, R.L., *Science*, 197, 1050–1056, 1977. With permission.)

11.2 ELECTROLYSIS AND ITS DERIVATIVE TECHNOLOGIES

Electrolysis of water is the decomposition of water (H_2O) into oxygen (O_2) and hydrogen (H_2) by the passage of electric current through it [11–21] (Fateev et al., 2012, pers. comm.; Laguna-Barcero et al., 2012, pers. comm.). This process requires a large amount of electrical energy that can be supplied by numerous sources such as hydropower, wind energy, solar energy, nuclear energy, geothermal energy, and electrical energy generated by fossil fuel and biomass. The electrical power needed can also be supplied by the energy stored in the form of hydrogen that is generated by other sources of renewable energy.

Electrolysis has been known to produce hydrogen since the early nineteenth century. It gives hydrogen at 99.99% purity. Bockris et al. [11] showed that the cost of energy generated by electrolysis can be expressed by the following formula:

$$\text{Cost of 1 GJ (\$)} = 2.29\, Ec + 3 \tag{11.2}$$

which assumes 100% Faraday efficiency. E is the potential difference across the electrodes to produce current density in the order 100–500 mA cm^{-2} and c is the cost of the electricity in cents per kilowatt-hour of electricity. The research carried out between 1970 and 1984 reduced the value of E from 2.2 V to around 1.6 V at current density of 500 mA cm^{-2}. Since 1995, all commercial electrolyzers have been operating with 1.6 V per cell. There are several indirect methods to improve electrolytic cell performance, and these are described by Bockris et al. [11]. In the recent years, hydrogen is also created by coal slurry electrolysis. Recent advances can also reduce the potential for the electrolysis to as low as 0.5 V.

If the electricity is obtained with a heat engine, Carnot efficiency limitation applies. Thus, if the electricity is obtained from coal, the normal efficiency is about 39% and the remaining 61% is lost as heat. This loss of thermal energy makes the electricity generated by wind or hydroelectric energy more efficient. The processes are often considered in combination with a nuclear or solar heat source. A high-temperature electrolysis (HTE) process may be favorable when high-temperature heat is available as waste heat from other processes. The use of such waste heat makes the overall process cost efficient.

11.2.1 Alkaline Electrolysis

Alkaline electrolyzers use an aqueous KOH solution (caustic) as an electrolyte that usually circulates through the electrolytic cells [1–11]. Alkaline electrolyzers are suited for stationary applications and are available at operating pressures up to 25 bar. Alkaline electrolysis is a mature technology allowing unmanned remote operation with significant operating experience in industrial applications. The following reactions take place inside the alkaline electrolytic cell:

$$\text{Electrolyte:}\quad 4H_2O \rightarrow 4H^+ + 4OH^- \tag{11.3}$$

$$\text{Cathode:}\quad 4H^+ + 4e^- \rightarrow 2H_2 \tag{11.4}$$

$$\text{Anode:}\quad 4OH^- \rightarrow O_2 + 2H_2O + 4e^- \tag{11.5}$$

$$\text{Sum:}\quad 2H_2O \rightarrow O_2 + 2H_2 \tag{11.6}$$

Commercial electrolyzers usually consist of a number of electrolytic cells arranged in a cell stack. The major research challenges for the future are the design and manufacturing of electrolyzer equipment at lower costs with higher energy efficiency and large turndown ratios.

11.2.2 HTE Process

HTE is more efficient economically than traditional room-temperature electrolysis because some of the energy is supplied by heat that is cheaper than electricity and the electrolysis reactions are more efficient at higher temperatures. While at 2500°C, thermal energy alone can dissociate water molecules, generally HTE systems operate between 100°C and 850°C [12,14,16–19] (Laguna-Barcero et al., 2012, pers. comm.). The efficiency of HTE process can be easily estimated by assuming that the heat required comes from heat engines and heat energy required for 1 kg of hydrogen (350 MJ) at 100°C gives the efficiency of 41%. Similar calculation at 850°C gives the efficiency of 64%.

The process requires a careful use of materials for electrodes and electrolyte. For a solid oxide electrolyzer cell (SOEC), numerous materials for electrodes and electrolytes have been tested. Recent studies have used yttria-stabilized zirconia (YSZ) electrolytes, nickel-cermet steam/hydrogen electrodes, and mixed oxide of lanthanum, strontium, and cobalt oxygen electrodes [18,19] (Laguna-Barcero et al., 2012, pers. comm.). Future advances in the HTE process will require materials that can withstand high temperature, high pressure, and corrosive environment. At the present time, the HTE process appears to be an inefficient way to generate hydrogen. The process will become more efficient if sources such as nuclear, solar, and hydro energy can be the source of thermal energy.

When the energy is supplied in the form of heat, such as by solar or nuclear energy, the production of hydrogen by HTE is very attractive. Unlike in low-temperature electrolysis, in HTE water converts more of initial thermal energy into chemical energy (like hydrogen) by increasing the conversion efficiency. Since energy in the HTE process is supplied in the form of heat, less of the energy must be converted twice (from heat to electricity and then to chemical form), and so less energy is lost and efficiency can be doubled up to 50%.

While the heat required for the HTE process can be obtained by solar energy or nuclear energy, the latter source is more reliable and is often used. The solar form of high-temperature heat is not consistent enough to bring down the capital cost of HTE equipment. More research into HTE and high-temperature nuclear reactors may eventually lead to hydrogen supply that is cost competitive with natural gas steam reforming. This concept of coupling a high-temperature electrolyzer and a high-temperature gas-cooled nuclear reactor (HTGR) has been demonstrated in a laboratory but not at a commercial scale, although Idaho National Laboratory is developing a commercial process based on this concept [17].

11.2.3 HPE Process

When electrolysis is conducted at high pressure, the produced hydrogen gas is compressed at around 120–200 bar (1740–2900 psi). By pressurizing the hydrogen in the electrolyzer, the need for an external hydrogen compressor is eliminated. The average energy consumption for internal compression is around 3% [13].

HPE is often carried out using a solid polymer electrolyte (SPE) membrane such as perflurosulfonic acid (Nafion) rather than classic liquid electrolyte (alkaline electrolyte) under high pressure. Laoun [13], LeRoy et al. [20,21], and Onda et al. [15] carried out a thermodynamic analysis of such a process and showed the importance of temperature and pressure on the entire efficiency of water electrolysis. Using the model and analysis of LeRoy et al. [20,21], Onda et al. [15] showed that a temperature change up to 250°C and pressure changes up to 70 atm can be carried out by polymer electrolytic membranes. They showed that an increase in pressure and a decrease in temperature deliver more power for water electrolysis. The increase is, however, found to be small at pressures above around 200 atm. They also found that hydrogen can be produced with about 5% less power using HPE than that required using atmospheric water electrolysis.

Fateev et al. (2012, pers. comm.) showed that water electrolysis using polymeric electrolyte membrane (PEM) has demonstrated its potential for high cell efficiency (energy consumption of about 4–4.2 kW/Nm^3 H_2) and gas purity of about 99.99%.

They studied the effects of increasing operating pressure up to several hundred bars for direct storage of hydrogen in a pressurized vessel. Their study showed that while PEM water electrolyzers operating at pressures up to 70 bar can be used to produce hydrogen and oxygen of electrolytic grade with high efficiencies; an increase in cross-permeation at higher pressure, can cause the hydrogen and oxygen mixture concentration to reach the critical level of explosive mixtures. The cross-permeation can be reduced by surface modification of solid electrolytes using low-permeability protective layers of coating. Contaminant concentration in the produced gases can also be reduced by adding catalyst gas recombiners either directly in the electrolytic cells or along the production line. Fateev et al. (2012, pers. comm.) showed that by using gas recombiners inside the electrolysis cell, it was possible to maintain the hydrogen content below 2 vol% at an operating pressure of 30 bar, with Nafion 117 as the solid electrolyte.

11.2.4 Photoelectrolysis

In this process, hydrogen and oxygen are separated in a light-driven electrolysis cell. Thus, the reactions that occur at the p-type cathode involve the evolution of hydrogen and those that occur at the n-type anode involve the evolution of oxygen. No external battery is used in the electrolysis process. While ideally the current between electrodes can be used as electricity and hydrogen produced from the process can be used as fuels, the efficiency of the overall process is about 1% [22–26] (Rajeshwar, 2012, pers. comm.). The progress in photoelectrolysis faces three major barriers: (1) There are no valid and significant theoretical analysis on the subject. The works of Scaife et al. [25], Scaife [26], and Ohashi et al. [22] appear to have some deficiencies. (2) The assumption made for years that Fermi level in solution as an important aspect of the conditions under which cells would work is proven not to be true. (3) The corrosion of semiconductor surfaces in contact with solution can be considerable. The corrosion is caused by heat as well as by photoelectrochemical reactions. Photoelectrochemical reaction efficiency is currently the same as that of photosynthesis. In the recent years, the increase in efficiency by photoelectrocatalysis has been achieved. Numerous metals such as TiO_2/pGaP, $SrTiO_2$/GaP, tin oxide, and other coatings of TiO_2 and CdS [22–26] (Rajeshwar, 2012, pers. comm.) on electrodes have been tested to improve the efficiency and life of the photoelectrolytic cell. The work of Szklarczyk and Bockris [23,24] showed that photoelectrocatalysis is directly related to electrocatalysis. The rate-determining step in photoelectrocatalysis is dependent on the transfer of charge at the metal–solution interface and not at the semiconductor–solution interface.

11.2.5 Photo-Aided Electrolysis

One method to improve efficiency is to have light falling upon an electrode. This can be achieved by applying a potential from an outside power source to the concerned electrodes. A 30%–40% efficiency in this case is not very impressive because the overall efficiency includes efficiencies for both light and electricity to hydrogen and not of light alone. The efficiency in this case can be improved by about 3%–4%

[27,28] (Rajeshwar, 2012, pers. comm.). This method is in general not preferred because not only it uses light as a source for hydrogen generation, but the overall process also requires electrical components.

11.2.6 Photovoltaic Electrolysis

One way to avoid corrosion problem in photoelecrolysis is to use the concept of photovoltaic cell working in air and electrolyzing a distant electrolyzer [11,28]. Here, semiconductors are not in direct contact with the solution so that corrosion problems cease to exist. This device can effectively be used with solar energy-generated electricity. The device contains two cells.

The best setup for efficiency is, however, recorded by Murphy–Bockris cell [28] using n-on-p gallium arsenide coated with ruthenium oxide and p-on-n gallium arsenide coated with platinum. Such a cell gave about 8% conversion of light to hydrogen production at current density in the range of 10 mA cm^{-2}. The cell life was also at least as good as that of the photovoltaics (PV) in air. Two advantages of Murphy–Bockris cell are that (1) cell is in solution so that the concentration of light upon the electrode that gives high temperatures can be used to provide household heat and (2) only one device is needed compared to two that are needed in PV cell in air. There are numerous ways to use PV cell in conjunction with electrolysis. One common method is described in Section 11.2.7.

11.2.7 Solar Electrolysis

The process of solar electrolysis involves generation of solar electricity via PV or concentrating solar power (CSP) followed by electrolysis of water [1,2,6,8,9,12]. This process is considered to be a benchmark for other thermochemical solar processes for water splitting that offers potential for energy-efficient large-scale production of H_2 (Figure 11.1). For solar electricity generated from PV cell and assuming solar thermal efficiencies at 15% or 20% and electrolyzer efficiency at 80%, the overall solar-to-hydrogen conversion efficiency will range from 12% to 16% [1,2,6,8,9,12]. If we assume solar thermal electricity cost of \$0.08/kWh, the projected cost of H_2 will range from \$0.15 to \$0.20/kWh, that is, from \$6 to \$8/kg H_2 [1,2,6,8,9,12]. For PV electricity, costs are expected to be twice as high. HTE process can significantly reduce electricity demand if it is operated at around 800°C–1000°C via SOEC. The high-temperature heat required for such a process can be supplied by the CSP system [1].

11.3 PHOTOCHEMICAL AND ITS DERIVATIVE TECHNOLOGIES

The dissociation of water can be assisted by photocatalysts that are directly suspended in water [29,30]. As shown below, a number of photocatalysts are possible. Early work by Gray et al. [31–34], Whitten et al. [35,36], and Maverick and Gray [37] showed that polynuclear inorganic complexes, excited metal complexes, and surfactant ruthenium complexes can help photochemical decomposition of water to produce hydrogen. Kiwi et al. [38] presented a review of homogeneous and heterogeneous photoproduction of hydrogen and oxygen from water. The majority of the photoredox systems (heterogeneous photolysis) involve a photosensitizer, an electron acceptor, and an electron donor, with the redox catalyst assisting in the gas evolution step. Excitation of the sensitizer (S) leads to an electron transfer:

$$S + A \rightleftarrows hv\, S + A \tag{11.7}$$

which is followed by the catalytic step:

$$A^{-} + H_2O \rightleftarrows \text{cat}\, A + OH^{-} + \frac{1}{2} H_2 \tag{11.8}$$

The back conversion of S^+ to S may be achieved by sacrificing a donor D added to the solution

$$S^{+} + D^{-} \rightarrow S + D^{+} \tag{11.9}$$

Koriakin et al. [39] used acridine dyes as sensitizers, Eu^{3+}, V^{2+} salicylates as electron acceptors, and "Adams" catalyst (PtO_2) as the redox catalysts. Numerous other sensitizers, electron acceptors, and redox catalysts are illustrated by Bockris et al. [11]. An efficiency of up to 30% at an elected wavelength for hydrogen production for a brief duration by photolytic process has been reported by Kalyanasundaram et al. [40].

11.3.1 Water Splitting on Semiconductor Catalysts (Photocatalysis)

Duonghong et al. [41] were the first to investigate the splitting of water by utilizing microsystems [41–65] (Correa, 2009, pers. comm.). In this system, the colloidal particles are made up of suitable conductor materials, for example, TiO_2. On these colloids are induced two metallic substances, for example, ruthenium oxide and platinum. When the system is irradiated, hydrogen is evolved on the platinum and oxygen on ruthenium oxide. Each colloidal particle is a micro photocell. Using small TiO_2 particles, a large area of TiO_2 can be exposed to light. The system needs to be heated to last more than several hours. There are some doubts whether or not equal production of hydrogen and oxygen is achieved and whether oxygen is engaged in side reactions. It is difficult to measure the efficiency of this system. In addition, hydrogen and oxygen come off from water together and their separations add extra cost. Furthermore, the simultaneous presence of oxygen and hydrogen in water can give rise to chemical catalysis and recombination to water [41–65].

11.3.1.1 Titanium Oxide Photocatalysts

TiO_2 was the first semiconductor used in water dissociation reaction [56]. A pure and powdered TiO_2, however, only absorbs UV fraction of solar light and thus not very effective for total absorption of solar light. The visible light response of TiO_2 was improved by chemical doping of TiO_2 with partially filled d-orbitals such as V_5^+, Cr_3^+, Fe_3^+, CO_2^+, and Ni_2^+ [44,61]. While these doping improved visible light response, they did not improve water dissociation reaction. Kato, Kudo, and coworkers [57–61] reported that TiO_2 co-doped with a combination of Sb_5^+ and Cr_3^+ became active for O_2 evolution under visible light from an aqueous solution using $AgNO_3$ as sacrificial agent. The physical doping of transition metal ions into TiO_2 by the advanced ion-implantation technique also allowed modified TiO_2 to work under visible light radiation. The ion implantation technique is, however, very expensive for commercial use. The visible light response can also be obtained by doping of anions such as N, S, or C [62–64] as substitutes for oxygen in the TiO_2 lattice. When TiO_2 is fused with metal oxides such as SrO, BaO, and Ln_2O_3, metal titanates and intermediate band gaps are obtained [44]. Materials such as $SrTiO_3$, $La_2Ti_2O_7$, and $Sm_2Ti_2O_7$ have shown some promising results. Promising results have also been shown by using $Sm_2Ti_2S_2O_5$, where sulfur anion is substituted for oxygen [44]. Under visible light radiation, the last material works as a stable photocatalyst for the reduction of H^+ to H_2 or the oxidation of H_2O to O_2 in the presence of sacrificial electron donor Na_2S–Na_2SO_3 or methanol or acceptor Ag^+ [44]. A new class of titanium semiconductors, titanium disilicide ($TiSi_2$) that absorbs a wide range of solar light, has recently been proposed as a prototype photocatalyst for the water dissociation reaction. More description of this catalyst is given in an excellent review by Navarro et al. [44].

11.3.1.2 Tantalates and Niobates

Layered and tunneling structures of oxides are considered as promising materials for water dissociation reaction. Tantalates and niobates oxides with corner-sharing octahedral MO_6 (M = Ta or Nb) have been examined as photocatalysts for water dissociation [44]. Kato and Kudo [42] observed that $MTaO_3$ (M = Li, Na, K) are effective photocatalysts for water dissociation under UV light. The oxides crystallize in pervoskite structure type. Lin et al. [43] showed that $NaTaO_3$ produced by sol–gel method gave higher activity for water dissociation than the same material prepared by the high-temperature solid-state synthesis. The most active photocatalysts were those that achieve higher nitrogen substitution, maintaining the original layered structure of $Sr_2Nb_2O_7$. More detailed discussion of these types of catalysts is given by Navarro et al. [44].

11.3.1.3 Transition-Metal Oxides, Nitrides, and Oxynitrides

Certain vanadium and tungsten compounds were found to be active in water dissociation reaction. $BiVO_4$ with scheelite structure and Ag_3VO_4 with pervoskite structure showed photocatalytic activity in visible light for oxygen evolution from an aqueous silver nitrate solution [61,65]. The WO_3 system also oxidizes water at moderately high rates in the presence of Ag^+ and Fe_3^+ ions [44]. Under visible light, Pt–WO_3 alone with $NaIO_3$ produces oxygen at high rate but produced no hydrogen [44]. Some other catalysts in this category are also examined by Navarro et al. [44].

Navarro et al. [44] also reported that nitrides and oxynitrides of transition metal cations with d10 electronic configurations (Ga_3^+ and Ga_4^+) constitute a class of photocatalysts suitable for water dissociation in visible light without sacrificial reagents. Among various cocatalysts examined, the largest improvement in activity was obtained when $(Ga_{1-x}Zn_x)(N_{1-x}O_x)$ was loaded with a mixed oxide of Rh and Cr [44]. This semiconductor evolves hydrogen and oxygen steadily and stoichiometrically under visible light from pure water in the absence of sacrificial agent. The solid solution between ZnO and $ZnGeN_2$ $(Zn_{1+x}Ge)$ – (N_2O_x) has also been found to be active oxynitride photocatalysts for pure water dissociation in visible light [44]. Finally, $(Zn_{1+x}Ge)(N_2O_x)$ solid solution loaded with nanoparticulate RuO_2 cocatalyst is also active under visible light, generating hydrogen and oxygen stoichiometrically from pure water [44].

11.3.1.4 Metal Sulfides

Navarro et al. [44] reported that while small band gaps in metal sulfides make them very attractive photocatalysts for water dissociation, they are unstable for water oxidation reaction under visible light. A common method for the reducing photocorrosion of the sulfides under irradiation is the use of suitable sacrificial agents such as Na_2S/Na_2SO_3 salt mixture [44]. CdS with wurtzite structure is the best-studied metal sulfide photocatalyst [66–68]. This catalyst property can be improved by improving preparation method that leads to CdS phases with good crystallinity and few crystal defects. Composite systems of CdS with TiO_2, ZnO, and CdO [69–71] also improved photoactivity. The incorporation of elements into the structure of CdS to make a solid solution is another strategy for improving the photocatalytic properties of CdS [44]. The substitution of ZnS into CdS structure improved the activity of the composite material [44].

ZnS was also another semiconductor investigated for photocatalytic activity [44]. The chemical doping of ZnS by Cu_2^+, Ni_2^+, and Pb_2^+ [59,60,72,73] allowed ZnS to absorb visible light. These doped ZnS photocatalysts showed high photocatalytic activity under visible light for hydrogen production from aqueous solutions using SO_3^{2-}/S^2 as electron donor reagents. Combining ZnS with $AglnS_2$ and $CulnS_2$ to produce solid solutions $(CuAgln)_xZn_{2(1-x)}S_2$ is another strategy for improving optical absorption in the visible light range [44,74–76]. Co catalysts such as Pt loaded on $(Agln)_{0.22}Zn_{1.56}S_2$ showed the highest activity for hydrogen evolution [44]. The ternary sulfides comprising ln^{3+} and one type of transition metal cation (Cd_2^+, Zn_2^+, Mn_2^+, Cu^+) found to have low efficiency for water dissociation in visible light. More description of sulfide photocatalysts is given by Navarro et al. [44]. An overview of recently developed photocatalysts for water splitting under visible light illumination is also summarized by Navarro et al. [44].

11.3.2 Photobiological Production of Hydrogen from Water

The water splitting can also be carried out photobiologically [49]. Biological hydrogen can be produced in an algae bioreactor. In the late 1990s, it was discovered that if the algae are deprived of sulfur, it will switch from the production of oxygen (a normal mode of photosynthesis) to the production of hydrogen. It seems that the production is now economically feasible by the energy efficiency surpassing 7%–10% [49].

Hydrogen can be produced from water by hydrogenase-catalyzed reduction of protons by the electrons generated from photosynthetic oxidation of water using sunlight energy. In the recent years, use of a variety of algae to produce hydrogen from water has been extensively investigated and reviewed [77,78]. These reviews mention the use of sulfur deprivation with *Chlamydomonas reinhardtii* to improve hydrogen production by algae [79,80]. In addition, certain polygenetic and molecular analyses were performed in green algae [81,82]. These methods, however, did not significantly improve the rate and the yield of algal photobiological hydrogen production. Solar-to-hydrogen energy conversion using algae has efficiency <0.1% [83]. The rate and yield of algal photobiological hydrogen production is limited by (1) proton gradient accumulation across the algal thylakoid membrane, (2) competition from carbon dioxide fixation, (3) requirement for bicarbonate binding at photosystem II (PSII) for efficient photosynthetic activity, and (4) competitive drainage of electrons by molecular oxygen. Recently, Lee [84–86] has outlined two inventions for more efficient and robust photobiological production of hydrogen from water: (1) designer proton channel algae and (2) designer switchable PSII algae. These two new inventions eliminate not only the four problems mentioned earlier but also oxygen sensitivity of algal hydrogenase and H_2–O_2 gas separation and safety issue. More work in this area is needed. The details of the two new inventions are described by Lee [49].

11.3.3 Plasma-Induced Photolysis

It has been suggested that plasma [87] can be used to produce photons of appropriate energy so that water can be dissociated in the gas phase. Thus, in a hypothetical fusion of hydrogen, it would be possible to produce a light in the region of 1800–950A by the addition of aluminum to the plasma [11,88]. The main gain from this method is that the thermal energy absorbed would be converted to electricity in a heat engine at about 30% efficiency. A gain in efficiency is obtained because hydrogen will be produced by both photolysis and electrolysis. At the present time, however, the production of high energy protons is only possible by the injections of aluminum into plasmas. The possibility of obtaining very high efficiency (up to 90% which is possible for electrolysis) is unlikely. Furthermore, the recombination of hydrogen and oxygen could be a major drawback of this process [11].

11.4 THERMAL AND THERMOCHEMICAL DECOMPOSITION OF WATER

The direct thermal dissociation has been examined since 1960s [1,11,89–140] (Funk, 2011, pers. comm.; Bamberger, 2011, pers. comm.). In direct thermal decomposition, the energy needed to decompose water is supplied by heat only. This requires a minimum temperature of at least 2200°C (even for partial decomposition) and as high as about 4700°C, and this makes the process somewhat unrealistic. At this temperature, about 3% of all water molecules are dissociated as H, H_2, O, O_2, and OH. Other reaction products like H_2O_2 or HO_2 remain minor. At about 3200°C, about half of the water molecules are dissociated. It is well known that an initiation of thermal splitting of water even at low pressure requires 2000 K (about 1730°C). As mentioned above, at an atmospheric pressure, 50% dissociation requires about 3500 K. This temperature can be reduced to less than 3000 K (about 2730°C) at 0.01 atm pressure. As will be discussed later, the catalysts can accelerate the dissociation at lower temperature. The lower total pressure favors the higher partial pressure of hydrogen, which makes the reactor to operate at pressures below an atmospheric pressure very difficult [1,11].

While a single-step thermolysis is conceptually simple, its realization is very challenging since it needs a high-temperature heat source above 2200°C for achieving a reasonable degree of dissociation and an effective technique to separate hydrogen and oxygen to avoid explosive mixture. The ideas proposed to separate hydrogen from the products include effusion separation and electrolytic separation. Membranes made of zirconia and other ceramics can withstand such high temperatures, but they fail to absorb severe thermal shocks that often occur when working under high-flux solar radiation. Other techniques that have been evaluated are rapid quench by injecting a cold gas, expansion in a nozzle, or submerging a solar-irradiated target in liquid water. The last technique is workable and simple, but a quench introduces a significant drop in energy efficiency and produces an explosive gas mixture. The efficiency can also be further decreased by reradiation, and the type of temperature (e.g., 2725°C for 64% dissociation at atmospheric pressure) required creates material limitations [1,11].

One of the problems for thermal dissociation of water is the materials that can stand temperatures in the excess of at least 2200°C–2500°C. Several materials such as tantalum boride, tantalum carbide, tungsten, and graphite are possible. However, at these temperatures, only oxides are stable. Graphite is chemically unstable in the presence of hydrogen and oxygen at these high temperatures. Tungsten and tungsten carbide get oxidized at these temperatures. The effect of hydrogen on oxide catalysts at these temperatures is not known. Ceramic materials such as boron nitride can also be useful if its oxidation can be controlled. Recent studies have shown that a low amount of dissociation is possible [11]. The separation of oxygen and hydrogen can be carried out in a semipermeable membrane of palladium or ZrO_2–CeO_2–Y_2O_3, which removes oxygen preferentially. Lede et al. [126,127] used a ZrO_2 nozzle through which steam is forced into a thermal stream and decomposed and unreacted water is quenched suddenly to remove water and oxygen. The resulting gas contained only a small amount (about 1.2 mol%) of hydrogen. Another possible solution is the use of heat-resistant membrane made of Pd or ZrO_2, both of which selectively permeate hydrogen. The gas can also be separated using a magnetic field. The source of heat is also an issue. Solar or nuclear sources are possibilities. They are, although at the early stages of development and at the present time, only possible on a smaller scale [11,135–139].

In the recent years, thermal dissociation of water is achieved using nuclear [90,91,93,94] (Funk, 2011, pers. comm.; Bamberger, 2011, pers. comm.) and solar energy [1,92,108–110]. Some prototype generation IV reactors operate at 850°C–1000°C, a temperature considerably higher than the existing commercial nuclear power plants. General Atomics predicts that hydrogen cost using HTGR would cost \$1.53/kg, a cost that compares well with \$1.40/kg costing by steam reforming mechanism. One advantage of nuclear reactor producing both electricity and hydrogen is that it can shift production between the two. For example, plant can produce electricity during the day and hydrogen during night by matching the variations in electricity demand. Thus, hydrogen can act as a storage unit from which electricity can be generated when needed. The peak demand of electricity can be handled by the energy stored in hydrogen.

The high temperature needed to split the water can also be provided by solar energy. In Spain, a 100-kW HYDROSOL II pilot plant is operated at the Plataforma

Solar de Almeria (PSA), which uses sunlight to get 800°C–1200°C to split water [1]. This plant has been in operation since 2008. A megawatt plant based on this concept can be built by having several parallel reactors operated by connecting the plant to heliostat fields (field of sun-tracking mirrors) of a suitable size [1,108,109,113]. H_2 power systems [1,110,113] have proposed a membrane system for solar dissociation of water at temperatures as high as 2200°C. The membrane separates hydrogen as soon as it is produced in the so-called solar water cracker. Such a cracker with 100 m^2 concentrator can produce almost 1 kg of hydrogen per hour during full sunlight conditions.

The required scale of thermal decomposition process such that it is economical remains questionable. Large volume may require exotic refractories. At present, the choice of thermal decomposition takes second place to the thermochemical cycles described later. At a laboratory scale, thermal decomposition has also been analyzed using solar energy as a source of heat. The overall efficiency of solar thermal process for hydrogen generation is considerably higher than that of PV/electrolysis [1,108–113]. As shown below, solar thermal splitting of water is aided by multiple chemical steps, but the following three principles govern the success of solar thermochemical reactions: (1) drive chemical reactions at the highest temperature possible, consistent with other pertinent constraints such as materials of construction and ability to concentrate light; (2) seek simple processes with as few steps as possible, preferably one (e.g., cracking); and (3) for multistep water splitting thermochemical cycles, seek processes involving a highly endothermic step driven using concentrated sunlight, followed by an exothermic step that is autothermal and can run continuously.

11.4.1 Thermochemical Decomposition of Water

Thermochemical cycles have been intensely investigated over the past more than four decades [1,11,89–139] (Funk, 2011, pers. comm.; Bamberger, 2011, pers. comm.). In this method, two-, three-, or four-step chemical reactions aided by a source of heat such as nuclear or solar can dissociate water and separate hydrogen and oxygen at temperatures around 800°C–900°C. The method has some inherent issues:

1. The original concept [1,11,107] was that since the method avoided the formation of electricity by the conversion of heat to mechanical work, it would avoid Carnot cycle, as this is the fundamental difficulty in reducing the price of hydrogen production by electrolysis method. The thermocycles were thought to produce hydrogen at a cost of about half of that for electrolytic method. This thinking was fallacious because the methods have to have reactions carried out at different temperatures in order that the entropic properties of the partial reactions in each cycle can be used to maximum advantage. Furthermore, when the individual reactions have a positive entropy change, it is desirable to carry out reactions at the highest temperature possible to minimize the overall free-energy change. Conversely, if the entropy changes are negative, the reactions should be carried out at the lowest temperature. However, this requirement of

changing the temperature of the reaction in various cycles gives rise to the requirements to change the pressure too, and so it would be necessary to pump gases from one temperature and one pressure to another and this is similar to the Carnot cycle.

2. With three to four cycles and need to change apparatus for each, plant capital costs for unit hydrogen production are likely to be more than those that occur in the electrolysis. Furthermore, at temperatures such as 800°C–900°C, the corrosion will cause the plant life to be short.
3. Generally, it is assumed [131,132] that the reaction would take place along the free-energy pathway, but in reality it takes place down a reaction rate pathway [133,134] and not necessarily on thermodynamic pathway. Also, because of possible side reactions, the final product may not be what was intended. Due to these reasons, if cyclicity in thermochemical steps fails even by 1%, a considerable amount of unwanted materials will build up and calculated economics based on the cyclical nature of the process is no longer valid.

In spite of these arguments, a considerable investigation on thermochemical cycles to produce hydrogen at the temperatures lower than one required for the thermal dissociation has been carried out. The moderate temperatures used in these cycles, in general, also cause less material and separation problems. More than 300 different types of chemical cycles have been proposed and tested. In this section, we evaluate some of the important thermochemical cycles.

Previously, thermochemical cycles were characterized as those that use process heat at temperatures <950°C. These are expected to be available from high-temperature nuclear reactors. These cycles required three or more chemical reaction steps, and they are challenging because of material problems and inherent inefficiency involved with heat transfer and product separation in each step. One example is hybrid sulfuric acid cycle that requires two steps incorporating one electrolysis step. The leading candidates for multistep thermochemical cycles include mainly three-step sulfur–iodine (S–I) cycle based on thermal decomposition of sulfuric acid at 850°C and four-step UT-3 cycle based on hydrolysis of calcium and iron bromide at 750°C and 600°C, respectively [87–131] (Funk, 2011, pers. comm.; Bamberger, 2011, pers. comm.).

Recent advancement in the development of optical systems for large-scale solar concentrations capable of achieving mean solar concentration ratio that exceeds 5000 suns allows high radiation fluxes capable of getting temperature >1200°C. Such high temperatures allowed the development of efficient two-step thermochemical cycles using metal oxide–redox reactions (see Figure 11.3). Some of the important cycles are briefly described in Sections 11.4.1.1 through 11.4.1.10.

11.4.1.1 The UT-3 Cycle

The UT-3 cycle is based on two pairs of chemical reactions [91,93,94] (Funk, 2011, pers. comm.). The first pair is as followa:

$$CaO + Br_2 \rightarrow CaBr_2 + \frac{1}{2}O_2 \qquad (550°C) \qquad (11.10)$$

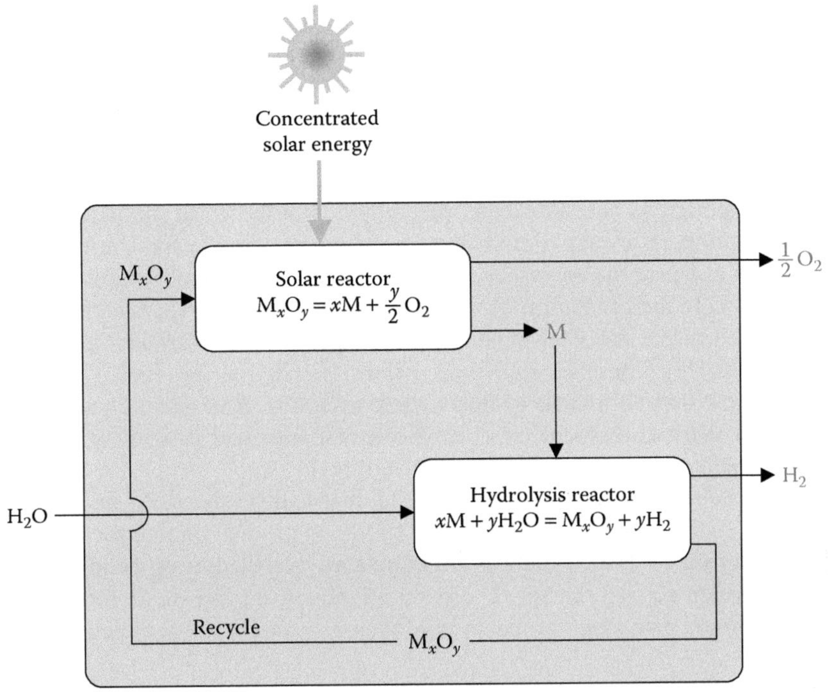

FIGURE 11.3 **(See color insert.)** Thermochemical route based on metal oxide–redox reactions. (From Meier, A. and Sattler, C., "Solar fuels from concentrated sunlight," SolarPACES, Solar Power and Chemical Energy Systems, IEA report, 2009.)

$$CaBr_2 + H_2O \rightarrow CaO + 2HBr \qquad (725°C) \tag{11.11}$$

According to these reactions, a production of hydrobromic acid is accompanied by the release of oxygen. The next set of two reactions is as follows:

$$Fe_3O_4 + 8HBr \rightarrow 3FeBr_2 + 4H_2O + Br_2 \qquad (250°C) \tag{11.12}$$

$$3FeBr_2 + 4H_2O \rightarrow Fe_3O_4 + 6HBr + H_2 \qquad (575°C) \tag{11.13}$$

It indicates the reduction of water by a bromide, accompanied by release of hydrogen. In the original concept, these two reactions operate separately and sequentially in two separate reactors, wherein heterogeneous reactions between gases and solids are carried out. The main difficulty encountered was the cycling behavior of these matrices. For example, in the first reactor during the first cycle, CaO is converted to $CaBr_2$; in the second cycle reverse transformation occurs; and so on. The design proved difficult to extrapolate to an industrial scale. Many design issues for commercial applications are still under investigation [91–94] (Funk, 2011, pers. comm.).

11.4.1.2 Zn/ZnO Cycle

One of the most researched metal oxide–redox pair is Zn/ZnO [1,91,95,110,111]. Since the product of ZnO decomposition at high temperature (namely, Zn and oxygen) readily recombines, the quenching of the product is necessary (Figure 11.3). Without heat recovery from the quench process, the estimated exergy efficiency [1] of this cycle is around 35%. The electrothermal process to separate Zn and oxygen at high temperatures has been experimentally demonstrated in small-scale reactors. Such high-temperature separation allows recovery of sensible and latent heats of the products to enhance the energy efficiency of the entire process. A high-temperature solar chemical reactor (Figure 11.4) was developed for this process, and solar tests were carried out at the Paul Scherrer Institute (PSI) solar furnace in Switzerland [1,95,107,110,111]. These tests allowed surface temperature to reach 1700°C in 2 s, with very low thermal inertia of the reactor system. In 2010, solar chemical reactor concept for thermal dissociation of ZnO was demonstrated in a 100-kW pilot plant in a larger solar research facility [1,95,107,110,111].

More recent work on this cycle showed that hydrolysis of Zn by the reaction $Zn + H_2O \rightarrow ZnO + H_2$ gave reasonable hydrogen production rate for the temperatures greater than 425°C. This was experimentally verified using nano-Zn particles and water in an aerosol reactor. The required molten Zn and steam for this process can be obtained using heat of reaction. Molten Zn can also be supplied by a quench

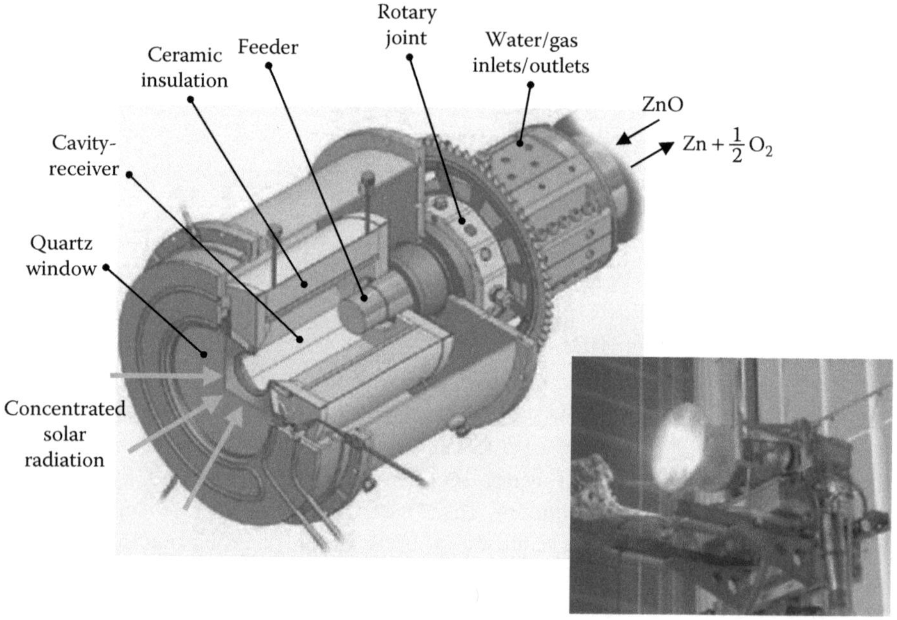

FIGURE 11.4 **(See color insert.)** Rotary solar reactor for the thermal dissociation of zinc oxide to zinc and oxygen at above 1700°C. (From Meier, A. and Sattler, C., "Solar fuels from concentrated sunlight," SolarPACES, Solar Power and Chemical Energy Systems, IEA report, 2009.)

unit of the nearby solar plant. The availability of Zn at the reaction site eliminates the storage and transportation need for the produced hydrogen. Attempts have also been made to store solar energy directly into Zinc–air batteries using Zn energy carrier from the process. The technology of redox batteries for solar energy storage is already commercially available [112,120–122].

11.4.1.3 SnO/SnO_2 Cycle

Another successful thermochemical cycle involves SnO/SnO_2 where exergy and energy efficiencies of 30% and 36%, respectively, can be obtained. The work carried out in 1 kW solar reactor at atmospheric and reduced pressure at Odeillo, France, has shown that SnO_2 reduction can be efficiently carried out at 1500°C and SnO hydrolysis can be carried out at 550°C [1,112].

11.4.1.4 Mixed Iron Oxide Cycle

Besides those mentioned above, manganese oxide, cobalt oxide, and iron-based mixed oxide–redox pairs have also been tested [90,93,94,106] (Funk, 2011, pers. comm.). The mixed iron oxide cycle was demonstrated at 10 kW level in the European Union's R&D project called "HYDROSOL" (2002–2005). The model for the monolithic solar thermochemical reactor (see Figure 11.5) was the catalyst converter used for automobile exhaust treatment. The multichanneled monoliths reactor with no moving parts absorbed solar radiation. The monolith channels were coated with mixed iron oxides–nanomaterials that can be activated by heating to 1250°C. The reactor dissociated water vapor and trapped oxygen allowing hydrogen to be released in the product stream at 800°C. Thus, a cyclic operation in a single closed receiver–reactor system separated produced oxygen and hydrogen. With the use of two or more reactor chambers in an alternate fashion, quasi-continuous stream of hydrogen was produced. "HYDROSOL II" (2005–2009) process tested 100 kW dual-chamber pilot reactor at PSA, Spain [1,90,93,94,106] (Funk, 2011, pers. comm.).

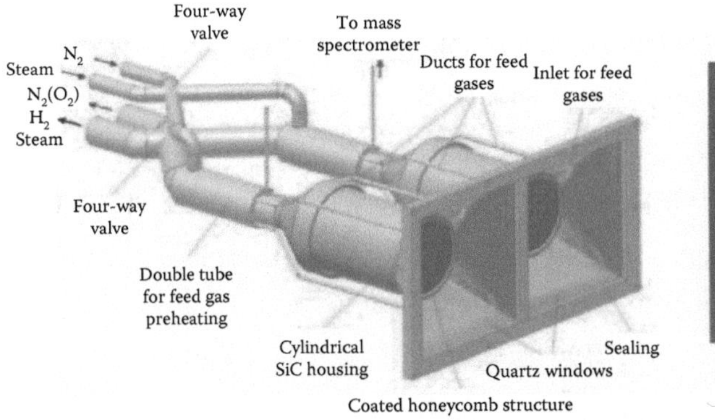

FIGURE 11.5 (See color insert.) Monolithic dual-chamber solar receiver reactor for continuous hydrogen production. (From Meier, A. and Sattler, C., "Solar fuels from concentrated sunlight," SolarPACES, Solar Power and Chemical Energy Systems, IEA report, 2009.)

Inoue et al. [89] examined mixed $ZnO/MnFe_2O_4$ system for two-step thermochemical cycle for the dissociation of water. This system among many other mixed oxide system is workable for producing hydrogen by thermochemical cycle. At 1000°C, the mixture of ZnO and $MnFe_2O_4$ reacted with water to generate hydrogen gas with 60% yield. The oxygen was produced around 1027°C completing a two-step cycle.

11.4.1.5 Carbothermal Reduction of Metal Oxides

In the recent years, under the European Union's R&D project SOLZINC (2001–2005), a 300-kW solar chemical reactor at the solar power research facility of the Weizmann Institute of Science (WIS) in Israel at temperatures ranging from 1000°C to 1200°C yielded up to 50 kg/h of 95% purity Zn and energy conversion efficiency of around 30% [1,90,95,106,121–124]. The process carried out carbothermal reduction of metal oxide (ZnO) using coke, natural gas, and other carbonaceous materials as reducing agents. This brings down the reduction of oxides even to lower temperatures. Carbothermal reductions of metal oxides such as iron oxide, manganese oxide, and zinc oxide with carbon and natural gas to produce the metals and the use of syngas were demonstrated in the solar furnaces. Such a solar chemical reactor concept—PSI's "two-cavity" solar reactor based on the indirect irradiation of ZnO and carbon (C) for producing Zn and carbon monoxide (CO)—was scaled up in the SOLZINC project [1,90,95,121–124].

11.4.1.6 Sulfur Family Thermochemical Water Splitting Cycles

All sulfur family thermochemical water splitting cycles (TCWSCs) depend on concentration and decomposition of sulfuric acid for the oxygen evolution step of the cycle [91,95–105,107,128–131]. The sulfuric acid decomposition step presents serious materials and catalyst deactivation challenges. The most active Pt catalysts deactivate very rapidly. Metal sulfate-based TCWSCs overcome this difficulty, but they use thermal input, thus degrading photonic energy. T-Raissi et al. [107] introduced FSEC's (Florida Energy Systems Consortium) metal sulfate–ammonia (MSO_4–NH_3) hybrid photochemical cycle/TCWSC that can be represented as follows:

$$\begin{gathered} SO_2(g) + 2NH_3(g) + H_2O\,(l) \rightarrow (NH_4)_2SO_3(aq) \\ \text{(Chemical absorption, 25°C)} \end{gathered} \tag{11.14}$$

$$\begin{gathered} (NH_4)_2SO_3(aq) + H_2O \rightarrow (NH_4)_2SO_4(aq) + H_2(g) \\ \text{(Solar photocatalytic, 80°C)} \end{gathered} \tag{11.15}$$

$$\begin{gathered} x(NH_4)_2SO_3 + M_2O_x \rightarrow 2xNH_3 + M_2(SO_4)_x + xH_2O \\ \text{(Solar thermocatalytic, 500°C)} \end{gathered} \tag{11.16}$$

$$\begin{gathered} M_2(SO_4)_x(s) \rightarrow xSO_2(g) + 2MO(s) + (x-1)\,O_2(g) \\ \text{(Solar thermocatalytic, 1100°C)} \end{gathered} \tag{11.17}$$

where:
M = Zn, Mg, Ca, Ba, Fe, Co, Ni, Mn, and Cu

Chemical equilibrium calculations for the reaction between ZnO and $(NH_4)_2SO_4$ indicate that both $ZnSO_4$ and $ZnO.2ZnSO_4$ can form stable reaction products. More than 20 sulfuric acid and/or metal sulfate decomposition-based TCWSCs have been reported. Major issue remains to be electrolytic oxidation of sulfur dioxide. The use of a depolarized electrolyzer as well as addition of a third process step such as S–I, S–Br, and S–Fe cycles has also been attempted. Some of these are described below [91,107]:

Ispra Mark 13 sulfur/bromine cycle [128]

$$Br_2(I) + SO_2 + 2H_2O(l) \rightarrow 2HBr(aq) + H_2SO_4(aq) \qquad 77°C \tag{11.18}$$

$$H_2SO_4(g) \rightarrow SO_2(g) + H_2O\,(g) + \frac{1}{2}O_2 \qquad 850°C \tag{11.19}$$

$$2HBr\,(aq) \rightarrow Br_2(aq) + H_2(\text{electrolytic}) \qquad 77°C \tag{11.20}$$

General Atomics' S–I cycle is described in Section 11.4.1.7 [129]. *Sulfur–iron cycle* can be described as follows:

$$Fe_2(SO_4)_3(aq) + SO_2 + 2H_2O \rightarrow 2FeSO_4(aq) + 2H_2SO_4 \qquad 25°C \tag{11.21}$$

$$H_2SO_4 \rightarrow SO_2(g) + H_2O\,(g) + \frac{1}{2}O_2 \qquad 850°C \tag{11.22}$$

$$2FeSO_4(aq) + H_2SO_4(aq) \rightarrow Fe_2SO_4(aq) + H_2 \qquad 25°C \tag{11.23}$$

To make the separation of HI and H_2O easier, Sato et al. [130] have proposed a nickel–iodine–sulfur version of S–I cycle. Others include the following:

$$CO + H_2O \rightarrow CO_2 + H_2 \qquad 550°C \tag{11.24}$$

$$CO_2 + SO_2 + H_2O \rightarrow H_2SO_4 + CO \qquad 500°C \tag{11.25}$$

$$H_2SO_4(g) \rightarrow H_2O(g) + SO_2(g) + \frac{1}{2}O_2 \qquad 900°C \tag{11.26}$$

$$SO_2 + H_2O + I_2 \rightarrow SO_3 + 2HI \qquad 200°C \tag{11.27}$$

$$SO_3 \rightarrow SO_2 + \frac{1}{2}O_2 \qquad 900°C \tag{11.28}$$

$$2HI \rightarrow H_2 + I_2 \qquad 450°C \tag{11.29}$$

$$2FeSO_4 + I_2 + 2H_2O \rightarrow 2Fe(OH)SO_4 + 2HI \qquad 20°C \tag{11.30}$$

$$2Fe(OH)SO_4 \rightarrow 2FeSO_4 + H_2O + \frac{1}{2}O_2 \qquad 100°C \tag{11.31}$$

$$3FeCl_2(s) + 4H_2O \rightarrow Fe_3O_4(s) + 6HCl(g) + H_2 \qquad 650°C \qquad (11.32)$$

$$Fe_3O_4(s) + Fe_2O_3(s) + 6HCl + 2SO_2 \rightarrow 3FeCl_2 + 2FeSO_4 + 3H_2O \quad 100°C \quad (11.33)$$

$$2FeSO_4 \rightarrow Fe_2O_3(s) + 2SO_2(g) + \frac{1}{2}O_2 \qquad 850°C \qquad (11.34)$$

Although these cycles address the issue of water solubility of SO_2, they have other issues of their own. For example, efficient separation of sulfuric acid from reaction products such as HI, HBr, and $FeSO_4$ is challenging. The determination and control of solution pH, particularly when other acids such as HI and HBr are formed, is a major issue. Abanades et al. [131] screened 280 TCWSCs and selected 30 as promising. There were nine metal sulfate-based TCWSCs in this selection because H_2SO_4 and M_SO_4 present an effective method for the heat-absorbing step of the TCWSCs. Some of these thermochemical cycles are also given by T-Raissi et al. [107].

The second approach is to introduce a metal oxide as a catalyst to convert low-concentration sulfuric acid to metal sulfate that is then decomposed to produce oxygen, sulfur dioxide, and metal oxide. Sulfur dioxide and water are sent to acid electrolysis unit for generation of hydrogen and sulfuric acid, thus closing the cycle. Introducing ZnO into the Westinghouse TCWSC, a new modified $ZnSO_4$ decomposition-based Westinghouse cycle can be written as follows [107]:

$$SO_2(g) + 2H_2O(l) = H_2 + H_2SO_4(aq) \qquad 77°C \text{ (electrolytic)} \qquad (11.35)$$

$$H_2SO_4(aq,\ 50\ wt\%) + ZnO(s) = ZnSO_4 \cdot H_2O(s) \qquad 80°C - 350°C \quad (11.36)$$

$$ZnSO_4 \cdot H_2O(s) = ZnSO_4(s) + H_2O(g) \qquad 450°C \qquad (11.37)$$

$$ZnSO_4(s) = SO_2(g) + \frac{1}{2}O_2 + ZnO(s) \qquad 850°C \qquad (11.38)$$

Similarly, metal oxide catalyst can be added to sulfur–bromine, S–I, and sulfur–iron cycles. These will give new modified metal-based TCWSCs. When energy input for these cycles is solar energy, they can utilize only the thermal energy, degrading the photonic portion of solar spectrum to lower grade heat.

11.4.1.7 S–I Cycle

The S–I cycle is one of the promising cycles for thermochemical hydrogen production [107,129,130]. It consists of three pure thermochemical steps that sum to the dissociation of water. These steps are as follows:

$$H_2O + SO_2 + I_2 = H_2SO_4 + 2HI \qquad (25°C - 120°C) \qquad (11.39)$$

$$H_2SO_4 = H_2O + SO_2 + \frac{1}{2}O_2 \qquad (11.40)$$

$$2HI = H_2 + I_2 \qquad (200°C - 400°C) \qquad (11.41)$$

The second reaction is a two-step reaction is as follows:

$$H_2SO_4 \rightarrow H_2O + SO_3 \qquad (400°C - 600°C) \tag{11.42}$$

$$SO_3 \rightarrow SO_2 + \frac{1}{2}O_2 \qquad (800°C - 900°C) \tag{11.43}$$

The first exothermic reaction is called Bunsen reaction and is operated at 120°C. The second endothermic reaction needs a temperature of about 850°C (in two steps as shown above). The last endothermic reaction runs at temperatures between 300°C and 450°C. Three reactors that are a part of the cycle are called Gibbs reactor, Bunsen reactor, and equilibrium reactor. The separation of H_2SO_4–HI mixture is the most critical part of the S–I cycle [107,129,130].

This cycle has been investigated by several research teams because the cycle involves only liquids and gases. General Atomics has discovered that it is possible to separate two acids in the presence of excess iodine and water. However, an efficient separation of HI from water and excess iodine at the outcome of Bunsen reaction still remains an issue. The high-temperature decomposition of acids is also an issue. The cycle was successfully tested in Japan to produce 45 l of hydrogen. It was also tested in France at the capacity of 50 l/h [107,129,130].

11.4.1.8 The Westinghouse Process

The Westinghouse process is one of the "sulfur family" of thermochemical cycles being considered for the generation of hydrogen [91,107,108]. It is a sulfur cycle using hybrid electrochemical/thermochemical process for decomposing water into hydrogen and oxygen. Sulfurous acid and water are reacted electrolytically to produce hydrogen and sulfuric acid. The resulting sulfuric acid is vaporized to produce steam and sulfur trioxide, which is subsequently reduced at higher temperatures into sulfur dioxide and oxygen. The process may be seen as a variant of the S–I process, in which iodine reactions are substituted for by sulfur dioxide electrolysis as follows:

$$SO_2 + 2H_2O \rightarrow H_2SO_4 + H_2 \qquad (20°C - 110°C,\ P = 2 - 10 \text{ bar}) \tag{11.44}$$

Following the separation of the water and sulfur dioxide for recycle to the electrolyzer, oxygen is available as a by-product. This has the advantage of requiring only one intermediate element. Sulfur was used because it is relatively inexpensive, its properties are well known, and it can assume a variety of valence states, thereby facilitating its use in oxidation–reduction reactions. The process requires electric energy that restricts its efficiency. Electrolysis is carried out in a strong acid medium, leading to corrosion issues. Moreover, this would require several compartments to restrict parasitic sulfur and H_2S production at the cathode [91,107].

11.4.1.9 Copper–Chlorine Cycle

The copper–chlorine (Cu–Cl) cycle is an important cycle due to its requirement for relatively low-temperature heat compared to other thermochemical water decomposition cycles [91,104,107]. It was identified by Atomic Energy of Canada Ltd. as a highly promising cycle for hydrogen production. The advantages of this cycle are (1) reduced construction materials, (2) inexpensive chemical agents, (3) minimal solids handling, and (4) reactions going to completion with few side reactions. It is well suited for energy supplied by the nuclear reactor. The important five steps in Cu–Cl cycle are as follows [91,104,107]:

Step 1: HCl production step

$$2CuCl_2(s) + H_2O(g) \rightarrow CuO * CuCl_2(s) + 2HCl(g) \qquad 400°C \qquad (11.45)$$

Step 2: Oxygen production step

$$CuO * CuCl_2(s) \rightarrow 2CuCl(l) + \frac{1}{2}O_2(g) \qquad 500°C \qquad (11.46)$$

Step 3: Electrochemical process

$$2CuCl(s) \rightarrow 2CuCl(aq) \rightarrow CuCl_2(aq) + Cu(s) \qquad \text{Ambient} \qquad (11.47)$$

Step 4: Flash drying

$$CuCl_2(aq) \rightarrow CuCl_2(s) \qquad > 100°C \qquad (11.48)$$

Step 5: Hydrogen production

$$2Cu(s) + 2HCl(g) \rightarrow 2CuCl(l) + H_2(g) \qquad 430 - 475°C \qquad (11.49)$$

This cycle is unusual in that it contains five chemical steps, although efforts have been made to reduce the number of chemical steps. Just like S–I cycle, Cu–Cl cycle has a significant potential due to lower temperature requirements. The literature has shown that the cost of hydrogen production by Cu–Cl cycle is better than electrolysis method at higher hydrogen production capacity (>30 tons per day) [91,104,107].

11.4.1.10 Copper–Sulfate Cycle

The copper/sulfate cycle involves two major steps: (1) hydrogen production from the reaction of water, $SO_2(g)$, and CuO(s) at room temperature and (2) the thermal decomposition of the products of the first step to form oxygen and to regenerate reagents for the first step [91,97–103]. The first step is performed electrolytically and the second step appears to be possible at a temperature of around 850°C. More complex versions of the copper/sulfate cycle called H-5 and H-7 involve four and six reactions. Law et al. [97–105] have given a very detailed accounting of this thermochemical cycle.

Brown et al. [90] examined efficiency of more than 100 thermocycles that can use high-temperature heat from advanced nuclear power stations. A basic requirement was the ability to deliver heat to the process interface heat exchanger at temperatures up to 900°C. They also developed a set of requirements and criteria considering design, safety, operational, economic, and development issues. Helium-cooled nuclear reactor was chosen to interface with the thermochemical cycles. The best two-, three-, and four-step cycles with the greatest commercial potential identified from their analysis are illustrated in Table 11.1 [92,140]. They also concluded that

TABLE 11.1
Best Two-, Three-, and Four-Step Thermochemical Cycles

Name/Major Compound	Temperature (°C)	Details of Cycles
		Two-Step Cycles
Tokyo Institute	1000	$2MnFe_2O_4 + 3Na_2CO_3 + H_2O \rightarrow 2Na_2MnFe_2O_6 + 3CO_2(g) + H_2(g)$
of Technology/ ferrite	600	$4Na_2MnFe_2O_6 + 6CO_2(g) \rightarrow 4MnFe_2O_4 + 6Na_2CO_3 + O_2(g)$
Westinghouse/	850	$2H_2SO_4\ (g) \rightarrow 2SO_2(g) + 2H_2O(g) + O_2(g)$
sulfur	77	$SO_2(g) + 2H_2O(a) \rightarrow H2SO_4(a) + H_2(g)$
Nickel ferrite	800	$NiMnFe_4O_6 + 2H_2O \rightarrow NiMnFe_4O_8 + 2H_2(g)$
	800	$NiMnFe_4O_8 \rightarrow NiMnFe_4O_6 + O_2(g)$
Hallett Air	800	$2Cl_2(g) + 2H_2O(g) \rightarrow 4HCl(g) + O_2(g)$
Products/chlorine	25	$2HCl \rightarrow Cl_2(g) + H_2(g)$
		Three-Step Cycles
Ispra Mark 13/	850	$2H_2SO_4(g) \rightarrow 2SO_2(g) + 2H_2O(g) + O_2(g)$
bromine/sulfur	77	$2HBr(a) \rightarrow Br_2(a) + H_2(g)$
	77	$Br_2(l) + SO_2(g) + 2H_2O(l) \rightarrow 2HBr(g) + H_2SO_4(a)$
Ispra Mark 8/	700	$3MnCl_2 + 4H_2O \rightarrow Mn3O_4 + 6HCl + H_2(g)$
manganese/	900	$3MnO_2 \rightarrow Mn3O_4 + O_2(g)$
chlorine	100	$4HCl + Mn3O_4 \rightarrow 2MnCl_2(a) + MnO_2 + 2H_2O$
Ispra/CO/Mn_3O_4	977	$6Mn_2O_3 \rightarrow 4Mn3O_4 + O_2(g)$
	700	$C(s) + H_2O(g) \rightarrow CO(g) + H_2(g)$
	700	$CO(g) + 2Mn3O_4 \rightarrow C + 3Mn_2O_3$
Ispra Mark 3/V/	850	$2Cl_2(g) + 2H_2O(g) \rightarrow 4HCl(g) + O_2(g)$
chlorine	170	$2VOCl_2 + 2HCl \rightarrow 2VOCl_3 + H_2(g)$
	200	$2VOCl_3 \rightarrow Cl_2(g) + 2VOCl_2$
Julich Center EOS/	800	$2Fe_3O_4 + 6FeSO_4 \rightarrow 6Fe_2O_3 + 6SO_2 + O_2\ (g)$
iron/sulfur	700	$3FeO + H_2O \rightarrow Fe_3O_4 + H_2(g)$
	200	$Fe_2O_3 + SO_2 \rightarrow FeO + FeSO_4$
Gaz de France/	725	$2K + 2KOH \rightarrow 2K_2O + H_2(g)$
KOH/K	825	$2K_2O \rightarrow 2K + K_2O_2$
	125	$2K_2O_2 + 2H_2O \rightarrow 4KOH + O_2(g)$
Aachen University	850	$2Cl_2(g) + 2H_2O(g) \rightarrow 4HCl(g) + O_2(g)$
Jülich 1972/Cr/Cl	170	$2CrCl_2 + 2HCl \rightarrow 2CrCl_3 + H_2(g)$
	800	$2CrCl_3 \rightarrow 2CrCl_2 + Cl_2(g)$
US-Chlorine/Cu/Cl	850	$2Cl_2(g) + 2H_2O(g) \rightarrow 4HCl(g) + O_2(g)$
	200	$2CuCl + 2HCl \rightarrow 2CuCl_2 + H_2(g)$
	500	$2CuCl_2 \rightarrow 2CuCl + Cl_2(g)$
Ispra Mark 9/Fe/Cl	420	$2FeCl_3 \rightarrow Cl_2(g) + 2FeCl_2$
	150	$3Cl_2(g) + 2Fe_3O_4 + 12HCl \rightarrow 6FeCl_3 + 6H_2O + O_2(g)$
	650	$3FeCl_2 + 4H_2O \rightarrow Fe_3O_4 + 6HCl + H_2(g)$

(*Continued*)

TABLE 11.1
(Continued) Best Two-, Three-, and Four-Step Thermochemical Cycles

Name/Major Compound	Temperature (°C)	Details of Cycles
LASL-U/uranium	25	$3CO_2 + U_3O_8 + H_2O \rightarrow 3UO_2CO_3 + H_2(g)$
	250	$3UO_2CO_3 \rightarrow 3CO_2(g) + 3UO_3$ 1
	700	$6UO_3(s) \rightarrow 2U_3O_8(s) + O_2(g)$
Ispra Mark 2 (1972)/Na/Mn	100	$Na_2O.MnO_2 + H_2O \rightarrow 2NaOH(a) + MnO_2$
	487	$4MnO_2(s) \rightarrow 2Mn_2O_3(s) + O_2(g)$
	800	$Mn_2O_3 + 4NaOH \rightarrow 2Na_2O.MnO_2 + H_2(g) + H_2O$
Sulfur–Iodine/S/I	850	$2H_2SO_4(g) \rightarrow 2SO_2(g) + 2H_2O(g) + O_2(g)$
	450	$2HI \rightarrow I_2(g) + H_2(g)$
	120	$I_2 + SO_2(a) + 2H_2O \rightarrow 2HI(a) + H_2SO_4(a)$
		Four-Step Cycles
Vanadium chloride	850	$2Cl_2(g) + 2H_2O(g) \rightarrow 4HCl(g) + O_2(g)$
	25	$2HCl + 2VCl_2 \rightarrow 2VCl_3 + H_2(g)$
	700	$2VCl_3 \rightarrow VCl_4 + VCl_2$
	25	$2VCl_4 \rightarrow Cl_2(g) + 2VCl_3$
Ispra Mark 4/Fe/Cl	850	$2Cl_2(g) + 2H_2O(g) \rightarrow 4HCl(g) + O_2(g)$
	100	$2FeCl_2 + 2HCl + S \rightarrow 2FeCl_3 + H_2S$
	420	$2FeCl_3 \rightarrow Cl_2(g) + 2FeCl_2$
	800	$H_2S \rightarrow S + H_2(g)$
Ispra Mark 6/Cr/Cl	850	$2Cl_2(g) + 2H_2O(g) \rightarrow 4HCl(g) + O_2(g)$
	170	$2CrCl_2 + 2HCl \rightarrow 2CrCl_3 + H_2(g)$
	700	$2CrCl_3 + 2FeCl_2 \rightarrow 2CrCl_2 + 2FeCl_3$
	420	$2FeCl_3 \rightarrow Cl_2(g) + 2FeCl_2$
Ispra Mark 1C/Cu/Ca/Br	100	$2CuBr_2 + Ca(OH)_2 \rightarrow 2CuO + 2CaBr_2 + H_2O$
	900	$4CuO(s) \rightarrow 2Cu_2O(s) + O_2(g)$
	730	$CaBr_2 + 2H_2O \rightarrow Ca(OH)_2 + 2HBr$
	100	$Cu_2O + 4HBr \rightarrow 2CuBr_2 + H_2(g) + H_2O$
UT-3 University of Tokyo/Fe/Ca/Br	600	$2Br_2(g) + 2CaO \rightarrow 2CaBr_2 + O_2(g)$
	600	$3FeBr_2 + 4H_2O \rightarrow Fe_3O_4 + 6HBr + H_2(g)$
	750	$CaBr_2 + H_2O \rightarrow CaO + 2HBr$
	300	$Fe_3O_4 + 8HBr \rightarrow Br_2 + 3FeBr_2 + 4H_2O$

Source: Brown, L.C., Besenbrauch, G.E., Schultz, K.R., Showalter, S.K., Marshall, A.C., Pickard, P.S., and Funk, J.F., *Spring National Meeting of AIChE, Nuclear Engineering Session THa01 139-Hydrogen Production and Nuclear Power*, New Orleans, LA, (2002). With permission; Schultz, K., *Presentation to the Stanford Global Climate and Energy Project*, General Atomics, San Diego, CA (2003). With permission.

S–I cycle was overall the best to interlink with the helium-cooled nuclear reactor. The use of solar and nuclear energy for direct thermolysis or thermochemical breakdown of water has also been extensively examined in the literature [135–138,140].

11.5 OTHER MISCELLANEOUS TECHNOLOGIES

Bockris et al. [11] described several novel methods for hydrogen production. Some of the methods described closely follow their description in Sections 11.5.2, 11.5.3, 11.5.6, and 11.5.7.

11.5.1 Chemical Methods

A number of materials react with liquid water or water containing acids to release hydrogen [11,139,141–148]. While these methods somewhat resemble steam reforming, they differ in that reactant is liquid water instead of gaseous water and the solids involved are not naturally occurring such as coal and shale oil but those that require a significant energy and efforts recovering such as zinc, aluminum, and iron.

In laboratory, zinc reacts with strong acids in Kipp's apparatus. In the presence of sodium hydroxide, aluminum and its alloys react with water to generate hydrogen [11,107]. This is, however, an expensive process due to the high cost of aluminum, and the process also results in a large amount of waste heat that must be disposed or recovered. In relative terms, aluminum is cheaper and safer than some other materials, and the produced hydrogen can be easily stored and transported than using other hydrogen storage materials such as sodium borohydride.

The reaction between water and aluminum follows the path:

$$\mathrm{Al} + 3\mathrm{H_2O} + \mathrm{NaOH} \rightarrow \mathrm{NaAl(OH)_4} + 1.5\mathrm{H_2} \qquad (11.50)$$

$$\mathrm{NaAl(OH)_4} \rightarrow \mathrm{NaOH} + \mathrm{Al(OH)_3} \qquad (11.51)$$

Overall reaction follows:

$$\mathrm{Al} + 3\mathrm{H_2O} \rightarrow \mathrm{Al(OH)_3} + 1.5\mathrm{H_2} \qquad (11.52)$$

The first two reactions are similar to the process that occurs inside an aluminum battery. The second reaction precipitates crystalline aluminum hydroxide. This process works well at a smaller scale, and every 1 kg of aluminum can produce up to 0.111 kg of hydrogen that can be very useful in the device such as fuel cell where released hydrogen can generate electricity. Aluminum along with $NaBH_4$ can also be used as compact storage devices for hydrogen. The above reaction is mildly exothermic, and hence the reaction is carried out under mild temperatures and pressures providing a stable and compact source of hydrogen. The process can be a backup process for remote or marine applications. The negative effect of passivation of aluminum can be minimized by changing the temperature, alkali concentration, physical form of aluminum, and solution composition.

11.5.2 Magmalysis

This process is another form of chemical method in which steam is injected on a magma that is near the surface [11,143]. According to Northrup et al. [143], the following reaction would occur:

$$2FeO + H_2O \rightarrow 2FeO_{1.5} + H_2 \quad (11.53)$$

Fresh basaltic lava contains on the order of 10 wt% ferrous oxide (FeO) and 1–2 wt% ferric oxide ($FeO_{1.5}$). These components exist as dissolved constituents within the melt and in the mineral suspended in the magma. Northrup et al. [143] calculated hydrogen concentration, which resulted from equilibration of water with a solid assemblage of hematite–magnetite for a total pressure of 100 MPa. The calculation agreed well with the measured data.

As water accumulates in the basaltic lava, most of FeO is converted into $FeO_{1.5}$, resulting in the drop of hydrogen production [144,145]. Northrup et al. [143] also estimated the hydrogen production at 1200°C. The estimates indicate that about 2.2×10^6 tons of hydrogen is potentially recoverable by water interacting with 1 km^3 of basalt at high temperatures at 1000 MPa. The exact calculations of hydrogen production rate requires the knowledge of available magma surface area and its cooling rate. Northrup et al. [143] estimated that about 10^5 km^3 of magma bodies in areas of the United States exist where hydrogen production by this method is possible.

11.5.3 Radiolysis

Radiolysis involves the injection of radioactive substances such as $UO_2(NO_3)_2$ into water which emits particles that have an energy in the region of 10^6 eV [11,142,149]. This energy will decompose some 10^5 water molecules per particle, and if there were no recombination, significant amounts of hydrogen and oxygen would be generated. When radioactive particles pass by water molecules, they strip a part of electron shells so that protons are produced and the oxygen becomes cationic. The conversion efficiency is, however, low; between 1% and 5% of the radioactive energy is translated in the productions of hydrogen and oxygen [141]. The efficiency can be improved by the use of salts such as B^{10} and Li^6 compounds. The process generates hydrogen and oxygen in a mixture, which can be separated using a fuel cell where hydrogen and oxygen are separated by anode and cathode, respectively.

The method can be valuable if the efficiency is improved to greater than 10% and the radioactive material used is waste. Gomberg and Gordus [142] improved the efficiency by using the nuclear fission either in a solid fuel configuration where the radiation energy/heat ratio can be about 1/4 or in a fluid fuel configuration where all the energy is available as radiation.

11.5.4 Shock Waves and Mechanical Pulses

Attempts to dissociate water using shock waves and mechanical pulses have also been made [11]. The use of shock wave to dissociate diatomic molecules and organic compounds has been successful [149]. It is possible to induce OH bond dissociation by introducing anharmonic oscillations in the molecule. A novel method could be the excitation of water molecules adsorbed on fiber optics that could be made conductive to allow part of the light wave being transmitted to interact with the adsorbed water.

11.5.5 Catalytic Decomposition of Water

Another approach to the thermolysis of water is to pass water through a "getter" that will remove oxygen [11]. The getter then needs to be regenerated after obtaining hydrogen. Kasal and Bishop [150,151] used zeolites for this purpose. They [40] also described a simple two-step cycle to decompose water by cycling water over chromium- and indium-substituted alumno silicates. For a two-step thermochemical process consisting of an endothermic step operating at lower temperature T_L and the second step operating at higher temperature T_H, the transition between these two steps will be accompanied by a large entropy change. A large entropy change can also be realized by resorting to a cycle consisting of many reaction steps or a single reaction involving many molecules. England [152] proposed a thermochemical cycle based on the results of Kasal and Bishop as follows:

$$Al_2O_3 + 4H_2O\,(g) + 2CrO \rightleftarrows Al_2O_3 \cdot 3H_2O + Cr_2O_3 + H_2(g) \tag{11.54}$$

at low temperatures with an entropy change of −128.5 eu and

$$Al_2O_3 + 3H_2O + Cr_2O_3 \rightleftarrows Al_2O_3 + H_2(g) + 2CrO + \frac{1}{2}O_2 \tag{11.55}$$

at high temperatures with an entropy change of 139.1 eu.

11.5.6 Plasmolysis

The direct thermal dissociation of water by thermal means at temperatures around 3000°C suffers from the lack of durable materials for the reactor at these high temperatures [11,139]. One method by which this difficulty may be avoided is to use electrically produced plasmas [139]. Electrical generation of the plasmas involves the transformation of the energy from an electric field (microwave, radio frequency, or d.c.) into kinetic energy of electrons, which is further transformed into molecular excitations and to the kinetic energy of heavy particles. These discharged plasmas are divided into either hot (thermal) or cold (nonthermal) plasmas. Both types of plasmas can result in electron temperature to be several thousand degrees. Although the difference in energy content is a function of temperature, the low-temperature discharge has sufficient energy to dissociate water.

11.5.7 Magnetolysis

The idea of producing high current and low voltage was abandoned for a long time due to the fact that resistance losses are less when electricity is transmitted at high voltages over a power line than when it is transmitted at low voltage and high current [148]. However, in an electrolyzer, what is needed is low voltage and very high currents. This can be achieved by the application of a homopolar generator conceived by Faraday [147]. Bockris and Gutmann [146] suggested that using this concept electrolysis can be carried out by generating the necessary potential difference by magnetic induction inside the electrolyzer [11,146–148]. Bockris et al. [11] examined the details of this method.

REFERENCES

1. Meier, A. and Sattler, C., "Solar fuels from concentrated sunlight," SolarPACES, Solar Power and Chemical Energy Systems, IEA report (August 2009).
2. Smitkova, M., Janicek, F., and Riccardi, J., "Analysis of the selected processes for hydrogen production," *WSEAS Transactions on Environment and Development*, 11 (4), 1026–1035 (2008).
3. Funk, J.E. and Reinstrom, R.M., "Energy requirements in the production of hydrogen from water," *Industrial & Engineering Chemistry Process Design and Development*, 5 (3), 336–342 (1966).
4. Funk, J.E., "Thermodynamics of multi-step water decomposition processes," *Proceedings of the Symposium on Non-Fossil Chemical Fuels*, ACS 163rd National Meeting, April, Boston, MA, 79–87 (1972).
5. Abe, I., "Hydrogen productions from water," *Energy Carriers and Conversion Systems*, 1, 1–3 (2001).
6. Taylan, O. and Berberoglu, H. "Fuel production using concentrated solar energy," in Rugescu, R. (ed.) *The Applications of Solar Energy*, INTECH, Rijeka, Croatia, 33–67 (2013).
7. Klausner, J., Petrasch, J., Mei, R., Hahn, D., Mehdizadeh, A., and Auyeung, N., "Solar fuel: Pathway to a sustainable energy future," Florida Energy Summit, University of Florida, Gainesville, FL (August 17, 2012).
8. Ris, T. and Hagen, E., "Hydrogen production-gaps and priorities," IEA Hydrogen Implementing Agreement (HIA), 1–111 (2005).
9. Suarez, M., Blanco-Marigorta, A., and Peria-Quintana, J., "Review on hydrogen production technologies from solar energy," International Conference on Renewable Energies and Power Quality, La Palmas de Gran Canaria, April 13–15, Spain (2011).
10. Nowotny, J., Sorrell, C., Sheppard, L., and Bak, T., "Solar-hydrogen: Environmentally safe fuel for the future," *International Journal of Hydrogen Energy*, 30 (5), 521–544 (2005).
11. Bockris, J., Dandapani, B., Cocke, D., and Ghoroghchian, J., "On the splitting of water," *International Journal of Hydrogen Energy*, 10 (30), 179–201 (1985).
12. Naterer, G.F., "Economics and synergies of electrolytic and thermochemical methods of environmentally benign hydrogen production," *Proceedings of the World Hydrogen Energy Conference*, May 16–21, Essen, Germany (2010).
13. Laoun, B., "Thermodynamics aspect of high pressure hydrogen production by water electrolysis," *Revue des Énergies Renouvelables*, 10 (3), 435–444 (2007).
14. Funk, J.E., "Thermochemical and electrolytic production of hydrogen," in Vezeroglu, T.N. (ed.), *Introduction to Hydrogen Energy*. International Association for Hydrogen Energy, Miami, FL, 19–49 (1975).
15. Onda, K., Kyakuno, T., Hattori, K., and Ito, K., "Prediction of production power for high pressure hydrogen by high pressure water electrolysis," *Journal of Power Sources*, 132, 64–70 (2004).
16. "High temperature electrolysis," Wikipedia, the free encyclopedia, 1–3 (2012).
17. McKellar, M., Harvego, E., and Gandrik, A., "System evaluation and economic analysis of a HTGR powered high temperature electrolysis production plant," *Proceedings of the High Temperature Reactor*, Paper 093, October 18–20, Prague, Czech Republic (2010).

18. Herring, J., O'Brien, J., Stoots, C., Lessing, P., and Anderson, R., "High temperature solid oxide elctrolyzer system," Hydrogen, Fuel Cells, and Infrastructure Technologies, DOE Progress Report, May 25, 2004, INEEL, Idaho Falls, ID, 1–5 (2003).
19. Herring, J., O'Brien, J., Stoots, C., Lessing, P., and Hartvigsen, J., "High temperature electrolysis for hydrogen production," Paper presented by Idaho National Laboratory at the Materials Innovations in an Emerging Hydrogen Economy Conference, February 26, Hilton Oceanfront, Cocoa Beach, FL (2008).
20. LeRoy, R.L., Bowen, C.T., and LeRoy, D.J., *Journal of the Electrochemical Society*, 127, 1954 (1980).
21. LeRoy, R.L., Janjua, M.B.I., Renaud, R., and Leuenberger, U., *Journal of the Electrochemical Society*, 126, 1674 (1979).
22. Ohashi, K., McCann, J., and Bockris, J.O.M., *Nature*, 266, 610 (1977).
23. Szklarczyk, M. and Bockris, J.O.M., *Applied Physics Letters*, 42, 1035 (1983).
24. Szklarczyk, M. and Bockris, J.O.M., *Journal of Physical Chemistry*, 88, 1808 (1984).
25. Scaife, D.E., Weller, P.F., and Fisher, W.G., "Crystal preparation and properties of cesium tin(II) trihalides," *Journal of Solid State Chemistry*, 9 (3), 308–314 (1974).
26. Scaife, D., "Oxide semiconductors in photoelectrochemical conversion of solar energy," *Solar Energy*, 25 (1), 41–54 (1980).
27. Shyu, R., Weng, F., and Ho, C., "Manufacturing of a micro probe using supersonic aided electrolysis process," DTIP of MEMS and MOEMS, April 9–11, Nice, France (2008).
28. Bockris, J. and Murphy, O., "One-unit photo-activated electrolyzer," US Patent No. US4790916 A (December 13, 1988).
29. Bockris, J., Gutmann, F., and Craven, W., "The economics of production of hydrogen," in Veziroglu, T., Van Vorst, W., Kelley, J., (eds.), *Hydrogen Energy Progress IV*. International Association of Hydrogen Energy, Miami, FL, 1475 (1982).
30. Okada, G., Guruswamy, V., and Bockris, J., *Journal of the Electrochemical Society*, 128, 2097 (1981).
31. Trogler, W., Geoffrey, G., Erwin, D., and Gray, H., *Journal of the American Chemical Society*, 100, 1160 (1978).
32. Erwin, D., Geoffrey, G., Gray, H., Hammond, G., Soloman, E., Trogler, W., and Zagers, A., *Journal of the American Chemical Society*, 99, 3620 (1977).
33. Tyler, D. and Gray, H., *Journal of the American Chemical Society*, 103, 1683 (1981).
34. Mann, K., Lewis, N., Miskowski, V., Erwin, D., Hammond, G., and Gray, H., *Journal of the American Chemical Society*, 99, 5525 (1977).
35. Sprintschnik, G., Sprintschnik, H., Kirsch, P., and Whitten, D., *Journal of the American Chemical Society*, 98, 2337 (1976).
36. Sprintschnik, G., Sprintschnik, H., Kirsch, P., and Whitten, D., *Journal of the American Chemical Society*, 99, 4947 (1977).
37. Maverick, A. and Gray, H., *Pure and Applied Chemistry*, 52, 2339 (1980).
38. Kiwi, J., Kalyanasundaram, K., and Gratzel, M., *Structure and Bonding*, 49, 37 (1982).
39. Koriakin, B., Dshabiev, T., and Shivlov, A., *Doklady Akademii Nauk SSSR*, 298, 620 (1977).
40. Kalyanasundaram, K., Micic, O., Pramauro, E., and Gratzel, M., *Helvetica Chimica Acta,* 62, 2432 (1979).
41. Duonghong, D., Borgarello, E., and Gratzel, M., *Journal of the American Chemical Society*, 103, 4685 (1981).
42. Kato, H. and Kudo, A., *Chemistry Letters*, 1207 (1999).
43. Lin, W., Cheng, C., Hu, C., and Teng, H., *Applied Physics Letters*, 89, 211904 (2006).
44. Navarro, Y.R., Alvarez-Galvan, M., del Valle, F., Villoria de la Mano, J., and Fierro, J., "Water splitting on semiconductor catalysts under visible light irradiation," *ChemSusChem*, 2 (6), 471–485 (2009).

45. Navarro, R., Yerga, J., and Fierro, G., "Photo catalytic decomposition of water," *ChemSusChem*, 2, 471–485 (2009).
46. Sato, J., Saito, N., Nishiyama, H., and Inoue, Y., "Photocatalytic water decomposition by RuO_2 loaded antimonates, $M_2Sb_2O_7$ (M = Ca, Sr), $CaSb_2O_6$ and $NaSbO_3$ with d10 configuration," *Journal of Photochemistry and Photobiology A: Chemistry*, 148, 85–89 (2002).
47. Zou, Z., Ye, J., Sayama, K., and Arakawa, H., "Photocatalytic hydrogen and oxygen formation under visible light irradiation with M-doped $InTaO_4$ (M = Mn, Fe, Co, Ni or Cu) photocatalysts," *Journal of Photochemistry and Photobiology A: Chemistry*, 148, 65–69 (2002).
48. "Water splitting," Wikipedia, the free encyclopedia (2012).
49. Lee, J. (ed.), "Designer transgenic, algae for photobiological production of hydrogen from water," in *Advanced Biofuels and Bioproducts*. Springer, New York, 371–404 (2012).
50. Sato, J., Kobayashi, H., Saito, N., Nishiyama, H., and Inoue, Y., "Photocatalytic activities for water decomposition of RuO_2 loaded $AInO_2$ (A = Li, Na) with d10 configuration," *Journal of Photochemistry and Photobiology A: Chemistry*, 158, 139–144 (2003).
51. Zou, Z. and Arakawa, H., "Direct water splitting into H_2 and O_2 under visible light irradiation with a new series of mixed oxide semiconductor photocatalysts," *Journal of Photochemistry and Photobiology A: Chemistry*, 158, 145–162 (2003).
52. Harda, H., Hosoki, C., and Kudo, A., "Overall water splitting by sonophotocatalytic reaction: The role of powdered photocatalyst and an attempt to decompose water using a visible light sensitive photocatalyst," *Journal of Photochemistry and Photobiology A: Chemistry*, 141, 219–224 (2001).
53. Abe, R., Hara, K., Sayama, K., Domen, K., and Arakawa, H., "Steady hydrogen evolution from water on Eosin Y-fixed TiO_2 photocatalyst using a silane coupling reagent under visible light irradiation," *Journal of Photochemistry and Photobiology A: Chemistry*, 137, 63–69 (2000).
54. Fujishima, A., Rao, T., and Tryk, D., "Titanium dioxide photocatalysis," *Journal of Photochemistry and Photobiology C: Photochemistry Reviews*, 1, 1–21 (2000).
55. Navarro, R., Sanchez-Sanchez, M., Alvarez-Galvan, M., del Valle, F., and Fierro, J., "Hydrogen production from renewable sources: Biomass and photocatalytic opportunities," *Energy & Environmental Science*, 2, 35–54 (2009).
56. Fujishima, A. and Honda, K., *Nature*, 238, 37 (1972).
57. Kato, H. and Kudo, A., *Journal of Physical Chemistry B*, 106, 5029 (2002).
58. Ishii, T., Kato, H., and Kudo, A., *Journal of Photochemistry and Photobiology A*, 163, 181 (2004).
59. Kudo, A. and Sekizawa, M., *Catalysis Letters*, 58, 241 (1999).
60. Kudo, A. and Sekizawa, M., *Chemical Communications*, 1371 (2000).
61. Konta, R., Ishii, T., Kato, H., and Kudo, A., *Journal of Physical Chemistry B*, 108, 8992 (2004).
62. Anpo, M. and Takeuchi, M., *Journal of Catalysis*, 216, 505 (2003).
63. Asahi, R., Morikawa, T., Ohwaki, T., Aoki, K., and Taga, Y., *Science*, 293, 269 (2001).
64. Umebayashi, T., Yamaki, T., Itoh, H., and Asai, K., *Applied Physics Letters*, 81, 454 (2002).
65. Kudo, A., Omori, K., and Kato, H., *Journal of the American Chemical Society*, 121, 11459 (1999).
66. Asokkumar, M., *International Journal of Hydrogen Energy*, 23, 427 (1998).
67. Kalyanasundaram, M., Graetzel, M., and Pelizzetti, E., *Coordination Chemistry Reviews*, 69, 57 (1986).
68. Meissner, D., Memming, R., and Kastening, B., *Journal of Physical Chemistry*, 92, 3476 (1988).
69. Fuji, H. and Guo, L., *Journal of Physical Chemistry B*, 110, 1139 (2006).

70. Spanhel, L., Weller, H., and Hanglein, A., *Journal of the American Chemical Society*, 109, 6632 (1987).
71. Navarro, R., del Valle, F., and Fierro, J., *International Journal of Hydrogen Energy*, 33, 4265 (2008).
72. Tsuji, I. and Kudo, A., *Journal of Photochemistry and Photobiology A*, 156, 249 (2003).
73. Kudo, A., Nishiro, R., Iwase, A., and Kato, H., *Chemical Physics*, 339, 104 (2007).
74. Tsuji, I., Kato, H., and Kudo, A., *Angewandte Chemie International Edition*, 117, 3631 (2005); *Angewandte Chemie International Edition*, 44, 3565 (2005).
75. Kudo, A., Tsuji, I., and Kato, H., *Chemical Communications*, 1958 (2002).
76. Tsuji, I., Kato, H., Kobayashi, H., and Kudo, A., *Journal of Physical Chemistry B*, 109, 7323 (2005).
77. Prince, R. and Kheshgi, H., "The photobiological production of hydrogen: Potential efficiency and effectiveness as a renewable fuel," *Critical Reviews in Microbiology*, 31, 19–31 (2005).
78. Ghysels, B. and Franck, F., "Hydrogen photo-evolution upon S deprivation stepwise: An illustration of microalgal photosynthetic and metabolic flexibility and a step stone for future biological methods of renewable H_2 production," *Photosynthesis Research*, 106, 145–154 (2010).
79. Mells, A., Zhang, L., Forestier, M., Ghirardi, M., and Seibert, M., "Sustained photobiological hydrogen gas production upon reversible inactivation of oxygen evolution in the green alga *Chlamydomonas reinhardtii*," *Plant Physiology*, 122, 127–135 (2000).
80. Ghirardi, M., Zhang, L., Lee, J., Flynn, T., Seibert, M., Greenbaum, E., and Melis, A., "Microalgae: A green source of renewable H_2," *Trends in Biotechnology*, 18, 506–511 (2000).
81. Nguyen, A., Thomas-Hall, S., Malnoe, A., Timmins, M., Mussgnug, J., Rupprecht, J., Krause, O., Hankamer, B., and Schenk, P., "Transcriptome for photobiological hydrogen production induced by sulfur deprivation in the green alga *Chlamydomonas reinhardtii*," *Eukaryotic Cell*, 7 (11), 1965–1979 (2008).
82. Timmins, M., Thomas-Hall, S., Darling, A., Zhang, E., Hankamer, B., Marx, U., and Schenk, P., "Phylogenetic and molecular analysis of hydrogen producing green algae," *Journal of Experimental Botany*, 60 (6), 1691–1702 (2009).
83. Berberoglu, H. and Pilon, L., "Maximizing the solar to H_2 energy conversion efficiency of outdoor photobioreactors using mixed cultures," *International Journal of Hydrogen Energy*, 35 (2), 500–510 (2010).
84. Lee, J., "Designer proton channel transgenic algae for photobiological hydrogen production," PCT International Patent Application WO2007/134340 A2 (2007).
85. Lee, J., "Designer proton channel transgenic algae for photobiological hydrogen production," US Patent No. 7932437 B2 (2011).
86. Lee, J., "Switchable photosystem-II designer algae for photobiological hydrogen production," US Patent No. 7642405 B2 (2010).
87. Eastlund, B. and Gough, W., Paper Presented at the 163th National Meeting of American Chemical Society, April 14, Boston, MA (1972).
88. McWhirter, R., "Spectral Intensities," in Huddlestone, R. and Leonard, S. (eds.), *Plasma Diagnostic Techniques*. Academic Press, New York, 201 (1965).
89. Inoue M., Uehara, R., Hasegawa, N., Gokon, N., Kaneko, H., and Tamaura, Y., "Solar hydrogen generation with $H_2O/ZnO/MnFe_2O_4$ system," ISES, Solar World Congress, Adelaide, Australia 1723–1729 (2001), www.aseanenergy.info/Abstract/32008845.pdf.
90. Brown, L.C., Besenbrauch, G.E., Schultz, K.R., Showalter, S.K., Marshall, A.C., Pickard, P.S., and Funk, J.F., "High efficiency generation of hydrogen fuels using thermochemical cycles and nuclear power," Spring National Meeting of AIChE, Nuclear Engineering Session THa01 139-Hydrogen Production and Nuclear Power, March 11–15, New Orleans, LA (2002).

91. Rosen, M.A., "Developments in the production of hydrogen by thermochemical water decomposition," *International Journal of Energy and Environmental Engineering*, 2 (2), 1–20 (2011).
92. Steinfeld, A. and Palumbo, R., "Solar thermochemical process technology," in Meyers, R. (ed.), *Encyclopedia of Physical Science and Technology*. Academic Press, New York, Vol. 15, 237–256 (2001).
93. Funk, J.E., Conger, W.L., and Cariy, R.H., "Evaluation of multi-step thermochemical processes for the production of hydrogen from water," *Proceedings of the THEME Conference*, March, Miami, FL, Sll-.ll (1974).
94. Chao, R.E., "Thermochemical water decomposition processes," *Industrial & Engineering Chemistry Process Design and Development*, 13 (2), 94–101 (1974).
95. Weimer, A., "II.1.2 fundamentals of a solar-thermal hydrogen production process using a metal oxide based thermochemical water splitting cycle," DOE Contract No. DE-FC36-05G015044, Annual Progress Report (2006).
96. Roeb, M. and Sattler, C., "HycycleS—Materials and components of hydrogen production by sulfur based thermochemical cycles," FPS-Energy-212470, German Aerospace Center (DLR), Solar research report (2010).
97. Law, V.J., Prindle, J.C., and Gonzales, R.B., "Analysis of the copper sulfate cycle for the thermochemical splitting of water for hydrogen production," Paper submitted for presentation at the 2007 AIChE National Meeting, November, Salt Lake City, UT (2007).
98. Law, V.J., Prindle, J.C., and Gonzales, R.B., "Level 1 and level 2 analysis of the copper sulfate cycle for the thermochemical splitting of water for hydrogen production," Contract No. 6F-003762, Argonne National Laboratory, Lemont, IL (July 2006).
99. Law, V.J., Prindle, J.C., and Gonzales, R.B., "Level 3 analysis of the copper sulfate cycle for the thermochemical splitting of water for hydrogen production," Contract No. 6F-003762, Argonne National Laboratory, Lemont, IL (August 2006).
100. Law, V.J., Prindle, J.C., Bang, R., Hoerger, K., and Ledbetter, J., "Progress report no. 1 experimental studies of the hydrogen generation reaction for the $CuSO_4$ cycle," Contract No. 6F-01144, Argonne National Laboratory, Lemont, IL (October 2006).
101. Law, V.J., Prindle, J.C., Bang, R., Hoerger, K., and Ledbetter, J., "Progress report no. 2, experimental studies of the hydrogen generation reaction for the $CuSO_4$ cycle," Contract No. 6F-01144, Argonne National Laboratory, Lemont, IL (January 2007).
102. Law, V.J., Prindle, J.C., Bang, R., Hoerger, K., and Ledbetter, J., "Progress report no. 3, experimental studies of the hydrogen generation reaction for the $CuSO_4$ cycle," Contract No. 6F-01144, Argonne National Laboratory, Lemont, IL (March 2007).
103. Law, V.J., Prindle, J.C., Bang, R., Hoerger, K., and Ledbetter, J., "Progress report no. 4, experimental studies of the hydrogen generation reaction for the $CuSO_4$ cycle," Contract No. 6F-01144, Argonne National Laboratory, Lemont, IL (May 2007).
104. Law, V.J., Prindle, J.C., and Lupulescu, A.I., "Aspen plus modeling of the three-reaction version of the copper-chloride thermochemical cycle for hydrogen production from water," Contract No. 6F-01144, Argonne National Laboratory, Lemont, IL (August 2007).
105. Law, V., Prindle, J., and Gonzales, R., "Analysis of the copper sulphate cycle for the thermochemical splitting of water for hydrogen production," Chemical Engineering Department, Tulane University, New Orleans, LA (2010).
106. Roeb, M., Sattler, C., Kluser, R., Monnerie, N., Oliveira, L., Konstandopoulos, A., Agrafiotis, C. et al., "Solar hydrogen production by a two step cycle based on mixed iron oxides," *Journal of Solar Energy Engineering*, 128, 125–133 (2006).
107. T-Raissi, A., Huang, C., and Muradov, N., "Hydrogen production via solar thermochemical water splitting," NASA/CR-2009-215441, Report of research for period March 2004–February 2008 (2009).

108. Steinfeld, A. and Palumbo, R., "Solar thermochemical process technology," *Encyclopedia of Physical Science and Technology*, 15 (1), 237–256 (2001).
109. Steinfeld, A. and Meier, A., *Solar Fuels and Materials*. Elsevier, Amsterdam, the Netherlands, 623–637 (2004).
110. Steinfeld, A., "Solar hydrogen production via a two-step water-splitting thermochemical cycle based on Zn/ZnO redox reactions," *International Journal of Hydrogen Energy*, 27 (6), 611–619 (2002).
111. Funke, H.H., Diaz, H., Liang, X., Carney, C.S., Weimer, A.W., and Li, P., "Hydrogen generation by hydrolysis of zinc powder aerosol," *International Journal of Hydrogen Energy*, 33 (4), 1127–1134 (2008).
112. Abanades, S., Charvin, P., Lemont, F., and Flamant, G., "Novel two-step SnO_2/SnO water-splitting cycle for solar thermochemical production of hydrogen," *International Journal of Hydrogen Energy*, 33 (21), 6021–6030 (2008).
113. Meier, A. and Steinfeld, A., "Solar thermochemical production of fuels," *Advances in Science and Technology*, 74 (1), 303–312 (2011).
114. Chueh, W.C. and Haile, S.M., "Ceria as a thermochemical reaction medium for selectively generating syngas or methane from H_2O and CO_2," *ChemSusChem*, 2 (8), 735–739 (2009).
115. Kappauf, T. and Fletcher, E.A., "Hydrogen and sulfur from hydrogen sulfide VI. Solar thermolysis," *Energy*, 14 (8), 443–449 (1989).
116. Zaman, J. and Chakma, A., "Production of hydrogen and sulfur from hydrogen sulfide," *Fuel Processing Technology*, 41 (2), 159–198 (1995).
117. Steinfeld, A., "Solar thermochemical production of hydrogen: A review," *Solar Energy*, 78 (5), 603–615 (2005).
118. Perret, R., "Solar thermochemical hydrogen production research (STCH), thermochemical cycle selection and investment priority," Report No. SAND2011-3622, Sandia National Laboratories, Livermore, CA (2011).
119. Almodaris, M., Khorasani, S., Abraham, J.J., and Ozalp, N. (eds.), "Simulation of solar thermo-chemical hydrogen production techniques," ASME/JSME 8th Thermal Engineering Joint Conference, March 13–17, ASME, Honolulu, HI (2011).
120. Aoki, A., Ohtake, H., Shimizu, T., Kitayama, Y., and Kodama, T., "Reactive metal-oxide redox system for a two-step thermochemical conversion of coal and water to CO and H_2," *Energy*, 25 (3), 201–218 (2000).
121. Osinga, T., Olalde, G., and Steinfeld, A., "Solar carbothermal reduction of ZnO: Shrinking packed-bed reactor modeling and experimental validation," *Industrial & Engineering Chemistry Research*, 43 (25), 7981–7988 (2004).
122. Osinga, T., Frommherz, U., Steinfeld, A., and Wieckert, C., "Experimental investigation of the solar carbothermic reduction of ZnO using a two-cavity solar reactor," *Journal of Solar Energy Engineering*, 126 (1), 633–637 (2004).
123. Epstein, M., Olalde, G., Santén, S., Steinfeld, A., and Wieckert, C., "Towards the industrial solar carbothermal production of zinc," *Journal of Solar Energy Engineering*, 130 (1), 014505 (2008).
124. Wieckert, C., Frommherz, U., Kräupl, S., Guillot, E., Olalde, G., Epstein, M., Santen, S., Osinga, T., and Steinfeld, A., "A 300 kW solar chemical pilot plant for the carbothermic production of zinc," *Journal of Solar Energy Engineering*, 129 (2), 190–196 (2007).
125. Kromer, M., Roth, K., Takata, R., and Chin, P., "Support for cost analyses on solar-driven high temperature thermochemical water-splitting cycles," Report No. DE-DT0000951, TIAX, LLC2011, Lexington, MA (February 22, 2011).
126. Lede, J., Lapicque, F., and Villermaux, J., *International Journal of Hydrogen Energy*, 8, 675 (1983).
127. Lapicque, F., Lede, J., Villermaux, J., Caler, B., Baumard, J., Anthony, A., Abdul-Aziz, G., Puechbertz, D., and Ledrix, M., *Entropie*, 19, 42 (1983).

128. Beghi, G., "A decade of research on thermochemical water hydrogen at the Joint Research Center, Ispra," *International Journal of Hydrogen Energy*, 11 (12), 761–771 (1986).
129. Besenbruch, G., "General atomic sulfur-iodine thermochemical water splitting process," *American Chemical Society, Division of Petroleum Chemistry*, preprint, 271, 48 (1982).
130. Sato, S., Shimizu, S., Nakajima, N., and Ikezoe, Y., "A nickel-iodine-sulfur process for hydrogen production," *International Journal of Hydrogen Energy*, 8 (1), 15–22 (1983).
131. Abanades, S., Charvin, P., Flamant, G., and Neveu, P., "Screening of water-splitting thermochemical cycles potentially attractive for hydrogen production by concentrated solar energy," *Energy*, 31, 2805–2822 (2006).
132. Marchetti, C., *Chemical Economy & Engineering Review*, 5, 7 (1973).
133. DeBeni, G. and Marchetti, C., "A chemical process to decompose water using nuclear heat," Paper presented at the 163rd National Meeting of the American Chemical Society, April 9, Boston, MA (1972).
134. Appleby, A. and Bockris, J., *International Journal of Hydrogen Energy*, 6, 1 (1981).
135. Pyle, W., Hayes, M., and Spivak, A., "Direct solar-thermal hydrogen production from water using nozzle/skimmer and glow discharge," IECEC96535, Report from H-ION Solar Inc., Richmond, CA (2010).
136. Baykara, S., "Experimental solar water thermolysis," *International Journal of Hydrogen Energy*, 29 (14), 1459–1469 (2004).
137. Harvey, W.S., Davidson, J.H., and Fletcher, E.A., "Thermolysis of hydrogen sulfide in the temperature range 1350–1600 K," *Industrial & Engineering Chemistry Research*, 37 (6), 2323–2332 (1998).
138. Perkins, C. and Weimer, A.W., "Solar-thermal production of renewable hydrogen," *AIChE Journal*, 55 (2), 286–293 (2009).
139. Venugopalan, M. and Jones, R., *Chemistry of Dissociated Water Vapor and Related Systems*. Wiley, New York (1968).
140. Schultz, K., "Thermochemical production of hydrogen from solar and nuclear energy," Presentation to the Stanford Global Climate and Energy Project, April 14, General Atomics, San Diego, CA (2003).
141. Kerr, W. and Majumdar, D., "Aqueous homogeneous reactor for hydrogen production," in Veziroglu, T. (ed.), *Hydrogen Energy, Part A*. Plenum Press, New York, 167 (1975).
142. Gomberg, H. and Gordus, A., *Journal of Fusion Energy*, 2, 319 (1982).
143. Northrup, C., Jr., Gerlach, T., Modreski, P., and Galt, J., *International Journal of Hydrogen Energy*, 3 (1) (1978).
144. Fudali, R., *Geochimica et Cosmochimica Acta*, 29, 529 (1948).
145. Kennedy, G., *American Journal of Science*, 246, 529 (1948).
146. Bockris, J. and Gutmann, F., *Applied Physics Communications*, 1, 121 (1981–1982).
147. Farady, M., *Diary*. Bell, London, Vol. 1, 381 (1932).
148. Appleton, A., "Super conducting DC machines," in Fouer, S. and Scwartz, B. (eds.), *Superconducting Machines and Devices*. Plenum Press, New York, 219 (1973).
149. Boyd, R. and Burns, G., in Lifshitz, A. (ed.), *Shock Waves in Chemistry*. Marcel Dekker, New York, 131 (1981).
150. Kasal, P. and Bishop, R., Jr., US Patent No. 3963830 (1976).
151. Kasal, P. and Bishop, R., Jr., *The Journal of Physical Chemistry*, 81, 1527 (1977).
152. England, C., in Veziroglu, T., Van Vorst, W., and Kelley, H. (eds.), *Hydrogen Energy Progress IV, Proceedings of the World Hydrogen Energy Conference IV*, June 13–17, Pasadena, CA, Vol. 2, p. 462, Pergamon Press, Oxford (1982).
153. Fletcher, E.A. and Moen, R.L., "Hydrogen and oxygen from water," *Science*, 197, 1050–1056 (1977).

12 Methane from Gas Hydrates

12.1 INTRODUCTION: WHAT IS GAS HYDRATE AND HOW IS IT FORMED?

Clathrate hydrates are solid crystalline "inclusion" compounds, which are formed when water is contacted with small hydrophobic molecules such as methane, ethane, H_2S, and CO_2 [1–7] (Harrison, 2010, pers. comm.) under certain pressure and temperature conditions. When the inclusion compound is a constituent of natural gas, clathrate hydrates are also referred to as gas hydrates [1–15] (Harrison, 2010, pers. comm.). The gas (or methane) hydrate composition is in general 5.75 mol of water for every molecule of methane, although this number does depend on the cage structure of the water ice. Various molecular structures of gas hydrate and clathrate are illustrated in Figure 12.1 [2]. The average density of methane hydrate is about 0.9 g/cc. Under standard conditions, the volume of methane hydrate will be 164 times less than that of methane gas [1–16] (Harrison, 2010, pers. comm.).

Gas hydrates are formed when natural gas and water are brought together under suitable conditions of low temperatures and elevated pressures. The formation depends on (1) the presence of sufficient amount of water, (2) the presence of hydrate former, and (3) the appropriate pressure and temperature conditions. In a gas hydrate reservoir, free gas, ice, water, and other components such as ethane, propane, hydrogen sulfide, and carbon dioxide can be found at different temperatures, pressures, and depth values. Two- and three-phase equilibria curves [5–7,13–16] (Harrison, 2010, pers. comm.) are used for correlation between phases where the amount of components present plays a significant role; very small and large amounts of water are not conducive to the formation of hydrates.

The gas hydrates are unstable compounds in which the water molecules form a sort of cage or lattice around the methane molecules, and the two establish weak chemical bonds with one another. Methane from methane hydrates must be released *in situ* due to the inherent instability of hydrate molecules. The temperature at which methane hydrate is stable depends on the prevailing pressure. For example, at 0°C, it is stable under a pressure of about 30 atm, whereas at 25°C, nearly 500 atm pressure is needed to maintain its integrity. The occlusion of other gases within the ice structure tends to add stability, whereas the presence of salts requires higher stabilizing pressures. Appropriate conditions of temperature/pressure exist on the earth in the upper 2000 m of sediments in two regions: (1) permafrost at high latitudes in polar regions where the surface temperatures are very low and (2) submarine continental slopes and rises where not only is the water cold but the pressures are high (>30 atm). Phase boundary of methane hydrates in permafrost and deep-sea regions

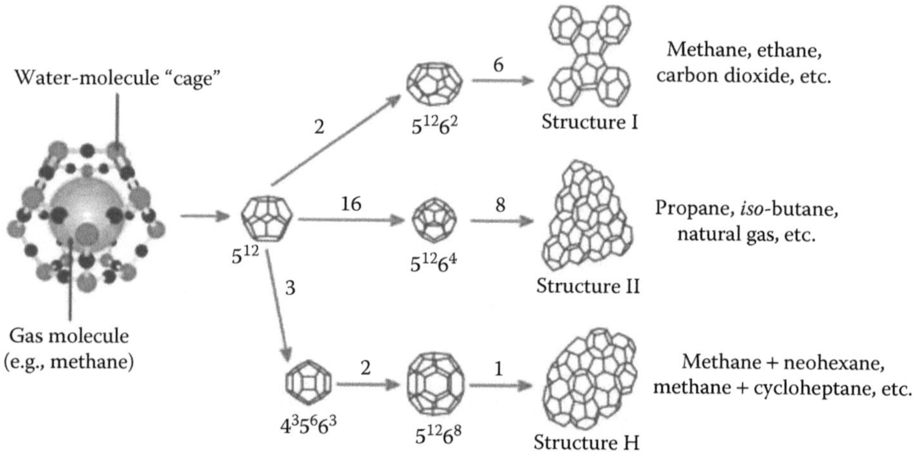

FIGURE 12.1 **(See color insert.)** Various molecular structures of gas hydrate and clathrate depending on guest molecules. (From "Methane hydrates," A communication by Center for Gas Hydrate Research, Heriot-Watt University, Edinburgh, The Hydrate forum Org., 2012. With permission.)

are graphically illustrated in Figure 12.2a and b (Tohidi, 2013, pers. comm.). These two figures show the estimations of regions where the stable hydrate formations are most likely to occur. The pressure–temperature phase diagram for methane hydrate is shown in Figure 12.3 [1,2,13–15] (Harrison, 2010, pers. comm.).

12.2 SOURCES, SIZES, AND IMPORTANCE OF GAS HYDRATE DEPOSITS

Gas hydrates were only discovered in the late twentieth century, and along with geopressurized zone gas, they are the best means of prolonging the carbohydrate age of energy [3,9,17–59] (Harrison, 2010, pers. comm.; USGS, 2012, pers. comm.). As mentioned earlier, vast quantities of methane gas hydrates can be discovered in sediments and sedimentary rocks within about 2000 m of the earth surface in polar and deep-water regions. Furthermore, the required conditions are found either in polar continental sedimentary rocks where surface temperature is <0°C or in oceanic sediment at water depths >300 m where the water temperature is around 2°C. Methane hydrates can also be formed in fresh water but not in salt water.

In 1995, the US Geological Survey (USGS) conducted a study to assess the quantity of natural gas hydrate (NGH) resources in the United States and found that the estimated quantity exceeded known conventional domestic gas resources [1]. The USGS estimates that methane hydrates may contain more carbon than world's coal, oil, and conventional natural gas combined. A comparison of estimated carbon in gas hydrates and other carbon sources on this earth is depicted

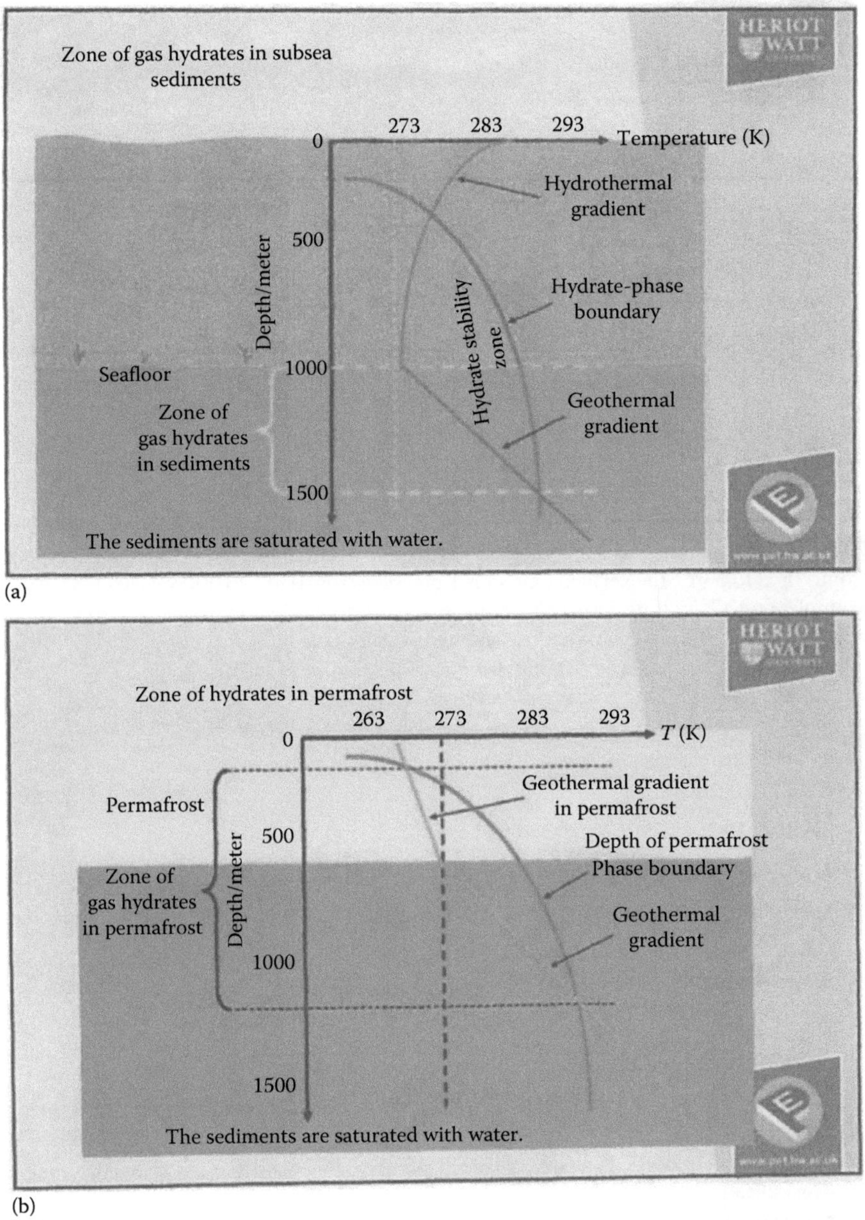

FIGURE 12.2 **(See color insert.)** Gas hydrate stability fields for (a) nominal marine settings and (b) permafrost settings. (From Tohidi, 2013, pers. comm. With permission.)

in Table 12.1 [1,60] (Tohidi, 2013, pers. comm., numerous works of Collet and coworkers at USGS). These data clearly show the dominance of gas hydrates as a source of carbon.

Types of methane hydrate deposits found on this earth are graphically illustrated in Figure 12.4 (Methane hydrate..., 2013, pers. comm.). Methane hydrates

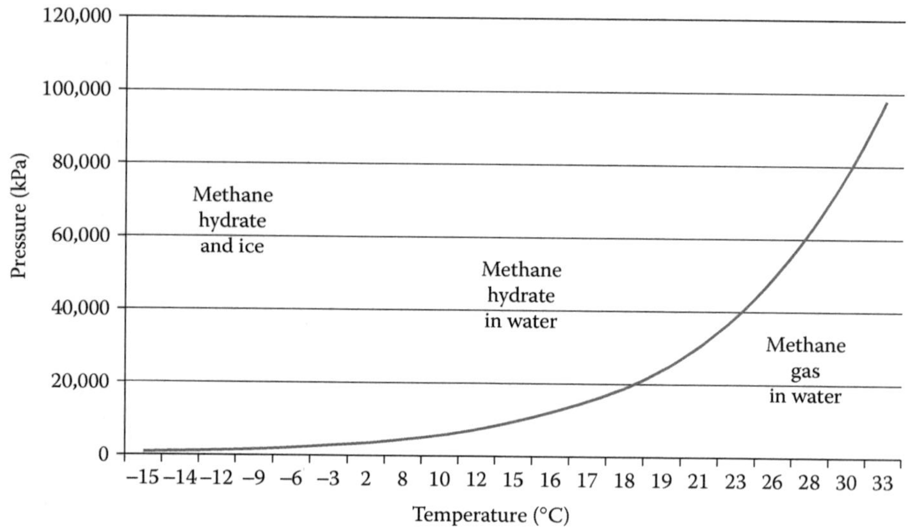

FIGURE 12.3 Methane hydrate phase diagram. The horizontal axis shows temperature from −15°C to 33°C, the vertical axis shows pressure from 0 to 120,000 kPa (0–1184 atm). For example, at 4°C, hydrate forms above a pressure of about 50 atm. (Adapted from "Methane hydrate phase diagram," Wikipedia, the free encyclopedia, 2010.)

TABLE 12.1
Distribution of Organic Carbon in the Earth

Source of Carbon	Amount (10^5 g of Carbon)[a]	Total Carbon (%)
Gas hydrates (onshore and offshore)	10,000	53.26
Recoverable and nonrecoverable fossil fuels (coal, oil, natural gas)	5,000	26.63
Soil	1,400	7.46
Dissolved organic matter in water	980	5.22
Land biota	830	4.42
Peat	500	2.68
Detrital organic matter	60	0.33
Atmosphere	3.6	0.0
Marine biota	3	0.0

Source: Englezos, P., *Industrial & Engineering Chemistry Research*, 32, 1251–1274, 1993; Kvenvolden, K.A., *Chemical Geology*, 71, 41–51, 1988. With permission; Tohidi, 2013, pers. comm.; Collet's work at USGS.

Note: This excludes dispersed organic carbon such as kerogen and bitumen, which equals nearly 1000 times the total amount shown in the table.

[a] These are best estimates.

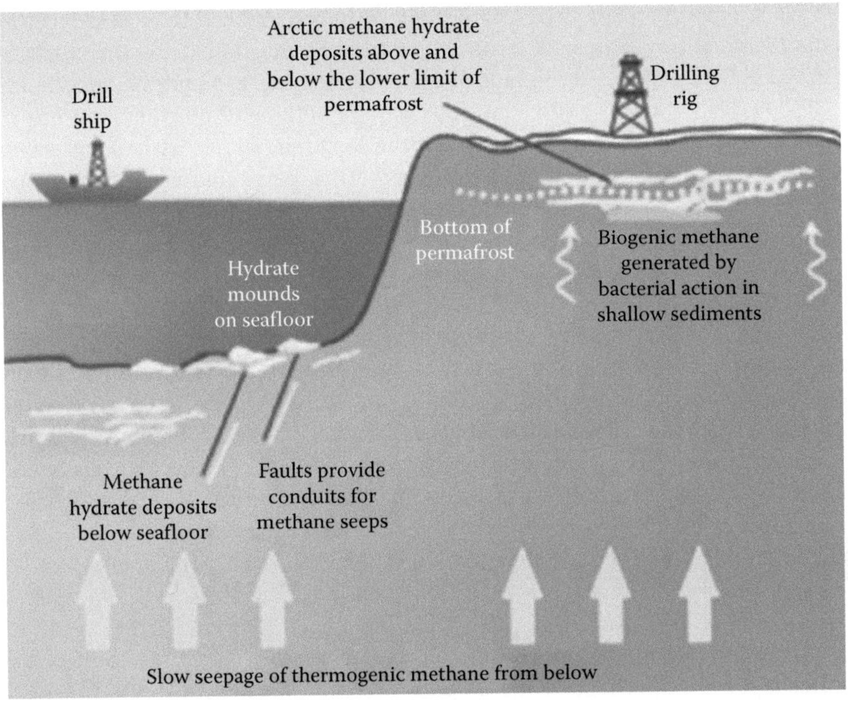

FIGURE 12.4 Types of methane hydrate deposits. (Adapted from "Methane hydrate—The world largest natural gas resource is trapped beneath permafrost and ocean sediments," Geology.com, a communication, 2013.)

are believed to be formed by the migration of gas from depth along geological faults, followed by precipitation, or crystallization on contact of the rising gas stream with cold seawater. Methane hydrates are also present in deep arctic sea cores and record a history of atmospheric methane concentrations dating to 800,000 years ago [17–27].

In polar regions, methane hydrates are found where temperatures are cold enough for onshore and offshore permafrost to be present. In offshore sediments, methane hydrates are found at water depths of 300–500 m, according to prevailing water temperatures. Continental deposits have been located in Siberia and Alaska in sandstone and siltstone beds at depth <800 m. Oceanic deposits seem to be widespread in the continental shelf and can occur within the sediments at depth or close to the sediment–water interface. They may cap even larger deposits of gaseous methane. In 2008, Canadian and Japanese researchers extracted a constant stream of natural gas from Mallik gas hydrate field in the Mackenzie River delta [17,36,39,43,48,49,51] (USGS, 2012, pers. comm.). This hydrate field was first discovered by Imperial Oil Co. in 1971–1972.

The occurrence of gas hydrates on the Alaska North Slope was confirmed in 1972 in the northwest part of the PBU (Prudhoe Bay Unit) field [17,46,48,49] (USGS, 2012, pers. comm.), and the North Slope now is known to contain several well-characterized gas hydrate deposits. The methane hydrate stability zone extends beneath most of the coastal plain province and has thicknesses >1000 m in the Prudhoe Bay, Kuparuk River, and Milne Point oil fields on the North Slope of Alaska. The estimated amount of gas within these gas hydrate accumulations is ~37–44 Tcf, which is equivalent to twice the volume of conventional gas in the Prudhoe Bay field [11]. More details on the locations of gas hydrate reservoirs in Alaska are given in various USGS reports (2012, pers. comm.). As mentioned in Refs. [3,9,17–54] (USGS, 2012, pers. comm.), besides Alaska, hydrate fields have been discovered in other countries of the world, which include Japan, China, India, Korea, Russia, and Canada. In the United States, hydrates have also been discovered in the Gulf of Mexico [3,9,40,47].

The size of the oceanic methane clathrate reservoir is poorly known. The recent estimates constrained by direct sampling suggest the global inventory occupies between 1 and 5 million cubic kilometers. This estimate corresponds to 500–2500 gigatons carbon that is substantially larger than 230 gigatons estimated for other natural gas resources. The reservoir in Arctic permafrost has been estimated at 400 gigatons, but no estimates for Antarctic reservoirs are available. Low concentrations at most sites imply that only small percentage of clathrate deposits may be economically recoverable [3,9,17–54] (USGS, 2012, pers. comm.).

There are two distinct types of oceanic deposits. The most common type is one where methane is contained in I clathrate and generally found in the depth of the sediment. This type is derived from microbial reduction of CO_2. These deposits are located within a mid-depth zone around 300–500 m thick in the sediments. The second less common type is found near the sediment surface. This type is formed by the thermal decomposition of organic matter. Examples of this type are found in the Gulf of Mexico and Caspian Sea. Some deposits have characteristics intermediate between the microbial and thermal source types, and they are considered to be formed from a mixture of two.

While the sedimentary methane hydrate reservoir probably contains 2–10 times the currently known reserves of conventional natural gas, the majority of the site's deposits are too dispersed to recover economically. The detection of viable sources is also problematic. The technology for extraction of methane gas from hydrate is also an issue. To date, Messoyakha Gas field in the Russian city of Norilsk is the only sustained commercial operation. Japan is planning to develop a commercial operation by 2016 [20,33,43], and China has invested $100 million over 10 years to study hydrates [45]. A possible economic reserve in the Gulf of Mexico may contain 10^{10} m^3 of gas [3,9,40,47].

Gas hydrates are of great importance for a number of reasons graphically illustrated in Figure 12.5 [54]. Naturally occurring methane gas clathrates contain an enormous amount of strategic energy reserve [37,39,46]. In offshore hydrocarbon drilling and production operations, gas hydrates can cause major and potentially hazardous flow assurance problems. The recovery of gas hydrates by carbon dioxide provides an opportunity to dispose carbon dioxide by sequestration [61–83]. Gas hydrates also provide an increasing awareness of the relationship between hydrate and subsea slope stability. Gas hydrates also pose a potential danger to deep-water drilling installations, pipelines, and subsea cables [55–59,84–116] (LaBelle, 2012, pers. comm.; Tohidi, 2012, pers. comm.). Finally, it poses a long-term concern

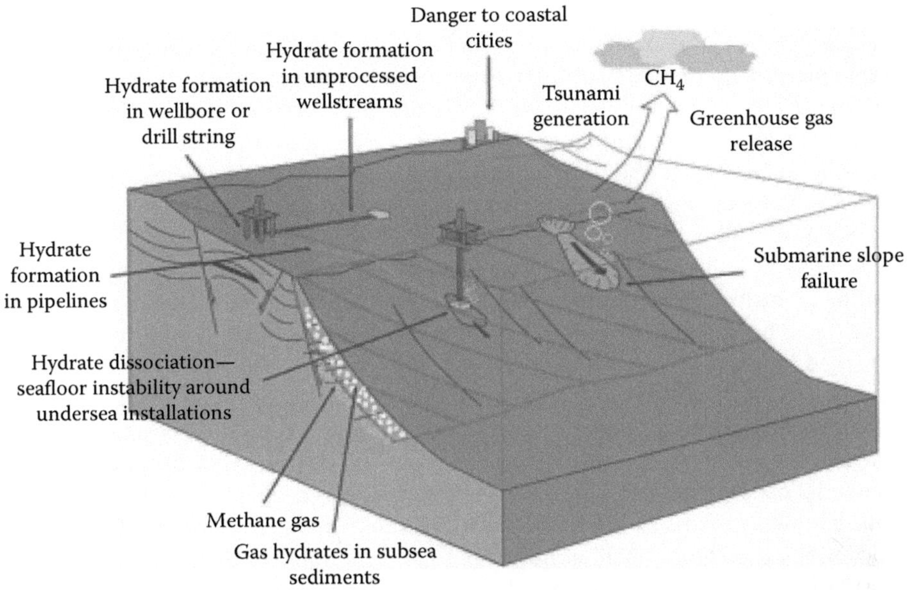

FIGURE 12.5 Reasons for the importance of methane hydrates. (From "Why are gas hydrates important," Institute of Petroleum Engineering, Heriot Watt University, Edinburgh, 2011. With permission.)

regarding hydrate stability and methane release and its subsequent effect on global climate change [117–180] (Kennett, 2012, pers. comm.). Some of these topics are briefly discussed in Sections 12.3 through 12.5.

12.3 IMPORTANCE OF GAS HYDRATES ON OFFSHORE OIL AND GAS OPERATIONS

The existence of gas hydrates affects both drilling and production of offshore oil and gas operations [55–59,84–116] (LaBelle, 2012, pers. comm.; Tohidi, 2012, pers. comm.). These effects are briefly described in Sections 12.3.1 through 12.3.3 [91–116] (LaBelle, 2012, pers. comm.; Tohidi, 2012, pers. comm.).

12.3.1 Drilling

Methane clathrates (hydrates) are commonly formed during natural gas production operations, when liquid water is condensed in the presence of methane at high pressure. It is known that larger hydrocarbon molecules such as ethane and propane can also form hydrates, although these are not as stable as methane hydrates. Once formed, hydrates can block pipeline and processing equipment. They are generally removed by (1) reduction of the pressure, (2) addition of heat, or (3) dissolving them using chemicals such as methanol and ethylene glycol. Care must be taken to ensure that the removal of the hydrates is carefully controlled, because as the hydrate undergoes phase transition, the release of water and methane can occur at very high rates. The rapid release of methane gas in a closed system can result in a rapid increase in pressure [104,105], which can be harmful to the drilling operation. In recent years, hydrate formation during drilling operation is controlled with the use of kinetic hydrate inhibitors [96–99,113–116], which dramatically slow the rate of hydrate formation and anti-agglomerates, which prevent hydrates from sticking together to block pipes and other parts of equipment.

When drilling in oil- and gas-bearing formations submerged in deep water [55–59,84,85], the reservoir gas may flow into the well bore and form gas hydrates owing to the low-temperature and high-pressure conditions found during deep-water drilling. The gas hydrates may then flow upward with drilling mud or other discharged fluids. As they rise, the pressure in the annulus decreases and the hydrates dissociate into gas and water. The rapid gas expansion ejects fluid from the well, reducing the pressure further, which leads to more hydrate dissociation and further fluid ejection. The resulting violent expulsion of fluid from the annulus is one potential cause or contributor to what is referred to as a "kick" [104,105], which can cause blowouts. This can cause serious well safety and control problems and create hazardous conditions such as flow blockage, hindrance to drill string movement, loss of circulation, and even abandonment of the well. Since gas hydrates contain 85% water, their formation can withdraw water from drilling fluids, changing the properties of the fluids, thus causing salt precipitation, an increase in fluid weight, or the formation of solid plug.

The condition of the hydrate formation during kick depends on the composition of the kick gas, temperature, and pressure. A combination of salts and chemical inhibitors can provide a required inhibition to avoid hydrate formation, particularly at water depths >1000 m [96–99,115–116].

12.3.2 Production by Enhanced Oil and Gas Recovery Methods

Enhanced oil and gas recovery methods increase the risk of the gas hydrate formation. Process equipment and multiphase transfer lines from wellhead to the production platform where low-temperature and high-pressure conditions exist are prone to hydrate formation. The following methods are generally adopted to reduce hydrate problems in hydrocarbon transfer lines and process facilities [86–93]:

1. Use high flow rates, which limit the time for hydrate formation in a volume of fluid, thereby reducing the kick potential [104,105]. Make careful measurement of line flow to detect incipient hydrate plugging [104,105], particularly at low gas production rate. Also, monitor the pressure rise in wellcasing after it is "shut in" (isolated). The hydrate formation will decrease the rate of pressure rise [104,105].
2. Additions of energy (e.g., the energy released by setting cement used in well completion) can raise the temperature and convert hydrates to gas, producing a "kick."
3. For a given pressure, operate at temperatures higher than the hydrate formation temperature. This can be done by insulation or heating of

the equipment. At fixed temperature, operate at pressure below hydrate formation pressure.

4. Reduce water concentration to avoid hydrate formation. Change feed composition.
5. Add compounds such as methanol, salts, or other kinetic inhibitors to prevent hydrate formation. Also prevent hydrate clustering by using hydrate growth modifiers or covering working surfaces with hydrophobic substances [86–93,104,105].

With conventional oil and gas exploration methods extending into progressively deeper waters, the potential hazards gas hydrates can pose to operation are becoming increasingly more important. Two possible events—the release of overpressurized gas (or fluids) trapped below the zone of hydrate stability and destabilization of *in situ* hydrates—can be hazardous. Care must be taken to avoid these incidences [96–102,113–116].

12.3.3 Natural Gas Hydrates versus Liquefied Natural Gas in Transportation

Since methane clathrates are stable at a higher temperature than liquefied natural gas (LNG) (–20°C vs. –162°C) [108], there is some interest in converting natural gas into clathrates rather than liquefying it when transporting it by seagoing vessels. A significant advantage would be that the production of NGH from natural gas at the terminal would require a smaller refrigeration plant and less energy than LNG would. Offsetting this, for 100 tons of methane transported, 750 tons of methane hydrate would have to be transported. Since this would require a ship of 7.5 times greater displacement, or require more ships, an application of this approach has not been economically attractive.

12.4 ENVIRONMENTAL IMPACTS OF GAS HYDRATES

Gas hydrates alter the physical properties of the sediment. In the absence of hydrates, fluids and gas migrate freely at seafloor. The solid hydrates reduce permeability and restrict sediment consolidation, fluid expulsion, and cementation. The hydrate dissociation leads to increased pore fluid pressure and underconsolidated sediments, with a reduced cohesive strength compared to overlying hydrate-bearing sediments, forming a zone of weakness. This zone of weakness could act as a site of failure in the event of increased gravitational loading or seismic activity. The link between seafloor failure and gas hydrate destabilization is a well-established phenomenon [1–15]. The exploration of hydrates from ocean floor by drilling through hydrate zones can create the problem of destabilizing support foundations for platforms and production wells. The disruption of ocean floor can also result in surface slumping or faulting, which can endanger work crews and the environment [1–15].

Since hydrates prevent sediment compaction, their *in situ* dissociation can also cause climate change and falling of sea level. If the hydrate breaks down, it will weaken the sediment and may cause submarine landslides and simultaneously release methane into the atmosphere. The methane released from the reservoir to the atmosphere can contribute to the climate change. Submarine landslides can cause tsunamis and catastrophic coastal flooding. The thickness of the gas hydrate stability zone (GHSZ) in continental margins depends on water depth (hydrostatic pressure), water temperature, geothermal gradient, and gas composition [1,60] (Tohidi, 2013, pers. comm.).

Methane is a powerful greenhouse gas. Despite its short atmospheric half-life of seven years, methane has a significant global warming potential [1–15] (Harrison, 2010, pers. comm.). Recent research carried out in 2008 in the Siberian Arctic has shown millions of tons of methane being released [153,162,166,168,169,176] (Kennett, 2012, pers. comm.), with concentrations in some regions reaching up to 100 times above normal [1–16] (Harrison, 2010, pers. comm.). Past and future climate changes can be linked to methane released from gas hydrates.

Currently, the link between stability of gas hydrates and global warming is being examined. Since methane warms the environment 15–20 times more than carbon dioxide, the release of methane can create a chain reaction for global warming, leading to more hydrate instability with additional release of methane. Methane release in air eventually (within 10 years) is converted to carbon dioxide, another greenhouse gas [117–180] (Kennett, 2012, pers. comm.).

The analysis of the link between gas hydrate and climate warming can be divided into five parts [117–180] (Kennett, 2012, pers. comm.):

Region 1: Thick (≥300 m) onshore permafrost. Gas hydrates that occur within or beneath thick terrestrial permafrost will remain largely stable even if climate warming lasts hundreds of years. The warming could, however, cause hydrates at the top of the stability zone, about 625 ft below the earth's surface to dissociate over thousands of years [117–180] (Kennett, 2012, pers. comm.). It contributes <1% of the total hydrates, and its effect on climate change will be minimal.

Region 2: Subsea permafrost on the circum-Arctic shelves. The shallow water continental shelves that circle the parts of the Arctic Ocean were formed when sea-level rise during the past 10,000 years inundated permafrost that was at the coastline. The methane hydrates in subsea permafrost that is thawing beneath these continental shelves is being released now. While this methane can rise to ocean surface and then to atmosphere, the amount is only considerably less than about 1% of the world gas hydrates [117–180] (Kennett, 2012, pers. comm.).

Region 3: Upper edge of stability (or deep-water marine hydrates at the feather edge of GHSZ). Gas hydrates on upper continental slopes beneath 1000–1600 ft of water lie at the shallowest water depth for which methane hydrates are stable. The upper continental slopes that ring all the continents could host gas hydrates in zones that are roughly 30 ft thick. Within the next 100 years, warm water can completely dissociate these hydrates, but they are more likely to be oxidized in water than released in the atmosphere. These hydrates contribute about 3.5% of the earth's total hydrates [117–180] (Kennett, 2012, pers. comm.).

Region 4: Deep-water gas hydrates. Ninety-five percent of earth's gas hydrates are at depths >3000 ft. Even with an increase in the ocean temperature, they are likely to stay stable over thousands of years. They also occur deep within the sediments, and the released methane will remain in the sediments, and if they move upward, they will form new hydrates or consumed by oxidation within water [117–180] (Kennett, 2012, pers. comm.).

Region 5: Seafloor gas hydrate mounds. At some marine seeps such as the Gulf of Mexico, massive relatively pure gas hydrate occurs in seafloor mounds. While seafloor gas hydrate mounds and shallow subsea floor gas hydrate constitute only a trace amount of the global gas hydrate inventory, they can dissociate rapidly due to the expulsion of warm fluids from the seafloor and release significant amount of methane to the atmosphere.

Based on the analysis of these five regions, a general consensus [117–180] (Kennett, 2012, pers. comm.) is that catastrophic widespread dissociation of methane gas hydrates will not be triggered by continued climate warming at a contemporary rate (0.2°C per decade) over a timescale of few hundred years. In spite of this conclusion, there has been an enormous interest in studying methane release from hydrates to the atmosphere and its effect on environment. The vast literature [117–180] (Kennett, 2012, pers. comm.) is cited here to demonstrate the significant interest on the subject.

12.5 PRODUCTION OF METHANE FROM GAS HYDRATE RESERVOIRS

Hydrates are known to occur at temperatures <295 K and pressure >3000 kPa. The dissociation of these hydrates occurs as

$$CH_4 \cdot 6H_2O(s) \rightarrow CH_4(g) + 6H_2O(l) \tag{12.1}$$

with enthalpy = 10–20 kcal/mol of gas dissociated [1–16] (Harrison, 2010, pers. comm.). This reaction requires an external energy source to propagate along the right-hand side [1–16] (Harrison, 2010, pers. comm.).

In conventional gas reservoirs, natural gas migrates to the recovery point via pressure gradients. For these reservoirs, the recovery rate is a function of the formation permeability and pressure gradients established between the reservoir and the extraction well(s). Production of methane from hydrate-bearing deposits requires additional energy to dissociate the crystalline water lattice that forms the gas hydrate structure. A variety of methods have been proposed for producing natural gas from hydrate deposits: (1) thermal stimulation, where the temperature is increased above the hydrate stability region; (2) depressurization, where the pressure is decreased below the hydrate stability region; (3) chemical injection of inhibitors, where the temperature and pressure conditions for hydrate stability are shifted; (4) CO_2 or mixed CO_2 and N_2 exchange, where CO_2 and N_2 replace CH_4 in the hydrate structure; and (5) enhanced gas hydrate recovery (EGHR) methods, where two-phase emulsion (of CO_2 and water) and other solution injection techniques are used to replace methane from hydrate structure. Each of these methods is briefly reviewed in Sections 12.5.1 through 12.5.6. This section also briefly reviews the numerical simulations that have been carried out for methane recovery from hydrates. Finally, production research that has been carried out for commercial sites is briefly assessed.

12.5.1 Thermal Stimulation

The recovery of methane gas from gas hydrates via thermal stimulation has been examined both experimentally [181–183] and theoretically [184–187]. Technologies for implementing thermal stimulation include steam injection, cyclic steam injection, fire flooding, hot brine injection and electromagnetic heating. The techniques of steam injection and cyclic steam injection are very similar to those used in the recovery of conventional and unconventional oils. Various possibilities for heating hydrates using steam or cyclic steam injections have been examined in the literature [182,187]. All of these techniques, however, suffer from high heat losses, and by-products of fire flooding can dilute the produced natural gas. The energy efficiency of electromagnetic heating is also low.

A more promising approach is to inject a saline aqueous solution at an elevated temperature into gas hydrate-bearing geological reservoir. In this method, the sensible heat carried by the brine solution is discharged to the gas hydrates by a convective heat-transfer mechanism. The dissolved salt depresses the dissociation temperature of the gas hydrate. The experimental evidences indicate that with the injection of brine, the hydrates become colloidal and migrate convectively with the brine [188–190]. Tang et al. [181,191] showed that the energy efficiency of the hot brine injection process is dependent on the brine temperature, injection rate, and initial hydrate saturation.

The energy efficiency is defined as the ratio of combustion heat of produced gas over the heat input of the brine. The study showed that a better energy efficiency was obtained at higher initial hydrate saturation and lower temperature and injection rates [181–188]. This higher energy efficiency is, however, accompanied by lower production rates. For moderate to high temperature and injection rate, about 50% of the recovered energy from methane is used to heat the brine solution. A modification of this approach was suggested by Chatterji and Griffith [78] who proposed an injection of two aqueous fluids that react and produce the heat required to release methane from the hydrates. This type of acidic and basic solutions reactions will yield a hot salt solution, and this will not require the external heating of brine solution, thereby improving the energy efficiency.

12.5.2 Depressurization

Gas hydrate production via depressurization is considered to be the most economically promising technology [190,192–200]. This method has been adopted in Messoyakha field in northern Russia, which contains both free natural gas and hydrates. This reservoir has been constantly producing natural gas because of dissociation of gas hydrates into gas due to depressurization. The production rate in this field is, however, controlled by the heat transfer toward the hydrate dissociation region.

Moridis et al. [197,201,202] and Moridis [203] numerically simulated the effect of depressurization at Mallik site assuming 0.03°C/m temperature gradient in the hydrate-bearing formation. The simulation showed a vertical drop in temperature in response to depressurization and hydrate dissociation. This temperature drop can be reversed by the injection of warmer water in the well, which provides the needed energy to sustain hydrate dissociation in the depressurized system. The simulation also indicated that, when steam or hot methane gas was injected from a second well, natural gas production was superior in terms of the ratios of produced gas to water and fraction of produced methane from hydrates.

Several other simulation studies showed that hydrate dissociation rates and associated gas productions are controlled by the far-field reservoir pressure and temperature, via energy supplied by natural gas conveyed from the far field to the dissociation front [203–212]. Few studies have reported experimental data of gas recovery by depressurization [194,195]. While depressurization is a viable option because of thermal self-regulation of gas hydrates, the method results in slow production rates. Sustained production requires a heat source, which at the Messoyakha field is supplied by thermal conduction and convection in the dissociation zone. This heat transfer ultimately controls the production rate.

There are three important mechanisms involved in the depressurization of the gas hydrates: (1) kinetics of dissociation, (2) conductive heat transfer, and (3) convective flow of fluids like gas and water. A significant theoretical work that uses a three-dimensional model of a porous media and simulates the exact conditions of a reservoir with regard to all the mechanisms involved has been reported [201–214]. However, to this date, conclusions of such analysis are only based on certain assumptions, whose validity needs to be experimentally verified. Often a two-well system involving a combination of depressurization at the production well and a thermal input (by hot fluid injection) at the injection well appears to be better than a single vertical system [190,192–200].

12.5.3 Inhibitor Injection

Sung et al. [214], Kawamura et al. [215], and Li et al. [216,217] showed that the thermodynamic inhibitors lower the hydrate formation temperature, which can result in hydrate dissociation when injected into a gas hydrate-bearing formation. The most important thermodynamic organic inhibitors are methanol, monoethylene glycol (MEG), and diethylene glycol (DEG) commonly referred to as glycol [218–222]. Dissolved salts such as NaCl, $CaCl_2$, KCl, and NaBr can also be inhibitors [191]. While gas hydrate inhibitors are an effective methodology for preventing hydrate formation in engineering applications, their use in the production of NGHs is restrictive due to environmental impact, prohibitive costs, and thermal self-regulation of gas hydrates. Of the inhibitors examined, methanol and glycols are the most successful ones [221]. The principles by which alcohol, glycols, and salts inhibit hydrates are the same. However, salts have corrosion problems, and they cannot be easily vaporized due to their low vapor pressures.

In adding inhibitors, besides temperature and pressure conditions, composition and amount of inhibitors are important. The inhibitor must be at or below its water dew point (i.e., must be water saturated). In addition, dehydration can be used as an alternative. An addition of an inhibitor can shift pressure–temperature diagram such that the temperature decreases at specific pressures, and this facilitates hydrate dissociation. After temperature depression due to an addition of an inhibitor, free gas will form and hydrate zone will shift to the left to lower the temperature side. Methanol has a high vapor pressure and infinite water solubility and can easily shift to the gas phase.

In most offshore applications, hydrate formation is controlled by injection of a thermodynamic hydrate inhibitor. Inhibitor injection at a given pressure will reduce the temperature at which hydrate is formed. Overall, ethylene glycol seemed to be the most useful inhibitor for the gas hydrates.

12.5.4 Gas Exchange

Exchanging CO_2 with CH_4 concept was first advanced by Ohgaki et al. [68]. Their experimental study showed that CO_2 be preferentially clathrated over CH_4 in the hydrated phase. They also demonstrated the possibility of producing CH_4 by injecting CO_2 gas. Ohgaki et al. [68] noted that during the exchange process, mole fraction of CO_2 in the hydrate phase was greater than that in the gas phase.

This effect was further studied quantitatively by Seo and Lee [69] and Seo et al. [70]. They showed that CO_2 concentration in the hydrate phase was >90% when gas-phase concentration of CO_2 in the hydrate formers (i.e., CO_2 and CH_4) was above 40%. Pure CH_4 and CO_2 form structure I (sI) type hydrates, and their mixtures also form sI type hydrates [61–75]. In forming mixed CH_4 and CO_2 hydrates, the CH_4 molecules occupy both the large and small cages of sI type hydrates, whereas CO_2 molecules only occupy the large cages. Without hydrate dissociation, there is an upper limit to the substitution of CO_2 for CH_4 in hydrates.

Lee et al. [218] showed that ~64% of CH_4 can be released by exchange with CO_2. In addition to equilibrium considerations, the heat of CO_2 hydrate formation is higher (−57.9 kJ/mol) than the heat of dissociation of CH_4 hydrate (−54.5 kJ/mol), making the overall process exothermic that favors the normal exchange of CO_2 with CH_4 hydrate.

While the exchange of CO_2 for CH_4 is thermodynamically a favorable process, the kinetics of exchange mechanism is slow [61–75,209], with induction time requiring several days. The original studies also did not address the rate of CO_2 gas penetration further into gas hydrate, beyond the first few hundred manometers at the interface [203]. The exchange of CO_2 with CH_4 at high pressure (with liquid CO_2) was also examined in the literature, but once again slow rate of exchange was observed. The use of nitrogen instead of CO_2 gave a much higher rate. For liquid CO_2 injection, thermodynamic conditions can either favor CO_2 or CH_4 cage occupation [76–83]. This transition occurs when the pure CO_2 and CH_4 temperature-versus-pressure equilibrium functions cross at the pressure above the gas–liquid CO_2 phase boundary.

Thermodynamic properties of hydrates depend on the pore size distribution in the geologic media; hydrate formation will occur in large pores first and then in small pores until equilibrium is achieved [205,212]. Porous media also affect other thermodynamic properties of hydrates. In geologic media that have distribution of pore sizes, hydrates would form and dissociate over a range of temperatures and pressures according to the distribution of pore radii and the impact of salts in the residual pore water [191]. Goel [67] and Goel et al. [209] indicated that in order to understand gas-exchange technology in porous media, quantitative estimates of formation and dissociation processes in a typical geologic media core samples are needed.

12.5.5 EGHR Method

As shown above, a strict gas exchange of CO_2 for CH_4 in bulk methane hydrate is slow by several orders of magnitude to be considered as an effective method of gas hydrate production. An EGHR process that involves injecting a two-phase emulsion of liquid CO_2 and water at proper volumetric ratio can considerably enhance (three times or higher) the production rate over injecting cool water (15°C) alone [76–83]. It is important to know the range of reservoir conditions where EGHR technique can be applied. Collett and coworkers [204,222,223] calculated these conditions for Alaska Northern slope (ANS) and concluded that EGHR method can be applied over a large fraction of ANS. They also found that CO_2 hydrate would be stable under almost any conditions on the ANS short of very near the ground surface. They also suggested that typical ANS reservoir conditions would inject liquid CO_2 with a density ~82%–94% of the water phase. ANS well log temperature data as well as carbon dioxide hydrate and vapor–liquid equilibrium data are described by Collet et al. [204,222,223].

The laboratory studies indicated that there are no signs of coagulation into macrodroplets as the emulsion moves away from the injector—a conclusion that needs to be tested at reservoir scale [73]. Another important restriction is that the temperature of the water–CO_2 emulsion remains above the equilibrium point where CO_2 hydrate could form in the wellbore or near the wellbore. Interruption of the supply of emulsion fluid during production for an extended period could result in the premature formation of CO_2 hydrate and plugging [73,77]. Provisions for temporary injection of heat may be needed to allow for flow interruptions, which are important for well maintenance.

The EGHR method has been tested in laboratory for continuous production of a suitable liquid carbon dioxide and water emulsion [73,76–82]. This test is largely one dimensional. A suitable downhole tool that can work in actual field needs to be developed. The injector tool design should be compatible with downhole conditions typical of gas hydrate formations. Wellbore completion requirements such as open hole, uncased, or perforated casing influence the design parameters of the injection tool. Injection of the liquid carbon dioxide and water emulsion in the target formation is the most important requirement. A new design to fit these requirements is depicted in Figure 12.6 [73]. Here, emulsion outlets are located on the side. Surface-warmed liquid carbon dioxide and water can be directed into such an injector from the high-pressure lines. Use of produced water to form emulsion would eliminate issues associated with disposal of these fluids in arctic conditions. Both rate and distance of formation penetration can

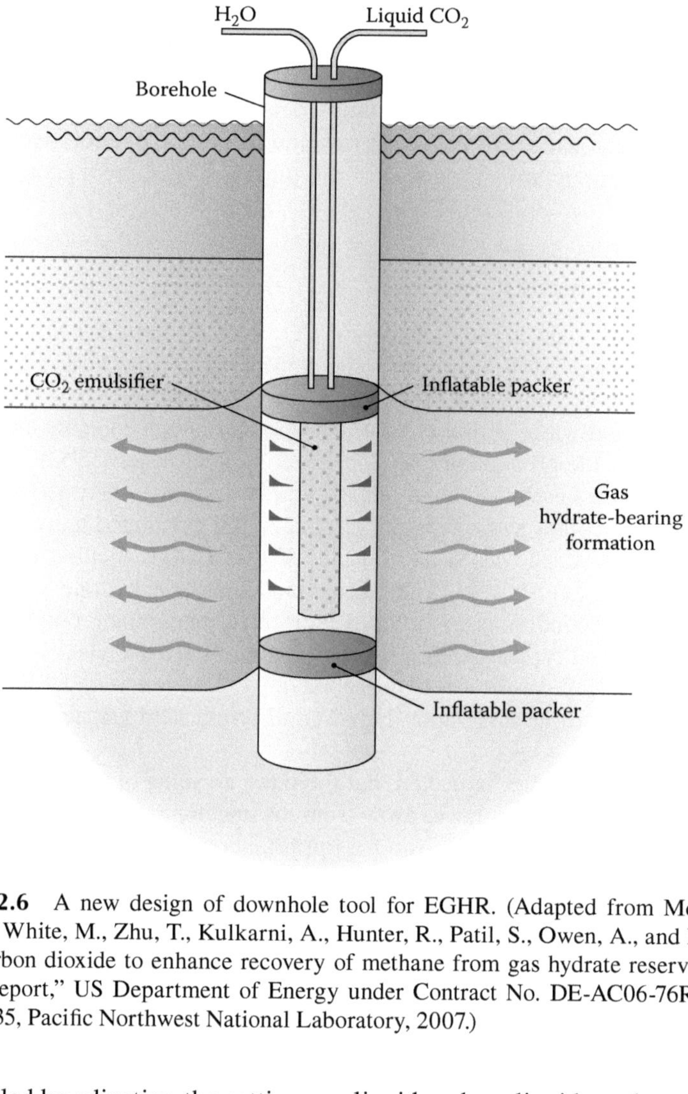

FIGURE 12.6 A new design of downhole tool for EGHR. (Adapted from McGrail, B., Schaef, H., White, M., Zhu, T., Kulkarni, A., Hunter, R., Patil, S., Owen, A., and Martin, P., "Using Carbon dioxide to enhance recovery of methane from gas hydrate reservoirs: Final summary report," US Department of Energy under Contract No. DE-AC06-76RLO 1830, PNNL 17035, Pacific Northwest National Laboratory, 2007.)

be controlled by adjusting the settings on liquid carbon dioxide and water pumps from the surface.

An EGHR technique is still being developed [73,77,81]. A number of questions such as placement of recovery wells including the distance from the injection site and spacing to maximize recovery of CH_4 gas need to be determined. Identification and delivery logistics of an economic supply of carbon dioxide for a given site also need to be ascertained. Both theoretical and experimental works that address these issues need to be pursued [73,77,81].

In sum, the EGHR process has several advantages: (1) Since the heat generated from the formation of CO_2 hydrate is ~20% greater than the heat consumed from the dissociation of methane, the replacement of methane with carbon dioxide in gas hydrate sediments is thermodynamically favorable. This net exothermic process allows the dissociation of hydrates to be carried out with only minimal requirement of an additional heat source. (2) Once CO_2-rich fluid fills pores vacated by methane, the subsequent formation of carbon dioxide hydrate would mechanically stabilize the formation, eliminating subsidence concerns in some production environment, and (3) the overall process is carbon neutral since methane is permanently replaced by carbon dioxide as gas hydrate. Produced water can also be used to form the emulsion, eliminating a problematic disposal issue in arctic settings [73,77,81].

12.5.6 Computer Simulation

There are some reported computer simulation studies of commercial production methods for gas hydrates, and most of them have examined conventional production concepts of depressurization coupled with some form of thermal stimulation [83,201–214,224–226]. An EGHR process that utilizes a microemulsion of liquid CO_2 and water to decompose methane hydrate *in situ* and produce free gas described earlier has been successfully demonstrated in laboratory-scale experiments with gas hydrate-bearing sediments. Since these laboratory-based studies were extremely encouraging, a reservoir modeling assessment that compared and contrasted the EGHR process with conventional methods of gas hydrate production was carried under a Department of Energy (DOE) project [73,210,212].

Within the DOE project [73,210], STOMP-HYD simulator was applied to a series of one- and two-dimensional simulations that investigated the production of CH_4 hydrates in geologic media using CO_2 injection. Effectively, the project considered two approaches to producing CH_4 hydrate in geologic media using CO_2 injection: (1) hydrate dissociation and reformation and (2) direct molecular exchange. In the hydrate dissociation and reformation approach, the injected CO_2 first dissociates CH_4 hydrate. This stage is followed by reformation of a mixed gas hydrate, which predominately comprises CO_2. In the direct molecular exchange approach, the injected CO_2 exchanges with the CH_4 in the hydrate structure, maintaining the hydrate integrity. The dissociation–reformation approach has the advantage of releasing CH_4 in both the small and large cages. In the direct-exchange approach, only the CH_4 in the large cages is released. Co-injection of CO_2 and N_2 has been shown to allow molecular exchange of CH_4 in both the small and large cages. Because the STOMP-HYD simulator did not track small- and large-cage occupancies, it is currently limited to CO_2 exchange with CH_4 in large cages. The principal conclusion from this series of simulations was that both CO_2 exchange approaches yielded faster production times, but lower CH_4 recoveries over pure water injections. Without consideration of the cage occupancies, the direct exchange yielded faster production times over the dissociation–reformation approach, with nearly equivalent CH_4 recoveries. The CO_2-to-water ratio in the injecting fluid primarily affected production rates, with higher ratios yielding faster productions.

STOMP-HYD simulation results also showed the following conclusions [73,213,224]:

1. Preliminary depressurization to a point above the freezing point of the aqueous phase opens pore space for injection of mobile fluids.
2. Kinetics of the direct exchange of hydrate formers (i.e., CO_2 with CH_4) are important.
3. Cage occupancies of the sI structure are expected to have significant impacts on the efficiency of direct gas-phase CO_2–CH_4 exchange.
4. Controlling secondary hydrate formation is critical to prevent pore plugging.
5. Heat transfer into the production zone is not required under properly controlled production conditions.

One critical finding of the above-described Battelle's simulation modeling work was that the formation of secondary CO_2 hydrate has the potential to halt the production process by inhibiting fluid migration. A complete exchange of CO_2 and CH_4 is possible without forming excessive secondary hydrate and while maintaining elevated hydrate saturations. The pore-water salinity may play a strong role in the inhibition of secondary hydrate formation beyond certain saturation levels, an observation in agreement with the published experimental results [73].

12.5.7 Commercial Applications

In the recent years, the above-described production methods and computer simulations have been applied to numerous practical sites [33,222,227–234]. The North Slope of Alaska and numerous sites in that region (such as Mallik field, Milne point) have been tested [33,222,227–229]. Nankai Trough [231] and Ulleung basin of the Korea [233] have also been examined. Several general production strategies have also been investigated [216,234–237]. More work on the applications (both theoretical and experimental) of various production methods to the commercial sites (both on land and in deep water) is needed. The successful commercial operations to recover methane from gas hydrates will significantly increase our energy resource. Once again, water is the cause for this important energy and fuel source.

REFERENCES

1. Englezos, P., "Clathrate Hydrates," *Industrial & Engineering Chemistry Research*, 32 (7), 1251–1274 (1993).
2. "What are gas hydrates?" a communication by Center for Gas hydrate Research, Heriot-Watt University, Edinburgh, The Hydrate forum Org. (2012).
3. Boswell, R. and Collett, T.S., "Current perspectives on gas hydrate resources," *Energy and Environmental Science*, 4, 1206–1215 (2011).
4. Collett, T.S., Johnson, A.H., Knapp, C.C., and Boswell, R. (eds.), "Natural gas hydrates: A review," in *Natural Gas Hydrates—Energy Resource Potential and Associated Geologic Hazards*, AAPG Memoir 89. AAPG, Tulsa, OK, 146–219 (2009).
5. McIver, R., "Gas hydrates," in Meyer, R. and Olson, J. (eds.), *Long-Term Energy Resources*. Pitman, Boston, MA, 713–726 (1981).
6. Collett, T.S. "Gas hydrates as a future energy resource," *Geotimes*, 49 (11), 24–27 (2004).
7. Sloan, E.D. and Koh, C., *Clathrate Hydrates of Natural Gases*, 3rd ed. Taylor & Francis, Boca Raton, FL (2008).

8. Ruppel, C., "Methane hydrates and the future of natural gas," MITEI Natural Gas Report, Supplement Paper No. 4 (2011).
9. Boswell, R. and Collett, T., "The gas hydrates resource pyramid," *Fire in the Ice*, 6 (3), 5–7 (2006).
10. Ruppel, C., "Ruppel: MITEI Natural Gas Report," Supplementary Paper on Methane Hydrates, 19 (2011).
11. Energy Information Administration (EIA), "Natural gas," in *International Energy Outlook 2010*. US Department of Energy, Washington, DC (2010).
12. Holder, G.D., Kamath, V.A., and Godbole, S.P., "The potential of natural gas hydrates as an energy resource," *Annual Review of Energy*, 9, 427–445 (1984).
13. "Methane hydrate phase diagram," Wikipedia, the free encyclopedia (May 10, 2010).
14. "Clathrate hydrate," Wikipedia, the free encyclopedia, 1–7 (2012).
15. "Methane clathrate," Wikipedia, the free encyclopedia (2012).
16. Makogon, Y.F., Holditch, S.A., and Makogon, T.Y., "Natural gas-hydrates—A potential energy source for the 21st Century," *Journal of Petroleum Science and Engineering*, 56 (1–3), 14–31 (2007).
17. Collett, T.S., "Energy resource potential of natural gas hydrate," *American Association of Petroleum Geologists Bulletin*, 86, 1971–1992 (2002).
18. Ruppel, C., "Ruppel: MITEI Natural Gas Report," Supplementary Paper on Methane Hydrates, 18 (2011).
19. Frye, M., "Preliminary evaluation of in-place gas hydrate resources: Gulf of Mexico Outer Continental Shelf," Minerals Management Service Report 2008-004, US Department of the Interior, Washington, DC (2008).
20. Fujii, T., Saeiki, T., Kobayashi, T., Inamori, T., Hayashi, M., Takano, O., Takayama, T. et al., "Resource assessment of methane hydrate in the Nankai Trough, Japan," Offshore Technology Conference, Paper 19310, Houston, TX (2008).
21. Gornitz, V. and Fung, I., "Potential distribution of methane hydrates in the world's oceans," *Global Biogeochemical Cycles*, 8, 225–347 (1994).
22. Milkov, A., "Global estimates of hydrate-bound gas in marine sediments: How much is really out there?" *Earth-Science Reviews*, 66, 183–197 (2004).
23. Trofimuk, A., Cherskiy, N., and Tsarev, V., "Accumulation of natural gases in zones of hydrate—Formation in the hydrosphere," *Doklady Akademii Nauk SSR*, 212, 931–934 (1973) (Russian).
24. Collett, T.S., "Gas hydrates as a future energy resource," *Geotimes*, 49 (11), 24–27 (2004).
25. Kvenvolden, K.A., Ginsburg, G.D., and Soloviev, V.A., "Worldwide distribution of subaquatic gas hydrates," *Geo-Marine Letters*, 13 (1), 32–40 (1993).
26. Kvenvolden, K.A., "Natural-gas hydrate occurrence and issues," *Sea Technology*, 36 (9), 69–74 (1995).
27. Sloan, E.D., Jr., *Clathrate Hydrates of Natural Gases*. Marcel Dekker, New York (1998).
28. Klauda, J.B. and Sandler, S.I., "Global distribution of methane hydrate in ocean sediment," *Energy & Fuels*, 19 (2), 459–470 (2005).
29. Borowski, W.S., "A review of methane and gas hydrates in the dynamic, stratified system of the Blake Ridge region, offshore southeastern North America," *Chemical Geology*, 205, 311 (2004).
30. Cherskiy, N.V., Tsaarev, V.P., and Nikitin, S.P., "Investigation and prediction of conditions of accumulation of gas resources in gas-hydrate pools," *Petroleum Geology*, 21, 65 (1982).
31. Collett, T., "Natural gas hydrates of the Prudhoe Bay and Kuparuk River area, North Slope, Alaska," *American Association of Petroleum Geologists Bulletin*, 77 (5), 793–812 (1993).
32. Collett, T. and Ginsburg, G., "Gas hydrates in the Messoyakha gas field of the West Siberian Basin—A re-examination of the geologic evidence," *International Journal of Offshore and Polar Engineering*, 8 (1), 22–29 (1998).

33. Dallimore, S.R. and Collett, T.S. (eds.), "Scientific results from the Mallik 2002 Gas Hydrate Production Research Well Program, Mackenzie Delta, Northwest Territories, Canada," Geological Survey of Canada Bulletin 585, USGS, Reston, VA (2005).
34. Uchida, T., Lu, H., Tomaru, H., and the MITI Nankai Trough Shipboard Scientists, "Subsurface occurrence of natural gas hydrate in the Nankai Trough area: Implication for gas hydrate concentration," *Resource Geology*, 54, 35–44 (2004).
35. Collett, T., Riedel, M., Boswell, R., Cochran, J., Kumar, P., Sethi, A., and Sathe, A., "International team completes landmark gas hydrate expedition in the offshore of India," *Fire in the Ice*, 6 (3), 1–16 (2006).
36. Park, K.P., "Gas hydrate exploration in Korea," *Proceedings of the 2nd International Symposium on Gas Hydrate Technology*, November 1–2, Daejeon, Korea (2006).
37. Boswell, R., Collett, T., McConnell, D., Frye, M., Shedd, B., Mrozewski, S., Guerin, G. et al., "Joint Industry Project Leg II discovers rich gas hydrate accumulations in sand reservoirs in the Gulf of Mexico," *Fire in the Ice*, 9 (3), 1–5 (2009).
38. Collett, T., Agena, W., Lee, M., Zyrianova, M., Bird, K., Charpentier, T., Houseknecht, D. et al., "Assessment of gas hydrate resources on the North Slope, Alaska, 2008," US Geological Survey Fact Sheet 2008-3073, USGS, Reston, VA, 4 (2002).
39. Collett, T., Riedel, M., Cochran, J.R., Boswell, R., Kumar, P., and Sathe, A.V., "Indian continental margin gas hydrate prospects: Results of the Indian National Gas Hydrate Program (NGHP) expedition 01," *Proceedings of the 6th International Conference on Gas Hydrates*, Vancouver, BC (2008).
40. Dai, J., Snyder, F., Gillespie, D., Koesoemadinata, A., and Dutta, N., "Exploration for gas hydrates in the deepwater northern Gulf of Mexico: Part I. A seismic approach based on geologic model, inversion, and rock physics principles," *Marine and Petroleum Geology*, 25, 830–844 (2008).
41. Dai, J., Banik, N., Gillespie, D., and Dutta, N., "Exploration for gas hydrates in the deepwater northern Gulf of Mexico: Part II. Model validation by drilling," *Marine and Petroleum Geology*, 25, 845–859 (2008).
42. Ryu, B.-J., Riedel, M., Kim, J.-H., Hyndman, R.D., Lee, Y.-J., Chung, B.-H., and Kim, I., "Gas hydrates in the western deep-water Ulleung Basin, East Sea of Korea," *Marine and Petroleum Geology*, 26, 1483–1498 (2009).
43. Tsuji, Y., Ishida, H., Nakamizu, M., Matsumoto, R., and Shimizu, S., "Overview of the MITI Nankai Trough wells: A milestone in the evaluation of methane hydrate resources," *Resource Geology*, 54, 3–10 (2004).
44. Tsuji, Y., Fujii, T., Hayashi, M., Kitamura, R., Nakamizu, M., Ohbi, K., Saeki, T. et al., "Methane-hydrate occurrence and distribution in the Eastern Nankai Trough, Japan: Findings of the Tokai-oki to Kumano-nada methane-hydrate drilling program," in Collett, T., Johnson, A., Knapp, C., and Boswell, R. (eds.), *Natural Gas Hydrates—Energy Resource Potential and Associated Geologic Hazards*, AAPG Memoir 89, AAPG, Tulsa, OK, 228–249 (2009).
45. Wu, N., Yang, S. et al., "Preliminary discussion on gas hydrate reservoir system of Shenhu area, north slope of South China Sea," *Proceedings of the 6th International Conference on Gas Hydrates*, July 6–10, Vancouver, 8 (2008).
46. Zhang, H., Yang, S., Wu, N., Su, X, Holland, M., Schultheiss, P., Rose, K., Butler, H., Humphrey, G., and GMGS-1 Science Team, "Successful and surprising results for China's first gas hydrate drilling expedition," *Fire in the Ice*, 7 (3), 6–9 (2007).
47. Paull, C., Reeburgh, W.S., Dallimore, S.R., Enciso, G., Green, S., Koh, C.A., Kvenvolden, K.A., Mankin, C., and Riedel, M., "Realizing the energy potential of methane hydrate for the United States," National Research Council Report (2010).
48. Cook, A., Goldberg, D., and Kleinberg, R., "Fracture-controlled gas hydrate systems in the Gulf of Mexico," *Marine and Petroleum Geology*, 25 (9), 932–941 (2008).

49. Collett, T.S., Lee, M.W. et al., "Permafrost associated natural gas hydrate occurrences on the Alaskan North Slope," *Marine and Petroleum Geology*, 28, 279–294 (2011).
50. Dallimore, S.R., Uchida, T., and Collett, T.S. (eds.), "Scientific results from the JAPEX/JNOC/GSC Mallik 2L-38 gas hydrate research well," Geological Survey of Canada Bulletin 544, Mackenzie Delta, Northwest Territories (1999).
51. Ruppel, C., "Ruppel: MITEI Natural Gas Report," Supplementary Paper on Methane Hydrates, 21 (2011).
52. Walsh, T., Stokes, P., Panda, M., Morahan, T., Greet, D., MacRae, S., Singh, P., and Patil, S., "Characterization and quantification of the methane hydrate resource potential associated with the barrow gas field," *Proceedings of the 6th International Conference on Gas Hydrates*, Vancouver, BC (2008).
53. Park, K.P., Bahk, J.J. et al., "Korean National Program expedition confirms rich gas hydrate deposits in the Ulleung Basin, East Sea," *Fire in the Ice*, 8 (2), 6–9 (2008).
54. Bahman, T.K., "Why are gas hydrates important," Institute of Petroleum Engineering, Heriot-Watt University, Edinburgh (2011).
55. Collett, T.S. and Ladd, J., "Detection of gas hydrate with downhole logs and assessment of gas hydrate concentrations (saturations) and gas volumes on the Blake Ridge with electrical resistivity log data," in Paull, C.K., Matsumoto, R., Wallace, P.J., and Dillion, W.P. (eds.), *Proceedings of the ODP, Science Results*, Vol. 164, 179–191 (2000).
56. Hato, M., Matsuoka, T., Inamori, T., and Saeki, T., "Detection of methane-hydrate-bearing zones using seismic attributes analysis," *The Leading Edge*, 25, 607–609 (2006).
57. Holbrook, W.S., Gorman, A.R., Hornbach, M., Hackwith, K.L., and Nealon, J., "Direct seismic detection of methane hydrate," *The Leading Edge*, 21, 686–689 (2002).
58. Hovland, M. and Gudmestad, O.V., "Potential influence of gas hydrates on seabed installations," in Paull, C. and Dillon, W. (eds.), *Natural Gas Hydrates—Occurrence, Distribution and Detection*. American Geophysical Union, Washington, DC, 307–315 (2001).
59. Lee, J.Y., Santamarina, J.C., and Ruppel, C., "Parametric study of the physical properties of hydrate-bearing sand, silt, and clay sediments: 1. Electromagnetic properties," *Journal of Geophysical Research*, 115, B11104 (2010).
60. Kvenvolden, K.A., "Methane hydrate-A major reservoir of carbon in the shallow geosphere," *Chemical Geology*, 71, 41–51 (1988).
61. Ersland, G., Husebo, J., Graue, A., Baldwin, B., Howard, J.J., and Stevens, J.C., "Measuring gas hydrate formation and exchange with CO_2 in Bentheim sandstone using MRI tomography," *Chemical Engineering Journal*, 158, 25–31 (2010).
62. Farrell, H., Boswell, R., Howard, J., and Baker, R., "CO_2–CH_4 exchange in natural gas hydrate reservoirs: Potential and challenges," *Fire in the Ice*, 10 (1), 19–21 (2010).
63. Graue, A., Kvamme, B., Baldwin, B.A., Steven, J., Howard, J., Aspenes, E., Ersland, G., Husebo, J., and Zornes, D., "MRI visualization of spontaneous methane production from hydrates in sandstone core plugs when exposed to CO_2," *SPE Journal*, 13, 146–152 (2008).
64. Jung, J.W., Espinoza, D.N., and Santamarina, J.C., "Properties and phenomena relevant to CH_4–CO_2 replacement in hydrate-bearing sediments," *Journal of Geophysical Research*, 115, B10102 (2010).
65. Lee, H., Seo, Y., Seo, Y.-T., Moudrakovski, I., and Ripmeester, J., "Recovering methane from solid methane hydrate with carbon dioxide," *Angewandte Chemie International Edition*, 42 (41), 5048–5051 (2003).
66. Park, Y., Kim, D., Lee, J., Huh, D., Park, K., Lee, J., and Lee, H., "Sequestering carbon dioxide into complex structures of naturally occurring gas hydrates," *Proceedings of the National Academy of Sciences of the United States of America*, 103–34, 12690–12694 (2006).

67. Goel, N. "In situ methane hydrate dissociation with carbon dioxide sequestration: Current knowledge and issues," *Journal of Petroleum Science and Engineering*, 51 (3/4), 169–184 (2006).
68. Ohgaki, K., Takano, K., Sangawa, H., Matsubara, T., and Nakano, S., "Methane exploitation by carbon dioxide from gas hydrates—Phase equilibria for CO_2–CH_4 mixed hydrate system," *Journal of Chemical Engineering of Japan*, 29 (3), 478–483 (1996).
69. Seo, Y.T. and Lee, H., "Multiple-phase hydrate equilibria of the ternary carbon dioxide, methane, and water mixtures," *Journal of Physical Chemistry B*, 105 (41), 10084–10090 (2001).
70. Seo, Y.T., Lee, H., and Yoon, J., "Hydrate phase equilibria of the carbon dioxide, methane, and water system," *Journal of Chemical & Engineering Data*, 46 (2), 381–384 (2001).
71. Smith, D.H., Seshadri, K., and Wilder, J., "Assessing the thermodynamic feasibility of the conversion of methane hydrate into carbon dioxide hydrate in porous media," *First National Conference on Carbon Sequestration*, National Energy Technology Laboratory (2001).
72. Uchida, T., Takeya, S., Ebinuma, T., and Narita, H., "Replacing methane with CO_2 in clathrate hydrate: Observations using Raman spectroscopy," in Williams, D.J., Durie, R.A., McMullan, P., Paulson, C.A.J., and Smith, A.Y. (eds.), *Proceedings of the 5th International Conference on Greenhouse Gas Control Technologies*. CSIRO Publishing, Collingwood, 523–527 (2001).
73. McGrail, B., Schaef, H., White, M., Zhu, T., Kulkarni, A., Hunter, R., Patil, S., Owen, A., and Martin, P., "Using carbon dioxide to enhance recovery of methane from gas hydrate reservoirs: Final summary report," US Department of Energy under Contract No. DE-AC06-76RLO 1830, PNNL 17035, Pacific Northwest National Laboratory, Richland, WA (September 2007).
74. White, M. and McGrail, P., "Numerical simulation of methane hydrate production from geologic formations via carbon dioxide injection," *Society of Petroleum Engineers, Offshore Technology Conference*, Paper No. OTC-19458 (2008).
75. Yezdimer, E., Cummings, P., and Chalvo, A., "Extraction of methane from its gas clathrate by carbon dioxide sequestration—Determination of the Gibbs Free Energy of gas replacement and molecular simulation," *Journal of Physical Chemistry A*, 106, 7982–7987 (2002).
76. "Comparative assessment of advanced gas hydrate production methods," DOE/NETL methane hydrates projects, DE-FC26-06NT42666 (September 2009).
77. Ota, M., Morohashi, K., Abe, Y., Watanabe, M., Smith, R., and Inomata, H., "Replacement of CH_4 in the hydrate by use of liquid CO_2," *Energy Conversion and Management*, 46 (11/12), 1680–1691 (2005).
78. Chatterji, J. and Griffith, J., "Methods of decomposing gas hydrates," Patent No. 5713416 (1998).
79. Hirohama, S., Shimoyama, Y., Wakabayashi, A., Tatsuta, S., and Nishida, N., "Conversion of CH_4-hydrate to CO_2-hydrate in liquid CO_2," *Journal of Chemical Engineering of Japan*, 29 (6), 1014–1020 (1996).
80. Cortis, A. and Ghezzehei, T., "On the transport of emulsions in porous media," *Journal of Colloid Interface Science*, 313 (1), 1–4 (2007).
81. Tegze, G., Gránásy, L., and Kvamme, B., "Phase field modeling of CH_4 hydrate conversion into CO_2 hydrate in the presence of liquid CO_2," *Physical Chemistry Chemical Physics*, 9 (24), 3104–3111 (2007).
82. Yan, L., Thompson, K., and Valsaraj, K., "A numerical study on the coalescence of emulsion droplets in a constricted capillary tube," *Journal of Colloid Interface Science*, 298 (2), 832–844 (2006).

83. Kurihara, M., Funatsu, K., Ouchi, H., Masuda, Y., Yasuda, M., Yamamoto, K., Numasawa, M. et al., "Analysis of the JOGMEC/NRCAN/Aurora Mallik gas hydrate production test through numerical simulation," *Proceedings of the 6th International Conference on Gas Hydrates*, Vancouver, BC (2008).
84. Lee, J.Y., Francisca, F.M., Santamarina, J.C., and Ruppel, C., "Parametric study of the physical properties of hydrate-bearing sand, silt, and clay sediments: 2. Small-strain mechanical properties," *Journal of Geophysical Research*, 115, B11105 (2010).
85. Moridis, G.J., Reagan, M.T., and Zheng, K., "On the performance of Class 2 and Class 3 hydrate deposits during co-production with conventional gas," *Offshore Technology Conference*, OTC 19435-MS (2008).
86. Paull, C.K., Matsumoto, R. et al., "Proceedings of the Ocean Drilling Program," Initial Reports 164, Ocean Drilling Program, College Station, TX (1996).
87. Rutqvist, J. and Moridis, G., "Evaluation of geohazards of in situ gas hydrates related to oil and gas operations," *Fire in the Ice*, 10 (2), 1–4 (2010).
88. Ruppel, C., Collett, T., Boswell, R., Lorenson, T., Buckzowski, B., and Waite, W., "A new global gas hydrate drilling map based on reservoir type," *Fire in the Ice*, 11 (1), 15–19 (2011).
89. Satyavani, N., Sain, K., Lall, M., and Kumar, B.J.P., "Seismic attribute study of gas hydrates in the Andaman, offshore India," *Marine Geophysical Research*, 29, 167–175 (2008).
90. Ruppel, C., "Tapping methane hydrates for unconventional natural gas," *Elements*, 3 (3), 193–199 (2007).
91. Ruppel, C., Boswell, R., and Jones, E., "Scientific results from Gulf of Mexico gas hydrates joint industry project Leg 1 drilling: Introduction and overview," *Marine and Petroleum Geology*, 25, 819–829 (2008).
92. Ruppel, C., "Ruppel: MITEI Natural Gas Report," Supplementary Paper on Methane Hydrates, 25 (2011).
93. Birchwood, R.A., Noeth, S., Tjengdrawira, M.A., Kisra, S.M., Elisabeth, F.L., Sayers, C.M., Singh, R. et al., "Modeling the mechanical and phase change stability of wellbores drilled in gas hydrates by the Joint Industry Participation Program (JIP) Gas Hydrates Project Phase II," *SPE Annual Technical Conference*, Anaheim, CA, SPE 110796, November 11–14 (2007).
94. Ameripour, S., "Prediction of gas-hydrate formation conditions in production and surface facilities," MS thesis, A&M University, College Station, TX, 79 (August 2005).
95. Dalmazzone, D., Kharrat, M., Lachet, V., Fouconnier, B., and Clausse, D., "DSC and PVT measurements—Methane and trichlorofluoromethane hydrate dissociation equilibria," *Journal of Thermal Analysis and Calorimetry*, 70, 493–505 (2002).
96. Edmonds, B., Moorwood, R.A.S., and Szczepanski, R., "A practical model for the effect of salinity on gas hydrate formation," *European Production Operations Conference & Exhibition*, SPE 35569, Stavanger, Norway (1996).
97. Grigg, R.B. and Lynes, G.L., "Oil-based drilling mud as a gas-hydrates inhibitor," *SPE Drilling Engineering*, 7, 32–38 (1992).
98. Kotkoskie, T.S., Al-Ubaidi, B., Wildeman, T.R., and Sloan, E.D., Jr., "Inhibition of gas hydrates in water-based drilling muds," *SPE Drilling Engineering*, 7, 130–136 (1992).
99. Lai, D.T. and Dzialowski, A.K., "Investigation of natural gas hydrates in various drilling fluids," *SPE/IADC Drilling Conference*, SPE 18637, February, New Orleans, LA (1989).
100. Tohidi, B., Østergaard, K.K., Danesh, A., Todd, A.C., and Burgass, R.W., "Structure-H gas hydrates in petroleum reservoir fluids," *The Canadian Journal of Chemical Engineering*, 79, 384–391 (2001).
101. Yousif, M.H., Dunayevsky, V.A., and Hale, A.H., "Hydrate plug remediation: Options and applications for deep water drilling operations," *SPE/IADC Drilling Conference*, SPE 37624, March, Amsterdam, the Netherlands (1997).

102. Kim, N., Bonet, E., and Ribeiro, P., "Study of hydrate in drilling operations: A review," *Proceedings of the 4th PDPETRO*, October 21–24, Campinas, SP (2007).
103. Bagirov, E. and Lerche, I., "Hydrate represent gas source, drilling hazard," *Oil & Gas Journal*, 95, 99–104 (1997).
104. Helgeland, L., Kinn, A., Kvalheim, O., and Wenaas, A., "Gas kick due to hydrates in the drilling for offshore natural gas and oil," Report by the Department of Petroleum Engineering and Applied Geophysiscs, NTNU, Trondheim, Norway (November 2012).
105. Skalle, P., "Pressure control during oil well drilling," in *Special Offshore Safety Issues*, 2nd ed. Ventus Publishing, http://bookboon.com/en/textbooks/geoscience/pressure-control-during-oil-well-drilling.
106. Amodu, A.A., "Drilling through gas hydrate formations: Possible problems and suggested solutions," MS thesis, Texas A&M University, Houston, TX (August 2008).
107. Dillon, W.P. and Max, M.D., "Oceanic gas hydrates," in Max, M.D. (ed.), *Natural Gas Hydrates in Oceanic and Permafrost Environments*, Kluwer, London, pp. 61–76 (2003).
108. Skalle, P., "Pressure control during oil well drilling," in *Schlumberger Oilfield Glossary, "Blowout."* http://www.glossary.oilfield.slb.com/Display.cfm?Term=blowout.
109. Khabibullin, T., Falcone, G., and Teodoriu, C., "Drilling through gas hydrate sediments: Managing wellbore stability risks," SPE-131332 (June 2006), http://dx.doi.org/10.2118/131332-MS.
110. Qadir, M.I., "Gas hydrates: A fuel for future but wrapped in drilling challenges," SPE-156516, *Paper presented at SPE/PAPG Annual Technical Conference*, November 22–23, Islamabad, Pakistan (2011).
111. Williamson, S.C., McConnell, D.R., and Bruce, R.J., "Drilling observations of possible hydrate-related annular flow in the deep water Gulf of Mexico and Implications on Well Planning," *Paper presented at the 2005 Offshore Technology Conference*, OTC 17279, May 2–5, Houston, TX (2005).
112. Hannegan, D., Todd, R.T., Pritchard, D.M., and Jonasson, B., "MPD—Uniquely applicable to methane hydrate drilling," SPE-91560, *Paper presented at the SPE/IADC Underbalanced Technology Conference and Exhibition*, October 11–12, Houston, TX (2004), http://dx.doi.org/10.2118/91560-MS.
113. Ebeltoft, H., Yousif, M., and Sægråd, E., "Hydrate control during deepwater drilling: Overview and new drilling-fluids formulations," SPE-68207, March (2001), http://dx.doi.org/10.2118/68207-PA.
114. Halliday, W., Clapper, D.K., and Smalling, M., "New gas hydrate inhibitors for deepwater drilling fluids," IADC/SPE-39316, *Paper presented at the 1998 SPE/IADC Drilling Conference*, March 3–6, Dallas, TX (1998).
115. Ravi, K. and Moore, S., "Cement slurry design to prevent destabilization of hydrates in deepwater environment," SPE-113631, *Paper presented at the 2008 Indian Oil and Gas Technical Conference and Exhibition*, March 4–6, Mumbai, India (2008).
116. Catak, E., "Hydrate dissociation during drilling through in-situ hydrate formations," MS thesis, Department of Petroleum Engineering, Louisiana State University, Baton Rouge, LA (May 2006).
117. Ruppel, C., "Methane hydrates and contemporary climate change," *Natural Education Knowledge*, 3 (10), 29 (2011).
118. Buffett, B. and Archer, D., "Global inventory of methane clathrate: Sensitivity to changes in environmental conditions," *Earth and Planetary Science Letters*, 227, 185–199 (2004).
119. Dutta, N.C., Utech, R.W., and Shelander, D., "Role of 3D seismic for quantitative shallow hazard assessment in deepwater sediments," *The Leading Edge*, 29, 930–942 (2010).
120. Ellis, M., Evans, R.L., Hutchinson, D., Hart, P., Gardner, J., and Hagen, R., "Electromagnetic surveying of seafloor mounds in the northern Gulf of Mexico," *Marine Petroleum Geology*, 25, 969–968 (2008).

121. Hadley, C., Peters, D., Vaughan, A., "Gumusut-Kakap project: Geohazard characterisation and impact on field development plans," *International Petroleum Technology Conference*, 12554-MS, 15pp (2008).
122. Weitemeyer, K., Constable, S., Key K., "Marine EM techniques for gas-hydrate detection and hazard mitigation," *The Leading Edge*, 25 (5), 629–632 (2006).
123. Bunz, S. and Meinert, J., "Overpressure distribution beneath hydrate-bearing sediments at the Storegga Slide on the Mid-Norwegian margin," *Proceedings of the International Conference on Gas Hydrates*, June 13–16, Trondheim, Norway, Vol. 3, Paper 3007, 755–758 (2005).
124. Archer, D. et al., "Ocean methane hydrates as a slow tipping point in the global carbon cycle," *Proceedings of the National Academy of Sciences of the United States of America*, 106, 20596–20601 (2009).
125. Biastoch, A. et al., "Rising arctic ocean temperatures cause gas hydrate destabilization and ocean acidification," *Geophysical Research Letters*, 38, L08602 (2011).
126. Bock, M. et al., "Hydrogen isotopes preclude marine hydrate CH_4 emissions at the onset of Dansgaard-Oeschger events," *Science*, 328, 1686–1689 (2010).
127. Bohannon, J., "Weighing the climate risks of an untapped fossil fuel," *Science*, 319, 1753 (2008).
128. Bowen, R.G. et al., "Geomorphology and gas release from pockmark features in the Mackenzie Delta, Northwest Territories, Canada," in Kane, D.L. and Hinkel, K.M. (eds.), *Proceedings of the 9th International Conference on Permafrost*, Institute of Northern Engineering, Fairbanks, AK, 171–176 (2008).
129. Dickens, G.R. et al., "Dissociation of oceanic methane hydrate as a cause of the carbon isotope excursion at the end of the Paleocene," *Paleoceanography*, 10, 965–971 (1995).
130. Dickens, G.R., "Down the rabbit hole: Toward appropriate discussion of methane release from gas hydrate systems during the Paleocene-Eocene thermal maximum and other past hyperthermal events," *Climate of the Past*, 7, 831–846 (2011).
131. Harvey, L.D.D. and Huang, Z. "Evaluation of potential impact of methane clathrate destabilization on future global warming," *Journal of Geophysical Research*, 100, 2905–2926 (1995).
132. Hesselbo, S.P. et al., "Massive dissociation of gas hydrate during a Jurassic anoxic event," *Nature*, 406, 392–395 (2000).
133. Hinrichs, K.-U. and Boetius, A., "The anaerobic oxidation of methane: New insights in microbial ecology and biochemistry," in Wefer, G. et al. (eds.), *Ocean Margin Systems*. Springer, Berlin, Germany, 457–477 (2002).
134. Hu, L. et al. "Methane fluxes to the atmosphere from deepwater hydrocarbon seeps in the northern Gulf of Mexico," *Journal of Geophysical Research* (2011).
135. Intergovernmental Panel on Climate Change (IPCC). *Climate Change: The Scientific Basis*. Cambridge University Press, New York (2001).
136. Intergovernmental Panel on Climate Change (IPCC). *Climate Change: The Physical Basis*. Cambridge University Press, New York (2007).
137. Jiang, G. et al., "Stable isotope evidence for methane seeps in Neoproterozoic postglacial cap carbonates," *Nature*, 426, 822–826 (2003).
138. Judge, A.S. and Majorowicz, J.A., "Geothermal conditions for gas hydrate stability in the Beaufort-Mackenzie area: The global change aspect," *Palaeogeography, Palaeoclimatology, Palaeoecology*, 98, 251–263 (1992).
139. Kennett, J.P. et al., *Methane Hydrates in Quaternary Climate Change: The Clathrate Gun Hypothesis*. American Geophysical Union, Washington, DC (2003).
140. Kessler, J.D. et al., "A persistent oxygen anomaly reveals the fate of spilled methane in the deep Gulf of Mexico," *Science*, 331, 312–315 (2011).
141. Krey, V. et al., "Gas hydrates: Entrance to a methane age or climate threat?" *Environmental Research Letters*, 4, 034007 (2009).

142. Lachenbruch, A.H., "Permafrost, the active layer, and changing climate," Open File Report 94-694, USGS, Reston, VA (1994).
143. Lachenbruch, A.H. and Marshall, B.V., "Changing climate: Geothermal evidence from permafrost in the Alaskan Arctic," *Science*, 234, 689–696 (1986).
144. Lammers, S. et al., "A large methane plume east of Bear Island (Barents Sea): Implications for the marine methane cycle," *Geologische Rundschau*, 84, 59–66 (1995).
145. Lelieveld, J. et al., "Changing concentration, lifetime and climate forcing of atmospheric methane," *Tellus*, 50B, 128–150 (1998).
146. Liro, C.R. et al., "Modeling the release of CO_2 in the deep ocean," *Energy Conversion and Management*, 33, 667–674 (1992).
147. Macdonald, G., "Role of methane clathrates in past and future climate," *Climatic Change*, 16, 247–281 (1990).
148. Macdonald, I.R. et al., "Thermal and visual time-series at a seafloor gas hydrate deposit on the Gulf of Mexico slope," *Earth and Planetary Science Letters*, 233, 45–59 (2005).
149. Macdonald, I.R. et al., "Gas hydrate that breaches the seafloor on the continental slope of the Gulf of Mexico," *Geology*, 22, 699–702 (1994).
150. Majorowicz, J.A. et al., "Onset and stability of gas hydrates under permafrost in an environment of surface climatic change-past and future," *Proceedings of the 6th International Conference on Gas Hydrates*, Vancouver, BC (2008).
151. Mascarelli, A.L., "A sleeping giant?" *Nature Reports Climate Change*, 3, 46–49 (2009).
152. Maslin, M. et al., "Linking continental-slope failures and climate change: Testing the clathrate gun hypothesis," *Geology*, 32, 53–56 (2004).
153. Maslin, M. et al. "Gas hydrates: Past and future geohazard," *Philosophical Transactions of the Royal Society A: Mathematical, Physical & Engineering Sciences*, 368, 2369–2393 (2010).
154. Mau, S. et al., "Dissolved methane distributions and air-sea flux in the plume of a massive seep field, Coal Oil Point, California," *Geophysical Research Letters*, 34, L22603 (2007).
155. McGinnis, D.F. et al., "Fate of rising methane bubbles in stratified waters: How much methane reaches the atmosphere?" *Journal of Geophysical Research*, 111, C09007 (2006).
156. Niemann, H. et al., "Novel microbial communities of the Haakon Mosby mud volcano and their role as a methane sink," *Nature*, 443, 854–858 (2006).
157. Paull, C. et al., "Tracking the decomposition of submarine permafrost and gas hydrate under the shelf and slope of the Beaufort Sea," *Proceedings of the 7th International Conference on Gas Hydrates*, Edinburgh, Scotland (2011).
158. Petrenko, V. et al., "$^{14}CH_4$ measurements in Greenland ice: Investigating last glacial termination CH_4 sources," *Science*, 324, 506–508 (2009).
159. Rachold, V. et al., "Near-shore arctic subsea permafrost in transition," *Eos, Transactions of the American Geophysical Union*, 88, 149–156 (2007).
160. Reagan, M.T. and Moridis, G.J., "Dynamic response of oceanic hydrate deposits to ocean temperature change," *Journal of Geophysical Research*, 113, C12023 (2008).
161. Reeburgh, W.S., "Oceanic methane biogeochemistry," *Chemical Reviews*, 107, 486–513 (2007).
162. Renssen, H. et al., "Modeling the climate response to a massive methane release from gas hydrates," *Paleoceanography*, 19, PA2010 (2004).
163. Röhl, U. et al., "On the duration of the Paleocene-Eocene thermal maximum," *Geochemistry, Geophysics, Geosystems*, 8, Q12002 (2007).
164. Ruppel, C. et al., "Degradation of subsea permafrost and associated gas hydrates offshore of Alaska in response to climate change," *Sound Waves*, 128, 1–3 (2010).
165. Schmidt, G.A. and Shindell, D.T., "Atmospheric composition, radiative forcing, and climate change as consequence of a massive methane release from gas hydrates," *Paleoceanography*, 18, 1004 (2003).

166. Semiletov, I. et al., "Methane climate forcing and methane observations in the Siberian Arctic Land-Shelf system," *World Resource Review*, 16, 503–543 (2004).
167. Shakhova, N. et al., "Geochemical and geophysical evidence of methane release over the East Siberian Arctic Shelf," *Journal of Geophysical Research*, 115, C08007 (2010).
168. Shakhova, N. et al., "Extensive methane venting to the atmosphere from sediments of the East Siberian Arctic Shelf," *Science*, 327, 1246–1250 (2010).
169. Solomon, E.A. et al., "Considerable methane fluxes to the atmosphere from hydrocarbon seeps in the Gulf of Mexico," *Nature Geoscience*, 2, 561–565 (2009).
170. Sowers, T., "Late quaternary atmospheric CH_4 isotope record suggests marine clathrates are stable," *Science*, 311, 838 (2006).
171. Suess, E. et al., "Sea floor methane hydrates at Hydrate Ridge, Cascadia Margin," in Dillon, W.P. and Paull, C.K. (eds.), *Natural Gas Hydrates—Occurrence, Distribution and Detection*. American Geophysical Union, Washington, DC, 87–98 (2001).
172. Treude, T. et al., "Anaerobic oxidation of methane at hydrate ridge (OR)," *Geochimica et Cosmochimica Acta*, 67, A491 (2003).
173. Tryon, M.D. et al., "Fluid and chemical flux in and out of sediments posting methane hydrate deposits on Hydrate Ridge, OR. II: Hydrological processes," *Earth and Planetary Science Letters*, 201, 541–557 (2002).
174. Walter, K.M. et al., "Methane bubbling from northern lakes: Present and future contributions to the global methane budget," *Philosophical Transactions of the Royal Society A: Mathematical, Physical & Engineering Sciences*, 365, 1657–1676 (2007).
175. Wang, J.S. et al., "A 3-D model analysis of the slowdown and interannual variability in the methane growth rate from 1988 to 1997," *Global Biogeochemical Cycles*, 18, GB3011 (2004).
176. Westbrook, G.K. et al., "Escape of methane gas from the seabed along the West Spitsbergen continental margin," *Geophysical Research Letters*, 36, L15608 (2009).
177. Yvon-Lewis, S.A. et al., "Methane flux to the atmosphere from the deepwater horizon oil disaster," *Geophysical Research Letters*, 38, L01602 (2011).
178. Zachos, J. et al., "Trends, rhythms, and aberrations in global climate 65 Ma to present," *Science*, 292, 686–693 (2001).
179. Zachos, J. et al., "Rapid acidification of the ocean during the Paleocene-Eocene thermal maximum," *Science*, 308, 1611–1615 (2005).
180. Ruppel, C. and Noserale, D., "Gas hydrates and climate warming," USGS Report, USGS, Reston, VA, 1–9 (January 2012).
181. Tang, L.G., Xiao, R., Huang, C., Feng, Z.P., and Fan, S.S., "Experimental investigation of production behavior of gas hydrate under thermal stimulation in unconsolidated sediment," *Energy & Fuels*, 19 (6), 2402–2407 (2005).
182. Kawamura, T., Ohtake, M., Sakamoto, Y., Yamamoto, Y., Haneda, H., Komai, T., and Higuchi, S., "Experimental study on steam injection method using methane hydrate core samples," *Proceedings of the 7th ISOPE Ocean Mining Symposium*, July 1–6, Lisbon, Portugal, 83–86 (2007).
183. Kamath, V.A., "Study of heat transfer characteristics during dissociation of gas hydrates in porous media," PhD dissertation, University of Pittsburgh, Pittsburgh, PA (1984).
184. Hancock, S., Collett, T.S., Dallimore, S.R., Satoh, T., Huenges, E., and Henninges, J., "Overview of thermal stimulation production test results for the Japex/JNOC/GSC Mallik Gas Hydrate Research Well," in Dallimore, S.R. and Collett, T.S. (eds.), *Scientific Results from Mallik 2002 Gas Hydrate Production Research Well Program*. Geological Survey of Canada Bulletin 585, Mackenzie Delta, NWT (2005).
185. Circone, S., Kirby, S., and Stern, L., "Thermal regulation of methane hydrate dissociation: Implications for gas production models," *Energy & Fuels*, 19 (6), 2357–2363 (2005).

186. Li, G., Li, X., Tang, L.-G., and Li, Q.-P., "Control mechanisms for methane hydrate production by thermal stimulation," *Proceedings of the 6th International Conference on Gas Hydrates*, July 6–10, Vancouver, BC (2008).
187. Computer Modeling Group, Steam, Thermal, and Advanced Processes Reservoir Simulator (STARS), www.cmgroup.com/software/stars.htm.
188. Ruppel, C., Dickens, G., Castellini, D., Gilhooly, W., and Lizzarralde, D., "Heat and salt inhibition of gas hydrate formation in the northern Gulf of Mexico," *Geophysical Research Letters*, 32 (4), L04605 (2005).
189. Bai, Y., Li, Q., Li, X., and Du, Y., "The simulation of nature gas production from ocean gas hydrate reservoir via depressurization," *Science in China Series E: Technological Sciences*, 51, 1272–1282 (2008).
190. Hong, H. and Pooladi-Darvish, M., "Simulation of depressurization for gas production from gas hydrate reservoirs," *Journal of Canadian Petroleum Technology*, 44 (11), 39–46 (2005).
191. Tang, L.G., Li, G., Hao, Y.M., Fan, S.S., and Feng, Z.P., "Effects of salt on the formation of gas hydrate in porous media," *Proceedings of the 5th International Conference on Gas Hydrate*, Trondheim, Norway, 155–160 (2005).
192. Ji, C., Ahmadi, G., and Smith, D., "Natural gas production from hydrate decomposition by depressurization," *Chemical Engineering Science*, 56 (20), 5801–5814 (2001).
193. Kono, H.O., Narasimhan, S., Song, F., and Smith, D.H., "Synthesis of methane gas hydrate in porous sediments and its dissociation by depressurizing," *Powder Technology*, 122 (2–3), 239–246 (2002).
194. Moridis, G.J., Kowalsky, M.B., and Pruess, K., "Depressurization-induced gas production from class 1 hydrate deposits," *SPE Reservoir Evaluation & Engineering*, 10 (5), 458–481 (2007).
195. Tang, L., Li, X., Feng, Z., Li, G., and Fan, S., "Control mechanisms for gas hydrate production by depressurization in different scale hydrate reservoirs," *Energy & Fuels*, 21 (1), 227–233 (2007).
196. Yousif, M.H., Li, P.M., Selim, M.S., and Sloan, E.D., "Depressurization of natural gas hydrates in Berea sandstone cores," *Journal of Inclusion Phenomena and Macrocyclic Chemistry*, 8, 71–88 (1990).
197. Moridis, G.J., Kowalsky, M., and Pruess, K., "Depressurization-induced gas production from class 1 hydrate deposits," *SPE Reservoir Evaluation & Engineering*, 10 (5), 458–481 (2007).
198. Fan, S.S., Zhang, Y.Z., Tian, G.L., Liang, D.Q., and Li, D.L., "Natural gas hydrate dissociation by presence of ethylene glycol," *Energy & Fuels*, 20 (1), 324–326 (2006).
199. Kamath, V.A.M.P.N., Sira, J.H., and Patil, S.L., "Experimental study of Brine injection and depressurization methods for dissociation of gas hydrate," *SPE Formation Evaluation*, 6 (4), 477–484 (1991).
200. Kawamura, T., Yamamoto, Y., Ohtake, M., Sakamoto, Y., Komai, T., and Haneda, H., "Dissociation experiment of hydrate core sample using thermodynamic inhibitors," *Proceedings of the 15th International Offshore and Polar Engineering Conference*, June 19–24, Seoul, South Korea (2005).
201. Moridis, G.J., Collett, T., Dallimore, S., Satoh, T., Hancock, S., and Weatherill, B., "Numerical studies of gas production from several CH_4 hydrate zones at the Mallik Site, Mackenzie Delta, Canada," *Journal of Petroleum Science and Engineering*, 43 (3–4), 219–238 (2004).
202. Moridis, G.J., "Numerical studies of gas production from class 2 and class 3 hydrate accumulations at the Mallik Site, Mackenzie Delta, Canada," *SPE Reservoir Evaluation & Engineering*, 7 (3), 175–83 (2004).

203. Moridis, G.J., Collett, T.S., Boswell, R., Kurihara, M., Reagan, M.T., Koh, C., and Sloan, E.D., "Toward production from gas hydrates: Current status, assessment of resources, and simulation-based evaluation of technology and potential," *SPE Reservoir Evaluation & Engineering*, 12 (5), 745–771 (2009).
204. Moridis, G.J. and Collett, T.S., "Strategies for gas production from hydrate accumulations under various geologic conditions," Report LBNL-52568, Lawrence Berkeley National Laboratory, Berkeley, CA (2004).
205. Moridis, G., Collett, T.S., Boswell, R., Kurihara, M., Reagan, M., Koh, C., and Sloan, E., "Toward production from gas hydrates: Current status, assessment of resources, and simulation-based evaluation of technology and potential," *SPE Unconventional Reservoirs Conference*, SPE 114163 (2008).
206. Moridis, G.J., "Numerical studies of gas production from methane hydrates," *SPE Journal*, 8 (4), 359–370 (2003).
207. Pooladi-Darvish, M., "Gas production from hydrate reservoirs and its modeling," *Journal of Petroleum Technology*, 56 (6), 65–71 (2004).
208. Sun, X., Nanchary, N., and Mohanty, K., "1-D modeling of hydrate depressurization in porous media," *Transport in Porous Media*, 58 (3), 315–338 (2005).
209. Goel, N., Wiggins, M., and Shah, S., "Analytical modeling of gas recovery from in situ hydrates dissociation," *Journal of Petroleum Science and Engineering*, 29 (2), 115–127 (2001).
210. Tsypkin, G.G., "Mathematical models of gas hydrates dissociation in porous media," *Gas Hydrates: Challenges for the Future*, 912, 428–436 (2000).
211. Moridis, G.J., "Numerical studies of gas production from methane hydrates," *SPE Journal*, 32 (8), 359 (2003).
212. Moridis, G.J., Collett, T., Dallimore, S., Satoh, T., Hancock, S., and Weatherhill, B., "Numerical studies of gas production from several methane hydrate zones at the Mallik Site, Mackenzie Delta, Canada," *Journal of Petroleum Science and Engineering*, 43, 219 (2004).
213. White, M.D. and Oostrom, M., *STOMP: Subsurface Transport over Multiple Phases, Version 4.0, User's Guide.* PNNL-15782, Pacific Northwest National Laboratory, Richland, WA (2006).
214. Sung, W.M., Lee, H., and Lee, C., "Numerical study for production performances of a methane hydrate reservoir stimulated by inhibitor injection," *Energy Sources*, 24 (6), 499–512 (2002).
215. Kawamura, T., Sakamoto, Y., Ohtake, M., Yamamoto, Y., Haneda, H., Yoon, J.H., and Komai, T., "Dissociation behavior of hydrate core sample using thermodynamic inhibitor," *International Journal of Offshore and Polar Engineering*, 16 (1), 5–9 (2006).
216. Li, X.S., Wan, L.H., Li, G., Li, Q.P., Chen, Z.Y., and Yan, K.F., "Experimental investigations into the production behavior of methane hydrate in porous sediment under ethylene glycol injection and hot brine stimulation," *Industrial & Engineering Chemistry Research*, 47 (23), 11 (2008). www.intechopress.com/download.pdf/11347.
217. Li, G., Tang, L., Huang, C., Feng, Z., and Fan, S., "Thermodynamic evaluation of hot brine stimulation for natural gas hydrate dissociation," *Huagong Xuebao/Journal of Chemical Industry and Engineering (China)*, 57 (9), 2033–2038 (2006).
218. Kawamura, T., Yamamoto, Y., Ohtake, M., Sakamoto, Y., Komai, T., and Haneda, H., "Experimental study on dissociation of hydrate core sample accelerated by thermodynamic inhibitors for gas recovery from natural gas hydrate," *The 5th International Conference on Gas Hydrate*, June 12–16, Trondheim, Norway (2005).
219. Li, G., Li, X., Tang, L., and Zhang, Y., "Experimental investigation of production behavior of methane hydrate under ethylene glycol stimulation in unconsolidated sediment," *Energy & Fuels*, 21 (6), 3388–3393 (2007).

220. Li, G., Li, X., Tang, L., Zhang, Y., Feng, Z., and Fan, S., "Experimental investigation of production behavior of methane hydrate under ethylene glycol injection," *Huagong Xuebao/Journal of Chemical Industry and Engineering* (China), 58 (8), 2067–2074 (2007).
221. Sira, J.H., Patil, S.L., and Kamath, V.A., "Study of hydrate dissociation by methanol and glycol injection," *Proceedings of the SPE Annual Technical Conference and Exhibition*, Society of Petroleum Engineers of AIME, Richardson, TX, 977–984 (1990).
222. Inks, T.L., Lee, M.W., Agena, W.F., Taylor, D.J., Collett, T.S., Zyrianova, M.V., and Hunter, R.B., "Seismic prospecting for gas hydrate and associated free gas prospects in the Milne Point area of northern Alaska," in Collett, T., Johnson, A., Knapp, C., and Boswell, R. (eds.), *Natural Gas Hydrates—Energy Resource Potential and Associated Geologic Hazards*. AAPG Memoir 89, AAPG, Tulsa, OK, 555–583 (2009).
223. Walsh, M.R., Hancock, S.H., Wilson, S.J., Patil, S.L., Moridis, G.J., Boswell, R., Collett, T.S., Koh, C.A., and Sloan, E.D., "Preliminary report on the commercial viability of gas production from natural gas hydrates," *Energy Economics*, 31, 815–823 (2009).
224. Pawar, R.J, Zyvoloski, G., Tenma, N., Sakamoto Y., and Komai, T., "Numerical simulation of gas production from methane hydrate reservoirs," *Proceedings of the 5th International Conference on Gas Hydrates*, Paper 1040, June 13–16, Trondheim, Norway, Vol. 1, 258–267 (2005).
225. Moridis, G.J., Kowalsky, M.B., and Pruess, K., "TOUGH+HYDRATE v1.0 user's manual: A code for the simulation of system behavior in hydrate-bearing porous media," Report LBNL-149E, Lawrence Berkeley National Laboratory, Berkeley, CA (2008).
226. Wilder, J.W., Moridis, G.J., Wilson, S.J., Kurihara, M., White, M., Masuda, Y., Anderson, B.J. et al., "An international effort to compare gas hydrate reservoir simulators," *Proceedings of the 6th International Conference on Gas Hydrates*, Vancouver, BC (2008).
227. Dallimore, S.R., Collett, T.S., Uchida, T., Weber, M., and Takahashi, H.; Mallik Gas Hydrate Research Team, "Overview of the 2002 Mallik Gas Hydrate Production Research Well Program," *Proceedings of the 4th International Conference on Gas Hydrates*, Vol. 1, 36–39 (2002).
228. Hancock, S., Collett, T., Pooladi-Darvish, M., Gerami, S., Moridis, G., Okazawa, T., Osadetz, K., Dallimore, S., and Weatherill, B., "A preliminary investigation on the economics of onshore gas hydrate production based on the Mallik Field discovery," *American Association of Petroleum Geologists Hedberg Conference Proceedings*, Vancouver, CA (2004).
229. Howe, S.J., "Production modeling and economic evaluation of a potential gas hydrate pilot production program on the North Slope of Alaska," MS thesis, University of Alaska Fairbanks, Fairbanks, AK, 138pp (2004).
230. Nakano, S., Yamamoto, K., and Ohgaki, K., "Natural gas exploitation by carbon dioxide from gas hydrate fields—High-pressure phase equilibrium for an ethane hydrate system," *Proceedings of the Institution of Mechanical Engineers*, 212, 159–163 (1998).
231. Ohgaki, K., Takano, K., and Moritoki, M., "Exploitation of CH4 hydrates under the Nankai Trough in combination with CO_2 storage," *Kagaku Kogaku Ronbunshu*, 20, 121–123 (1994).
232. Takahashi, H., Fercho, E., and Dallimore, S.R., "Drilling and operations overview of the Mallik 2002 Production Research Well Program," in Dallimore, S.R. and Collett, T.S. (eds.), *Scientific Results from Mallik 2002 Gas Hydrate Production Research Well Program, Mackenzie Delta, Northwest Territories, Canada*, Vol. Bulletin 585. Geological Survey of Canada, Vancouver, BC (2005).
233. Moridis, G.J., Reagan, M.T., Kim, S.J., Seol, Y., and Zhang, K., "Evaluation of the gas production potential of marine hydrate deposits in the Ulleung basin of the Korean East Sea," *SPE Journal*, 14 (4), 759–781 (2009).

234. Sung, W.M., Lee, H., Kim, S., and Kang, H., "Experimental investigation of production behaviors of methane hydrate saturated in porous rock," *Energy Sources*, 25 (8), 845–856 (2003).
235. Yousif, M.H., Abass, H.H., Selim, M.,S., and Sloan, E.D., "Experimental and theoretical investigation of methane-gas-hydrate dissociation in porous media," *SPE Reservoir Engineering*, 6 (4), 69–76 (1991).
236. Moridis, G.J. and Reagan, M., "Strategies for production from oceanic Class 3 hydrate accumulations," OTC Paper 18865 (2007).
237. Tohidi, B., Anderson, R., Clennell, M.B., Burgass, R.W., and Biderkab, A.B., "Visual observation of gas-hydrate formation and dissociation in synthetic porous media by means of glass micromodels," *Geology*, 29 (9), 867–870 (2001).

13 Power and Energy Directly from Water

13.1 INTRODUCTION

In Chapters 2 through 12, we examined (1) the benign role water plays for fuel production and energy carrier; (2) water (in the form of steam, water, or supercritical water) as a chemical solvent, reactant, or catalyst to generate fuels; and (3) the direct role it plays to generate hydrogen and methane. In this chapter, we briefly examine the direct role of water for the generation of power and electricity.

Water can directly generate energy and power in three different ways: hydropotential energy (or hydroelectricity), hydrokinetic energy (or a mixture of hydrokinetic energy and hydropotential energy like in tidal wave), and the use of ocean thermal energy conversion (OTEC) technologies. Here, we examine all three methods for generating power with the direct use of water.

The use of water dams to generate hydroelectricity has been practiced for a long time. This is a very clean method for power generation since it has a very little effect on greenhouse gas (GHG) production. Along with hydroelectricity, in recent years, hydrokinetic energy that uses the kinetic energy stored in tidal waves, sea and ocean shore waves, and undercurrents and inland waterways has been harnessed to generate electricity with the numerous different types of devices. The energy can also be harnessed from the temperature difference between the surfaces of the ocean (warm) and deep water (cold) using numerous OTEC devices. All three methods solely use water to generate power. This chapter briefly describes our current state of art on this subject.

13.2 HYDROELECTRIC POWER BY WATER DAMS

Water has been used for a long time to directly generate energy and power through hydroelectricity [1–32] (Zainuddin et al., 2012, pers. comm.). In this process, electricity is generated by hydropower, the production of electrical power through the use of the gravitational force of falling or flowing water. It is the most widely used form of renewable energy for power generation and accounts for 16% of global electricity consumption. This method has generated about 3427 TWh of electricity in 2010 [1]. Hydropower is produced in 150 different countries, with the Asia-Pacific region generating 32% of global hydropower in 2010 [1]. The top 10 countries for hydroelectricity generation in 2009 are listed in Table 13.1 [1]. As summarized in Table 13.1, China is the largest producer of hydroelectricity. Major new projects that are under construction worldwide are listed in Table 13.2 [1].

TABLE 13.1
Ten Largest Hydroelectricity Producers as of 2009

Country	Annual Hydroelectricity Production (TWh)	Installed Capacity (GW)
China	652.05	196.79
Canada	369.5	88.974
Brazil	363.8	69.080
United States	250.6	79.511
Russia	167.0	45.000
Norway	140.5	27.528
India	115.6	33.600
Venezuela	85.96	14.622
Japan	69.2	27.229
Sweden	65.5	16.209

Source: "Hydroelectricity," Wikipedia, the free encyclopedia, 1–7, 2012.

TABLE 13.2
Major Global Hydroelectricity Projects (≥3000 MW) under Construction

Country	Name	Maximum Capacity (MW)	Construction (Start/Completion)
China	Xiluodu Dam	12,600	2005/2015
Brazil	Belo Monte Dam	11,181	2011/2015
India	Upper Siang HE project	11,000	2009/2024
Burma	TaSang Dam	7,100	2007/2022
China	Xiangjiaba Dam	6,400	2006/2015
China	Nuozhadu Dam	5,850	2006/2017
China	Jinping-II hydropower station	4,800	2007/2014
Pakistan	Diamer-Bhasha Dam	4,500	2011/2023
China	Jinping-I hydropower station	3,600	2005/2014
Brazil	Santo Antônio Dam	3,150	2008/2013
Brazil	Jirau Dam	3,300	2008/2013
China	Pubugou Dam	3,300	2004/2010
China	Goupitan Dam	3,000	2003/2011
China	Guanyinyan Dam	3,000	2008/2015
China	Lianghekou Dam	3,000	2009/2015
Russia	Boguchany Dam	3,000	1980/2013

Source: "Hydroelectricity," Wikipedia, the free encyclopedia, 1–7, 2012.

The cost of producing hydroelectricity is relatively low, making it competitive with other renewable sources of energy. The average cost of electricity from a hydroplant larger than 10 MW is about 3–5 cents per kWh [1]. Hydroelectricity is also a very flexible source because plants can be ramped up and down very quickly to adapt the changing energy demands. While hydroelectricity creates very little waste and

GHG emissions, the creation and operation of dams have some environmental effects such as changing ecosystems and displacing people and wildlife, and these issues need to be considered in the development of the new hydroelectricity projects. There are numerous methods for obtaining hydroelectricity. These are briefly described in Sections 13.2.1 through 13.2.3 and in Section 13.2.6 [1–13].

13.2.1 Conventional Dams

Most hydroelectric power comes from the potential energy of dammed water driving a water turbine and generator [1–13]. The power extracted from the water depends on the height of the dam and volume of waterfall. The potential energy is proportional to the height of the dam. A large pipe called penstock delivers water to the turbine. A typical cross section of dam, turbine, and generator is graphically illustrated in Figure 13.1.

13.2.2 Pumped Storage

In many situations, water at the lower level reservoir is pumped back to the higher level reservoir [1–13]. Thus, when there is a higher demand, water is released back into the lower reservoir through a turbine. The pumped-storage scheme provides the most commercially important means of large-scale grid energy storage and improves the daily capacity factor of the generation system. It also supplies extra energy during emergency needs.

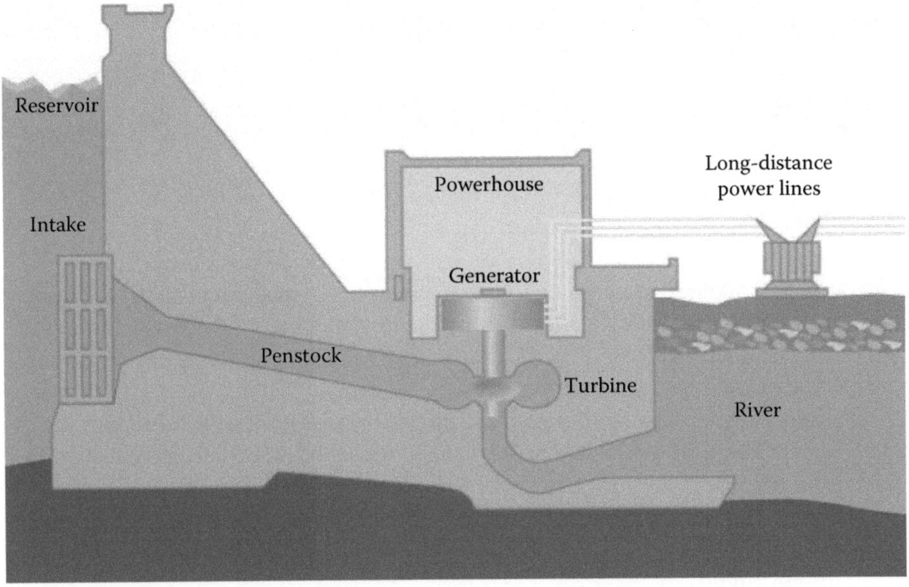

FIGURE 13.1 **(See color insert.)** Cross section of a conventional hydroelectric dam. (Adapted from "Hydroelectricity," Wikipedia, the free encyclopedia, 1–7, 2012.)

13.2.3 Other Methods

Numerous other methods are also used to generate hydroelectricity [1–32] (Zainuddin et al., 2012, pers. comm.). Some of these combine the principles of hydrokinetic energy to generate power and electricity. If there is small or no reservoir capacity, the water from the upstream is used for generation according to the need and bypassing the dam. This method is called run-of-the-river hydroelectric station. Hydroelectricity can also be generated using rise and fall of ocean water due to tides. The conditions of tides are predictable, and if reservoirs can be constructed, it can supply hydroelectricity during high demand periods. Sometimes tides are so high (>40 ft [33]) that they provide both potential and kinetic energy for the power generation. This subject is further discussed in Section 13.3.

An underground power station can make use of a large natural difference between two waterways such as waterfall or a mountain lake. An underground tunnel can also be constructed to take the water from high reservoir to the power-generating hall built in an underground cavern near the lowest point of the water tunnel and a horizontal water pipe taking the water away to the lower waterway. The size of the hydroelectricity generated by these methods can be large, small [14–16], micro (<100 kW) [17–25], or pico (<5 kW) [26–32] (Zainuddin et al., 2012, pers. comm.). These different levels of hydropower generations are further discussed in Section 13.2.6.

13.2.4 Advantages and Disadvantages of Hydroelectric Power

Hydroelectric power offers various advantages and disadvantages, some of which are listed in Sections 13.2.4.1 and 13.2.4.2.

13.2.4.1 Advantages

Hydroelectric power offers the following advantages:

1. Hydroelectricity is flexible and relatively inexpensive.
2. The operating costs for dams are usually low because of automation and low manual labor requirement during normal operations.
3. A new plant for specific purposes can be added with relatively low construction cost. The life of dam can be 50–100 years.
4. Hydroelectric plants can also be suitable for specific industrial purposes. For example, dedicated hydroelectric plant provides a substantial amount of electricity needed for aluminum electrolytic plants. There are numerous examples of the use of hydroelectricity in aluminum plants in the United States and New Zealand.
5. Hydroelectricity does not produce carbon dioxide emissions, although some CO_2 emissions can be produced during the manufacture and construction of the project.
6. Hydroelectric dams can also provide water sports, an attraction for tourists, aqua culture, and irrigation support to agriculture industry with constant water supply.
7. Large dams can control floods that can affect people living downstream of the dams.

13.2.4.2 Disadvantages

While hydroelectricity offers many advantages as stated above, it also has some drawbacks. Some of these drawbacks are briefly stated as follows

1. Hydroelectric projects can change the ecosystem and loss of the land. Dams can have a serious effect on salmon and fish populations around the dams.
2. When water flows, it has the ability to transport heavier particles downstream. This has a negative effect on the operation of dams and subsequently their power stations.
3. High siltation can fill a reservoir and reduce its capacity to control floods along with causing additional horizontal pressure on the upstream portion of the dam. This can ultimately result in the failure of the reservoir due to accumulation of large sediments [10,11].
4. When a dam is on the river, the change in river flow can affect the amount of power supply to the neighborhood. This can be important as the temperature of the water and rainfall changes under different climates.
5. In tropical regions, reservoirs of power plants may produce a substantial amount of methane. This is because of the decay of plant materials in anaerobic environment that results in methane production by microbiological reactions. This type of emission is higher when the reservoir is large compared to its electricity-generating capacity, and the forests surrounding the dams are not cleared [10,11].
6. The new dam projects also displace people. The capital costs for new dam projects are high.
7. For large dams, failure of dam due to poor construction can be the largest man-made disasters in the history. The Banqiao Dam failure in Southern China resulted in the deaths of 26,000 people, with another 145,000 from epidemics, and millions were left without homes [1–13,34].

13.2.5 Environmental Issues

Hydroelectric power is environment friendly. Some evidences toward this fact are as follows

1. Unlike in coal- or gas-driven power plants, hydroelectricity eliminates flue gas emissions, including pollutants such as sulfur dioxide, nitric oxide, carbon monoxide, dust, and mercury. It also eliminates emissions of GHG like methane and carbon dioxide.
2. It eliminates all the health hazards related to coal mining and coal dust emissions.
3. It also eliminates all the negative consequences of "fracking process" to recover unconventional gas for gas-driven power plants.
4. Compared to nuclear power, hydroelectricity generates no nuclear waste and has none of the dangers associated with uranium mining nor nuclear leaks.
5. Unlike coal-, gas-, or uranium-driven power plants, hydroelectricity is a renewable and carbon-free energy source.

Due to all these environmental benefits, the use of hydroelectricity for power generation will grow worldwide over the next several decades.

13.2.6 Size and Capacities of Hydroelectric Power Facilities

While there are no official breakdowns of the type of hydroelectric power plants, they are generally categorized based on the amount of power generation into four parts:

1. Large power plants (generally >10 GW)—Currently, there are three such plants in the world: Three Gorges dam (22.5 GW), Itaipu dam (14 GW), and Guri dam (10.2 GW) [1–13].
2. Small power plants (generally with capacity up to 10–30 MW) with significant growth in China, Japan, the United States, and India [14–16].
3. Micropower plants (generally with a capacity up to 100 kW) provide power to isolated home or community [17–25].
4. Picopower plants (generally with power capacity under 5 kW) used in remote community requiring small amount of electricity [26–32] (Zainuddin et al., 2012, pers. comm.).

Since we already examined some details on large hydroelectric power generation, in Sections 13.2.6.1 through 13.2.6.3, we briefly assess some of the characteristics of small-, micro-, and picopower plants.

13.2.6.1 Small Hydropower Plants

Small hydropower plants are generally used in a small community or an industrial plant [14–16]. They may be connected to conventional electrical distribution networks as a source of low-cost renewable energy. They are also often built in remote areas where it would be uneconomical to provide electricity from national electrical distribution network. They have low environmental impact. In a typical installation, water is fed from a reservoir through a channel or pipe into a turbine. The pressure of the flowing water on the turbine blades makes the shaft to rotate. This rotating shaft is connected to the electric generator, which converts the motion of the shaft into electrical energy. Small hydropower plants can be further subdivided into mini-hydro, which has production capacity of less than 1000 kW [1,14–16].

13.2.6.2 Microhydropower Plants

These type of power plants produce electricity up to 100 kW using flow of water [17–25]. The installation is often just a small dammed pool at the top of a waterfall, with several hundred feet of pipe leading to a small generator housing. They are useful to provide electricity to a small community. This type of power plant is frequently accomplished with a Pelton wheel for high-head, low-flow water supply. Construction details of microhydropower plant are site specific. The production range of such systems is often calculated in terms of “head” and “flow”; the higher each of these are, more power can be generated. The construction of such a power station requires an “intake” structure where water is diverted from natural stream, river, or waterfall. Such a structure screens out floating debris, fish, and other large objects. Microhydropower plants are very popular in poor countries for local power supply [17–25].

13.2.6.3 Picohydropower Plants

This is generally built in a remote community that requires only small amount of electricity (<5 kW) [26–32] (Zainuddin et al., 2012, pers. comm.). These types of setups typically are run-off stream, meaning that dams are not used but pipe diverts some of the flow and drops this down a gradient and through the turbine before being exhausted back to the stream. Two examples of picohydropower stations are in the towns of Kithamba and Thimba in Kenya [26]. Just like microhydropower stations, these are very popular in poor countries. Both micro- and picohydropower stations can be improved by custom-designed power-generation systems [31,32] (Zainuddin et al., 2012, pers. comm.).

13.3 HYDROKINETIC ENERGY AND POWER GENERATION

Fundamentally, hydrokinetic energy is the energy generated from moving water [34–71].[*,†] The power of tidal, river, and ocean currents and ocean waves is tremendous, and the basic concept behind hydrokinetic power is not new. For a century, people have harnessed the power of river currents by installing water wheels of various sorts to turn shafts or belts [35,49,50].[‡] Modern tidal/river/ocean current hydrokinetic machines use new technology that is designed to operate in high amplitude waves and fast currents. These emerging technologies have the potential to provide significant amounts of affordable electricity with low environmental impact given proper care in their deployment and operation [53–60] (Bertsch, 2012, pers. comm.).

13.3.1 Why Hydrokinetic Energy?

It is estimated that the amount of hydrokinetic energy that can be feasibly captured can power 67 million homes [50–60] (Bertsch, 2012, pers. comm.). It is expected that by 2025, 13,000 MW of power can be generated using hydrokinetic energy [50–60] (Bertsch, 2012, pers. comm.). This can displace 22 new coal-fired power plants [57,60]. Just like hydroelectricity, hydrokinetic energy has very little effect on air and global climate change. They generate power only from the kinetic energy of moving water (current). This power is a function of the density of water and speed

[*] Hydrokinetic energy was included as an eligible renewable energy resource by the Energy Policy Act of 2005. Various funding authorizations for research and development were also included in this Act as well as the Energy Independence and Security Act of 2007.

[†] Bertsch, D.J., Juris Doctoral candidate, The University of South Dakota School of Law, 2011; Congress defined hydrokinetic energy as "electrical energy from waves, tides, and currents in oceans, estuaries, and tidal areas; free flowing water in rivers, lakes, and streams, or man-made channels; and differentials in ocean temperature (ocean thermal energy conversion)," The Energy Independence and Security Act of 2007, 42 USC §17211 (2006).

[‡] Hydrokinetic energy was included as an eligible renewable energy resource by the Energy Policy Act of 2005. Various funding authorizations for research and development were also included in this Act as well as the Energy Independence and Security Act of 2007.

of the current cubed. The available hydrokinetic power thus depends on the speed of the river, ocean, or tidal current. The operation of the hydrokinetic device requires a minimum current and water depth. As water flows through a turbine or other device, the kinetic energy of the flowing river, tidal fluctuations, or waves is converted into electricity by the appropriate converting device.

Hydrokinetic energy is different from hydroelectricity in the fact that it does not require a change in elevation. Also, unlike traditional hydropower projects, hydrokinetic energy projects do not require impoundments or diversions of water [34,61–65] (Bertsch, 2012, pers. comm.).* Instead, these projects harness the power of moving water in waves, currents, and tidal channels [34,61–65] (Bertsch, 2012, pers. comm.).† Hydrokinetic technologies can thus be distinguished based on these three major sources for harnessing hydrokinetic energy. Waves, currents, and tidal channels can be either from oceans, rivers, or inland waterways. Surface wave energy is generally harnessed near sea or ocean shores while energy from undercurrents is harnessed using technologies installed below the water surface [34–47,61–65] (Bertsch, 2012, pers. comm.; Dixon et al., 2008, pers. comm.).‡,§ Tidal power is harnessed by new tidal power-harnessing technologies, such as tidal barrage, tidal lagoons, and new axial or cross-flow turbine technologies [53,58].

While hydrokinetic energy can be obtained in a number of different ways, capturing the energy contained in near and offshore waves is thought to have the greatest energy production potential among various options. The rise and fall of ocean waves is driven by winds and influenced by oceanic geology. The extraction of only 15% of the energy in coastal waves would generate as much electricity as we currently produce in conventional hydroelectric dams [36–40] (Bertsch, 2012, pers. comm.; Dixon et al., 2008, pers. comm.).¶ Much of this wave potential is found along our Pacific coast, near big cities and towns. Besides waves, ocean tides hold promise as an energy resource. Each change in the tide creates a current, called tidal stream. Regular tidal streams have the potential to provide a reliable new source of electricity without building dams and barrages. Ocean currents, such as Gulf stream, also offer hydrokinetic energy. These result from winds and equatorial solar heating. Free flowing rivers (without dams) and constructed waterways such as irrigation canals also allow the use of hydrokinetic

* Bertsch, D.J., Juris Doctoral candidate, The University of South Dakota School of Law, 2011; Congress defined hydrokinetic energy as "electrical energy from waves, tides, and currents in oceans, estuaries, and tidal areas; free flowing water in rivers, lakes, and streams, or man-made channels; and differentials in ocean temperature (ocean thermal energy conversion)," The Energy Independence and Security Act of 2007, 42 USC §17211 (2006).

† Ibid.

‡ Hydrokinetic energy was included as an eligible renewable energy resource by the Energy Policy Act of 2005. Various funding authorizations for research and development were also included in this Act as well as the Energy Independence and Security Act of 2007.

§ Bertsch, D.J., Juris Doctoral candidate, The University of South Dakota School of Law, 2011; Congress defined hydrokinetic energy as "electrical energy from waves, tides, and currents in oceans, estuaries, and tidal areas; free flowing water in rivers, lakes, and streams, or man-made channels; and differentials in ocean temperature (ocean thermal energy conversion)," The Energy Independence and Security Act of 2007, 42 USC §17211 (2006).

¶ Hydrokinetic energy was included as an eligible renewable energy resource by the Energy Policy Act of 2005. Various funding authorizations for research and development were also included in this Act as well as the Energy Independence and Security Act of 2007.

energy. Stream-based hydrokinetic energy is not as much researched as wave energy, although it is suggested that these resources can provide electricity needs of 23 million homes [34–47,61–65] (Bertsch, 2012, pers. comm.; Dixon et al., 2008, pers. comm.).[*,†] In lower southwest where wind energy has low potential, stream hydrokinetic energy can be a very valuable resource. Waves and ocean currents can provide a continuous power [34,61–65] (Bertsch, 2012, pers. comm.), which is not possible for solar, wind, and tidal stream energy.[‡] Since the kinetic energy from a stream is proportional to the cube of the speed of the current, site location is very important.

The hydrokinetic energy projects are renewable and emission free. The United States can avoid emitting 250 million metric tons of carbon dioxide per year if hydrokinetic energy represented 9% of the US energy consumption [34,61–65] (Bertsch, 2012, pers. comm.).[§] Electric Power Research Institute (EPRI) estimates that hydrokinetic resources can provide about 10% of today's electric consumption in the United States [44] (Dixon et al., 2008, pers. comm.). The sources of hydrokinetic energy are generally predictable and unaffected by weather variability. Wave patterns can be accurately forecast several days in advance [34,61–65] (Bertsch, 2012, pers. comm.) and tides will always reoccur every 12 h 25 min because they are connected to the moon's gravitational pull (Bertsch, 2012, pers. comm.).[¶] Both wave and tidal energy can provide base load power, eliminating the needs for backup sources [34–65] (Bertsch, 2012, pers. comm.).[**,††] River currents, however, fluctuate seasonally and are susceptible to wet and dry years, and this makes them difficult to predict from year to year [36].

13.3.2 Hydrokinetic versus Hydroelectric Energy: Potentials and Issues

Unlike hydroelectric energy projects, hydrokinetic energy projects have little or no effects on the local aesthetics [36–40] (Bertsch, 2012, pers. comm.; Dixon et al., 2008, pers. comm.).[‡‡] They are generally underwater or a little removed

[*] Hydrokinetic energy was included as an eligible renewable energy resource by the Energy Policy Act of 2005. Various funding authorizations for research and development were also included in this Act as well as the Energy Independence and Security Act of 2007.

[†] Bertsch, D.J., Juris Doctoral candidate, The University of South Dakota School of Law, 2011; Congress defined hydrokinetic energy as "electrical energy from waves, tides, and currents in oceans, estuaries, and tidal areas; free flowing water in rivers, lakes, and streams, or man-made channels; and differentials in ocean temperature (ocean thermal energy conversion)," The Energy Independence and Security Act of 2007, 42 USC §17211 (2006).

[‡] Ibid.

[§] Ibid.

[¶] Ibid.

[**] Hydrokinetic energy was included as an eligible renewable energy resource by the Energy Policy Act of 2005. Various funding authorizations for research and development were also included in this Act as well as the Energy Independence and Security Act of 2007.

[††] Bertsch, D.J., Juris Doctoral candidate, The University of South Dakota School of Law, 2011; Congress defined hydrokinetic energy as "electrical energy from waves, tides, and currents in oceans, estuaries, and tidal areas; free flowing water in rivers, lakes, and streams, or man-made channels; and differentials in ocean temperature (ocean thermal energy conversion)," The Energy Independence and Security Act of 2007, 42 USC §17211 (2006).

[‡‡] Hydrokinetic energy was included as an eligible renewable energy resource by the Energy Policy Act of 2005. Various funding authorizations for research and development were also included in this Act as well as the Energy Independence and Security Act of 2007.

from shore for visual observations [36–40,61–65] (Bertsch, 2012, pers. comm.; Dixon et al., 2008, pers. comm.).[*,†] They can be installed wherever energy is needed [61–65] (Bertsch, 2012, pers. comm.).[‡] Since many high-demand urban areas are located near water, hydrokinetic energy projects can be easily integrated into existing grid. It can also provide energy to rural and remote areas where other means of power and electricity may not be possible.

In general, hydrokinetic energy differs from traditional hydropower [34–65] in three ways[§,¶]:

1. As mentioned earlier, hydrokinetic energy projects do not require impoundments or diversions of water. Hydropower projects drastically alter the surrounding land. They can increase the likelihood of flooding upstream of the dam [1–13]. They also cause an increase in sedimentation in the reservoir [10,11]. It also causes an accelerated erosion of the riverbed caused by sediment-free water released downstream at a high velocity (Bertsch, 2012, pers. comm.).
2. Hydrokinetic projects do not displace a large number of people that hydropower projects do, particularly when the projects are large. For people staying near dam, there is a continuous threat of dam failure due to earthquake or other natural disasters [1–13] (Bertsch, 2012, pers. comm.).
3. Hydrokinetic energy projects can avoid impacts on surrounding wildlife, although it can have an effect on underwater ecosystem if not properly installed. The temperature and sea level fluctuations caused by hydrokinetic power projects can have an effect on the fisheries and other sea life. In general, if properly installed, hydrokinetic energy projects are more environment friendly than hydropower projects.

In this country, waterways in Alaska are well suited for tapping hydrokinetic energy. One of the challenges in hydrokinetic energy is the presence of glacial silt (in Alaska waters). Over time, silt and other sediments in the water flowing through hydrokinetic turbines can erode the machinery. In addition to this, the migration of fish

---

* Hydrokinetic energy was included as an eligible renewable energy resource by the Energy Policy Act of 2005. Various funding authorizations for research and development were also included in this Act as well as the Energy Independence and Security Act of 2007.

† Bertsch, D.J., Juris Doctoral candidate, The University of South Dakota School of Law, 2011; Congress defined hydrokinetic energy as "electrical energy from waves, tides, and currents in oceans, estuaries, and tidal areas; free flowing water in rivers, lakes, and streams, or man-made channels; and differentials in ocean temperature (ocean thermal energy conversion)," The Energy Independence and Security Act of 2007, 42 USC §17211 (2006).

‡ Ibid.

§ Hydrokinetic energy was included as an eligible renewable energy resource by the Energy Policy Act of 2005. Various funding authorizations for research and development were also included in this Act as well as the Energy Independence and Security Act of 2007.

¶ Bertsch, D.J., Juris Doctoral candidate, The University of South Dakota School of Law, 2011; Congress defined hydrokinetic energy as "electrical energy from waves, tides, and currents in oceans, estuaries, and tidal areas; free flowing water in rivers, lakes, and streams, or man-made channels; and differentials in ocean temperature (ocean thermal energy conversion)," The Energy Independence and Security Act of 2007, 42 USC §17211 (2006).

and marine mammals, ice and other debris, as well as river and ocean bed stability can significantly affect the performance of hydrokinetic energy machinery [34–65] (Bertsch, 2012, pers. comm.).*,†

13.3.3 Hydrokinetic Power Devices

While tidal/river/ocean current energy and wave energy converters (WECs) are sometimes categorized separately [51–60], both types of technology can be categorized under the general term "hydrokinetic power devices." Another marine energy technology, OTEC, is covered later in detail. It is mostly applicable to tropical areas [72–77].

Modern ocean wave energy conversion machines use new technology that is designed to operate in high amplitude waves, and modern tidal/river/ocean current hydrokinetic machines use new technology that is designed to operate in fast currents. Both of these emerging technologies have the potential to provide significant amounts of affordable power with low environmental impact. While each individual device can give a limited amount of power, often, a farm of devices are used to increase the total power generation capacity. All designs require careful thoughts and implementations of their deployment and operation.

The devices that generate energy from waves and currents are called hydrokinetic energy conversion devices. *Hydrokinetic* energy devices typically use vertical- or horizontal-axis turbines similar to those developed for wind generation or old water mills. Since water is approximately 850 times denser than air, the amount of energy generated by a hydrokinetic device is much greater than that produced by a wind turbine of equal diameter. In addition, river and tidal flow do not fluctuate as dramatically from moment to moment as wind does. This predictability benefit is particularly true for tidal energy. It can be predicted years in advance and is not affected by precipitation or evaporation.

Hydrokinetic energy conversion devices are broken into two categories: WECs and rotating devices [50,51]. WECs are generally installed at the surface of the water, while rotating devices are generally installed beneath the water surface. While the industry is still at the growing stage, there are a number of cost-effective and environmentally acceptable devices being developed, constructed, and installed. Some of these devices (and their modifications) have been tested at the pilot- and commercial-scale levels. The industry wants to build "wave parks" and turbine arrays that are capable of delivering clean and renewable electricity from different forms of wave energy.

* Hydrokinetic energy was included as an eligible renewable energy resource by the Energy Policy Act of 2005. Various funding authorizations for research and development were also included in this Act as well as the Energy Independence and Security Act of 2007.

† Bertsch, D.J., Juris Doctoral candidate, The University of South Dakota School of Law, 2011; Congress defined hydrokinetic energy as "electrical energy from waves, tides, and currents in oceans, estuaries, and tidal areas; free flowing water in rivers, lakes, and streams, or man-made channels; and differentials in ocean temperature (ocean thermal energy conversion)," The Energy Independence and Security Act of 2007, 42 USC §17211 (2006).

13.3.3.1 Wave Energy Converters

In its basic design, WEC utilizes the principle of motion of two or more bodies relative to each other [50,51]. One body, called displacer, is acted on by the waves. The second body, called reactor, moves in response to the displacer. While there are a number of designs and configurations of WECs, the four most common ones are graphically depicted in Figure 13.2a through d [50,51].

In the oscillating water column, waves enter and exit in a partially submerged collector from below, causing the water column inside the collector to rise and fall. The changing water level acts as a piston as it drives the trapped air in the device above the water into turbine. The turbine movement generates electricity via a coupled generator. This device is graphically illustrated in Figure 13.2a [50,51].

In overtopping device, shown in Figure 13.2b, a floating reservoir is formed as waves break over the walls of the device. This reservoir creates a head of water higher than that of the surrounding ocean surface. The pressure necessary to turn hydroturbine is provided by this head of water. The water leaves the bottom of the device to return back into the sea [50,51]. The device can be used in tidal currents.

In an attenuator WEC (shown in Figure 13.2c), long, multiple sectioned, floating structures, which are joined and anchored at each end, are aligned parallel to the wave direction and they generate electricity by riding the waves. These heavy-surge devices utilize the passing waves to set each section into the rotational motion relative to the next segment. This relative motion, which is concentrated at the joint between the two consecutive segments, is used to pressurize a hydraulic piston that drives fluids through a motor. The hydraulic piston in turn drives the coupled generator and produces electricity [50,51].

Finally, the point absorber depicted in Figure 13.2d drives a turbine by using waves from all directions at a single point by using the vertical motion of the waves to act as a pump that pressurizes seawater or an internal fluid. While this type of device has many configurations, one of which is the hose pump point absorber that consists of a surface-floating buoy anchored to the seafloor with the turbine as a part of the vertical connection. The wave-induced vertical motion of buoy causes the connection to expand and contract, thereby producing pumping action. The captured energy and the resulting electricity generation by this device can be optimized [50,51] by operating the device and wave in resonance.

13.3.3.2 Commercial Applications of WEC

A brief summary of commercial applications of various types of WECs outlined above is given in Table 13.3 [60]. Locations of the devices outlined in Table 13.3 can be shoreline, nearshore, and offshore. Types of power takeoff include hydraulic ram, elastomeric hose pump, pump to shore, hydroelectric turbine, air turbine, and linear electrical generator [60]. Some of these designs incorporate parabolic reflectors to increase the wave energy at the point of capture. These capture systems use the rise and fall of motion of waves to capture energy. Once the energy is captured, power is transmitted to its use or to the electrical grid by transmission power cables [60]. More details on each device outlined in Table 13.3 are given in Ref. [60]. A more detailed

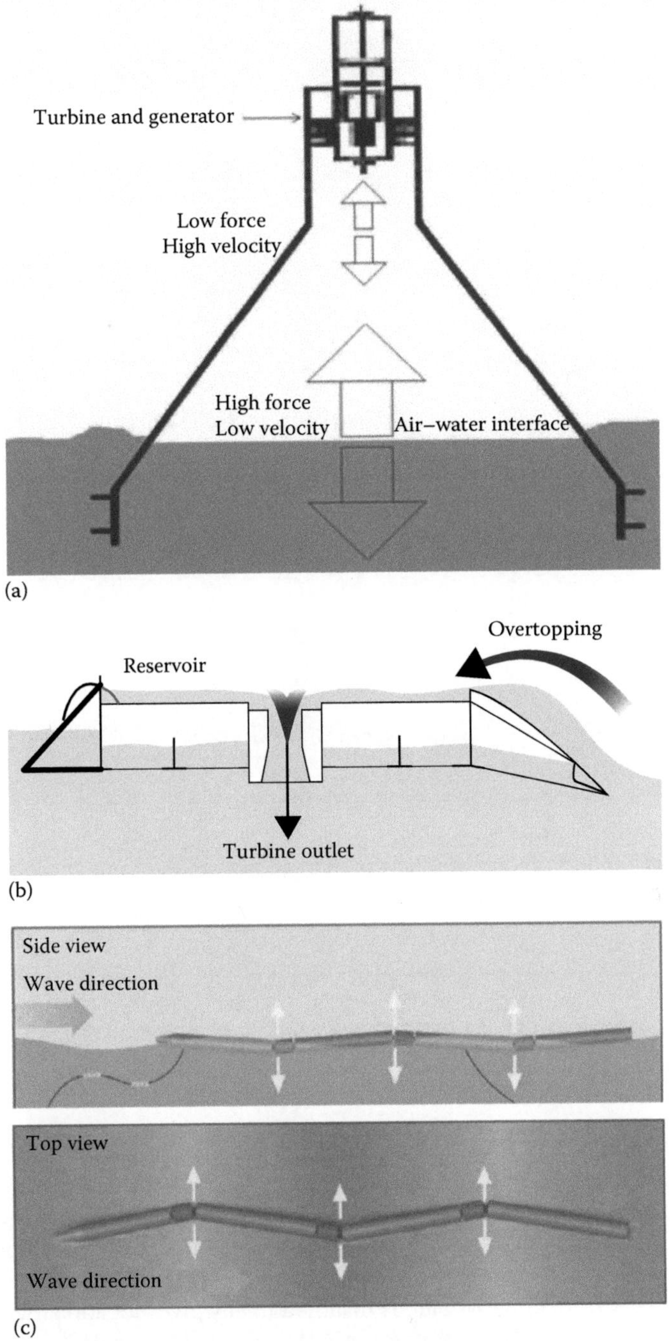

FIGURE 13.2 Types of WECs: (a) oscillating water column; (b) overtopping WEC; (c) attenuator WEC; (Adapted from "How hydrokinetic energy works?" Union of Concerned Scientists, 1–5, 2012.)

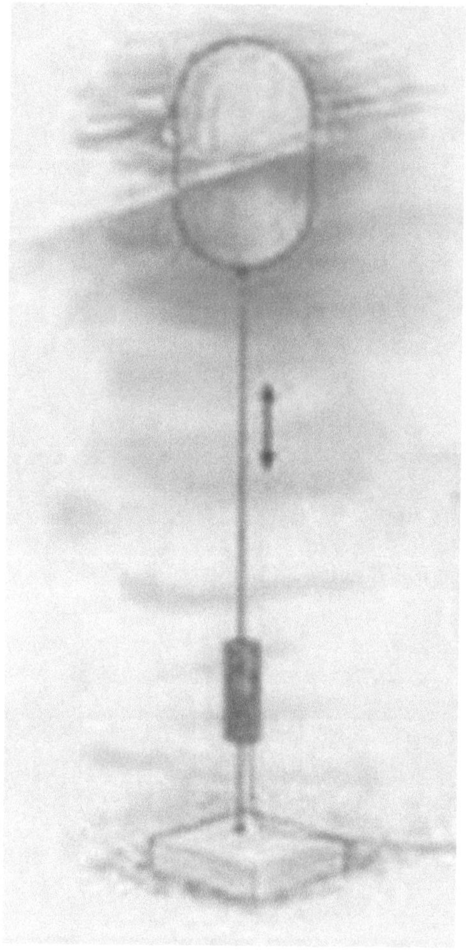

(d)

FIGURE 13.2 (Continued) Types of WECs: (d) point absorber. (Adapted from "How hydrokinetic energy works?" Union of Concerned Scientists, 1–5, 2012.)

assessment of WECs is also given in an excellent review by Drew et al. [78]. As can be seen from Table 13.3, WECs are used all around the world.

One commercial device, the tethered floating buoy [33,54,79,81], converts the energy in the rise and fall of the passing waves into electricity (often via hydraulics) (Figure 13.8a). This device uses the principle of point absorber depicted in Figure 13.2d. A farm of buoys can be installed in any given location (Section 13.3.4), and these are tied to the bottom of ocean and connected to the electrical grid system.

Other machines have chambers that, when filled and emptied by rising and falling wave water, compress and decompress air to drive an electric generator. One such design—"WaveRoller"—is illustrated in Figure 13.3. It uses the principle of

TABLE 13.3
Large-Scale Application of Various WECs

Device/Year	Proponent	Country	Capture Method
	Commercial Application of Point Absorber		
AquaBuOY/2003	Finavera Wind Energy SSE Renewable Ltd.	Ireland/Canada/ Scotland	Buoy
CETO wave power/1999	Carnegie	Australia	Buoy
FlanSea/2010	FlanSea	Belgium	Buoy
Lysekil Project/2002	Uppsala University	Sweden	Buoy
Oceanlinx/1997	Oceanlinx	Australia	Buoy
OE buoy/2006	Ocean Energy	Ireland	Buoy
PowerBuoy/1997	Ocean Power Technologies	United States	Buoy
SDE Sea Waves Power Plant/2010	SDE Energy Ltd.	Israel	Buoy
SeaRaser/2008	Alvin Smith (Dartmouth Wave Energy)/Electricity	United Kingdom	Buoy
Unnamed Ocean Wave Powered Generator/2004	SRI International	United States	Buoy
Wavebob/1999	Wavebob	Ireland	Buoy
Wave Star/2000	Wave Star A/S	Denmark	Multipoint Absorber
	Commercial Application of Attenuator		
Anaconda Wave Energy Converter/2008	Checkmate Sea Energy	United Kingdom	Surface following attenuator
AWS-III/2010	AWS Ocean Energy	United Kingdom (Scotland)	Surface following attenuator
Pelamis Wave Energy Converter/1998	Pelamis Wave Power	United Kingdom (Scotland)	Surface following attenuator
R38/50 kW, R115/150 kW/2010	40 South Energy	United Kingdom	Underwater attenuator
	Commercial Applications of Oscillating Water Column		
Islay LIMPET/1991	Islay LIMPET	Scotland	Oscillating water column
Oyster Wave Energy Converter/2005	Aquamarine Power	United Kingdom, Scotland, Irish	Oscillating wave surge converter
Wave Roller/1994	AW-Energy Oy	Finland	Oscillating wave surge converter
Cycloidal Wave Energy Converter/2006	Atargis Energy Corp.	United States	Fully submerged wave termination device
	Commercial Application of Overtopping Device		
Wave Dragon/2003	ErikFrils-Madsen	Denmark	Overtopping device

Source: "Wave power," Wikipedia, the free encyclopedia, 2013.

FIGURE 13.3 **(See color insert.)** WaveRoller wave energy farm installation in Peniche, Portugal. (Adapted from "Wave power," Wikipedia, the free encyclopedia, 2013.)

oscillating water column described in Figure 13.2a. This device has been in operation in Finland since 1994.

Yet another type of machine looks like a giant sea snake (or a farm of snakes) with floating pontoons that heave and sway on the ocean surface, driving hydraulic pumps to power an electric generator [64,81–89]. This Pelamis "snake," an offshore machine, consist of five tube sections that float on the surface and use the motion of the waves to generate electricity [60,64,81]. They use the principle of attenuator WEC (Figure 13.2c) described above. When the tube sections flex, hydraulic arms move in opposite directions and turn a generator that produces power. The amount of power required will dictate the number of such snakes in a given farm and to some extent the length of each snake (that can be as long as 600 ft). This device is also anchored to the seabed and must withstand marine environments. Waves powerful enough to drive these generators are often found off coasts with large oceans to their west (providing long wind fetch) and strong prevailing winds such as the west coasts of the United States, Chile, Australia, and in the North Sea among many others [46–60]. This technology is graphically illustrated in Reference [35] and extensively described in the Pelamis wave power website enquiries@pelamiswave.com. The snakes are being commercially used off the shores in Portugal and Scotland [35,64,81–90].

Finally, one device that uses the above-described "overtopping" principle (Figure 13.2b) is "Wave Dragon" (shown in Figure 13.4), which is being used in Denmark since 2003.

13.3.3.3 Rotating Hydrokinetic Devices

The kinetic energy of flowing tidal stream, ocean current, or river can also be captured by the rotating device [47–60] shown in Figures 13.5 [50]. Such a device is generally installed underwater to harness maximum energy from the currents.

FIGURE 13.4 **(See color insert.)** Wave Dragon seen from reflector. (Adapted from "Wave power," Wikipedia, the free encyclopedia, 2013.)

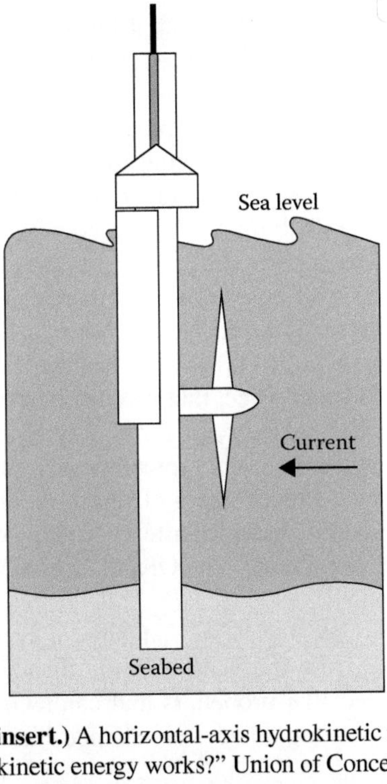

FIGURE 13.5 **(See color insert.)** A horizontal-axis hydrokinetic rotating device, tidal turbine. (Adapted from "How hydrokinetic energy works?" Union of Concerned Scientists, 1–5, 2012.)

FIGURE 13.6 (See color insert.) Cross-flow turbine used in Alaska Rivers: ORPC's TidGen™ power system. [Adapted from "Hydrokinetic energy (in river, tidal, and ocean current)," *Alaska Energy Wiki*, Alaska Center for Energy and Power, 1–4, 2012.]

In this device, the rotational energy created by the rotation of the blades drives turbine and creates electricity by a generator. This device is very similar to the wind turbine used for gathering wind energy for electricity. This similarity has helped a faster movement in its development. Some rotational device designs rotate around horizontal axis just like wind turbines, while others are either oriented around vertical axis or use a design resembling egg beaters. As shown later, this design has been successfully installed at a commercial scale at the bottom of New York City's East River.

The rotating device can also be a cross-flow device depicted in Figure 13.6 [51]. This type of cross-flow turbine is used in Alaska's rivers. The figure shows the Ocean Renewable Power Company (ORPC)'s Beta TidGen™ power system [51]. This design looks similar to old water wheel that was used to drive boats and barges. Such cross-flow turbines typically have a rotor formed by mounting two or more blades substantially parallel to a shaft that is typically vertical or horizontal. Horizontal cross-flow rotors can capture kinetic energy from flows in two directions (e.g., flood and ebb) without an orientation change, while a vertical-axis rotor can be omnidirectional depending on river conditions. The ORPC turbine shown in Figure 13.6 is an example of such a device [51].

Rotating devices also take a variety of forms in commercial applications but in general capture energy from the water flowing through or across a rotor. Some of these devices are shaped like propellers and can swing, or yaw, to face changing tidal currents. Other rotating devices are shaped like a jet engine, having many vanes turning within a fixed outer ring [35,47–51,53,55–60]. Many other designs for rotating devices are presented in an excellent review by Ortega-Achury et al. [91].

One commercial design—"the open-center turbine"—is designed to be deployed directly on the seabed, and its installation is silent and invisible from the surface. It is located at depth and presents no navigational hazard. Farms of open-center turbines can provide a significant and undetectable supply of clean, predictable, renewable power. More details of this type of turbine are given in "OpenHydro" company website.

Fast currents, like those in the Missouri and Mississippi rivers, in tidal channels such as the Puget Sound, or in ocean currents such as the Gulf Stream off Florida, have enough power to turn large rotating devices. Since the power from a hydrokinetic machine is proportional to the cube of the current velocity, faster currents are better, and sites with current velocities reaching 3 m/s are desirable.

13.3.3.4 Devices to Harness Tidal Power

Harnessing tidal power has traditionally suffered from high cost and limited availability of sites, with sufficiently high tidal ranges or flow velocities. The topic has, however, gained more attention due to new technological developments in design such as dynamic tidal power (DTP) or tidal lagoons and new turbine technology such as new axial flow and cross-flow turbines. Unlike in wave power, in tidal power, both the kinetic energy of the moving water and the potential energy difference between high and low tides can be used.

Tidal stream generator makes use of the kinetic energy of the moving water to power turbines just as wind turbines use wind to power turbines. Some tidal generators can be built in the structures of existing bridges. The world's first commercial scale and grid-connected *tidal stream generator*—SeaGen—was built in Strangford Lough [92]. This power generator is illustrated in Figure 13.7.

Tidal barrage makes use of the potential energy difference in height between high and low tides. Tidal barrage technology takes advantage of predictable ocean tides. A barrage, or dam across an estuary or tidal channel, traps tidal flows and then releases them through turbines as tides fall [47–60]. When using tidal barrages to generate power, the potential energy from a tide is seized through the strategic placement of specialized dams. When the sea level rises and the tide begins to come in, the temporary increase in tidal power is channeled into a large basin behind the dam, holding a large amount of potential energy. With the receding tide, this energy is then converted into mechanical energy as the water is released through large turbines that create electrical power through the use of generators [93]. Barrages are essentially dams across the full width of a tidal channel.

DTP device exploits an interaction between potential and kinetic energies in tidal flows. These are very long dams (about 20–30 miles long) built from coasts straight out into the sea or ocean, without enclosing an area. High and low tidal phase differences are introduced across the dam, leading to a significant water-level differential in shallow coastal seas. These types of tidal currents are often found in the countries like the United Kingdom, China, and Korea. Besides tidal power, tidal lagoons are being rapidly developed and deployed.

FIGURE 13.7 (See color insert.) The world's first commercial scale and grid-connected tidal stream generator—SeaGen—in Strangford Lough. (Adapted from "Tidal power," Wikipedia, the free encyclopedia, 2013.)

On a global scale, several countries have long pursued tidal power R&D activities [34,61–63,65,94] (Bertsch, 2012, pers. comm.).* The 240 MW La Rance barrage dam in France was the world's first tidal power station. This station was opened in 1966 and it has 24 turbines that generate electricity [33]. The Sihwa Lake Power station in Korea that began operating last year is the world's largest tidal power station with a generating capacity of 254 MW [33,90,95,96]. The Annapolis power plant at the Bay of Fundy in Canada which was built in 1984 also generates 20 MW of electricity from the Bay's record 43 ft tides [33].

13.3.3.5 Hydrokinetic Power Barges

These are designed for use in river and ocean currents with a horizontal-axis turbine in which a vertically submerged blade has performance characteristics similar to a horizontally mounted cross-flow turbine [52]. The turbine blades are concave such that the leading edge offers reduced resistance while the trailing edge is aerodynamically optimized to reduce the flat dynamic effect. The rotational speed of turbine is low. Since hydrokinetic power is proportional to the cube of velocity,

* Bertsch, D.J., Juris Doctoral candidate, The University of South Dakota School of Law, 2011; Congress defined hydrokinetic energy as "electrical energy from waves, tides, and currents in oceans, estuaries, and tidal areas; free flowing water in rivers, lakes, and streams, or man-made channels; and differentials in ocean temperature (ocean thermal energy conversion)," The Energy Independence and Security Act of 2007, 42 USC §17211 (2006).

turbine blade can be designed to accommodate flow rate; it could be long and broad for slow-moving deep currents or it could be short and thin for fast-moving shallow currents.

The turbine is horizontally mounted on a barge and partially submerged into the water flowing beneath the barge. The barge on which turbines are mounted is able to cope with fluctuations in water levels, substantial velocity increases, and direct impact from large and fast-moving debris. The power output from the turbine would decrease only when the water flow rate underneath the barge goes down. The barge generates 1 MW and produces 8760 MWh electricity annually at a maximum rating through a synchronous AC induction generator.

The power barge has very low maintenance cost and downtime and life span of about 20 years. These barges can be deployed in rivers such as Mississippi, Amazon, and Nile. Different designs of the barges from different countries (the United Kingdom, Belgium, and Australia) are illustrated by alternative energy website news publication [52].

13.3.3.6 Criteria for Choice of a Device and Its Location

Generally, the following criteria and considerations are used to decide on the choice and size of a device and its location [34–65] (Bertsch, 2012, pers. comm.; Dixon et al., 2008, pers. comm.)*,†:

1. The use of hydrokinetic energy devices needs to carefully consider local environmental implications, economics, and competing users of the site chosen. The needs of users such as fishermen, shipping vessel operators, recreational boaters, and coastal citizen groups need to be factored in choosing the appropriate site for the recovery of hydrokinetic energy.
2. The device may affect the habitats of benthic animals and plants like oysters, clams, and sea grass; the potential for fish strikes or impingement on device; and the effect of noise on movement and migration of aquatic animals or even alteration of hydrologic and sediment regimes. With careful selection of location, these impacts can, however, be minimized.
3. The cost of electricity produced by these devices depend on the power density of stream (kW/m^2) or wave crest (kW/m crest height), the distance that electricity must be transmitted to reach consumers, access of the site for maintenance and monitoring and availability of the site for federal subsidy, project financing, or guarantee market for the produced electricity.
4. In general, stronger currents and large wave heights will reduce the cost of the hydrokinetic electricity. The hydrokinetic devices require minimum

* Hydrokinetic energy was included as an eligible renewable energy resource by the Energy Policy Act of 2005. Various funding authorizations for research and development were also included in this Act as well as the Energy Independence and Security Act of 2007.

† Bertsch, D.J., Juris Doctoral candidate, The University of South Dakota School of Law, 2011; Congress defined hydrokinetic energy as "electrical energy from waves, tides, and currents in oceans, estuaries, and tidal areas; free flowing water in rivers, lakes, and streams, or man-made channels; and differentials in ocean temperature (ocean thermal energy conversion)," The Energy Independence and Security Act of 2007, 42 USC §17211 (2006).

current and water depth. The minimum current required to operate a hydrokinetic device is typically 2–4 knots, with an optimum requirement being 5–7 knot range. Water depth is an important factor in the total energy that can be extracted from a site since rotor diameter is dependent on adequate water level above the installed device.

5. Hydrokinetic devices are ideally installed in locations with relatively steady flow throughout the year and in locations not prone to serious flooding events, turbulence, or extended period of low water level. In cold weather like Alaska, glacier silt, silt, and other sediments within the water can be harmful to the hydrokinetic devices.

13.3.4 Recent Commercialization Examples in the United States

While the technology for the hydrokinetic power devices is still being continuously improved, and the current prices make hydrokinetic energy somewhat less competitive compared to other methods for power generation, there is a significant momentum to commercialize this technology [33,79–89,94–102]. Compared to wind and solar, this renewable technology is more permanent in nature because river currents and ocean tides are more predictable. In recent years, several new commercial installations have taken place within the United States. We site five examples, which are as follows:

1. Wind power buoys (Figure 13.8a) capture the energy in the up and down movement of waves and generate power, which is transmitted by an underwater cable to the electric grid onshore. While several types of buoys are under development, Ocean Power Technologies' Reedsport Wave Park power station commercializes this technology (approved by Federal Energy Regulatory Commission [FERC] in August 2012) and they have 10 large

(a) (b)

FIGURE 13.8 First commercial projects for hydrokinetic power generation: (a) ocean wave power buoy off the Oregon coast; (b) underwater turbine in New York City's East River. (Adapted from "Regulators approve first commercial hydrokinetic projects in the United States," *Today in Energy*, US Energy Information Administration, Washington, DC, 2012.)

buoys installed which collectively generate 1.5 MW of electricity. The power wave station is located 2.5 miles off the Oregon coast and is connected to the electric grid by an underwater cable. The construction of this power station is completed [33,79,80,101].

2. Underwater turbines (Figure 13.8b) use water currents to spin underwater blades and generate electricity. These technologies depend on the unconstrained currents found in rivers, tidal areas, or the open ocean. Vedant Powers' Roosevelt Island Tidal Energy project (approved by the FERC in January 2012) installed 30 three-blade hydrokinetic generators on the bottom of New York City's East River to produce about 1.0 MW electricity. With some initial ups and downs, the project is now completed with blades made out of fiberglass and plastic [33,98–100].
3. Tidal power harnesses the water flowing between low and high tides, turning a turbine to generate power. There are only 40 sites known in the world that have the required difference in water levels between the tides needed to produce electricity. Alaska has a significant potential for hydrokinetic development in both rivers and tidal basins. Most inland communities in Alaska are situated along navigable waterways that could host hydrokinetic installations, and Alaska, with 90% of the total US tidal energy resources, is a home of some of the best tidal energy resources in the world.

   In 2008, 5-kW, and, in 2009, 100-kW turbines were installed in Yukon River by New Energy Corporation. The New Energy Corp. EnCurrent machine with 5, 10, 25, 125, and 250 kW capabilities were also developed [33,94,100,102]. The water flow in Alaskan rivers is, however, season dependent, dropping of in winter compared to summer. This can create some challenging issues on constant power supply.
4. A company called Hydro Green Energy [102] is developing hydrokinetic power turbine arrays that are composed of truly modular, interchangeable, zero-head, current-based turbines. Hydro Green Energy's dual-duct, axial-flow, interchangeable hydrokinetic array of current-driven turbines operate in river (in-stream, free-flow, open-river, or hydrokinetic run of river), ocean (ocean power), and tidal settings (tidal power). The capacity of the Hydro Green Energy design is 98 kW per unit (at 3.5 m/s) with a rotor diameter of 12 ft.

   Due to a surface suspension system, there are inherent operational maintenance and safety advantages of this device. An on-board gantry allows for raising and lowering of individual generating units in the hydrokinetic turbine array. The floating raft provides a platform for operation and maintenance activities. In general, the current-based hydrokinetic energy device provides the following advantages [102]:
   a. High capacity factor (approx. 90%), maximum net energy, and highly predictable base load power for in-stream river and ocean current applications; peak power generation in tidal energy applications.
   b. Large projects consist of robust and simple metal construction. Small projects consist of reinforced plastic construction. It utilizes conventional moving systems and the installation is simple and safe using existing marine vessels.

   c. Scalable to large power-generating stations (100+ MW)-utility scale power production; this turbine has been installed at numerous locations in Alaska.
5. In the summer of 2012, a hydrokinetic power-generation project was completed off the coast of Maine. This project was a result of collaboration between the department of energy and local community. The project will provide electricity to the local community. The project used the cross-flow turbine design similar to that shown in Figure 13.6, and it will generate power using the energy generated from underwater currents off the coast of Maine [97]. The electric power will be supplied using underwater power grid to the local town.

Besides the above five examples of commercialization, a number of hydrokinetic generation technologies are moving beyond pilot or demonstration stages [58–63] (Bertsch, 2012, pers. comm.).* In 2011, the United States had less than 1 MW of installed hydrokinetic generation capacity, as compared to more than 77,000 MW of conventional hydroelectric generation capacity [40,42,44] (Dixon et al., 2008, pers. comm.). As of June 2011, the FERC had issued 70 preliminary permits for hydrokinetic projects (27 tidal, 8 wave, and 35 inland) with 9306 MW of generation capacity [37,42,44] (Dixon et al., 2008, pers. comm.). Preliminary permits were pending for an additional 147 projects with 17,353 MW of capacity [37,42,44,90].

The development and implementation of new commercial projects will require strong public/private financial backing, local political support for guaranty use of generated power, an efficient licensing approval system along with all the environmental considerations mentioned above [33,37,49,50,57–63] (Bertsch, 2012, pers. comm.).† Areas in the United States with good wave energy potentials include most of the continental US west coast, Hawaii, and Alaska. For tidal energy, good sites exist in the Puget Sound, San Francisco, a variety of east coast tidal channels, and Alaska. For river hydrokinetic energy, large inland rivers such as the Mississippi, Missouri, and Yukon have promising potential power [33,66–71,90].

13.4 OCEAN THERMAL ENERGY CONVERSION (OTEC)

This technology uses the temperature difference between warmer surface of the ocean and cooler deep water to run a heat engine to produce electricity. The heat cycle commonly used in the OTEC process is a Rankin cycle using a low-pressure turbine. While the attempt to develop and refine the OTEC technology started in the 1880s [92], one of the first successful plants generating 22 kW electricity was built in Matanzas, Cuba, in 1930 [75].

Japan is the major contributor to the OTEC technology. Tokyo Electric Power Company started building a 100-kW closed-loop cycle OTEC plant on the island of

---

* Bertsch, D.J., Juris Doctoral candidate, The University of South Dakota School of Law, 2011; Congress defined hydrokinetic energy as "electrical energy from waves, tides, and currents in oceans, estuaries, and tidal areas; free flowing water in rivers, lakes, and streams, or man-made channels; and differentials in ocean temperature (ocean thermal energy conversion)," The Energy Independence and Security Act of 2007, 42 USC §17211 (2006).

† Ibid.

Nauru [93] in 1970. While this plant became operational in 1981, it only generated net 30 kW power for a school and other systems [92]. Many earlier efforts made in India and U.S. [75] were not completely successful. Only successful effort was made in 1993 by National Energy Laboratory in Hawaii, which generated 255 kW energy and lasted for six years [75]. Currently only operating OTEC plant is the one built by Saga University with support of a various Japanese industries in March 2013 [75].

Ocean Thermal Energy Corporation has plans to install two 10 MW OTEC plants in the US Virgin Islands and 5–10 MW OTEC facility in the Bahamas [75]. Numerous projects in Hawaii, Hainan, and Japan have been proposed and are being pursued [75,92]. Basic operating principles, operating sites, other usages of OTEC process and barriers to its implementations are well described in Ref. [75]. Sections 13.4.1 through 13.4.4 briefly summarize the descriptions presented in this reference.

13.4.1 Operating Principles

A heat engine gives a higher efficiency when it is run with a higher temperature difference. The tropical area provides the largest temperature difference between ocean surface and deep water around 20°C–25°C. While OTEC can in principle provide 10–100 times more energy than wave power, its thermodynamic efficiency of 1%–3% (with old technology) compared to the theoretical maximum of 6%–7% for 20°C–25°C temperature difference has limited its use. Modern technologies, however, approach to the theoretical maximum efficiency. One approach that has worked is to pump vaporized low boiling fluid into the depths to be condensed, which reduces pumping volumes, technical and environmental problems, and costs.

The heat engine cycle can be operated in three different ways: close, open, and hybrid. Closed-cycle systems use fluid with a low boiling point, such as ammonia (having a boiling point around –33°C at atmospheric pressure), to power a turbine to generate electricity. Ammonia is used because of its superior transport properties, easy availability and low cost. Other fluids (such as CFC, HCFC, etc.) are possible but they have harmful environmental effects. The ammonia is vaporized and condensed by warm and cool water, respectively, with the use of two separate heat exchangers. The power is generated by the expanding vapor.

In an open-cycle OTEC process, warm surface water is converted to steam by passing it into a low-pressure vessel. The expanding steam can drive a low-pressure turbine. The steam is then converted to purified water by exposure to cold temperatures from deep-ocean water. This method thus produces water that is suitable for drinking, irrigation, or aquaculture [72]. The expanding (and rising) steam can also be used in a gas lift technique to lift water to significant heights. Depending on the local circumstances, this technique can generate power with the use of a hydroelectric turbine [73].

A hybrid cycle combines the features of the closed- and open-cycle systems. In a hybrid process, warm seawater enters a vacuum chamber and is flash evaporated. The steam then vaporizes the ammonia, which like in a closed cycle drives a turbine to produce electricity. The steam condenses within the heat exchanger and provides desalinated water. The hybrid cycle thus serves multiple purposes at the same time.

13.4.2 Operating Sites

While OTEC has the potential to produce a large amount of power and hydrogen (jointly with electrolysis), it is an expensive technology. OTEC plants require a long, large-diameter intake pipe, which is submerged a kilometer or more into the ocean's depths, to bring cold water to the surface. The operating site for OTEC can be land based, shelf based, or floating.

Land-based and near-shore facilities can be installed in sheltered areas so that they are relatively safe from storms and heavy seas. Electricity, desalinated water, and cold, nutrient-rich seawater could be transmitted from near-shore facilities via trestle bridges or causeways. In addition, land-based or near-shore sites allow plants to operate jointly with desalination or aquaculture industries. Land-based or near-shore sites can also support chilled water agriculture. Tanks or lagoons built on shore allow workers to monitor and control miniature marine environments and allow easy transport of the products to the markets.

Favored locations include those with narrow shelves (volcanic islands), steep (15°–20°) offshore slopes, and relatively smooth seafloors. These sites minimize the length of the intake pipe. A land-based plant could be built well inland from the shore, offering more protection from storms, or on the beach, where the pipes would be shorter. In either case, convenient access for construction and operation helps lower costs.

There are few disadvantages to land-based operations. The prolonged turbulent wave action in the surf zone and storms can damage discharge pipes. In addition, the mixed discharge of cold and warm seawater may need to be carried several hundred meters offshore to reach the proper depth before it is released, requiring additional expense in construction and maintenance. This can be avoided by building OTEC system just offshore in waters ranging from 10 to 30 m deep which use shorter intake and discharge pipes. The plant itself, however, would require protection from the marine environment, and the plant output would need to be transmitted to shore [74].

To avoid the turbulent surf zone as well as to move closer to the cold-water resource, the OTEC plants can be mounted to the continental shelf at depths up to 100 m. In general, however, the stress of open-ocean conditions, difficulty in product delivery and higher expenses for its construction and building a power delivery system to reach land make this approach less attractive [74].

Floating OTEC facilities operate offshore. Although potentially optimal for large systems, floating facilities present several difficulties. The difficulty of mooring plants in very deep water complicates power delivery. Cables attached to floating platforms are more susceptible to damage, and for depths >1000 m they are difficult to maintain and repair. The system needs to be connected to the sea floor by riser cables without entanglement [74]. Both warm-water intake and vertically suspended cold-water pipe can be damaged by major storms and heavy seas. This problem can be alleviated with the use of flexible polyethylene materials for pipes, which can be uncoupled from the plant during storm. Surface water can also be drawn directly into the platform. Precautions must be taken to reduce damage and interruptions to intake flow by heavy seas [74]. Connecting a floating plant to power delivery cables requires the plant to remain relatively stationary. Mooring is an acceptable method, but current mooring technology is limited to depths of about 2000 m (6600 ft). Even at shallower depths, the cost of mooring may be prohibitive.

13.4.3 Other Usages of OTEC

One of the attractions of the OTEC technology is its use for numerous other industries. Both open- and hybrid-cycle plants using surface condensers can desalinate seawater into potable water. A system analysis indicates that a 2-MW plant could produce about 150,000 cu ft of desalinated water each day [75].

The 41°F (5°C) cold seawater made available by an OTEC system creates an opportunity to provide large amounts of cooling to industries and homes near the plant. In 2010, Copenhagen Energy opened a district cooling plant in Copenhagen, Denmark. The plant delivered cold seawater to commercial and industrial buildings and reduced their electricity consumption by 80% [76]. OTEC technology supports chilled-soil agriculture. When cold seawater flows through underground pipes, it chills the surrounding soil. The temperature difference created by this method allows plants that require temperate climates to be grown in the subtropics [75].

Aquaculture is the best-known by-product of OTEC because it reduces the financial and energy costs of pumping large volumes of water from the deep ocean. Nonnative species such as salmon, lobster, abalone, trout, oysters, and clams can be raised in pools supplied by OTEC-pumped water. This extends the variety of fresh seafood products available for nearby markets and provides a low-cost refrigeration that can be used to maintain the quality of harvested fish [75]. In Kona, Hawaii, aquaculture companies working with an OTEC plant (Natural Energy Laboratory of Hawaii Authority [NELHA]) generate about $40 million annually, a significant portion of Hawaii's gross domestic product (GDP) [77]. Deep-ocean water can also be combined with surface water to deliver water at an optimal temperature.

Hydrogen can be produced via electrolysis using the OTEC electricity. The OTEC process-generated steam with an addition of electrolyte compounds to improve efficiency is a relatively pure medium for hydrogen production [75]. While OTEC can be scaled to generate large quantities of hydrogen, this method is as yet not competitive to other methods of hydrogen production [75].

The OTEC technology can also be used to recover a large number (>50) of trace salt elements, uranium, and other materials from ocean. Japanese investigators are pursuing this approach [75].

A schematic of the OTEC process with its other usages is graphically illustrated in Figure 13.9 [75].

13.4.4 Barriers to Implementation

The OETC technology faces several political, economical, and technical barriers. The stationary surface platforms of the technology may affect the United Nations

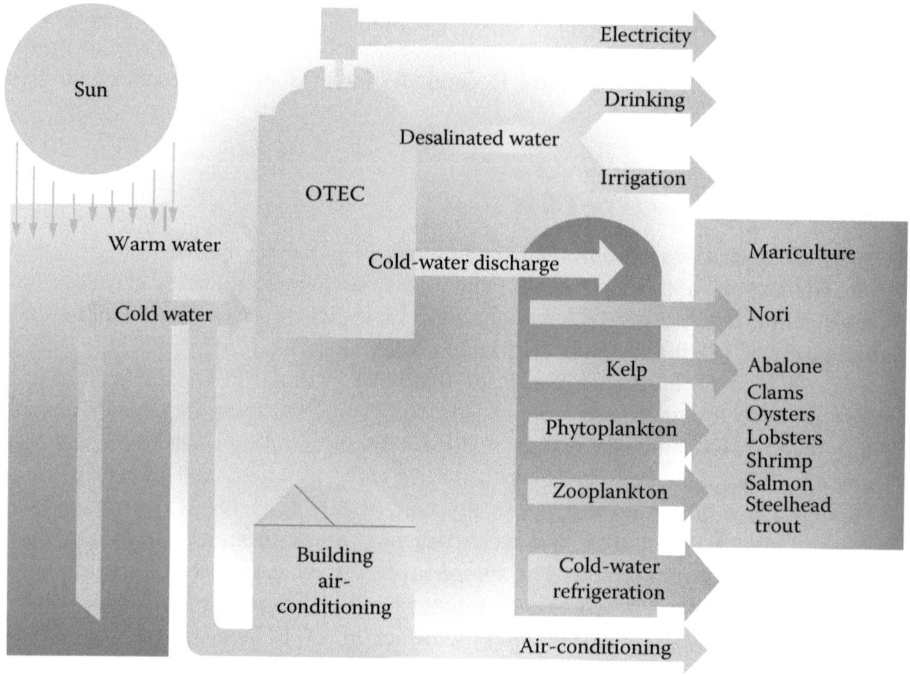

FIGURE 13.9 **(See color insert.)** A schematic of OTEC process with applications. (Adapted from "Ocean thermal energy conversion," Wikipedia, the free encyclopedia, 2013.)

convention on the law of the sea treaty [75]. They can affect fisheries and seabed mining operations. While the technology creates no waste and fuel consumption, its cost estimates are uncertain. It can be as low as 7 cents per kW depending on the cycle efficiency [75]. The technology is mostly applicable within 20° of the equator.

The technology faces several technical difficulties such as dissolved gases, microbial fouling, sealing, and parasitic power consumption by exhaust compressor [75]. The drop in pressure in an intake pipe can evolve gas, which can cause problems to the direct contact condensers. This issue may require installation of a deaeration unit [75]. The deposition of biofouling microbial layer from water in the heat exchanger wall may degrade its performance. Although the layer can be removed by brushing at short times, it may be difficult at longer times. This microbial layer may harden over time requiring more expensive treatment process [75]. The evaporator, turbine, and condenser operate in partial vacuum ranging from 3% to 1% of atmospheric pressure. The system must be carefully sealed to prevent in-leakage of atmospheric air that can degrade or shut down operation. The exhaust compressor parasitic power loss could be significant, and efforts must be made to reduce this loss and improve overall economics [75].

This technology can be in principle extended to cool air/water conversion in Arctic location where this temperature difference can be as high as 40°C [75].

13.5 GROWTH OF HYDROKINETIC ENERGY AND OTEC INDUSTRIES AND COST OF HYDROKINETIC AND OTEC POWER

A 2005 report by the EPRI estimated that some US utility-scale wave power projects could produce electricity for about 10 cents per kWh once the technology has fully matured [37,40,44,90] (Dixon et al., 2008, pers. comm.). They indicated that the present state of technology makes hydrokinetics a long-term investment opportunity, with potentially significant but highly uncertain returns.

A recent report by Pike Research [90] is, however, much more optimistic about the future growth of marine and hydrokinetic energy industry and pricing of hydroelectric power. Based on their own analysis, they made following assertions on the five chosen technologies:

1. *Tidal stream turbines.* These projects comprise over 90% of today's marine kinetic energy projects. However, majority of them depend on first-generation "barrage systems" that still rely on storage dams. The cost of power generation using these technologies is predicted to be 17 cents per kWh for 10 MW industry to as low as 4–9 cents per kWh for 100 MW industry. The target for this industry is 5 cents per kWh [90].
2. *Ocean wave energy technologies.* These "metal snake technologies" can span 600 ft floating on ocean wave horizontally. The generators can also be erected vertically akin to a buoy. Any western coastline in world has this wave energy potential. The cost of power generation for these technologies is 30 cents per kWh for 10 MW industry and 5–32 cents per kWh for 100 MW industry. The target for the industry is 5 cents per kWh [90].
3. *River hydrokinetic technologies.* This relies on the kinetic energy of moving water and it can be enhanced by tidal waves particularly at the intersections of river with sea or ocean. Alaska rivers are well suitable for these technologies. The cost of power generation by these technologies is <65 cents for 10 MW industry and about 18 cents per kWh for 100 MW industry. The target price for these technologies is 7–10 cents per kWh [90].
4. *Ocean current technologies.* This applies to deeper ocean currents near the shoreline. As mentioned in the Section 13.3, they are getting more attention in the recent years. The cost of power production for these technologies is about 20–40 cents for 10 MW industry. The data for the larger power plants using these technologies are not available. The target price for these technologies is 5 cents per kWh [90].
5. *OTEC technologies.* These technologies capture the energy from the difference in temperature between the ocean surface and lower depths. They can deliver power 24 h a day. The cost of power production for these technologies is >40 cents per kWh for 10 MW industry and >20 cents per kWh for 100 MW industry. The target price for these technologies is 15 cents per kWh [90].

Pike Research [90] in their two quarterly reports in 2012 projects very upbeat growth projections for marine and hydrokinetic energy productions. As shown in Figure 13.10,

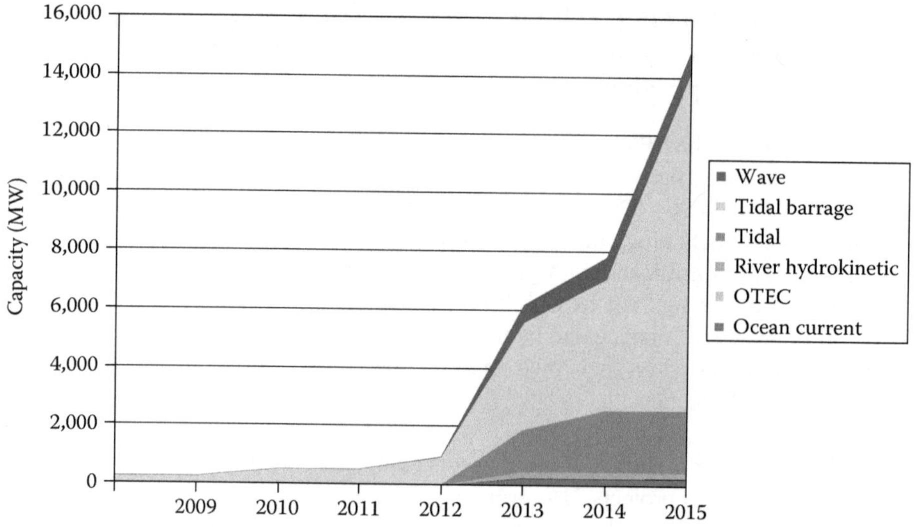

FIGURE 13.10 **(See color insert.)** Cumulative marine and hydrokinetic energy installed capacity by technology, world market: 2008–2017. (Adapted from Gauntlett, D. and Asmus, P., "Executive summary: Hydrokinetic and Ocean Energy; Renewable power generation from ocean wave, tidal stream, river hydrokinetic, ocean current, and ocean thermal technologies," Research report by Pike Research, Cleantech Market Intelligence, Boulder, CO, 2012.)

their growth projections for hydrokinetic energy capacity up to 2015 indicate more than ten-fold increase mainly due to two large projects—a 14 GW tidal barrage in the United Kingdom and a 2.2 GW tidal fence in Philippines—both may or may not be complete by 2015. Their growth projections for wave, river hydrokinetic, and ocean current energies during this period are modest. The figure predicts a negligible growth in energy by the OTEC projects during this period. Pike Research [90] sees Europe as a global leader for hydrokinetic energy producer and sees very significant global growth in wave, tidal stream, and tidal barrage energy by 2025.

REFERENCES

1. "Hydroelectricity," Wikipedia, the free encyclopedia, 1–7 (2012).
2. REN21, "Renewables 2011 global status report," Hydropower, REN21, Paris (July 11, 2011).
3. "History of hydropower," US Department of Energy (2012).
4. "Hydroelectric power," *Water Encyclopedia* (2012).
5. "Hydroelectric power—Energy from falling water," Clara.net (2012).
6. Robbins, P., "Hydropower," in *Encyclopedia of Environment and Society*, Sage Publications, Thousand Oaks, CA, Vol. 3 (2007).
7. World Watch Institute, "Use and capacity of global hydropower increases" (January 2012).
8. Atkins, W., "Hydroelectric power," *Water: Science and Issues*, 2, 187–191 (2003).
9. The Economist, "Binge and purge 98–99% of Norway's electricity comes from hydro-electric plants," *The Economist* (January 22, 2009).

10. International Rivers, "Sedimentation problems with dams" (July 16, 2010), Internationalrivers.org.
11. James, P. and Chansen, H., *Teaching Case Studies in Reservoir Siltation and Catchment Erosion*. TEMPUS Publications, Great Britain, 265–275 (1998).
12. Sentürk, F., "*Hydraulics of Dams and Reservoirs* (reference ed.). Water Resources Publications, Highlands Ranch, CO, 375 (1994).
13. "DAM," Wikipedia, the free encyclopedia, 1–19 (May 2013).
14. "Small hydro," Wikipedia, the free encyclopedia (2012).
15. Crettenand, N., "The facilitation of mini and small hydropower in Switzerland: Shaping the institutional framework. With a particular focus on storage and pumped-storage schemes," École Polytechnique Fédérale de Lausanne (EPFL), PhD thesis No. 5356 (2012), http://infoscience.epfl.ch/record/176337?ln=en.
16. Nachman-Hunt, N., "Small hydropower systems: Energy efficiency and renewable energy clearinghouse," Report DOE/GO-102001-1173, Merrifield, VA, FS217 (July 2001).
17. "Micro hydro," Wikipedia, the free encyclopedia (2012).
18. "How a micro hydro system works," US DOE (November 28, 2010).
19. "Micro hydropower systems," US DOE (November 28, 2010).
20. "Micro hydroelectric systems," Oregon DOE (December 1, 2010).
21. "Determining a potential micro hydropower site's flow," US DOE (November 28, 2010).
22. "Micro hydro," Research Institute for Sustainable Energy (December 9, 2010).
23. "Micro-hydro," The Ashden Awards for Sustainable Energy (November 20, 2010).
24. "Microhydropower," US DOE (November 20, 2010).
25. "Micro hydro power—Pros and cons," Alternative Energy News Network (November 24, 2010).
26. "Pico hydro," Wikipedia, the free encyclopedia (2012).
27. "Pico hydro power," T4cd.org (July 16, 2010).
28. Thomas, B., Jordan, B., and McGhee, R., "Pico-hydropower franchising: A test bed in rural honduras," National Collegiate Inventors and Innovators Alliance (January 24, 2013).
29. "Ashden award for pico hydro power in Kenya," Ashden Awards (September 6, 2010).
30. Khandker, S., Barnes, D., Samad, H., and Huu Minh, N., "Welfare impacts of rural electrification: Evidence from Vietnam," World Bank, Washington, DC (2008).
31. Redfield, S., *Five Gallon Bucket Hydroelectric Generator Build Manual*. Engineering for Change, Appropriate Infrastructure Development Group, Weston, MA (2012).
32. Howey, D., "Axial flux permanent magnet generators for pico-hydropower," Community of Practice: Energy, A conference hosted by Royal Academy of Engineering, February 20, Imperial College London, London (2006).
33. "Regulators approve first commercial hydrokinetic projects in the United States," *Today in Energy*, US Energy Information Administration, Washington, DC (October 2, 2012).
34. Sautter, J.A., "The clean development mechanism in China: Assessing the tension between development and curbing anthropogenic climate change," *Villanova Environmental Law Journal*, 27, 91, 107 (2009).
35. Center for Climate and Energy Solutions, *Hydrokinetic Electric Power Generation*. Center for Climate and Energy Solutions, Arlington, VA, 1–6 (2012).
36. Bedard, R., Previsic, M., Hagerman, G., Polagye, B., Musial, W., Klure, J., Jouanne, A., et al., "North American Ocean energy status—March 2007," *Proceedings of the 7th European Wave and Tidal Energy Conference*, September 11–13, Porto, Portugal (2007).
37. FERC, "Issued hydrokinetics projects preliminary permits," http://www.ferc.gov/industries/hydropower/indus-act/□hydrokinetics/permits-issued.asp.
38. Minerals Management Service, "Technology whitepaper on ocean current energy potential on the US outer continental shelf," Renewable Energy and Alternate Use Program, US Department of the Interior, Washington, DC, 3 (2006), http://ocenergy.anl.gov.

39. Cada, G., Ahlgrimm, J., Bahleda, M., Bigford, T., Stavrakas, S.D., Hall, D., Moursund, R., and Sale, M., "Potential impacts of hydrokinetic and wave energy conversion technologies on aquatic environments," *Fisheries*, 32 (4), 174–181 (2007).
40. Previsic, M., Polagye, B., and Bedard. R., "System level design, performance, cost and economic assessment—San Francisco tidal in-stream power plant, EPRI. EPRI-TP-006-SF CA (2006), http://oceanenergy.epri.com/streamenergy.html#reports.
41. Previsic, M., Moreno, A., Bodard, R., Polagye, B., Collar, C., Lockard, D., Toman, W., et al., "Hydrokinetic energy in the United States—Resources, challenges and opportunities," *Proceedings of the 8th European Wave and Tidal Energy Conference*, Uppsala, Sweden, 76–84 (2009).
42. Bedard, R., Previsic, M., Hagerman, G., Polagye, B., Musial, W., Klure, J., et al., "North American ocean energy status," EWTEC, Conference Paper, September 11–14, Porto, Portugal (2007).
43. Previsic, M. et al., "US Ocean energy resources," *Presentation at Hydrovision* (2008).
    Referred by Previsic, M., Moreno, A., Bodard, R., Polagye, B., Collar, C., Lockard, D., Toman, W., et al., "Hydrokinetic energy in the United States—Resources, challenges and opportunities," *Proceedings of the 8th European Wave and Tidal Energy Conference*, Uppsala, Sweden, 76–84 (2009).
44. "EPRI ocean energy reports, http://oceanenergy.epri.com/.
45. "Research at the intersection of marine/hydrokinetic energy and the environment," NSF Workshop, October 5–7, St. Anthony Falls Laboratory, University of Minnesota, Minneapolis, MN (2011).
46. Lalander, E., *Modelling Hydrokinetic Energy Resource for In-Stream Energy Converters*. Division of Electricity, Department of Engineering Sciences, Uppsala Universitet, Uppsala, Sweden, (2010).
47. US Department of Energy Wind and Water Power Program, *Marine and Hydrokinetic Technologies*. US Department of Energy, Washington, DC, 1–2, DOE/GO-102010-3038 (2010).
48. Bertsch, D., "Hydrokinetic energy: Trying to navigate the energy and water law framework to develop new renewable energy technology," The University of South Dakota Report, Vermillion, SD, 1–25 (2011).
49. "Hydrokinetic energy (in river, tidal and ocean current)," *Alaska Energy Wiki*, Alaska Center for Energy and Power, 1–6 (2012).
50. "How hydrokinetic energy works?" Union of Concerned Scientists, 1–5 (2012).
51. "Hydrokinetic energy (in river, tidal, and ocean current)," *Alaska Energy Wiki*, Alaska Center for Energy and Power, 1–4 (July 20, 2012).
52. "Hydrokinetic power barges," Future energy, hydro power inventions, alternative energy (August 18, 2009).
53. "Tidal power," Wikipedia, the free encyclopedia (May 9, 2013); also Tidal Power at Cook Inlet; a communication by Ground Truth Trekking (May 9, 2013).
54. Ocean Energy Council, "Tidal energy: Pros for wave and tidal power" (2011).
55. Dorf, R., *The Energy Factbook*. McGraw-Hill, New York (1981).
56. "Tidal energy, ocean energy," Racerocks.com (April 5, 2011).
57. BOEM, "Ocean wave energy," Bureau of Ocean Energy Management, Washington, DC (May 9, 2013).
58. "Hydrokinetic projects," FERC, US Administration, 74, 61–67 (August, 2013).
59. "Low head hydro power," Wikipedia, the free encyclopedia (May 9, 2013).
60. "Wave power," Wikipedia, the free encyclopedia (May 9, 2013).
61. International Energy Agency, *World Energy Outlook Executive Summary* (2009), http://iea.org/Textbase/npsum/WEO2009SUM.pdf.
62. Walsh, M.B., "A rising tide in renewable energy: The future of tidal in-stream energy conversion (TISEC)," *Villanova Environmental Law Journal*, 19, 193, 196 (2008).

63. Koch, L., "The promise of wave energy," *Villanova Environmental Law Journal*, 2, 162, 165 (2008).
64. Pelamis Wave Power, *Statement on Portuguese Aguçadoura Project* (2010), http://www.pelamiswave.com/media/statement_on_aguadoura_project.pdf.
65. PEW Center on Global Climate Change, *Hydrokinetic Electric Power Generation* (2009), http://www.pewclimate.org/docUploads/Hydrokinetic%2009%2012%2004.pdf.
66. Miller, C., "A brief history of wave and tidal energy experiments in San Francisco and Santa Cruz" (August 2004).
Referred by Vosough, A. "Wave Energy," *International Journal of Multidisciplinary Sciences and Engineering*, 2 (7), 60 (2011).
67. "Wave energy potential on the US outer continental shelf," US Department of the Interior, Washington, DC (2008).
68. Holthuijsen, L.H., *Waves in Oceanic and Coastal Waters*. Cambridge University Press, Cambridge (2007).
69. Clément, A., McCullen, P., Falcao, A., Fiorentino, A., Gardner, F., Hammarlund, K., Lemonis, G., et al., "Wave energy in Europe: Current status and perspectives," *Renewable & Sustainable Energy Reviews*, 6 (5), 405–431 (2002).
70. "The development of wave power," (December 18, 2009).
71. "Wave energy research and development at JAMSTEC," (December 18, 2009).
72. Vega, L.A., "Open cycle OTEC," *OTEC News*, The GreenOcean Project (1999).
73. Lee, C.K.B., and Ridgway, S., "Vapor/droplet coupling and the mist flow (OTEC) cycle," *Journal of Solar Energy Engineering*, 105, 181–186 (1983).
74. "Design and location," *What is Ocean Thermal Energy Conversion?* National Renewable Energy Laboratory, Washington, DC (January 22, 2012).
75. "Ocean thermal energy conversion," Wikipedia, the free encyclopedia (2013).
76. Green Tech, "Copenhagen's sea water cooling delivers energy and carbon savings," Forbes, New York (October 24, 2012).
77. Ponia, B., "Aquaculture updates in the Northern Pacific: Hawaii, Federated Sates of Mirconesia, Palau and Saipan," *SPC Fisheries Newsletter*, July 2006 (2013).
78. Drew, B., Plummer, A., and Sahinkaya, M., "A review of wave energy converter technology," *Proceedings of the IMechE: Journal of Power and Energy*, 223 Part A, 887–902 (2009).
79. "Agreement to develop wave power park in Oregon," (2008), renewableeneregyaccess.com.
80. Reedsport OPT Wave Park, "Reedsport OPT Wave Park FERC Project No. 12713 Application for a Major License," Federal Energy Regulatory Commission, Washington, DC (February 15, 2010).
81. "Anaconda WEC," *Science Daily* (July 7, 2008).
82. Siegel, S.G., Jeans, T.L., and McLaughlin, T.E., "Deep ocean wave energy conversion using a cycloidal turbine," *Applied Ocean Research*, 33 (2), 110–119 (2011).
83. Siegel, S.G., Fagley, C., and Nowlin, S., "Experimental wave termination in a 2D wave tunnel using a cycloidal wave energy converter," *Applied Ocean Research*, 38, 92–99 (2012).
84. Kraemer, S., "Wave Roller uses swinging door for underwater wave energy," Scientific American, Stevenson Ranch, CA (December 9, 2010).
85. Cruz, J., Gunnar, M., Barstow, S., Mollison, D., and Cruz, J. (eds.), *Green Energy and Technology, Ocean Wave Energy*. Springer, Berlin, Germany, 93 (2008).
86. Jha, A., "Making waves: UK firm harnesses power of the sea ... in Portugal," *The Guardian* (London) (October 9, 2008).
87. "Pelamis sinks Portugal wave power," cleantech.com (2009).
88. Lima, J., "Babcock, EDP and Efacec to collaborate on wave energy projects," *Bloomberg Television* (September 24, 2008).
89. Fyall, J., "600ft 'sea snake' to harness power of Scotland," *The Scotsman* (Edinburgh), 10–11 (May 19, 2010).

90. Gauntlett, D. and Asmus, P., "Executive summary—Hydrokinetic and Ocean Energy: Renewable power generation from ocean wave, tidal stream, river hydrokinetic, ocean current, and ocean thermal technologies," Research report by Pike Research, Cleantech Market Intelligence, Boulder, CO (2012).
91. Ortega-Achury, S., McAnally, W., Davis, T., and Martin, J., "Hydrokinetic power review," Prepared for US Army Corps of Engineers Research development Center, Vicksburg, Mississippi, by Bagley College of Engineering, Mississippi State University, Starkville, MS (April 2, 2010).
92. Douglas, C.A., Harrison, G.P., and Chick, J.P., "Life cycle assessment of the SeaGen marine current turbine," *Proceedings of the Institution of Mechanical Engineers, Part M: Journal of Engineering for the Maritime Environment*, 222 (1), 1–12 (2008).
93. Evans, R., *Fueling Our Future: An Introduction to Sustainable Energy*. Cambridge University Press, New York (2007).
94. "China endorses 300 MW ocean energy project," Renewableenergyworld.com (April 5, 2011).
95. "Korea's first tidal power plant built in Uldolmok, Jindo," Korea.net, Gateway to Korea (2012).
96. "$3-B tidal power plant proposed near Korean islands," Korea.net, Gateway to Korea (2012).
97. Sharp, D., "1st tidal power delivered to US grid off Maine," *CBS MoneyWatch* (September 14, 2012).
98. "Turbines off NYC East River will create enough energy to power 9,500 homes," US Department of Energy, Washington, DC (February 13, 2012).
99. Johnson, K., "Project aims to harness the power of waves," *New York Times* (September 3, 2012).
100. "Renewable power from the ocean's waves," CETO Wave Power (November 9, 2010).
101. Lockheed Martin, "Woodside, Ocean Power Technologies in wave power project," Portland Victoria Wave Farm (2012).
102. "Hydrokinetic technology," Hydro Green Energy, Westmont, IL (December 7, 2012).

Index

Note: Locators "*f*" and "*t*" denote figures and tables in the text

C

D

E

F

G

H

P

R

S

W

X

Y